MECHANICAL AND ELECTRICAL SYSTEMS
in Construction
and Architecture

Third Edition

MECHANICAL AND ELECTRICAL SYSTEMS in Construction and Architecture

FRANK R. DAGOSTINO

Registered Architect
Architectural and Construction Consultant

PRENTICE HALL
Englewood Cliffs, New Jersey   Columbus, Ohio

Library of Congress Cataloging-in-Publication Data

Dagostino, Frank R.
    Mechanical and electrical systems in construction and architecture
    Frank R. Dagostino.—3rd ed.
        p.   cm.
    Includes index.
    ISBN 0–13–181462–1
    1. Buildings—Mechanical equipment.   2. Buildings—Electrical
equipment.   I. Title.
TH6010.D33   1995
696—dc20   94–43946
                                                           CIP

Acquisitions editor: Ed Francis
Editorial/production supervision and interior design: WordCrafters Editorial Services, Inc.
Cover design: Brian Deep
Cover photo: Steve Dunwell, The Image Bank
Prepress/manufacturing buyer: Pamela D. Bennett
Marketing manager: Debbie Yarnell

This book was set in Times Roman by BookMasters and was printed and bound by Semline. The
cover was printed by Phoenix Color Corp.

 © 1995, 1991, 1978 by Prentice Hall, Inc.
A Simon & Schuster Company
Englewood Cliffs, NJ 07632

The following figures are extracted from the 1993 National Standard Plumbing Code with permission of The
National Association of Plumbing Heating Cooling Contractors (PHCC): 1-17, 1-19, 2-2, 2-38, 2-39, 2-40,
2-41, 2-42, 2-43, 2-44, 2-45, 3-43, 3-44, 3-45, 3-46, 3-47, 3-48, 4-26, 4-27, 4-28, 4-29, 4-30, 5-17, 5-18, 5-19,
5-20, 5-21.
The following figures are reprinted with permission from ASHRAE, *Fundamentals Handbook, 1993*: 8-45, 8-46,
8-47, 8-48, 8-49, 8-50, 8-51, 8-52, 8-53, 9-31, 9-32, 9-33, 9-34, 10-35, 10-36, 12-8, 12-12, 13-21, 13-22, 14-4.
The following figures are reprinted with permission from ASHRAE, *Systems and Equipment Handbook, 1992*:
9-36, 10-34.
The following figures are reprinted with permission from NFPA 70–1993, the *National Electrical Code®*,
Copyright © 1992, National Fire Protection Association, Quincy, MA 02269: 15-16, 16-16, 16-20, 16-21, 16-22,
16-23, 16-24, 16-25, 16-26. This reprinted material is not the complete and official position of the National Fire
Protection Association, on the referenced subject which is represented only by the standard in its entirety. National Electrical Code® and NEC® are registered trademarks of National Fire Protection Association, Inc.,
Quincy, MA 02269.
The following figures are reprinted with permission from IES, *IES Lighting Handbook, 1993 Reference and
Application Handbook*: 17-5, 17-13, 17-14, 17-15, 17-16, 17-17, 17-18, 17-19, 17-20.

Printed in the United States of America
10   9   8   7   6   5

ISBN 0-13-181462-1

Prentice-Hall International (UK) Limited, *London*
Prentice-Hall of Australia Pty, Limited, *Sydney*
Prentice-Hall Canada Inc., *Toronto*
Prentice-Hall Hispanoamericana, S.A., *Mexico*
Prentice-Hall of India Private Limited, *New Delhi*
Prentice-Hall of Japan, Inc., *Tokyo*
Simon & Schuster Asia Pte. Ltd., *Singapore*
Editora Prentice-Hall do Brasil, Ltda., *Rio de Janeiro*

Contents

Preface

This text is geared for use by the entire construction industry—from those interested in the actual design of the systems to those who realize they must know and understand the mechanical systems in order to successfully design, draw or build a building or project.

The third edition has been revised to include metrics. It has all of the latest code and associated design revisions in each area of the text. The designer must check the code requirements in the area where the building will be constructed and use that code in designing the system.

The mechanical equipment covered in this book is an integral part of the design and construction of all buildings and projects. Its *successful* integration into the design depends upon the close cooperation of the designer, the drafters, the mechanical systems designer, the general construction contractor, and each of the contractors who may install a portion of the mechanicals.

To be successful, each of these people or groups of people must be familiar with the requirements of the mechanical systems. Each should at least be familiar with the basic design procedures used, flexibilities in each system, space required, and the time at which such work must be done on the job so it is fully coordinated. To get an idea of the coordination, consider each step briefly.

The designer must consider whether the layout provides space for the equipment required by the heating system (ducts, pipes, radiation devices, etc.), the water supply and drainage systems, and the electrical system.

The architectural (general construction) drafter must be certain that the details allow for the inclusion of the mechanicals, that the walls are wide enough for pipes to pass through, that there is enough space for the ducts to run, and that soffits are provided where required to conceal pipes.

The mechanical systems designers must work closely with all of the others. During the design stage, they must work with the designer and the owner to analyze the requirements of the project and to be certain that the designer incorporates the systems required in the design. Next, they must coordinate the mechanical systems designed with the drafter who will draw the system up. This drafter may be working there with them or may be in a separate part of the architectural designer's office. If this is a different drafter from those working on the general construction drawings, their efforts must be coordinated. Once the project is under construction, they will be involved in the coordination of the mechanical drawings, the general contractor, and the mechanical contractor.

The mechanical drafter may or may not be the same drafter doing the general construction drawings. The drafter must be careful to check against the general construction drawings to be certain that there is sufficient space for all of the mechanicals required.

The general construction contractor, and all the workers, must be certain that the mechanical contractors have adequate notice and opportunity to install their work as the project proceeds. There will be pipe sleeves in foundation walls, conduit and pipes in concrete floors, conduit and boxes in masonry walls, cable and boxes in stud walls before the wall boards are installed—any many, many other installations that require coordination.

Mechanical contractors are specialists in the field, tested and licensed by the government. Their staff will do an estimate on the mechanicals, submit bids and, if they are successful bidders, install the mechanical systems. Generally, these contractors are separated into three groups: plumbing; electrical; and heating, ventilation and air conditioning contractors. There may be separate contractors for each of these, and quite often a firm may bid plumbing and heating, ventilation and air conditioning.

Because all of these people are involved, and because the mechanicals must fit into the construction of the building, it is important that the entire construction "team" be familiar with the mechanical systems and their principles of design.

The basic format of this text is to discuss the chapter topic, describe the materials available and then show, step by step, how to approach a design problem. In those chapters that describe how to design a system, an actual design has been worked out. For easy reference, the charts and tables used in the designs have been located at the end of each chapter.

It has been the goal of this book to explain, step by step, the process of designing the mechanical systems and how to accumulate and tabulate the design information. Where applicable, the text not only discusses how to design a system but also what to look out for—what the variables are.

Frank R. Dagostino

Metric Notes

The metric system adopted by Canada, and to be adopted by the United States, is the *Systeme International d'Unites,* which is referred to as the SI metric system. In this book all dimensions will be given in the traditional U.S. system (such as feet and inches), with the metric conversion also given. Those learning the construction business at this time will find it necessary to become proficient in both the U.S. system and the metric system. The use of the traditional system will die out slowly, and both systems of measurement will be in use for many years.

Conversion of the U.S. system to metrics comes in two forms, soft and hard. *Soft conversion* occurs when the actual size of the product does not change; it is still available in its standard size, but is labeled in metrics. For example, a 1 quart bottle is labeled 0.95 liters. Similarly, a 4 inch pipe is called 102 millimeters (mm). In each case, the actual size of the product has not changed—only the method of measurement.

Hard conversion occurs when the actual size or quantity changes. For example, the 1 quart bottle is replaced with a 1 liter bottle (which actually holds 1.057 quarts). Another hard conversion will occur when a 100 mm pipe replaces the 4 inch pipe currently in use.

Note: At this time both hard and soft conversions are available in many products. In some other countries the 100 mm pipe is available, while in the United States and Canada the 4 inch (102 mm) pipe is in use. For all practical purposes it makes no difference to the designer whether the pipe is 4 inches (102 mm) or 100 mm. Design tables and charts may show 102 mm or 100 mm depending on the source. Once again, do not be confused; there is no practical difference.

METRIC CONVERSION

Metric conversion in the construction industry will involve primarily linear, liquid, weight, and quantity of heat measurements. Common conversions are listed in Appendix E.

Linear Measurements

Using the SI metric system, 1 millimeter is the basic linear measurement. A millimeter is small, equal to 0.0394 inch, and about the same thickness as a paper match. One inch actually equals 25.4 mm, but in the soft conversion 1 inch equals 25 mm. When the conversion to metrics is complete, all of the materials used in construction will be available in 100 mm increments. This should eventually reduce the time needed to design, draw, and actually construct the building. This 100 mm module will even become a design module used in planning individual room sizes.

For large dimensions, such as lot dimensions and overall building dimensions, the meter is used. One meter is equal to 39.37 inches, 1.1 yards, and 1,000 millimeters. Room areas will be given in square meters (m^2), as will such items as carpet and paint and stain coverage.

Volumes will be measured in cubic meters. This includes items such as cement and concrete that are now sold by the cubic foot or cubic yard. This measurement will also be used for the volume of rooms and other spaces.

Linear soft conversions include lumber sizes, such as keeping the nominally sized 2 × 4 an actual 1½ × 3½ inches and referring to it as a 38 × 89 mm stud, which is the same as its actual size. Hard conversions will include such items as walls being 2400 mm high, or about 1½ inches lower than the standard 8-foot height currently in use.

Liquid Measurements

In the SI metric system, the liter is the basic unit of liquid measurement. A liter is equal to 1.057 quarts. This will apply to such items as paints, stains, thinners, and any other liquids used. For large liquid measurements, a kiloliter may be used. One kiloliter is equal to 10 liters.

Weight Measurements

In the SI metric system, the gram and kilogram are the basic units of measurement. A gram equals 0.035 ounce, 457 grams equal 1 pound, 1,000 grams equal 1 kilogram, and 1 kilogram equals 2.2 pounds. The kilogram will be used for most weights such as aggregates and cement used for concrete, weight of a bag of cement, and weights of such items as reinforcing bars and structural steel members. Large amounts will be measured in megagrams. A megagram equals 1000 kilograms and is the equivalent of 2205 pounds. Large quantities of reinforcing bars, structural steel, and fine and coarse aggregates (used as fill and for concrete) are sometimes sold by the ton. One megagram equals 1.1 tons.

Temperature Measurement

In the SI metric system, the temperature is measured in Kelvins, which is a thermodynamic measurement. However, wide use is made of Celsius (formerly referred to as centigrade), and a one-degree increment in Celsius is exactly the same as a one-degree increment in Kelvin. Celsius is related to Kelvin as:

$$\text{Kelvin} = \text{Celsius} + 273.8° \text{ (exactly)}$$

Thus, a Celsius temperature of 20° is equal to a Kelvin temperature of 293.8°.

Quantity of Heat

In the SI system, the quantity of heat is expressed in joules and kilojoules instead of the Btu (British thermal unit) currently used. By definition, one joule equals the amount of heat required to heat one kilogram of water by one degree Kelvin. Joules per second (J/s) are equal to watts (W). To determine the number of watts required, the joules per second must be divided by the number of seconds in an hour (3600). Therefore, a building that requires 80,000 BTUH (British thermal units per hour) requires 84,384 kilojoules (kJO) and 23,440 watts, or 23.44 kilowatts. Other heat measurement values are:

$$C = J/K$$
$$A = m^2$$
$$\Delta T = K$$
$$\Delta x = \text{material thickness}$$
$$k = W \text{ (m.K)}$$

Force Measurement

The SI measurement for force is given in pascals, kilopascals, and megapascals. One pascal is the pressure or stress produced when one newton is applied to an area of one square meter. One newton is the amount of force required to move a one kilogram mass a distance of one meter per second squared. One pascal is quite small, about 0.000145 pounds per square inch (psi), and one psi equals about 6.9 kilopascals (kPa).

CHAPTER 1

Water Supply

1-1   WATER

Basically, water may be *potable* (suitable for human drinking) or *nonpotable* (not suitable for human drinking). While an abundant supply of water is vital to a prosperous economy, on an individual basis a supply of potable water is even more important to survival than food. This potable water must be supplied, or be available, for drinking and cooking. Nonpotable water may be used for flushing water closets (toilets), watering grass and gardens, washing cars, and for any use other than drinking or cooking.

At this time, potable water is used for many activities that could be done with nonpotable water. As potable water becomes scarcer and costs increase, more communities will require the use of nonpotable water wherever possible. These communities are installing separate water mains to provide nonpotable water to homes and businesses to preserve their supply of potable water. In some communities, the cost of potable water is so high that many of the residents have put in shallow wells that provide them with water for their lawns and gardens and for washing their cars.

1-2   WATER SOURCES

Rain is the source of most of the water available for our use, and it is classified as *surface water* or *groundwater*. Surface water is the rain that runs off the surface of the ground into streams, rivers, and lakes. Groundwater is the water that percolates (seeps) through the soil, building the supply of water below the surface of the earth.

Surface water readily provides much of the water needed by cities, counties, large industry, and others. However, this source is dependent on rain, and during a drought the flow of water may be significantly reduced. Most surface water will probably have to be treated to provide the potable water required. Where nonpotable water may be used, no treatment of the water may be necessary.

Also classified as surface water is rain, which may be collected in a small reservoir or tank (cistern) (Fig. 1-1) as it drains from the roof of a building. This water is then pumped into the supply line of the building for use. The need for water is so critical on certain islands that the government has covered part of the land surface (usually the side of a mountain or a hill facing the direction from which the rains usually come) with a plastic so that rain may be collected and stored for later use.

As groundwater percolates through the soil, it forms a water level below the surface of the earth. This water level is referred to as the *water table*. The distance from the ground surface to the water table (referred to as the *depth* of the water table) varies considerably; generally the more rainfall an area gets, the higher the water table will be. During a dry spell the water table will usually go down, while during a rainy season it will probably rise.

Since the water table is formed by an accumulation of water over an impervious stratum (a layer of earth, usually rock, that the water cannot pass through), the flow of the water follows the irregular path of the stratum, sometimes moving close to the surface

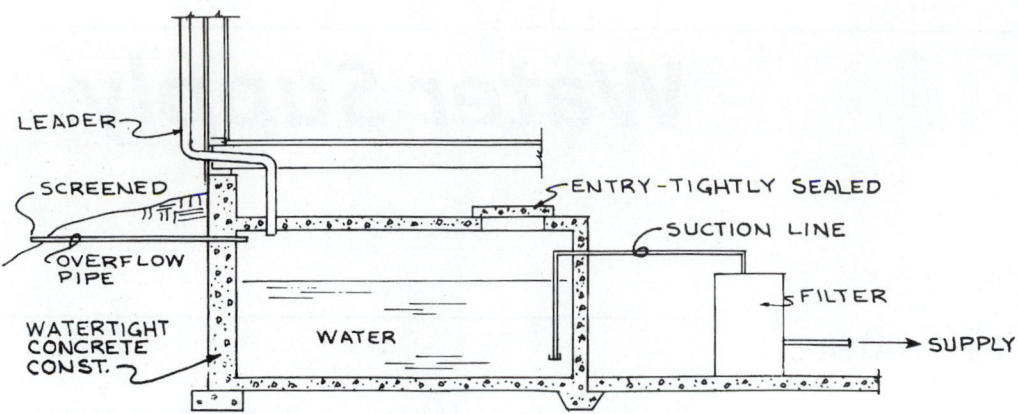

FIGURE 1-1 Cistern

while dropping off nearby (Fig. 1-2). This underground supply of water flows horizontally, and if it reaches a low spot in the ground surface, it may flow as a spring or seep out, creating a swampy area. Or, if the flowing underground water becomes confined between impervious strata, enough pressure may be built up in the water that if a stratum is opened (by drilling through the top stratum or by a natural opening in the stratum), it will create an artesian well.

Groundwater may require treatment to provide potable water, *but* often it does not. When treatment is required, it is generally less treatment than is required when making surface water potable.

The increased use (and misuse) of our potable water supply has forced the development of additional sources of water. This need for potable water has led to the desalination (taking the salt out) of water from the oceans and the purification of waste (sewage) water to be returned to the water system for reuse. To date, these methods involve a great deal of additional cost compared with the use and treatment of surface and groundwater.

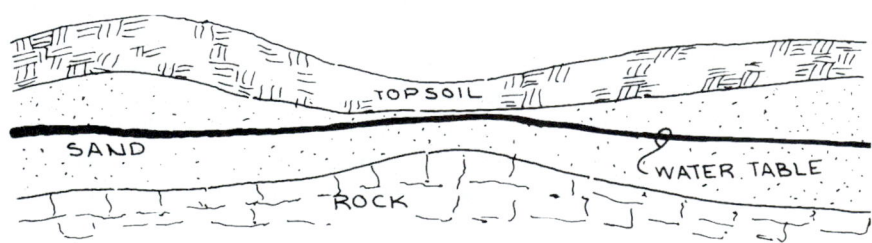

FIGURE 1-2 Water table

1-3 IMPURITIES

All water sources contain some impurities. It is the type and amount of these impurities which may affect the water's suitability for particular use.

As surface water runs over the ground, it may pick up various organic matter such as algae, fungi, bacteria, vegetable matter, animal decay and wastes, garbage wastes, and sewage.

As groundwater percolates down through the soil, it dissolves minerals such as calcium, iron, silica, sulphates, fluorides, and nitrates, and it may also entrap gases such as sulfide, sulphur dioxide, and carbon dioxide. It may also pick up contamination from public or private underground garbage and sewage wastes. Generally, as it percolates, it will filter out any organic matter which may have been accumulated at the surface or in the ground.

The impurities in the water may be harmful, of no importance, or even beneficial to a person's health. To determine what is in the water, it must be tested.

1-4 TESTS

All potable water supplies should be tested before being put in use and periodically checked during their use. It is assumed that whatever agency of a city, municipality, etc., controls the supply of water to a community regularly tests its water to be certain it is potable. Private water supplies, such as wells and streams, should always be checked before the water system is put into use and periodically thereafter.

Such tests are usually performed free of charge, or at a very low cost, by the local government unit in charge of public health; the governing unit (town hall, city hall, county health department) will put you in touch with the proper authorities, or it will refer you to a private testing laboratory.

The test for potable water provides a chemical analysis of the water, indicating the parts per million (ppm) of each chemical found in the water. A separate test is made for bacteriological quality of the water, providing an estimate of the density of bacteria in the water supply. Of particular concern in this test is the presence of any coliform organisms, which indicate that the water supply may be contaminated with human wastes (perhaps seepage from a nearby septic tank field). Since the test reports mean little to most people, a written analysis of the test or a standardized form is included with the test results saying whether the water is potable or not.

Water may have an objectionable odor and taste, even be cloudy and slightly muddied or colored in appearance, and yet the test may show it to be potable. This may not mean you *want* to drink it, but it does mean that it is drinkable. Such problems are often overcome by use of water-conditioning equipment, such as filters. As any traveler can quickly tell you, water varies considerably from place to place, depending on the water source of the area, the chemical and bacteria contents of the water, and the amount and type of treatment given the water before it is put into the system.

1-5 WATER SYSTEMS

The design of any water supply begins with a check of the water system from which the water will be obtained. Basically, water is available through systems which serve the community or through private systems.

Community Systems

Systems which provide water to a community may be government-owned, as in most cities, or privately owned, such as in a housing development where the builder or real estate developer provides and installs a central supply of water to serve the community. The water for these systems may have been obtained from any of the water sources listed in Sec. 1-2, and quite often it is drawn from more than one source. For example, part of the water may be taken from a river, and it may be supplemented by deep wells.

Before proceeding with the design of the water supply, the following information should be obtained:

1. What is the exact location of the water main (pipe) in relation to the property being built on?

2. If the main is on the other side of the street from the property, what procedures must be followed to get permission (in writing) to cut through the street, set up barricades, and patch the street? Also, what permits are required from local authorities, how much do the permits cost, and who will inspect the work and when? If available, obtain the specifications (written requirements) concerning the cutting and patching of the street.

3. If the water main does not run past the property, can it be extended from its present location to the property, who pays for the extension, and how long will it take?

4. Is there a charge to connect (tap) onto the community system? Many communities charge just to tap on, and the charge is often hundreds of dollars, sometimes thousands.

5. What is the water pressure in the main at the property? If it is too low for a residence [less than 30 psi (pounds per square inch) or about 200 kPa (kilopascal)], a storage tank and pump may be required to raise the pressure. Such a system is often used on commercial and industrial projects where the pressure may have to be quite high to meet the water demands. Water pressure that is too high (above 60 psi or about 400 kPa for a residence) will probably require a pressure-reducing valve in the system to cut the pressure to an acceptable level.

   Since plumbing fixtures are manufactured to operate efficiently with water pressures from about 30 to 60 psi (200 to 400 kPa), higher pressure may result in poor operation of the fixtures, rapid wearing out of the washers and valves, and noises in the piping. Low pressure may cause certain fixtures to operate sluggishly, especially showers, flush valve water closets (toilets), and garden hoses. The required water pressure at various fixtures and the water pressure from the main to the fixtures are discussed in detail in Chapter 2.

6. What is the cost of the water? Typically, a water meter is installed, either out near the road or somewhere in the project, and there is a charge for the water used. After determining what the charges are, a cost analysis may show that it is cheaper to put in a private system. Some areas do not allow private systems for potable water, but quite often it will be desirable to put in a well to provide nonpotable water for sprinkling the lawn and garden and for washing the car. Where costs for potable water are extremely high, it may be feasible to use separate potable and nonpotable water supply systems within the project (especially industrial and commercial projects).

Private Systems

Private systems may also use any of the water sources discussed in Sec. 1-2. Large industrial and commercial projects may draw all of their supply from one source, or they may draw part of their supply from one source (such as a stream) and supplement the supply with another source (such as a well). Such systems often include treatment plants, water storage towers, and sometimes even lakes or reservoirs to store the water.

Small private systems, such as those used for residences, usually rely on a single source of water to supply potable water through the system. Installing a well is the most commonly used method of obtaining water, and springs may be used when one is available.

Experts (usually consulting mechanical engineers, soil engineers, or water supply and treatment specialists) should be consulted early in the planning for any large project requiring its own private water system. Such specialists can make tests, interpret what the tests mean to the project, and make recommendations as to the quality and amount of water available.

1-6   WELLS

Most private water systems use wells to tap the underground water source. Wells are classified according to their depth and the method used to construct the well.

Depth	Construction Method
Shallow (25 ft or less) (7.6 meters or less)	Dug Driven Drilled
Deep (in excess of 25 ft) (in excess of 7.6 m)	Drilled Bored

The depth of the well is determined by the depth of the water table and the amount of water which can be pumped. This flow of water is considered the *yield* or capacity of the well. Once the water *demand* (the amount of water required) has been calculated (Sec. 2-10), it can be determined whether one well is enough or whether other wells will be needed to provide the required water for the project. Where the water table is high, it may not be necessary to go 25 ft (7.6 m) deep, but it is not unusual for wells to be 100 ft (30.5 m) deep, and in some areas well depths of several hundred feet are required to provide an adequate supply of water.

Dug wells (Fig. 1-3) should be 3 to 6 ft (1 to 2 m) in diameter and not more than 20 to 25 ft (about 7.6 m) deep. To minimize the chances of surface contamination, the well should have a watertight top and walls. The top should be either above the ground (Fig. 1-4) or sloped so that surface water will run away from it and not over it. The watertight walls should extend at least 10 ft (3 m) down. The walls may be concrete block, poured concrete, clay tile or precast concrete tile.

The water will flow into the well through the bottom of the well (and the water in the well will rise to about the level of the water table). Some wells also allow water to seep through the walls by use of porous construction near the bottom of the wall. This porous construction may be seepage pits (Fig. 1-5) or concrete block placed without mortar (normally used to hold the blocks together and make them watertight).

The placing of washed gravel in the bottom of the well, and on the sides of the well when porous walls are used, will reduce the sand particles or discoloration in the well water. Washed gravel is gravel (stone) that has been put through a wash (water sprayed over the stone) to remove much of the sand or clay from the stone. To further protect the water

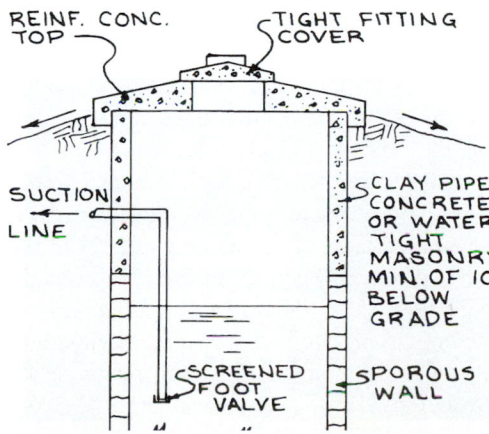

FIGURE 1-3   Dug well

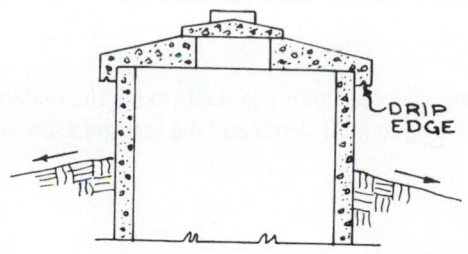

FIGURE 1-4   Dug well—top above ground

from possible contamination, tightly seal around the suction line pipe where it passes through the wall. Don't take water out with a bucket or other container since the container may have contamination on it, and by dipping such a container into the water, the contaminant is transferred to the water supply.

Shallow wells may also be driven. To drive a well, first attach a well point to a drive pipe and drive cap. Then, by means of an impact loading device, such as a small pile driver or even a sledge hammer for very shallow wells in soft, sandy soil, the well point is driven into the ground until it is into the water table. The well point has holes or slots in the side, allowing water to be sucked up to the surface by a shallow-well pump (Sec. 1-6). As the point is driven, additional lengths of pipe may be attached (usually 5-ft (1.5 m) lengths are used) by the use of a coupling. Driven wells will not pass through rock formations, and the maximum diameter commonly available is 2 in. (50 mm).

Shallow wells may have to be drilled if it is necessary to pass through rock to get to the water table.

Drilling and boring methods are used for deep wells. A well-drilling rig is used to form the well hole. Drilled and bored wells differ in that drilled wells have the holes formed by using rotary bits, while bored wells have the holes formed by using augers. Only the drilling method is effective in passing through rock.

As the hole is formed, a casing (pipe) is lowered into the ground. This steel or wrought-iron pipe (usually 3 to 6 in. (75 to 150 mm) in diameter) protects the hole against cave-ins where unstable soil conditions are encountered and keeps out surface drainage and possible surface or underground contamination. To further protect against surface drainage and contamination, a concrete apron, sloping away from the well, is poured around the casing at the surface (Fig. 1-6).

Well location and construction are often controlled by government regulations that set minimum distances between the well and any possible ground contaminant. When certain types of construction methods are used, these regulations may even require that licensed well drillers install the well. Various authorities and government regulations require different minimum distances, and there is no single set of standards used. The table in Fig. 1-7 shows minimum distances required in the *National Standard Plumbing Code*. It is important that local regulations be checked for each project.

FIGURE 1-5   Seepage pit

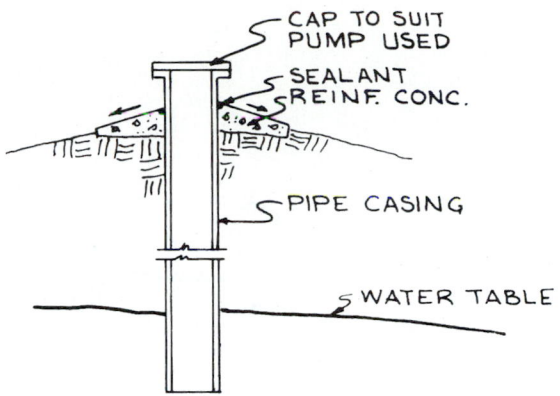

FIGURE 1-6   Drilled well

Possible well contamination from an underground flow of contaminants through rock formations which allow free-flowing groundwater to travel long distances is always possible, especially through strata of eroded limestone. Constant testing of water quality is required wherever there is a possibility of such contamination. For the well contaminated in this manner, three methods used to eliminate the problem are water treatment, relocation of the well, and elimination of the source of the contamination.

Before planning the well, local conditions should be checked to provide some background information. For example, existing local wells should be checked for depth and yield of water. This information can be obtained from local well drillers and government agencies and, if possible, verified by testing existing or just-completed wells.

Where insufficient information on well yields is available, and especially where large projects will require substantial water supplies, it may be necessary to have test wells made so that the yield can be checked. The well(s) should be tested by the driller to determine the yield, and a sample should be taken so that the quality of the water can be analyzed. It is important that this be done at an early stage in the design so that the size of the water storage tank can be determined and so that any water treatment equipment required can be designed and space allowed in the design of the project to locate the tank and equipment.

When a large supply of water is required for the continuous operation of the project, it may be necessary to put in other wells to be certain that the water yield will be sufficient to meet the projected demand (Sec. 2-10). For example, if one well provides adequate water, it may be a good investment to have a second well put in to act as a "back-up" in case the first well should fail in some way. This is usually not done for residences, but for industries or businesses which need water to operate (such as a car wash, farm, or apartment complex), it is a wise investment.

When more than one well is used, they must be spaced so that the use of one well will not lower the water table in the other well. Generally, deep wells must be 500 to 1,000 ft (150 to 300 m) apart, while shallow wells must be 20 to 100 ft (6 to 30 m) apart. Due to soil

Type of system	Distance from well
Building sewer	50′ (15 m)
Septic tank	50′ (15 m)
Distribution box	50′ (15 m)
Disposal field	100′ (30 m)
Seepage pit	100′ (30 m)
Drywell	50′ (15 m)

FIGURE 1-7   Well locations

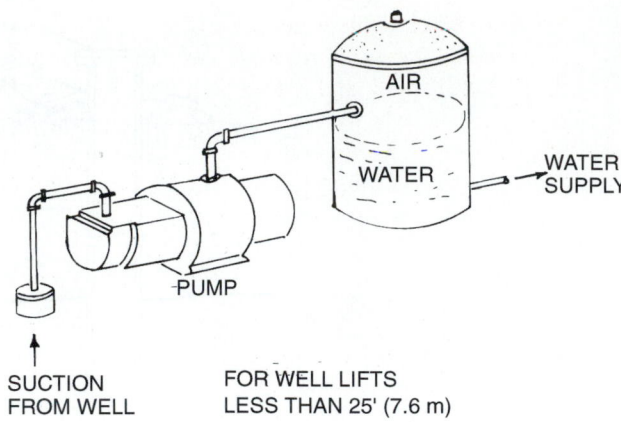

FIGURE 1-8   Shallow-well pump

variables, the minimum distance between wells can be determined only by testing (usually trial and error).

Many industries and businesses which draw their water from community systems have private systems that can be put into operation in case of a water shortage due to a breakdown in the system or a prolonged drought. As an example, during a drought in Raleigh, North Carolina, a local ordinance was passed prohibiting the washing of cars. This meant that all car washes served by the community system had to close down. Imagine the relief of one owner who had a well as an alternate source of water; he was in business while all the other car washes closed. The local newspaper even carried a story on it, providing the owner with free advertising.

1-7   PUMPS

Pumps used to bring well water to the surface are referred to as *shallow well* and *deep well,* depending on the type of well.

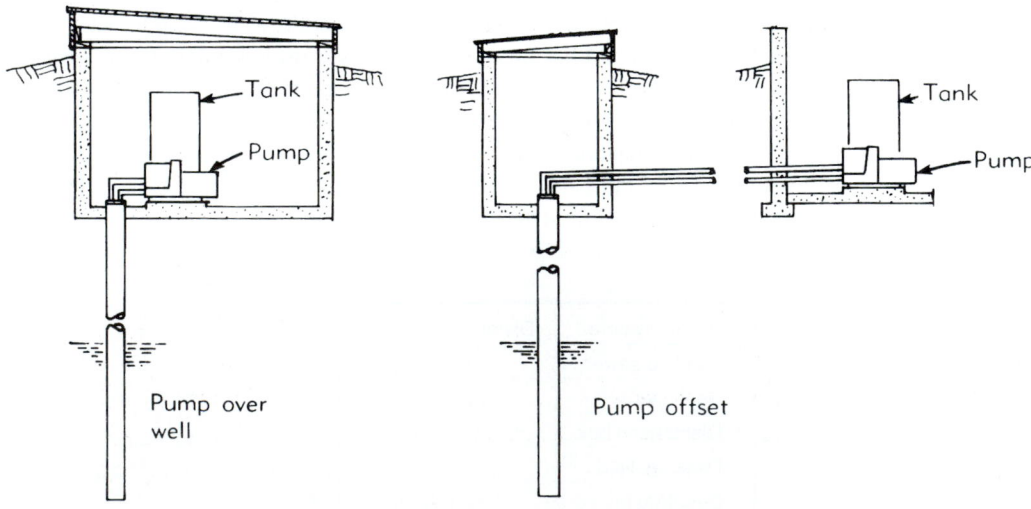

FIGURE 1-9   Deep well pump

Shallow-well pumps are located above the ground, and a suction line extends into the well below the water table (Fig. 1-8). The pump cannot lift or pull the water up more than about 25 ft (7.6 m), so any well with the water table deeper than 25 ft (7.6m) is considered a deep well, and a deep-well pump is used. The shallow-well pumps commonly used are the shallow-well jet, rotary, and reciprocating piston pumps.

The deep-well pumps most commonly used are the jet and submersible pumps. Jet pumps are located above ground, either directly over or offset from the well (Fig. 1-9). Submersible pumps have a waterproof motor and are placed in the well below the water table.

1-8   PLUMBING FIXTURES

The plumbing fixtures may be selected by the designer of the plumbing system, the architect, the owner, or a combination of these people. It is important that the designer of the plumbing system know what fixtures will be used (and even the manufacturer and model number, if possible) in order to do as accurate a job as possible in the system design.

The fixtures are the only portion of the plumbing system that the owners or occupants of the building will see regularly, since most of the plumbing piping is concealed in walls and floors. All fixtures should be carefully selected since they will be in use for years, perhaps for the life of the building.

The available sizes for each fixture should be carefully checked in relation to the amount of space available. Most manufacturers supply catalogs which show the dimensions of the fixtures they supply.

Whoever selects the fixtures should check with the local supplier to be certain that those chosen are readily available; if not, they may have to be ordered far in advance of the time they are required for installation. In addition, most of the fixtures are available in white or colors, so the color must also be selected.

Fixtures are grouped according to their use: water closets, urinals, bidets, bathtubs, showers, lavatories, kitchen sinks, and service sinks.

Water Closets

Water closets are made of solid vitrified china cast with an integral trap (also Fig. 3-3 and Sec. 3-3). Water closets are available as flush tank or flush valve fixtures.

A flush tank water closet (Fig. 1-10) has a water tank as a part of the fixture. As the handle (or button) is pushed, it lifts the valve in the tank, releasing the water to "flush out" the bowl. Then, when the handle is released, the valve drops and the tank fills through a tube attached to the bottom of the tank. This type of fixture cannot be effectively flushed again until the tank is refilled. There are several types of flushing action available on water closets, as illustrated in Fig. 1-11.

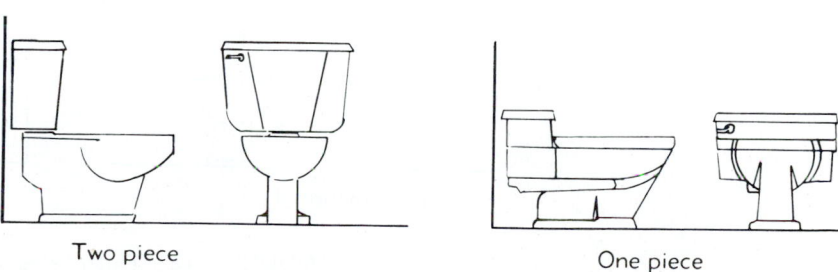

Two piece   One piece

FIGURE 1-10   Typical flush tank water closets

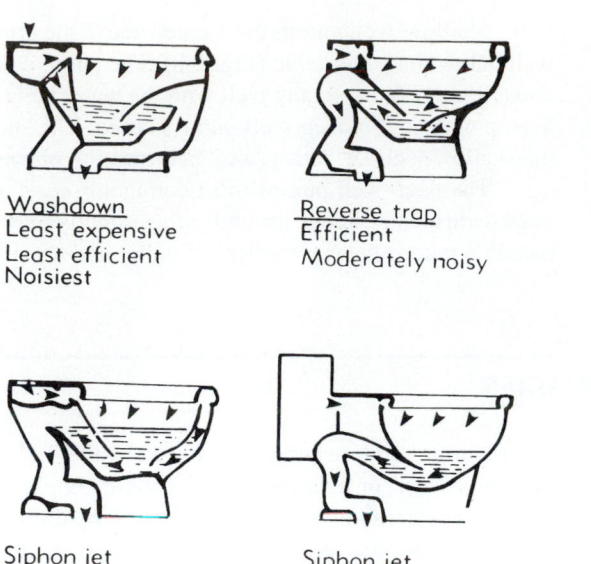

Washdown
Least expensive
Least efficient
Noisiest

Reverse trap
Efficient
Moderately noisy

Siphon jet
Efficient
Fairly quiet

Siphon jet
Quietest
Most expensive

FIGURE 1-11   Types of flushing action

Flush tank models range from those having the tank as a separate unit set on the closet bowl to those having a low tank silhouette with the tank cast as an integral part of the water closet. Generally, this low-slung appearance is preferred by clients, but it is considerably more expensive.

Flush valve water closets (Fig. 1-12) have no tank to supply water. Instead, when the handle is pushed, the water to flush the bowl comes directly from the water supply system at a high rate of flow. When used, it is important that the water supply system be designed to supply the high flow required. While most of the fixtures operate effectively at 20 psi (140 kPa) pressure, the manufacturer's specifications should be checked, since higher pressure is often required and must be considered in the design.

Water closets may be floor- or wall-mounted, as shown in Figs. 1-12 and 1-13. The floor-mounted fixture is much less expensive in terms of initial cost, but the wall-mounted fixture allows easier and generally more effective cleaning of the floor. Wall-mounted fixtures are considered desirable for public use, and some codes even require their use in public places. When wall-mounted fixtures are used in wood stud walls, a wider wall will be required than is sometimes used with floor-mounted fixtures.

Urinals

Urinals are commonly used in public restrooms, where it is desirable to reduce any possible contamination of the water closet seats. They are commonly available in vitreous

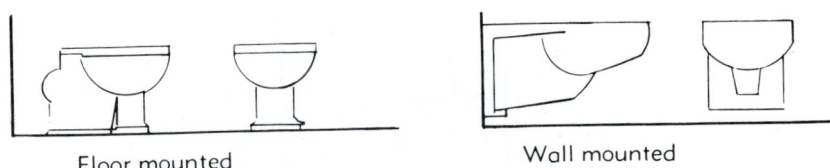

Floor mounted

Wall mounted

FIGURE 1-12   Flush valve water closets

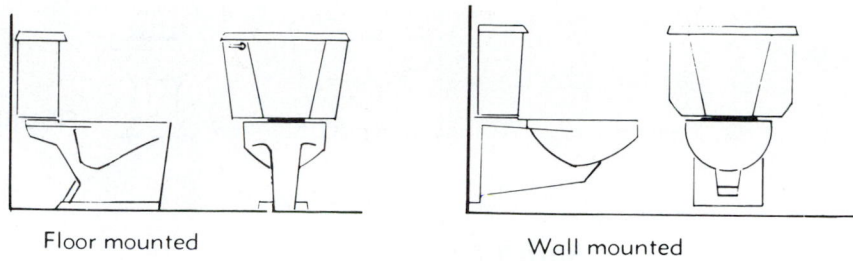

Floor mounted   Wall mounted

FIGURE 1-13   Floor- and wall-mounted water closets

china and sometimes in enameled iron. They may be flush tank or flush valve and are available in three basic styles—wall, stall, and pedestal—as shown in Fig. 1-14.

Bidets

Bidets (Fig. 1-15) are designed to wash the perineal area after using the water closet. The bidet is used extensively in Europe and South America and is enjoying increased usage in Canada and the United States. It is designed for use by the entire family and is installed beside the water closet. The user sits on the fixture facing the wall (and the water controls) and is cleansed by a rinsing spray. It is available in vitreous china.

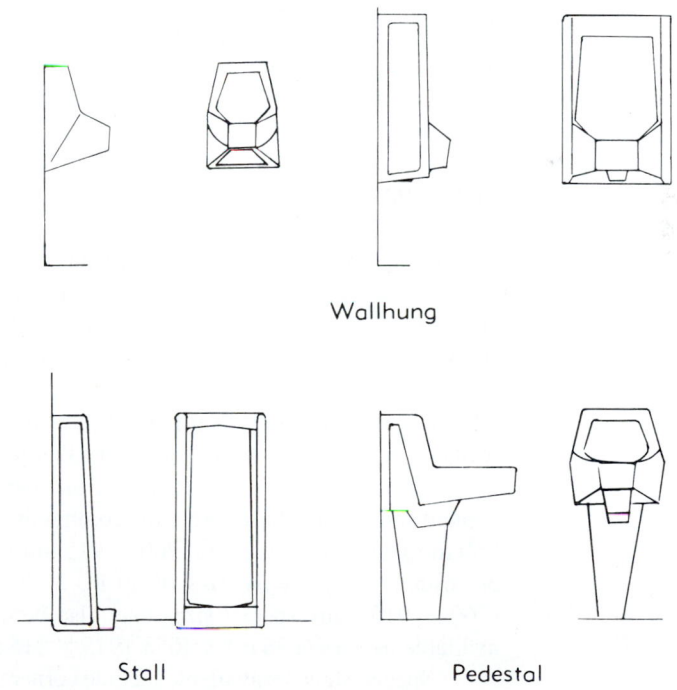

Wallhung

Stall   Pedestal

FIGURE 1-14   Urinals

FIGURE 1-15   Bidets

Bathtubs

Bathtubs are available in enameled iron, cast iron, or fiberglass. Tubs are available in quite a variety of sizes, the most common being 30 or 32 in. (760 or 810 mm) wide; 12, 14, or 16 in. (300, 350 or 400 mm) high; and 4 to 6 ft (1.2 to 1.8 m) long.

Enameled iron tubs (formed steel with a porcelain enamel finish) are generally available in lengths of 4½ and 5 ft (1.37 and 1.53 m) widths of 30 to 31 in. (760 to 785 mm), and typical depths of 15 to 15½ in. (375 to 387 mm).

The most commonly available length of fiberglass bathtubs is 5 ft (1.5 m), and it takes 34 to 36 in. (865 to 915 mm) of width to install. Generally, the units are cast in a single piece. Many include three walls (eliminating the need for ceramic tile around the tub). It is this single-piece feature, with no cracks or sharp corners to clean, which makes the fiberglass tub so popular with clients. The size of the unit makes it almost impossible to fit it through the standard bathroom door; it must therefore be ordered and delivered early enough to be set in place before walls and doors are finished. In wood frame buildings, these units are usually delivered to the job and put in place before the plaster or gypsum board is put on the walls or the doors installed. When selecting fiberglass tubs, be certain to specify only manufacturers who are widely known and respected, with long experience in the plumbing fixture field. Off-brands often give unsatisfactory results in that the fiberglass "gives" as it is stepped on, making a slight noise. In addition, some may be far more susceptible to scratching and damage.

Bathtub fittings may be installed on only one end of a tub, and the tub is designated by the end at which they are placed. As you face the tub, if the fittings are placed on the left, it is called a *left-handed* tub, and if placed on the right, it is *right-handed.*

Showers

Showers are available in units of porcelain enameled steel or fiberglass. They may be built in with a base (bottom) of tile, marble, cement, or molded compositions, and walls may be any of these finishes or porcelain enameled steel. Showers have overhead nozzles which spray water down on the bather. Shower fittings may be placed over bathtubs instead of having a separate shower space; this is commonly done in residences, apartments, and motels. However, it is important that when a shower head is used with a bathtub fixture, the walls be of an impervious material (one that will not absorb water).

Showers of tile, concrete, or marble may be built to any desired size or shape. Preformed shower stall bases are most commonly available in sizes of 30 in. × 30 in. (760 × 760 mm) and 30 in. × 36 in. (760 × 915 mm); other sizes may be ordered. Steel showers are usually available in sizes of 30 in. × 30 in. (760 × 760 mm) and 30 in. × 36 in. (760 × 915 mm); special sizes may also be ordered. Fiberglass showers are commonly available in sizes of 36 in. × 36 in. (915 × 915 mm) and 36 in. × 48 in. (915 × 1220 mm).

Special showers available include corner units and gang head units. Gang head showers are commonly used in institutions, schools, factories where workers must shower after work, and other situations where large numbers of people must shower. The code sets a minimum shower size (except as permitted herein) of at least 1,024 sq. in. (.66 sq. m) of interior cross-sectional space with a minimum interior dimension of 30 in. (760 mm). The

only exception is a prefabricated one-piece shower designed to accommodate a 32 in. ×
32 in. (800 × 800 mm) roughed-in opening, provided it has at least 900 sq. in. (.56 sq. m)
of interior area.

Lavatories

Lavatories (Fig. 1-16) are generally available in vitreous china or enameled iron, or they
may be cast in plastic or a plastic compound with the basin an integral part of the countertop. They are available in a large variety of sizes, and the shapes are usually square, rectangular, round, or oval (and even shell-shaped). The lavatory may be wall-hung, set on legs
or on a stand, or built into a cabinet. Lavatory styles are usually classified as flush-mount,
self-rimming, under-the-counter, or integral, or as units which can be wall-hung or supported on legs.

Special fittings for lavatories include foot controls (often used in institutions such as
hospitals and nursing homes), and self-closing faucets, which are commonly used in public facilities (especially on hot-water faucets) to conserve water.

Kitchen Sinks

Kitchen sinks are most commonly made of enameled cast iron or stainless steel. Sinks are
usually available in a single-or a double-bowl arrangement; some even have a third bowl
which is generally much smaller. Quite often a garbage disposal is connected to one of the
sinks. Kitchen sinks are generally flush-mounted into a plastic laminate or into a composi-
tion plastic counter.

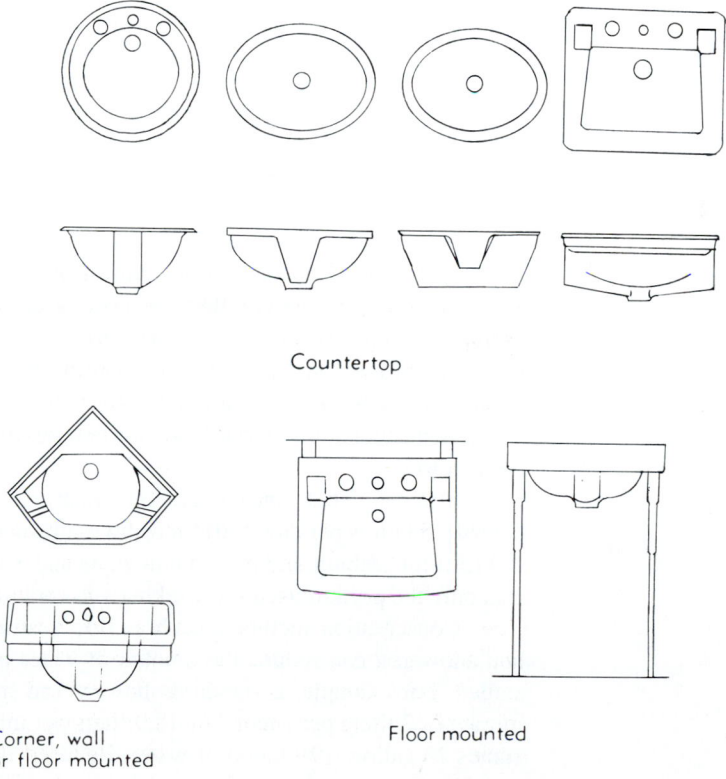

Countertop

Corner, wall
or floor mounted

Floor mounted

FIGURE 1-16   Lavatories

Service Sinks

Service sinks are made of enameled cast iron or vitreous china, and they are often called *slop sinks.* Most service sinks have high backs, and there may be two or as many as three bowl compartments. Other sinks commonly used are laundry trays, pantry sinks, bar sinks, and surgeon's sinks.

Minimum Requirements

The codes generally set the minimum number of fixtures that must be installed on a project according to the type of occupancy (Fig. 1-19). For example, a theatre with 300 seats must have two water closets for men (of which one may be a urinal), two water closets for women, and one lavatory in each bathroom. When designing any commercial, industrial, or institutional project, this minimum fixture chart must be checked.

Minimum Clearances

The code sets a variety of minimum fixture clearances (Fig. 1-17). These set the minimum distances between the fixtures and any walls and the spacing between fixtures and clearances in front of a fixture, either to another fixture or to a wall. Keep in mind that these are minimums; typically, larger clearances are used.

The Americans with Disabilities Act sets a variety of requirements to provide accessibility and usability for people with physical disabilities. In the case of plumbing fixtures, it means they must be accessible for their use. The American National Standards Institute (ANSI) has developed guidelines for minimum dimensions for plumbing fixture installation; the dimensions are available as the ANSI A1117.1 standards. Some typical sizes are shown in Fig. 1-17.

1-9   CONSERVATION

As people understand that the amount of potable water is not an inexhaustible resource (and becoming expensive), they are making efforts to cut back on its use. The introduction of running water and waste systems in buildings is a rather new experience, occurring only in the last hundred years. During this time, we have progressed from taking a bath once a week (the Saturday night bath) to bathing daily. Bathing more than once a day is not uncommon when we are hot, having just mowed the lawn, worked out, or gotten dirty.

It is estimated that the average water use is about 14 gallons per person per day (g/ppd) (53 liters per day, L/d) for dishwashing and clothes washing, with another 21 g/ppd (80 L/d) for bathing and personal hygiene and 3 g/ppd (12 L/d) for drinking and cooking. It is only the portion used for drinking and cooking that has to be potable.

Conservation methods, such as flow restricters on all water outlets (such as sinks and showers), can reduce the amount of water used by 50%, depending on the type installed. For example, a standard shower head may have a water flow of 5 gallons per minute (.32 liters per second or 18.9 liters per minute), so just a five-minute shower consumes 25 gallons (95 liters) of water. Reduced-flow shower heads are available from 1.8 to 2.5 gallons per minute (6.8 to 9.5 liters) flow. This not only saves water but also a great deal of heated water.

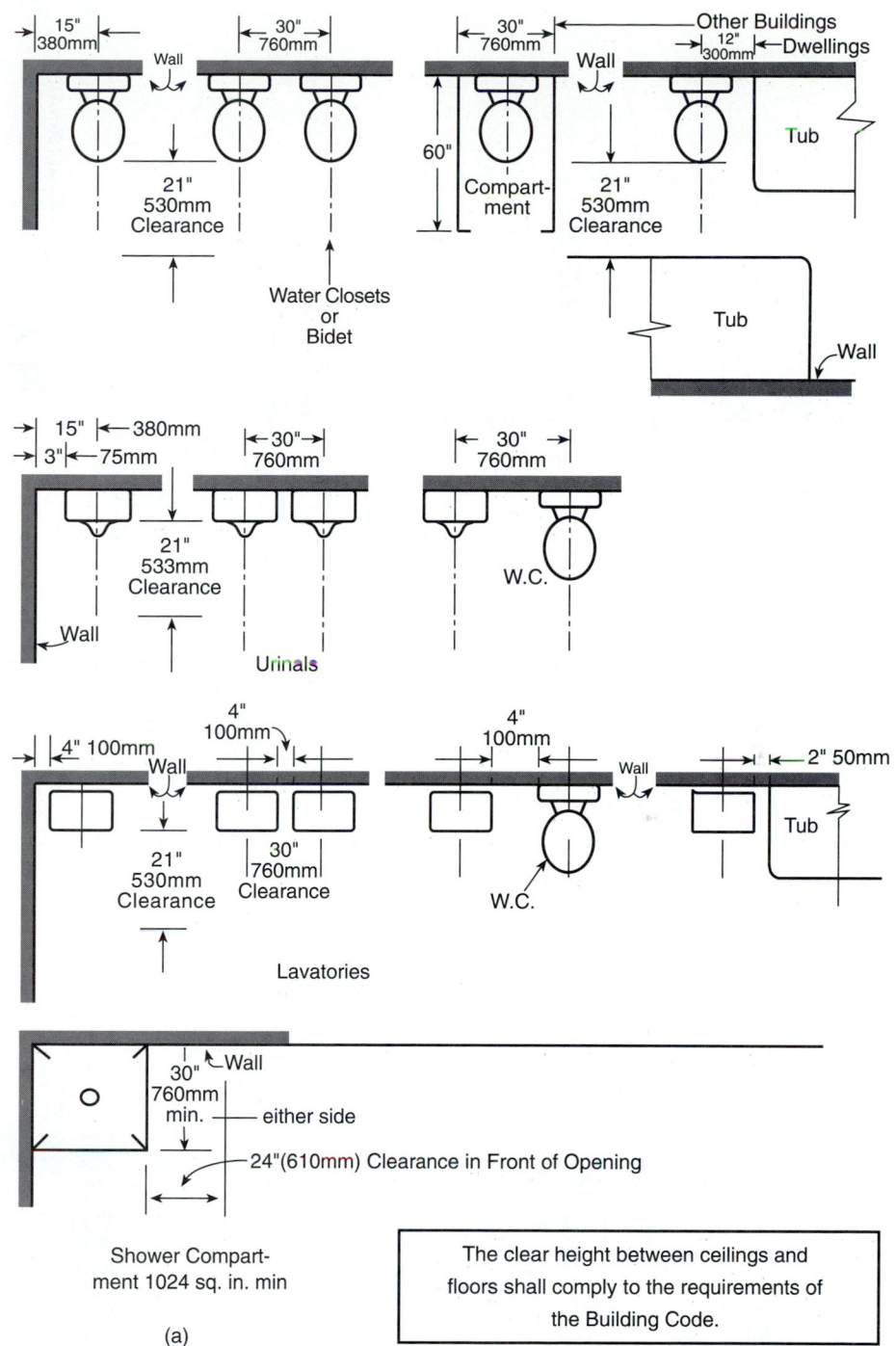

FIGURE 1-17 Minimum clearances (continued next page)

Similar savings can be realized by using water-saving toilets. Originally, the water tank used to flush a toilet bowl was mounted high on the wall so the water would gain enough velocity to wash away the waste. This could be accomplished with between 1 and 1½ gallons (3.8 to 5.7 liters) of water. Over the years, it became fashionable to put the tank lower, and finally units were designed that were single low-profile units. All of these improvements meant that more water volume was necessary to wash away the waste, and each flush used from 5 to 7 gallons of water.

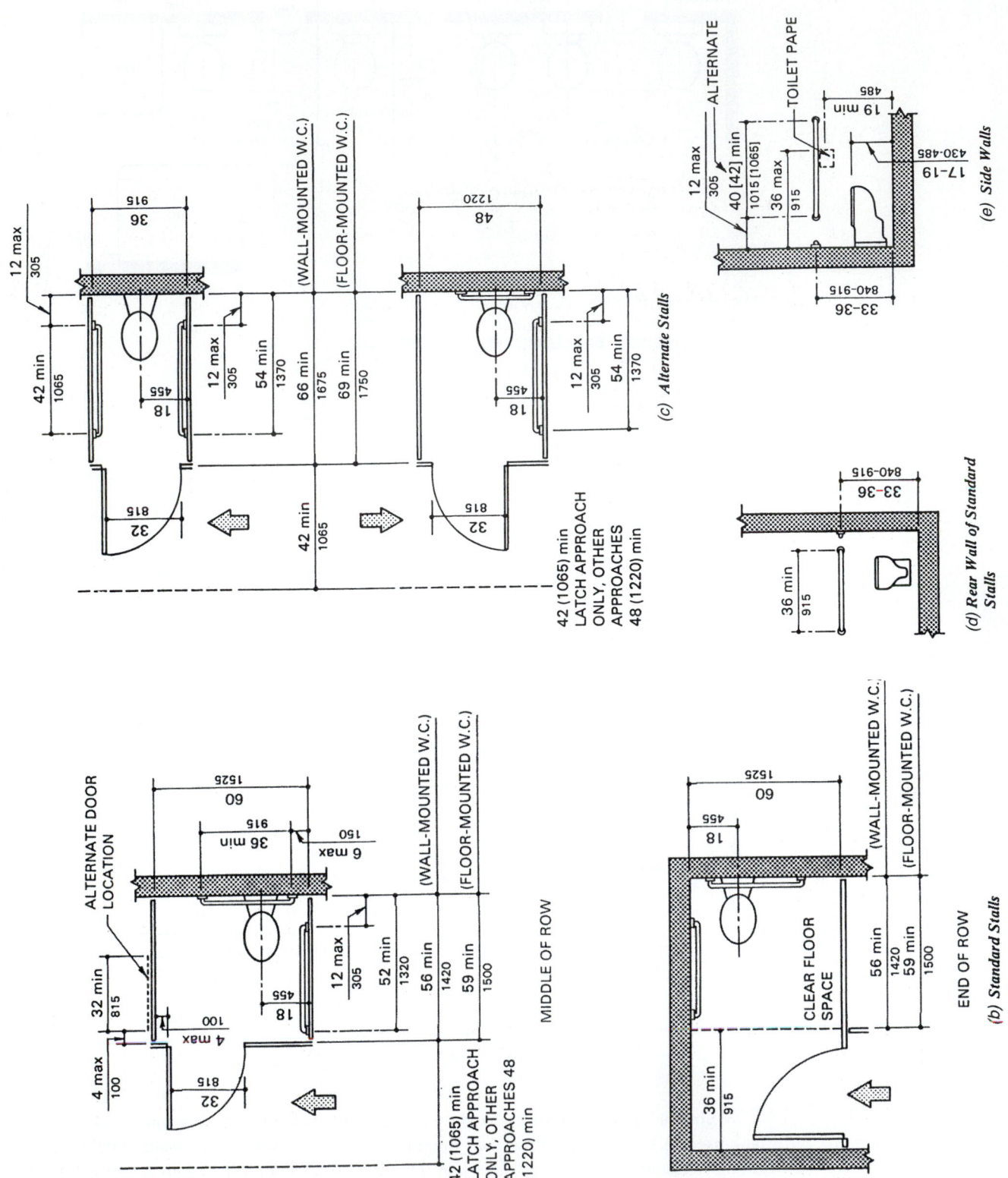

(c) Alternate Stalls

(WALL-MOUNTED W.C.)
(FLOOR-MOUNTED W.C.)

42 (1065) min
LATCH APPROACH
ONLY, OTHER
APPROACHES
48 (1220) min

(e) Side Walls

ALTERNATE

TOILET PAPE

(d) Rear Wall of Standard Stalls

ALTERNATE DOOR LOCATION

(WALL-MOUNTED W.C.)
(FLOOR-MOUNTED W.C.)

MIDDLE OF ROW

42 (1065) min
LATCH APPROACH
ONLY, OTHER
APPROACHES 48
(1220) min

CLEAR FLOOR SPACE

(WALL-MOUNTED W.C.)
(FLOOR-MOUNTED W.C.)

END OF ROW

(b) Standard Stalls

FIGURE 1-17   *continued*

Currently the U.S. government is allowing the manufacture of toilets that use no more than 1.6 gallons (6 liters) per flush. In addition, there are a number of devices available for use in existing water closet tanks to cut the use of water by 50% and more.

These installations would save significant amounts of potable water by themselves. In addition, they would greatly reduce the amount of water that must be treated by sewage treatment plants and reduce the need for additional plants.

Another approach to conserving potable water is a water reuse system. This system involves the processing of household waste water for reuse.

In the design of the system, a typical family of four is used with typical household appliances. The water from the bathtub or shower and the washing machine is run into a collection tank instead of going into the sewer lines. From the collection tank the water is filtered and chlorinated and then reused as water to flush the toilets. A schematic illustrating the flow of the water is shown in Fig. 1-18. This water reuse system cuts water consumption by one-half. Of course, the potable water system is kept completely separate from the reuse portion of the system, and all waste from water closets goes directly into the sewer.

In another experiment, all of the household water, except for that from the garbage disposal and water closets, is processed for multiple reuse in the system. This results in savings of up to 70% of overall household water consumption.

In addition to the reduced amount of water required, savings from such a system result in:

1. Smaller community or private sewer systems.

2. Smaller community treatment plants required to treat sewage.

3. Smaller community treatment plants required to treat supply water (when required).

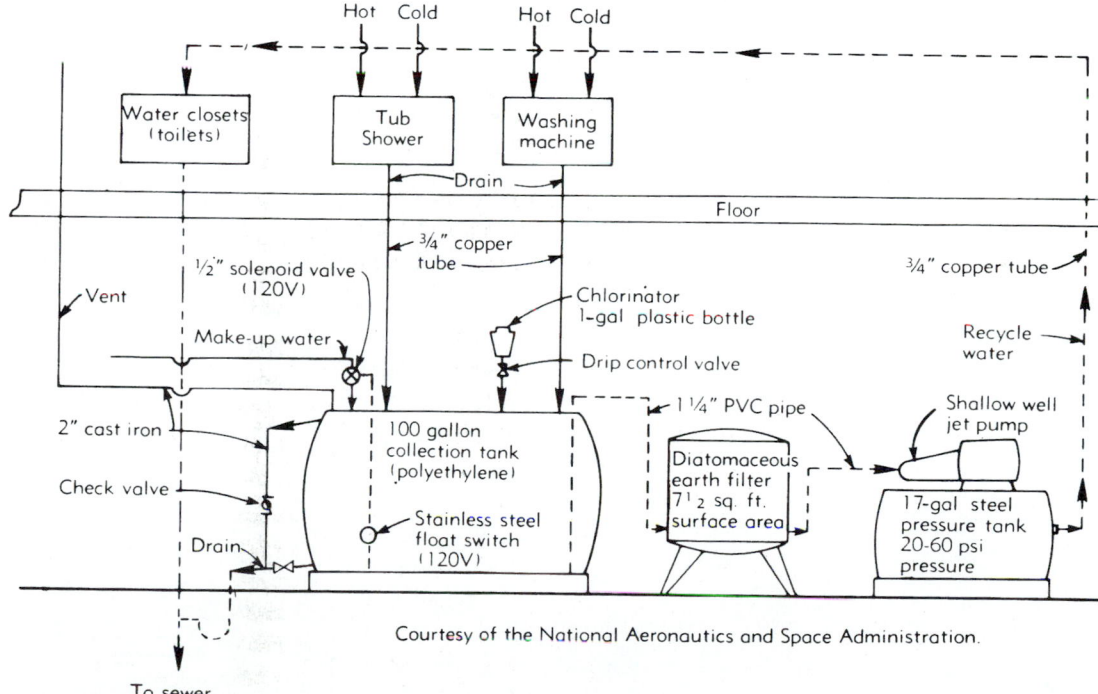

Courtesy of the National Aeronautics and Space Administration.

FIGURE 1-18   Water reuse system

QUESTIONS

1-1. What is the difference between potable and nonpotable water? For what purposes may each be used?

1-2. What sources of water supply may be available to a city? To a private individual?

1-3. Why should any source of water be tested before the water is used?

1-4. What is the basic difference between a community and a private water supply system, and what are the advantages and disadvantages of each?

1-5. When a project (building) being designed is going to connect to a community water supply system, what information about the system must be obtained?

1-6. How are wells classified, and what methods of construction may be used for each type?

1-7. Show with a sketch how a well may be protected from surface water contaminants.

1-8. How are wells protected from possible underground contaminants from sewage disposal fields?

1-9. What two types of pumps are used, and what are the limitations of each?

1-10. What is a water reuse system, how does it work, and why might such a system be desirable?

Design Exercise

1-11. Determine the minimum number of plumbing fixtures required for an auditorium planned for a total of 12,500 people (assume 50% of each sex).

1-12. Determine the minimum number of plumbing fixtures required for a restaurant planned for 250 seatings (50% of each sex).

1-13. Determine the minimum number of plumbing fixtures required for a higher education school building planned to accommodate 750 people (50% of each sex).

MINIMUM NUMBER OF REQUIRED PLUMBING FIXTURES[1]

Building Use Group	Water closets (Urinals)[5] No. of Persons of each sex	Male[4]	Female[4]	Lavatories	Drinking Water Facilities	Bathtub or Shower	Other	Note
Assembly and Educational								
A. Assembly Churches, auditoriums, museums, theaters, libraries and other similar uses.	1-50 51-300 ea. add'l 300 over 300	1 2 add 1	1 2 add 2	1/2 no. req. water closets	1/1000 people		1 service sink/floor	6, 9, 12
B. Assembly Arenas, stadiums, ballparks, passenger terminals, convention halls and other similar uses.	1-100 101-200 201-400 ea. add'l 300 over 400	2 3 4 add 1	2 3 4 add 2	1/2 no. req. water closets	1/1000 people		1 service sink/floor	2, 3, 6, 9, 12, 13
C. Assembly Restaurants and nightclubs where seating is provided.	1-25 26-50 51-100 ea. add'l 200 over 100	1 2 3 add 1	1 2 3 add 2	1/2 no. req. water closets	1/200 people		1 service sink/floor	2, 3, 6, 9, 12, 13
D. Assembly Recreational facilities: includes health spas, golf courses, public swimming pools and similar uses.	1-40 ea. add'l over 40	1 add 1	1 add 2	1/2 no. req. water closets	1/75 people	1 shower/15 people over 150 add 1/30 people	1 service sink/floor	2, 3, 6, 9, 12

FIGURE 1-19   Minimum number of required plumbing fixtures

Building Use Group	Water closets (Urinals)[5]			Lavatories	Drinking Water Facilities	Bathtub or Shower	Other	Note
	No. of Persons of each sex	Male[4]	Female[4]					
E. Educational								
1. Preschool/Day Care	1-15	1	1	1/2 no. req. water closets	1/100 people		1 service sink/floor	6, 9
	ea. add'l 15	add 1	add 1					
	ea. add'l 30	add 1	add 2					
	over 90							
2. Elementary	1-25	1	1	1/2 no. req. water closets	1/100 people		1 service sink/floor	6, 9
	ea. add'l 25	add 1	add 1					
	ea. add'l 50	add 1	add 2					
	over 100							
3. Secondary/Higher Education	1-30	1	1	1/2 no. req. water closets	1/100 people		1 service	6, 9
	ea. add'l 30	add 1	add 1					
	ea. add'l 60	add 1	add 2					
	over 120							

Notes for Table

1. This table shall be used unless superseded by the Building Code. For handicap requirements see local, state or national codes. Additional fixtures may be required where environmental conditions or special activities may be encountered.

2. Drinking fountains are not required in restaurants or other food service establishments if drinking water service is available. Drinking water is not required for customers where normal occupancy is short term. A kitchen or bar sink may be used for employee water drinking facilities.

3. In food preparation areas, fixture requirements may be dictated by local Health Codes.

4. Whenever both sexes are present in approximately equal numbers, multiply the total census by 60% to determine the number of persons for each sex to be provided for. This regulation only applies when specific information, which would otherwise affect the fixture count, is not provided.

5. Not more than 50% of the required number of water closets may be urinals.

6. In building with multiple floors, accessibility to the fixtures shall not exceed one vertical story.

9. Fixtures accessible only to private offices shall not be counted to determine compliance with this section.

12. Requirements for employees and customers may be met with a single set of restrooms. The required number of fixtures shall be the greater of the required number of employees, or the required number of customers.

13. If the design number of customers in food handling establishments exceeds 100, separate facilities for employees and customers are required.

FIGURE 1-19   *continued*

MINIMUM NUMBER OF REQUIRED PLUMBING FIXTURES[1]

Building Use Group	Water closets (Urinals)[5] No. of Persons of each sex Male[4]	Female[4]	Lavatories	Drinking Water Facilities	Bathtub or Shower	Other	Note
Workplaces							
A. Employees Stores, shopping centers, offices, banks, light industrial and similar uses.			1/2 no. req. water closets	1/100 people		1 service sink/floor	6, 8 12, 13
1-15	1	1					
16-40	2	2					
41-75	3	3					
over 75 ea. add'l 60	add 1	add 2					
B. employees Heavy industrial and service occupancies such as those where employees are subject to high heat or skin contamination			1/2 no. req. water closets	1/75 people	1 shower/10 when exposed to skin contamination	1 service sink/floor 1 emergency shower & eye wash	6, 9, 12
1-10	1	1					
11-25	2	2					
26-50	3	3					
51-75	4	4					
76-100	5	5					
over 100 ea. add'l 50	add 2	add2					

Notes for Table

1. This table shall be used unless superseded by the Building Code. For handicap requirements see local, state or national codes. Additional fixtures may be required where environmental conditions or special activities may be encountered.

4. Whenever both sexes are present in approximately equal numbers, multiply the total census by 60% to determine the number of persons for each sex to be provided for. This regulation only applies when specific information, which would otherwise affect the fixture count, is not provided.

5. Not more than 50% of the required number of water closets may be urinals.

6. In buildings with multiple floors, accessibility to the fixtures shall not exceed one vertical story.

8. In stores with floor areas of 150 sq. ft. or less, the requirements to provide facilities for use by employees may be met by providing central facilities located accessible to several stores. The maximum distance from entry to any store from this facility shall not exceed 300 ft.

9. Fixtures accessible only to private offices shall not be counted to determine compliance with this section.

12. Requirements for employees and customers may be met with a single set of restrooms. The required number of fixtures shall be the greater of the required number of employees, or the required number of customers.

13. If the design number of customers in food handling establishments exceeds 100, separate facilities for employees and customers are rquired.

FIGURE 1-19 *continued*

MINIMUM NUMBER OF REQUIRED PLUMBING FIXTURES[1]

Building Use Group	Water closets (Urinals)[5] No. of Persons of each sex	Male[4]	Female[4]	Lavatories	Drinking Water Facilities	Bathtub or Shower	Other	Note
Mercantile/Business								
Customers in stores, banks shopping centers, office buildings and carry-out food establishments where seating is not provided.	1-50	1	1	1/2 no. req. water closets	1/1000 people			2, 6, 7, 12
	51-300	2	2					
	ea. add'l 300 over 300	add 1	add 2					

Notes for Table

1. This table shall be used unless superceded by the Building Code. For handicap requirements see local, state or national codes. Additional fixtures may be required where environmental conditions or special activities may be encountered.

2. Drinking fountains are not required in restaurants or other food handling establishments if drinking water service is available. Drinking water is not required for customers where normal occupancy is short term. A kitchen or bar sink may be considered the equivalent of a drinking fountain for employees.

4. Whenever both sexes are present in approximately equal numbers, multiply the total census by 60% to determine the number of persons of each sex to be provided for. This regulation only applies when specific information, which would otherwise affect the fixture count, is not provided

5. Not more than 50% of the required number of water closets may be urinals.

6. In buildings with multiple floors, accessibility to the fixtures shall not exceed one vertical story.

7. Fixtures for public use as required by this section may be met by providing a centrally located facility accessible to several stores. The maximum distance from entry to any store to this facility shall not exceed 500 feet.

12. Requirements for employees and customers may be met with a single set of restrooms. The required number of fixtures shall be the greater of the required number of employees, or the required number of customers.

FIGURE 1-19   *continued*

MINIMUM NUMBER OF REQUIRED PLUMBING FIXTURES[1]

Building Use Group	Water closets (Urinals)[5] No. of Persons of each sex	Male[4]	Female[4]	Lavatories	Drinking Water Facilities	Bathtub or Shower	Other	Note
Dwelling Units								
A. Single		1 wc/unit		1/unit		1/unit	1 kitchen sink	
B. Multiple		1 wc/unit		1/unit		1/unit	1 kitchen sink/ unit/10; 1 laundry tray/ 100	
C. Dormitories Boarding House	1-20 ea. add'l 20 over 40	2 add 1	2 add 2	1/2 no. req. water closet	1/100 or 1 per floor	2 ea. add'l 20 add 1	1 service sink per floor; 1 laundry tray per 100	6, 9, 10
D. Hotel/Motel		1 wc/unit		1/unit		1/unit	1 service sink per floor	

Notes for Table

1. This table shall be used unless superceded by the Building Code. For handicap requirements see local, state or national codes. Additional fixtures may be required where environmental conditions or special activities may be encountered.

4. Whenever both sexes are present in approximately equal numbers, multiply the total census by 60% to determine the number of persons of each sex to be provided for. This regulation only applies when specific information, which would otherwise affect the fixture count, is not provided

5. Not more than 50% of the required number of water closets may be urinals.

6. In buildings with multiple floors, accessibility to the fixtures shall not exceed one vertical story.

9. Fixtures accessible only to private offices shall not be counted to determine compliance with this section.

10. Multiple dwelling or boarding houses without public laundry rooms shall not require trays.

FIGURE 1-19 *continued*

MINIMUM NUMBER OF REQUIRED PLUMBING FIXTURES[1]

Building Use Group	Water closets (Urinals)[5] No. of Persons of each sex	Male[4]	Female[4]	Lavatories	Drinking Water Facilities	Bathtub or Shower	Other	Note
Institutional								
A. Hospital	1-8 patients ea. add'l 8	1 add 1	1 add 1	1/2 no. req. water closets	1/100 people	1 per 20 people	1 service sink/floor	
B. Hospital rooms private or semi-private	1 wc/room			1 per room		1 per room		
C. Penal								
1. Penal short term detention	1 per cell or 1 per 4 inmates			1 per cell or 1 per 4 inmates		1/6 inmates	1 service sink/floor	14
2. Long term correctional	1 per cell or 1 per 8 inmates			1 per cell or 1 per 8 inmates		1/15 inmates	1 service sink/floor	

Notes for Table

1. This table shall be used unless superseded by the Building Code. For handicap requirements see local, state or national codes. Additional fixtures may be required where environmental conditions or special activities may be encountered.

4. Whenever both sexes are present in approximately equal numbers, multiply the total census by 60% to determine the number of persons of each sex to be provided for. This regulation only applies when specifiec information, which would otherwise affect the fixture count, is not provided.

5. Not more than 50% of the required number of water closets may be urinals.

14. Water closet and lavatory may be a combination fixture. All showers and lavatories shall have thermostatic control and timing devices.

FIGURE 1-19   *continued*

CHAPTER 2

Water Supply Design

2-1 INTRODUCTION

Plumbing systems are used to performed the two primary functions of water supply and waste disposal. The water supply portion of the system consists of the piping and fittings which supply hot and cold water from the building water supply to the fixtures, such as lavatories, bathtubs, water closets, dishwashers, clothes washers, and sinks, in the building or project. Only water supply is included in this chapter. The waste disposal portion of the system, which consists of the piping and fittings required to take that water supplied to the fixtures out of the building and into the sewer line or disposal field, is discussed in Chapter 3.

2-2 CODES

Building codes are the regulations which govern the private actions of those who build or modify buildings. They are for the protection of public health, safety, and welfare. The codes establish certain minimum requirements for the construction and subsequent occupancy of buildings.

Plumbing codes may be a part of the general building code or, more commonly, a separate code. The code in force in any locale is determined by the municipality involved (each individual area, or government unit, selects its own code). The most commonly used codes are the national, regional, and state codes which may be used intact (complete) or with changes to meet the local needs and requirements. It should also be noted that the government unit also has a right to decide to have no code at all.

Since the code in a given locale is a law, it must be complied with in all of the buildings constructed under its jurisdiction. For this reason, it is important that all of the people involved in the planning, design, and construction phases of the project become familiar with the code in effect in the locale in which the building will be constructed.

The National Standard Plumbing Code has been used as a basis for this text since this is the code most often used, referred to, or adapted. But it is important for you, when using the text and learning about the construction industry, to find out what code is used in your locale and to get a copy. You will probably find it very much in agreement with the *National Standard Plumbing Code,* which should also be a part of your library.

In general, plumbing codes limit the types of materials and the sizes of pipe used in the system. The codes generally also form the basis for regulating installation methods.

Multiple Governing Codes

A proposed project may be regulated by several codes, covering the same items at the same time. This situation occurs quite often when the government or a government agency is providing some or all of the financing on a project, and they have certain "rules" or "guidelines" which must be followed. This situation also sometimes occurs when doing business with large corporations. In such cases, the designers involved will have to become familiar with all of the applicable codes and regulations and, when they are in conflict, use the more stringent requirements.

Administration

Where codes are in force, there will probably be some form of Building Department or Department of Building in the government. The local administration and enforcement of the codes are performed by a building inspector or an engineer, who usually reviews the proposed contract documents (drawings and specifications) for compliance with the codes and then checks for compliance during construction. It is important to know just what work must be inspected. Inspections should be scheduled so that work which must be inspected will not be covered before being inspected; otherwise, the inspector may require it to be uncovered so it can be checked.

Such responsibilities require qualified personnel—those who are experienced, informed, and objective. Many levels of government offer courses for the inspectors to provide them with great experience and to make the latest technical information available to them.

2-3 PARTS OF THE SYSTEM

The parts of a typical water supply system are shown in Fig. 2-1. They include the building main, riser, horizontal fixture branch, fixture connection, and a meter in community systems.

Building Main
Connects to the community or private source and extends into the building to the furthest riser. The building main is typically run (located) in a basement, crawl space, or below the concrete floor slab.

Riser
Extends from the building main vertically in the building to the furthest horizontal fixture branch. It is typically run vertically in the walls.

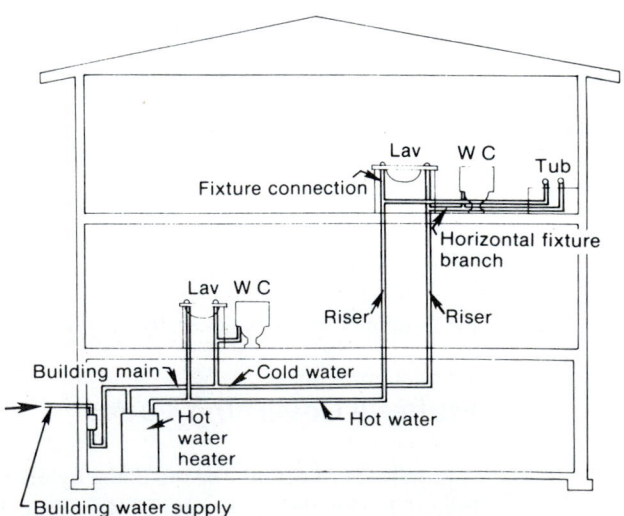

FIGURE 2-1   Parts of a typical water supply system

Horizontal Fixture Branch

Extends horizontally from the riser to the furthest fixture to be connected. It is usually run in the floor or in the wall behind the fixtures.

Fixture Connection

Extends from the horizontal fixture branch to the fixture.

Meter

Required by most community water supply systems to measure and record the amount of water used. It may be placed in a meter box located in the ground, near the street, or inside the building.

2-4 PIPING MATERIALS

Only piping used for water supply is included in this section. Drainage, waste, and vent piping are covered in Sec. 3-3. Piping most commonly used for water supply includes copper, wrought-iron, steel, plastic, and occasionally brass.

Copper is one of the most popular water supply pipes. The pipe types available are K, L, and M, with K having the thickest walls, then L, and finally M with the thinnest walls of this group (DWV copper tubing is used for drainage, waste, and vent piping). The thin walls of copper pipe are usually soldered to the fittings. This allows all of the pipes and fittings to be set into place before joints must be "finished" (in this case, by soldering). This advantage generally allows faster installation of copper pipe. Compared with iron or steel pipe, copper pipe also has the advantage of not rusting and of being highly resistant to any accumulation of scale (particles) in the pipe.

Type K copper tube is available either rigid (hard temper) or flexible (soft temper). It is used primarily for underground water service in water supply systems. Soft temper tubing 1 in. and smaller is usually available in coils 60 or 100 ft (18.3 or 30.5 m) long while 1¼ and 1½ in. tubing is available in 40- or 60-ft (12.2 or 18.3 m) coils. Hard temper is available in 12- and 20-ft (3.7 and 6.1 m) straight lengths. Type K copper tubing is color-coded in green for quick visual identification.

Type L copper tube is also available in either hard or soft temper in coils and straight lengths. The soft temper tubing is often used as replacement plumbing because the flexibility of the tube allows easier installation. Hard temper tubing is often used for new installations, particularly in commercial work. Type L copper tubing is color-coded blue. This type of tubing is most popular for use in water supply systems.

Type M copper tube is made in hard temper only and is available in straight lengths of 12 and 20 ft (3.7 and 6.1 m). It has the thinnest wall and is used for branch supplies where water pressure is not too great, but it is not used for risers and mains. It is also used for chilled water systems, exposed lines in hot water heating systems, and drainage piping. Type M copper tubing is color-coded red.

Copper tubing has a lower friction loss than wrought-iron or steel, providing the designer with an additional advantage. Also, the outside dimensions of the fittings are smaller, which makes a neater, better-looking job. With wrought-iron and steel pipe, the bigger outside dimensions of the fittings sometimes require that wider walls be used in the building.

Red brass piping, consisting of 85% copper and 15% zinc, is also sometimes used as water supply piping. The pipe is threaded for fitting connections, but this requires thicker walls to accommodate the threading, making installation and handling more difficult than for copper. In addition, its relatively higher total cost, installed on the job, limits its usage.

Plastic pipe is also available for water supply systems. Its economy and ease of installation make it increasingly popular, especially on projects such as low-cost housing or apartments, where "cost economy" is most important. Available in 10-ft (3 m) coils or lengths, it is lighter than steel or copper and requires no special tools to install. While many plumbing subcontractors, engineers, and architects still prefer copper, the use of plastic pipe will continue to increase. It is important to check the plumbing code in force in your locale since some areas still do not allow the use of plastic pipe for water supply systems. This dates back to early concerns about possible toxicity (poisoning) resulting from the use of plastic pipe; this concern has long since been proved groundless. Plastic pipe used for water supply should carry the NSF (National Sanitation Foundation) seal. However, not all plastic pipe available should be used for water supply; much of it has been manufactured

MATERIALS FOR POTABLE WATER (1) (2) (3)	WATER SERVICE PIPING	COLD WATER DISTRIBUTION	HOT WATER DISTRIBUTION
ABS Plastic Pipe, SDR (ASTM D2282)	●	●	
ABS Plastic Pipe, schedule 40 or 80 (ASTM D1527)	●	●	
Brass Pipe (ASTM B43)	●	●	●
Copper Pipe (ASTM B42)	●	●	●
Copper Water Tube, Type K or L (ASTM B88)	●	●	●
Copper Water Tube, Type M (ASTM B88)	●	●	●
CPVC Plastic Pipe, Schedule 40,80 (ASTM F441)	●	●	
CPVC Plastic Pipe, SDR (ASTM F442)	●	●	
CPVC Plastic Water Distribution Systems (ASTM D2846)	●	●	●
Ductile Iron Pipe, cement-lined (ASTM A377, ANSI/AWWA C151/A21.51)	●		
Fiberglass Pressure Pipe (AWWA C950)	●		
Galvanized Steel Pipe (ASTM A53)	●	●	●
PB Plastic Pipe, SDR (ASTM D3000)	●	●	
PB Plastic Pipe, SIDR (ASTM D2662)	●	●	
PB Plastic Tubing (ASTM D2666)	●	●	
PB Plastic Water Distribution Systems (ASTM D3309)	●	●	●
PB Plastic Pressure Pipe and Tubing (AWWA C902)	●		
PE Plastic Pipe, Schedule 40 (ASTM D2104)	●	●	
PE Plastic Pipe, Schedule 40,80 (ASTM D2447)	●	●	
PE Plastic Pipe, SDR (ASTM D3035)	●	●	
PE Plastic Pipe, SIDR (ASTM D2239)	●	●	
PE Plastic Tube (ASTM D2737)	●	●	
PE Plastic Pressure Pipe and Tubing (AWWA C901)	●		
PVC Plastic Pressure Pipe, AWWA C900	●		
PVC Plastic Pipe, Schedule 40,80,120 (ASTM D1785)	●	●	
PVC Plastic Pipe, SDR (ASTM D2241)	●	●	

(1) Piping for potable water shall be water pressure rated for not less than 160 psig at 73F.

(2) Piping for hot water shall be applied within the limits of its listed standard and the manufacturer's recommendations.

(3) Plastic piping materials shall comply with NSF 14.

FIGURE 2-2   Plastic pipe and tubing for potable water

for use in the drainage portion of the plumbing system. The chart in Fig. 2-2 shows plastic materials and their usual use in the water supply system.

Wrought-iron pipe is available in diameters from ⅜ in. to 24 in. (10 to 600 mm). Lightweight wrought-iron pipe, designated standard (or schedule 40), is the type most commonly used for water supply systems. The wrought-iron pipe used is most commonly galvanized to add extra corrosion resistance. Occasionally, it is used as the service main from the community main to the riser. Wrought-iron pipe is threaded for connection to the fittings, and it can be identified by a red spiral stripe on the pipe. The higher cost of wrought-iron pipe limits its increased use. Wrought-iron pipe also has a higher friction loss than copper.

Steel pipe is available in diameters from ⅜ in. to 12 in. (10 to 300 mm). Plain steel pipe is usually used only when the water is not corrosive. Galvanized steel pipe is moderately corrosion-resistant and suitable for mildly acid water. It is not used extensively in water supply systems; most plumbers and engineers prefer copper tubing because of its superior resistance to corrosion. Steel pipe is connected to its fittings with threaded connections. Steel pipe also has a higher friction loss than copper.

2-5   FITTINGS

A variety of fittings must be used to install the piping in a project and make all the pipe turns, branch lines, joinings on the straight runs, and stops at the end of the runs. Fittings for steel and wrought-iron pipe are threaded and made of malleable iron and cast iron. The fittings for plastic, copper, and brass pipe are made of the same materials as the pipe being connected.

The 45° and 90° elbows are used to change the direction of the pipe. Unions and couplings are used to join straight runs of pipe. A clamping piece on the coupling allows it to be more easily disengaged for uncoupling of the pipes when future piping revisions are expected at a given point. Tees are used when branch lines must be made; the reducing tee allows different pipe sizes to be joined together. Adapters are used where threaded pipe is being connected to copper or plastic. Adapters have one end threaded to accommodate the steel pipe.

Meters are required in all community systems that charge for water usage. The meter measures the amount of water that passes through it, and then the user is billed for that amount. In cold climates, the meter is usually put in where it is least likely to freeze, typically in a basement (Fig. 2-3). The amount of water is transmitted to a recording device outside the building so it can be read at any time. In warmer climates, the water supply line is

FIGURE 2-3   Water meter— cold climate

FIGURE 2-4   Water meter—warm climate

near the surface of the ground, and the meter is often located in a small box near the road side (Fig. 2-4).

2-6   VALVES

Valves are used to control the flow of the water throughout the system. There are usually valves at risers (vertical pipe serving the building), branches (horizontal pipe serving the fixtures), and any pipes to individual fixtures or equipment. The proper location of valves simplifies repairs to the system, fixtures, or equipment being serviced.

The *globe valve* (Fig. 2-5) is a compression-type valve, commonly used where there is occasional or periodic use, such as lavatories (faucets) and hose connections (called hose bibbs). This type of valve usually closes the flow of water and is partially or fully opened only periodically to allow the water to flow. In reviewing Fig. 2-5, the handle is turned and a washer on the bottom of the stem is forced against the metal seat, which stops the flow of water. To allow the water to flow, the handle is turned and the washer separates from the seat; the more flow desired, the more the valve is opened. The design of the globe valve is such that the water passing through is forced to make two 90° turns, which greatly increases the friction loss in this valve compared with that in a gate valve.

The *angle valve* (Fig. 2-6) is similar in operation to the globe valve, utilizing the same principle of compressing a washer against a metal seat to cut the flow of water. It is com-

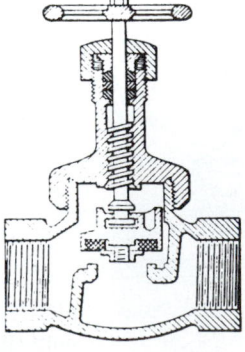

FIGURE 2-5   Globe valve

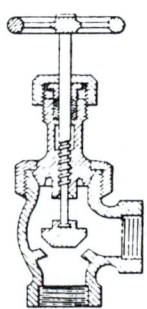

FIGURE 2-6   Angle valve

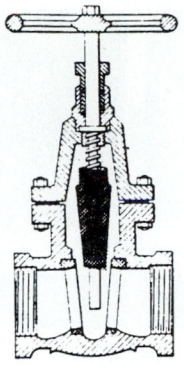

FIGURE 2-7   Gate valve

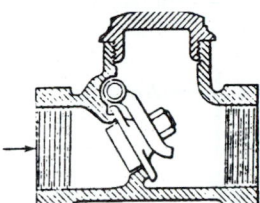

FIGURE 2-8   Check valve

monly used for outside hose bibbs. The angle valve has a much higher friction loss than the gate valve and about half the friction loss of the globe valve.

The *gate valve* (Fig. 2-7) has a wedge-shaped leaf which, when closed, seals tightly against two metal seats which are set at slight angles. This type of valve is usually used where the flow of the water is left either completely opened or closed for most of the time. Because the flow of water passes straight through the valve, there is very little water pressure lost to friction. The gate valve is usually used to shut off the flow of water to fixtures and equipment when repairs or replacement must be made.

The *check valve* (Fig. 2-8) has a hinged leaf which opens to allow the flow of water in the direction desired (indicated by an arrow in the illustration). But the leaf closes if there is any flow of water in the other direction. This eliminates any possible flow of water in a direction other than that desired, or required, by the designer. The check valve works automatically so there is no need for a handle. This valve is used in such places as the water feed line to a boiler (heating unit) where the water from the boiler might pollute the system if it backed up. The inside of the valve is made accessible for repairs by removing the cover (see Fig. 2-8).

Valves referred to as *standard weight* will withstand pressures up to 125 psi (860 kPa); high-pressure valves are also available. Most small valves used have bronze bodies, while large valves (2 in. [50 mm] and larger) have iron bodies with noncorrosive moving parts and seats which may be replaced. They are available threaded or soldered to match the pipe or tubing used.

2-7   WATER SHOCK

Water pressure surges from the quick closing of water valves (faucets) may cause the water system to be noisy. This abrupt closing of the valve causes the fast-flowing water to stop quickly and makes the pipes rattle. A length of pipe, installed above the water connection, will act as a cushion or shock absorber as it controls the pressure surge of the

FIGURE 2-9   Air cushion

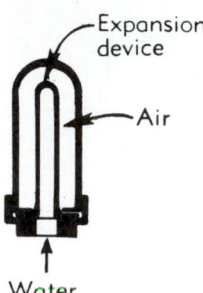

FIGURE 2-10   Shock absorber

water (Fig. 2-9). Special shock absorbers are also available (Fig. 2-10). Oftentimes the noise in the system is referred to as *water hammer.*

2-8   EXPANSION ALLOWANCES

No matter what type of pipe is used in the water supply system, some expansion in the pipe will occur, and this expansion must be considered in the design of the system. The amount of expansion will depend on the type of piping used and the range of temperatures to which the pipe will be subjected.

The piping for hot water will have to withstand a temperature range from about 70°F (21°C), the average indoor temperature, to about 180°F (82°C), the temperature of the wa-

Increase in temperature Deg. F	Elongation in inches per 100 ft. of pipe or tube		
	Steel pipe	Wrought iron pipe	Copper tubing
1°	.0076	.0079	.0112
°C	Elongation in millimeters per 30 meters of pipe or tube		
1°	.342	.355	.512

Note: Obtain plastic elongation characteristics from the manufacturer for the specific type of plastic being used. Plastic expands as much as five times as copper tubing.

FIGURE 2-11   Pipe and tubing expansion (elongation in inches)

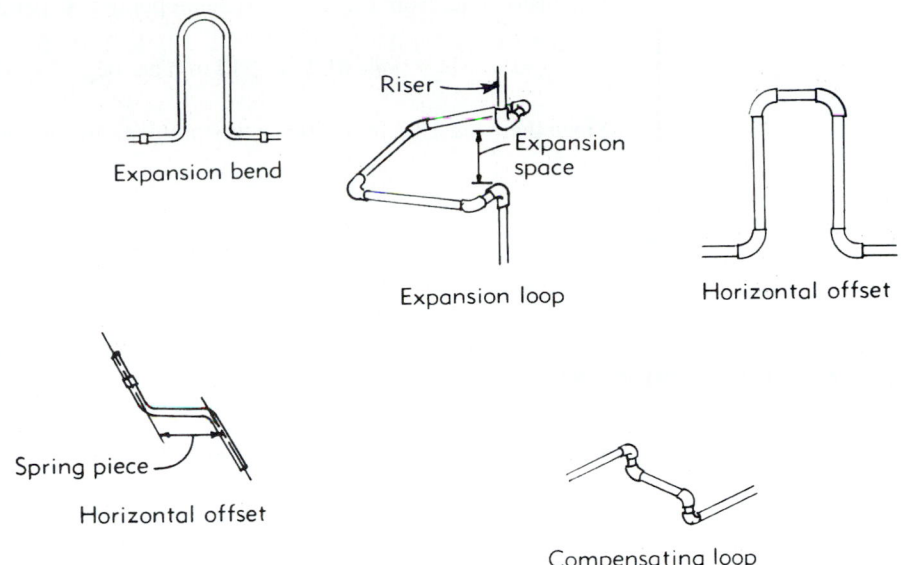

FIGURE 2-12   Expansion allowance

ter. This range will vary, sometimes considerably, on projects and must be checked. Cold water piping will be subjected to a much smaller temperature range, usually with a low of 40°F and a high of 75°F. This range may vary if the piping is placed very close to the hot water line, heating line, or other heat source which may raise the temperature.

The amount of expansion of pipe and tubing which will occur due to temperature change is shown in Fig. 2-11. A review of the expansion figures shows that it would be minimal in the average residence, but it must be considered in large homes and commercial projects. The expansion is allowed for in the system by using one of the methods shown in Fig. 2-12.

Example   *Given:* A four-story apartment building; water service (riser) height of about 40 ft.
Problem: What would the expansion be in a copper tube system if the temperature increases from 70°C to 180°F?

With a temperature difference of 180° − 70° = 110°F, Fig. 2-11 shows the following expansion figures:

0.112 in. per 100 ft. for 100°F temperature rise

0.112 × 1.1 (110°F) = 1.232 in. per 100 ft.

Since there are 40 ft of riser in this problem, the answer is 40/100, 40% or 0.4 of 1.232 in.:

1.232 in. × 0.4 = 0.493 in. of expansion

SI   Solve the expansion problem using SI units

Height: 12.2 m

Temperature increase: 21°C to 82°C

Temperature difference: 61°C

Material: copper tubing

Expansion rate, from Fig. 2-11: .512 mm per deg. C per 30 m

$$61°C \times .512 \text{ mm per deg. C} = 17.08 \text{ mm}$$

Since there are 12.2 m of riser in this problem, the answer is 12.2/30, or 40.67% of 17.08 mm

$$17.08 \text{ mm} \times .4067 = 12.7 \text{ mm of expansion}$$

2-9 WATER SUPPLY SYSTEMS

The two basic types of water supply system used inside the building (or project) are the *up-feed* and the *downfeed* systems.

Since the water pressure in community mains averages about 50 to 60 psi (345 to 410 kPa), and this is also considered about the upper limit for private systems, this places limits on how far the water can be moved in an *upfeed* system (Fig. 2-13). The pressure will be used up in friction losses as the water passes through the building main pipe, the meter, and the various fittings. In addition, part of the pressure is used to overcome *static head*— the pressure required to push water up vertically (up the riser). Also, there must be sufficient pressure left at the top floor to provide proper operation of the fixtures (a shower requires 12 psi, a bathtub 5 psi, etc.).

It requires 0.434 psi to push the water up 1 ft. Pushing the water up 20 ft will require 20×0.434 psi = 8.68 psi. Even if the entire 50 psi were available for static head, the maximum height would be 50 psi ÷ 0.434 psi per ft = 115.2 ft. (It should be noted that 0.434 psi per ft is the same as saying 1 psi can raise water 2.3 ft and 50 psi × 2.3 ft of head per psi = 115 ft.) Depending on the exact floor-to-floor height, this would raise the water 10 to 15 stories (floors) *if there were no friction loss or fixture operation to consider.* Practical limitations set about 60 ft as the usual maximum height, and 40 ft as the preferred height.

 To overcome static head requires 9.8 kPa to push water up one meter. Pushing water up seven meters will require 68.6 kPa (9.8 kPa × 7 m). Even if the entire 345 kPa was available for static head, the maximum height would be 345 kPa divided by 9.8 kPa per meter, or 35.1 meters.

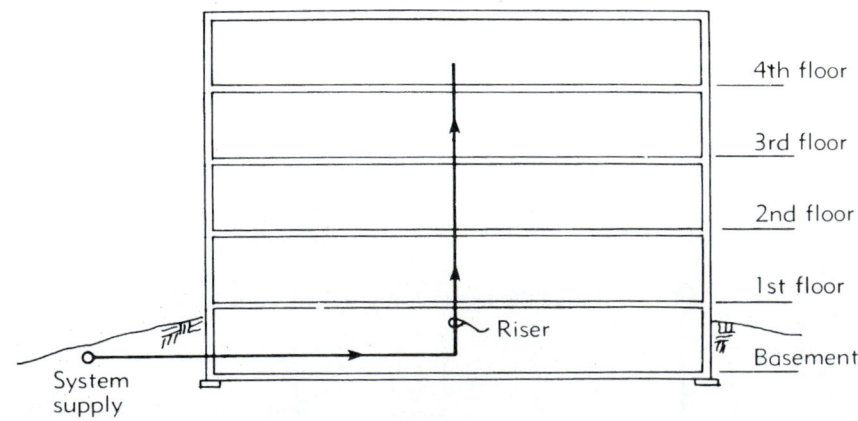

FIGURE 2-13   Upfeed system

In buildings which cannot be adequately serviced to the top floor by an upfeed system, the water is pumped to elevated storage tanks in, or on, the building, and the water is fed down into the building by gravity. This gravity system, fed from the upper stories to the lower, is called a *downfeed* system (Fig. 2-20).

2-10   WATER DEMAND

The amount of water required for the operation of the fixtures installed in a project will depend on the number and kind of fixtures installed and on the probable simultaneous use of the fixtures (for example, it would be highly unlikely that every sink, dishwasher, water closet, bathtub, shower, clothes washer, and garden hose in a residence would be used at one time). The amount of water required is referred to as the *demand load*.

An apartment building with a floor plan such as that shown in Fig. 2-14 would require two risers. When there is more than one riser, it is important that the demand be subtotaled for each riser and then totaled for the building.

Step-By-Step Approach

1. The first step in determining the demand load is to list the plumbing fixtures required on the project. Using the small apartment house shown in Appendix A, the following information is gathered:

   4-story apartment building

   3-in. (75 mm) service main

   Street main pressure; 50 psi (345 kPa)

   Fixtures *per* floor:

      2 flush valve water closets (toilets)

      2 tubs with shower heads

      2 lavatories (bathroom sinks)

      2 kitchen sinks

      4 hose bibs on the first floor (⅜ in. (10 mm) supply, general use)

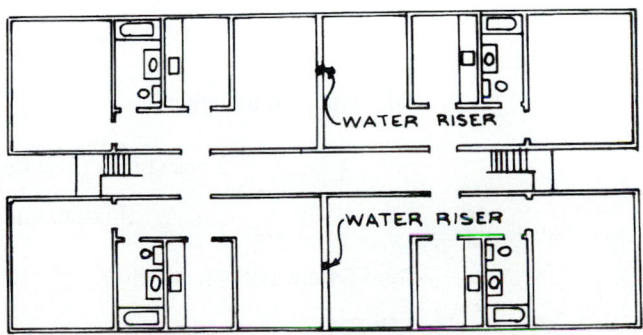

FIGURE 2-14   Water riser location

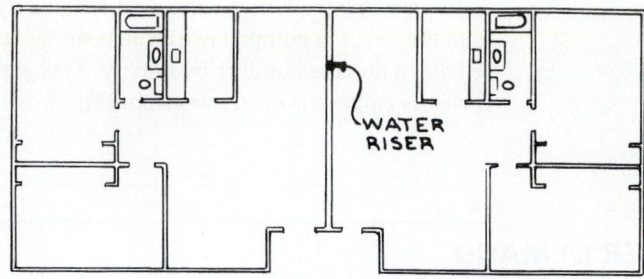

FIGURE 2-15   Single water riser

It should be noted at this point that the apartment building being sized for a water system has a repetitive floor plan for each floor. This allows the entire apartment to be serviced from a single water pipe (riser) going up the building (Fig. 2-15).

2. Next, the demand for each plumbing fixture is listed. This demand is measured in *water supply fixture units (w.s.f.u.)*, and the fixture-unit ratings for various commonly used plumbing fixtures are shown in Figs. 2-40. The w.s.f.u. provides a means of comparing the water supply demands for the fixtures since they are "relative to each other," meaning that a private bathtub with faucet rated at 2 water supply fixture units requires twice as much water as a private lavatory with faucet which is rated at 1 water supply fixture unit.

As the table in Fig. 2-40 shows, the type of occupancy of the building (public or private use) has an effect on the w.s.f.u. value since the greater use of the fixtures in public buildings increases the amount of simultaneous use. Since the water demand supplies the water for both hot and cold requirements, use the total w.s.f.u. column in Fig. 2-40.

Now, calculate the water supply fixture units for the entire apartment building (Appendix A):

Total fixture units per floor:

2 bathroom groups, flush valve, private 8 w.s.f.u. × 2 = 16 w.s.f.u.

2 kitchen sinks 2 w.s.f.u. × 2 = 4 w.s.f.u.

20 w.s.f.u. per floor

Apartment fixture units on riser:

20 f.u. (per floor) × 4 (floors) = 80 w.s.f.u.

Additional fixture units:

2 hose bibbs (first floor) 2 f.u. × 2 = 4 w.s.f.u.

(Fixture units from Fig. 2-41)

Total fixture units on this riser:

80 f.u. (apts.) + 4 f.u. (hose bibbs) = 84 w.s.f.u.

3. The demand for water in gallons per minute can now be determined by using the table in Fig. 2-42 for various water supply fixture unit loads.

$$84 \text{ w.s.f.u.} = 40 \text{ gpm}$$

SI $$84 \text{ w.s.f.u.} = 2.52 \text{ L/s (liters per second)}$$

2-11 UPFEED SYSTEM DESIGN

Water supply pipes must be of sufficient size to provide adequate pressure to all fixtures in the system at a reasonable cost. Selection of economical sizes for the piping and the meter is based on the total demand for water (Sec. 2-10); this process must take into account the available water pressure (street main pressure), the pressure loss due to static head (pressure loss to raise the water up the pipe), the pressure loss due to friction in piping and fittings, and the pressure required to operate the fixture requiring the most pressure on the top floor.

Street main pressure must be adequate to supply:

Required fixture flow pressure
+
Static head
+
Friction in pipe and meter

Step-By-Step Approach

{ Incorrect! in Book. }

The accumulated information to this point (using the apartment building discussed in Sec. 2-10 and shown in Appendix A) is as follows:

4-story apartment building
3-in. (75 mm) service main
Street main pressure: 50 psi (345 kPa)
Predominately flush *tank*

20 w.s.f.u. per floor (per riser)
84 w.s.f.u. (per riser), 168 w.s.f.u. total *min.*
Water demand: 40 gpm (2.52 L/s) per riser
59 gpm (3.72L/s) for main *40gpm*

1. The first step in sizing the supply is to determine the pressure losses. The pressure loss due to static head is figured by multiplying 0.434 psi per ft (9.8 kPa per meter) times the vertical distance of the riser from the service main into the building to the top branch water line (Fig. 2-16).

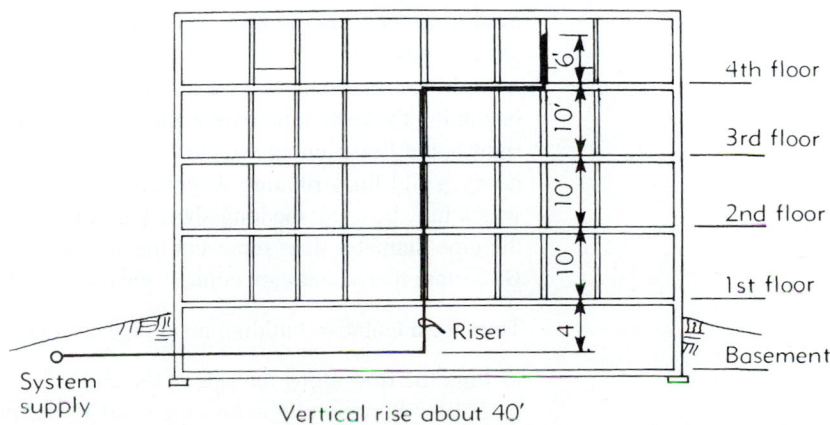

FIGURE 2-16   Vertical riser

$$\text{Static head loss} = 0.434 \text{ psi} \times \text{Vertical rise of water (in ft)}$$

SI

$$\text{Static head loss} = 9.8 \text{ kPa} \times \text{Vertical rise of water (in meters)}$$

From Fig. 2-16, the vertical rise in this four-story building is about 40 ft (12.2 m) from the top of the main to the top of the riser.

$$\text{Static head loss} = 0.434 \text{ psi} \times 40 \text{ ft} = 17.36 \text{ psi}$$

SI

$$\text{Static head loss} = 9.8 \text{ kPa} \times 12.2 \text{ m} = 120 \text{ kPa}$$

2. Next, find the pressure required to operate the fixture on the top floor, requiring the most pressure. The rate of water flow and the required pressure to operate various fixtures properly are shown in Fig. 2-37.

At this time only the pressure is needed, and the maximum pressure required on the top floor is that for the water closet. Of course, the design should also leave some excess pressure so that more than one fixture can be operated properly at the same time.

Water closet: 15 psi (check with manufacturer)

SI

Water closet: 100 kPa (check with manufacturer)

3. The pressure loss due to friction in the main and riser pipes, the fittings, and the meter must be found next. Add the static head and the pressure required for the top fixture; then subtract the total from the available street main pressure to find the pressure left, which is the maximum amount that can be lost to friction.

Street main pressure	50.00 psi
Static head + fixture pressure (17.36 + 15)	−32.36 psi
Pressure left to overcome friction	17.64 psi

Street main pressure	345 kPa
Static head + fixture pressure (120 + 100)	−220 kPa
Pressure left to overcome friction	125 kPa

4. Sizing the piping is often a matter of trial and error, even for experienced engineers; the process involves first selecting a pipe size for the building main, which runs from the water system to the riser(s), and then determining the friction loss for the pipe used from the charts in Figs. 2-43 and 2-44. The chart used will depend on the type of pipe used.

Since these charts have many lines and numbers, use them with care. Before beginning the selection, review what information is on each chart. Along the left and right is the flow, and along the bottom and top is the friction loss in the pipe. The heavy, solid lines running diagonally on the chart represent the diameters of pipe which may be used; the long-short-long lines going perpendicular (at a 90° angle) to the pipe diameter lines represent the velocity of the water in a pipe of a given size. Be certain that you do not confuse the diagonal lines.

To make a tentative building main pipe selection:

a. Find the flow along the side of the chart.
b. Follow horizontally across the chart to the pipe diameters (solid line) and note their sizes.

Going from the left side toward the right, the first pipe size servicing 59 gpm is 4 in. Since this is much larger than the street main, keep moving to the right to the 2½-in. pipe. Where the horizontal line from 59 gpm touches the line representing 2½-in., draw a vertical line to the bottom (or top), and the friction loss reads about 1.26 psi per 100 ft.

Continuing to the right, the following friction readings are taken:

2½-in. pipe = 1.26 psi per 100 ft

2-in. pipe = 3.20 psi per 100 ft

1½-in. pipe = 13.0 psi per 100 ft

1¼-in. pipe = 28.0 psi per 100 ft

1-in. pipe = 74.0 psi per 100 ft (type L copper)

SI

Going from the left side to the right, the first pipe size servicing 3.72 L/s is 100 mm. Since this is much larger than the street main, keep moving to the right to a smaller pipe (65 mm). Where the horizontal line from 3.72 L/s touches the line representing 65 mm, draw a vertical line to the top and the friction loss reads about 28.4 kPa per 100 meters.

Continuing to the right, the following readings are taken:

65 mm pipe = 28.4 kPa per 100 m

50 mm pipe = 72.3 kPa per 100 m

40 mm pipe = 293.8 kPa per 100 m

32 mm pipe = 632.8 kPa per 100 m

25 mm pipe = 1672 kPa per 100 m (type L copper)

5. Since all friction losses are given per 100 ft (100 m) of pipe length, the distance from the street main (or point of service, such as a storage tank) to the base of the riser must be checked on the plot plan.

Referring to Appendix A, the horizontal building main distance is estimated at 60 ft. So, for example, the friction loss in the building main pipe will be 60/100, 3/5, 0.6 or 60% of the figures recorded.

Friction losses for 60 ft of pipe are as follows:

2½-in. pipe = 1.26 psi per 100 ft = 1.26 psi (0.6) = 0.76 psi

2-in. pipe = 3.20 psi per 100 ft = 3.20 psi (0.6) = 1.92 psi

1½-in. pipe = 13.0 psi per 100 ft = 13.0 psi (0.6) = 7.8 psi

1¼-in. pipe = 28 psi per 100 ft   = 28 psi (0.6)   = 16.8 psi

1-in. pipe = 74 psi per 100 ft   = 74 psi (0.6)   = 44.4 psi

By observation, the 1-in. pipe is eliminated since its friction loss is slightly greater than the pressure left in the calculations (17.64 psi). The 1¼-in., 1½-in., and 2½-in. pipe are all possibilities at this point.

SI

Referring to Appendix A, the horizontal building main distance is estimated at 18.3 m. So, for this example, the friction loss in the building main pipe will be 18.3 / 100 or 18.3% of the figures recorded.

Friction losses for 18.3 m of pipe are as follows:

65 mm pipe = 28.4 kPa per 100 m   = 28.4 (0.183)   = 5.2 kPa

50 mm pipe = 72.3 kPa per 100 m   = 72.3 (0.183)   = 13.2 kPa

40 mm pipe = 293.8 kPa per 100 m = 293.8 (0.183) = 53.8 kPa

32 mm pipe = 632.8 kPa per 100 m = 632.8 (0.183) = 115.8 kPa

25 mm pipe = 1672 kPa per 100 m   = 1672 (0.183)   = 306 kPa

By observation, the 25 mm pipe is eliminated since its friction loss is higher than the pressure left for friction loss in the calculations (125 kPa). All of the other pipes are still possibilities at this time.

6. The next step is to determine the pressure loss as the water passes through the meter (if one is required; if not, this step is eliminated). The chart in Fig. 2-47 gives the pressure loss for various meter sizes with the flow in gpm along the bottom and the pressure loss on the left.

    When the main size is 1½ in. or greater, the most common main-meter selection will have the meter one size smaller than the main.

    For this problem, locate the gpm along the bottom—in this case, 59 gpm—and go vertically to the first diagonal line (3-in. meter); at this point move horizontally to the pressure loss reading (0.8 psi). Referring to Fig. 2-47, the pressure loss for a 2-in. pipe meter is about 3.8 psi.

3-in. meter = 0.8 psi

2-in. meter = 3.8 psi

1½-in. meter = 9.7 psi

75 mm meter = 5.56 kPa

50 mm meter = 26.2 kPa

40 mm meter = 66.9 kPa

Totaling the pressure for main and meter:

2½-in. pipe and 2-in. meter   = 0.76 + 3.8 = 4.56 psi

2½-in. pipe and 1½-in. meter = 0.76 + 9.7 = 10.46 psi

2-in. pipe and 2-in. meter   = 1.92 + 3.8 = 5.72 psi

2-in. pipe and 1½-in. meter   = 1.92 + 9.7 = 11.62 psi

1½-in. pipe and 1½-in. meter = 16.8 + 9.7 = 26.5 psi

Totaling the pressure for main and meter:

65 mm pipe and 50 mm meter = 5.2 + 26.2   = 31.4 kPa

65 mm pipe and 40 mm meter = 5.2 + 66.9   = 72.1 kPa

50 mm pipe and 50 mm meter = 13.2 + 26.2 = 39.4 kPa

50 mm pipe and 40 mm meter = 13.2 + 66.9 = 80.1 kPa

40 mm pipe and 40 mm meter = 293.8 + 66.9 = 360.7 kPa

Referring back in the notes, there are 17.64 psi left to overcome pressure losses due to friction in the main, meter, and riser. The totals above are pressure losses for the main and meter; the riser pressure loss must still be calculated.

Since the most economical solution involves the use of the smallest pipe sizes which will provide an adequate flow of water throughout the system, at this point a preliminary selection of 2½-in. pipe and 1½-in. meter is made. This means that of the 17.64 psi that are left to overcome friction, 10.46 psi will be used in the 2½-in. pipe and the 1½-in. meter. This leaves 7.18 psi as the maximum friction loss in the riser.

$$17.64 \text{ psi available for friction loss}$$
$$\underline{-10.46 \text{ psi friction loss in 2½-in. main and 1½-in. meter}}$$
$$7.18 \text{ psi available for riser friction loss}$$

Using a 2-in. main and 1½-in. meter would leave (17.64 − 11.62) = 6.02 psi for riser friction loss. Using the 2½-in. main and 2-in. meter would leave (17.64 − 4.56) = 13.08.

$$125 \quad \text{kPa available friction loss}$$
$$\underline{- \quad 72.1 \text{ kPa friction loss in 65 mm pipe and 40 mm meter}}$$
$$52.9 \text{ kPa available for riser friction loss}$$

Using a 50 mm main and a 40 mm meter would leave (125 − 80.1) = 44.9 kPa for friction loss. Using the 65 mm main and a 50 mm meter would leave (125 − 31.4) = 93.6 kPa.

7. The next step will be to select the possible pipe sizes for the riser. Using the friction-loss pipe chart (Fig. 2-43), find the 40 gpm (2.52 L/s) flow on the side, the following friction-loss readings are then taken:

$$2\text{-in. pipe} = 1.6 \text{ psi per 100 ft}$$

$$1½\text{-in. pipe} = 6.0 \text{ psi per 100 ft}$$

$$1¼\text{-in. pipe} = 13 \text{ psi per 100 ft}$$

$$1\text{-in. pipe} = 34 \text{ psi per 100 ft}$$

$$50 \text{ mm pipe} = \quad 36.2 \text{ kPa per 100 m}$$

$$40 \text{ mm pipe} = 135.6 \text{ kPa per 100 m}$$

$$32 \text{ mm pipe} = 293.8 \text{ kPa per 100 m}$$

$$25 \text{ mm pipe} = 768.4 \text{ kPa per 100 m}$$

Once again, note that the friction losses are given per 100 ft or 100 m of pipe length. This means that the vertical distance from the main to the highest and most remote fixture must be estimated. On some projects, this may mean a meeting with the architects to be certain of how high the ceilings in the building are and of how thick the floor construction is or of what the floor-to-floor height is.

For this apartment building, with floor-to-floor heights of *about* 10 ft-0 in. (3.0 m), the riser height will be approximately 40 ft (12.2 m). In such piping there is an additional friction loss for fittings (such as valves and tees). When the entire system has been designed and all pipes sized and fittings located, it is possible to determine the friction loss for the fittings. An example of how to calculate fittings is given later in this section. In reviewing Fig. 2-45 in order to determine the friction loss for fittings, the table lists the fitting across the top, the diameter required along the left and the equivalent length of pipe for each type and size. This equivalent length of pipe means that, for example, when water passes through a 2-in. (50 mm)-diameter standard 90° ell, the same friction loss occurs as when water passes through 2 lineal ft of 2-in. (.6 m of 50 mm) pipe. So the equivalent length of the 2-in.-diameter 90° elbow is 2 ft. (50 mm diameter 90° elbow is .6 m).

Since, during this stage of the design, the fittings have not been determined, it is common practice to allow an additional 50% of the piping length to account for the friction loss in the fittings.

In this case, the pipe length is 40 ft, and, adding 50% for friction loss in the fittings, the total *equivalent feet of piping* is 40 + 20 = 60 ft.

The friction loss for riser pipes and fittings is calculated next.

Using the friction loss for various pipe sizes per 100 ft (previously taken from Fig. 2-43), convert the friction loss to the equivalent length—in this case, 60 ft or 60/100, 0.60 or 60% of the figures recorded.

$$2\text{-in. pipe} = 1.6 \text{ psi per 100 ft} = 1.6 \text{ psi } (0.60) = 0.96 \text{ psi}$$

$$1\tfrac{1}{2}\text{-in. pipe} = 6.0 \text{ psi per 100 ft} = 6.0 \text{ psi } (0.60) = 3.60 \text{ psi}$$

$$1\tfrac{1}{4}\text{-in. pipe} = 13 \text{ psi per 100 ft} = 13 \text{ psi } (0.60) = 7.8 \text{ psi}$$

$$1\text{-in. pipe} = 34 \text{ psi per 100 ft} = 34 \text{ psi } (0.60) = 20.4 \text{ psi}$$

SI

$$50 \text{ mm pipe} = 36.2 \text{ kPa per 100 m} = 36.2 \text{ } (.183) = 6.62 \text{ kPa}$$

$$40 \text{ mm pipe} = 135.6 \text{ kPa per 100 m} = 135.6 \text{ } (.183) = 24.8 \text{ kPa}$$

$$32 \text{ mm pipe} = 293.8 \text{ kPa per 100 m} = 293.8 \text{ } (.183) = 53.8 \text{ kPa}$$

$$25 \text{ mm pipe} = 768.4 \text{ kPa per 100 m} = 768.4 \text{ } (.183) = 140.62 \text{ kPa}$$

These friction losses are now compared with the preliminary main size selections, and a final selection for main and riser is made. The choices are as follows:

a. Using a 2½-in. main and 1½-in meter which has 10.46 psi friction loss, leaving 7.18 psi for riser friction loss, the choice is limited to 1½-in. and 2-in. riser pipe.
b. Using a 2-in. main and 1½-in. meter which has 11.62 psi friction loss, leaving 6.02 psi for riser friction loss, the riser choices are only the 1½-in., and 2-in. pipe. The final selection may also depend on whether there is any possibility of adding any length to the riser (by adding more stories to the building, making it taller).

The final selection would depend on other variables—in particular whether the main will serve any other building and whether there is any chance of further development on the site (perhaps another apartment building).

There is no *one* correct answer; in this case, the solution which seems most economical is:

2½-in. main and 1½-in. meter with 1½-in. riser

SI

50 mm main and 40 mm meter with a 40 mm riser

Another approach to finding riser pipe size is often used. In this approach, once a main size has been tentatively selected, the riser may be more directly selected from the chart in Fig. 2-43. Using a main size of 2½-in. with a 1½-in. meter, the friction loss available for the riser is 7.18 psi. The following formula will convert the psi left for friction loss and the equivalent length of riser into the friction loss per 100 ft (using a 100-ft constant), so that the chart in Fig. 2-43 may be used more easily.

$$\frac{\text{Available psi} \times 100 \text{ ft}}{\text{Equivalent length (ft)}} = \text{psi loss per 100 ft}$$

In this case:

$$\frac{(7.18)(100)}{(40 + 20)} = 11.97 \text{ psi loss per 100 ft}$$

SI

$$\frac{\text{Available kPa} \times 100 \text{ m}}{\text{Equivalent length}} = \text{kPa per 100 m}$$

In this case:

$$\frac{(52.9)(100)}{(12.2 + 6.1)} = 289 \text{ kPa per 100 m}$$

Now the chart in Fig. 2-43 is used in a slightly different way. The flow is found on the side, and a horizontal line is made. The available friction loss per 100 ft (m) is found and a vertical line made. The smallest riser size which can be used is selected by finding the intersection of the horizontal and vertical lines and taking the first pipe size to the left.

In an actual layout, the equivalent length of the fittings should be checked against the assumed length. Assuming the following fittings are in the riser, the equivalent lengths are taken from Fig. 2-45.

Length Per
Fitting
—————

3 tees, branch 1½ = 5 = 15
3 tees, run, 1½ = 1 = 3
1 gate valve, 1½ = 2 = 2

SI

3 tees, branch, 40 mm = 2.1 = 6.3 m

3 tees, run, 40 mm = 0.5 = 1.5 m

1 gate valve, 60 mm = 0.3 = 0.8 m

In this case the equivalent length added (20 ft or 6.1 m) is just about what is needed. This actual equivalent length will sometimes be more than allowed and sometimes less. When there is considerable difference, the riser size selected may have to be recalculated.

At this point, the information on the upfeed system should be tabulated. The required information includes the fixture units per floor, the accumulated fixture units (which are the units that the riser must serve at any given point), the demand flow and the pipe size. The tabulated information for this problem is shown in Fig. 2-17.

	WATER SUPPLY			
	UPFEED ZONE, APT. BLDG. PROJ. #20–1			
Floor	Accumulated w.s.f.u.	Demand load (gpm)	Friction loss (psi per 100[1])	Pipe size
1	84	40		
2	60	33		
3	40	25		
4	20	14		

SI UNITS				
WATER SUPPLY				
UPFEED ZONE, APT. BLDG. PROJ. #20–1				
Floor	Accumulated w.s.f.u.	Demand load (L/s)	Friction loss (kPa per 100 m)	Pipe size (mm)
1	84	2.52		
2	60	2.08		
3	40	1.58		
4	20	0.88		

FIGURE 2-17   Upfeed data

Now the riser for the upper floors can be sized. First, the accumulated fixture units are figured:

2nd floor . . . . 20 w.s.f.u. . . . . 60 accumulated w.s.f.u.
3rd floor . . . . 30 w.s.f.u. . . . . 40 accumulated w.s.f.u.
4th floor . . . . 20 w.s.f.u. . . . . 20 accumulated w.s.f.u.

Next, the demand load is determined for each floor by using Fig. 2-42 and the accumulated fixture units for each floor (use line 2, predominantly flush tank):

2nd floor . . . . 60 accumulated w.s.f.u. . . . . 33 gpm
3rd floor . . . . 40 accumulated w.s.f.u. . . . . 25 gpm
4th floor . . . . 20 accumulated w.s.f.u. . . . . 14 gpm

The riser pipe sizes are now selected by using the demand load (gpm) for each floor, the pressure available for friction loss in the riser and the chart in Fig. 2-43.

2nd floor . . . . 33 gpm . . . . 1½-in. pipe
3rd floor . . . . 25 gpm . . . . 1½-in. pipe
4th floor . . . . 14 gpm . . . . 1-in. pipe

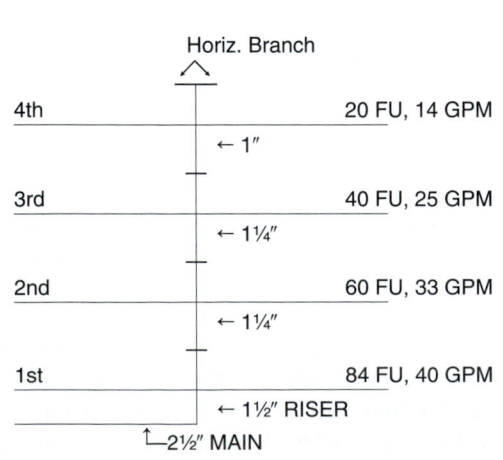
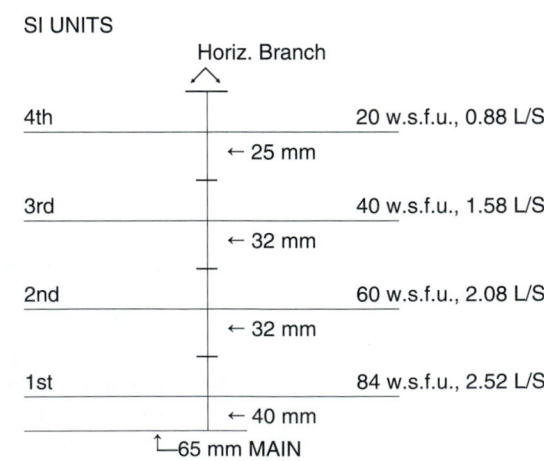

FIGURE 2-18   Upfeed diagram

WATER SUPPLY				
UPFEED ZONE, APT. BLDG. PROJ. #20–1				
Floor	Accumulated w.s.f.u.	Demand load (gpm)	Friction loss (psi per 100[1])	Pipe size
1	84	40	1.97	1½"
2	60	33		1¼"
3	40	25		1¼"
4	20	14		1"

SI UNITS				
WATER SUPPLY				
UPFEED ZONE, APT. BLDG. PROJ. #20–1				
Floor	Accumulated w.s.f.u.	Demand load (L/s)	Friction loss (kPa per 100 m)	Pipe size
1	84	2.52	289	40
2	60	2.08	289	32
3	40	1.58	289	32
4	20	0.88	289	25

FIGURE 2-19   Upfeed data

2nd floor . . . . 2.08 kPa . . . . 32 mm pipe
3rd floor . . . . 1.58 kPa . . . . 32 mm pipe
4th floor . . . . 0.88 kPa . . . . 25 mm pipe

Once all of this information is tabulated as shown in Fig. 2-19, the design is complete.

2-12 DOWNFEED SYSTEM DESIGN

For all tall buildings a downfeed system will be used (Fig. 2-20).

Step-by-Step Approach

1. The first step in the design is to list all the information about the project.

20-story public building
50 psi at community main

Floors 1–18:   Floors 19 and 20:

3 flush valve water closets   3 flush tank water closets
2 urinals, flush valve   2 urinals, flush tank
2 lavatories   2 lavatories
1 service sink   1 service sink

Floor-to-floor height: 11 ft (3.4 m)
Distance from community main to riser: 40 ft (12.2 m)

2. In order to determine the main, meter, and riser sizes for the system, the next step is to determine the fixture units per floor and the total fixture units in the building.

Floors 1–18 (Fixture units from Fig. 2–40):

3 flush valve water closets   10 w.s.f.u. each = 30
2 urinals, flush valve (¾")   5 w.s.f.u. each = 10
2 lavatories   2 w.s.f.u. each = 4
1 service sink   3 w.s.f.u. each = 3
                                               47 w.s.f.u. per floor

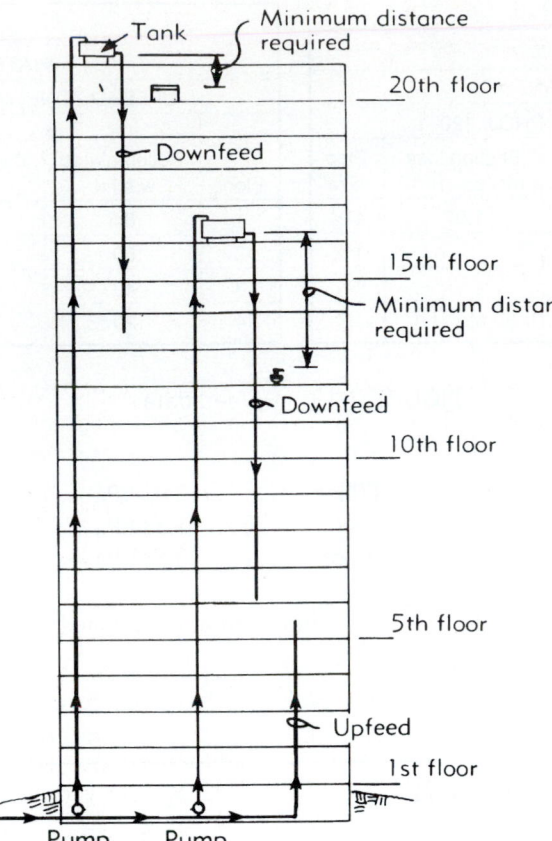

FIGURE 2-20   Downfeed system

47 w.s.f.u. (per floor) $\times$ 18 (floors) = 846 w.s.f.u. for floors 1–18

Floors 19 and 20:

3 flush tank water closets	5 f.u. each =	15
2 urinals, flush tank	3 f.u. each =	6
2 lavatories	2 f.u. each =	4
1 service sink	3 f.u. each =	3
		28 w.s.f.u. per floor

28 w.s.f.u. (per floor) $\times$ 2 (floors) = 56 w.s.f.u. for floors 19 and 20

Total fixture units:

$$846 \text{ w.s.f.u. (floors 1–18)}$$
$$+ \quad 56 \text{ w.s.f.u. (floors 19–20)}$$
$$902 \text{ w.s.f.u. total}$$

3. The next step is to determine the demand load for water in gpm (from Fig 2-42).

The system is predominantly flush valve, and the approximate demand load is:

$$902 \text{ w.s.f.u.} = 201 \text{ gpm}$$

$$902 \text{ w.s.f.u.} = 12.68 \text{ L/s}$$

SI

4. In this problem, the bottom floors of the building will be serviced by upfeed zones (Fig. 2-20 and Fig. 2-21) and the upper floors by downfeed zones.

    The first step will be to make a preliminary decision on how many floors each zone will supply and the number of zones required. The upfeed zone (refer to Sec. 2-11) is limited to a practical maximum height of about 60 ft (20 m). With a floor-to-floor height given as 11 ft (3.4 m), the upfeed system will serve five to six floors (stories), depending on how deep the community main is (a deep main would limit the zone to five floors).

    Since on this project the main is shallow (about 4 ft (1.2 m) below ground level), it will be possible to have an upfeed zone of six floors, if necessary.

5. Downfeed zones also have maximum heights (distances they can serve) without using special pressure reducing and control valves. First, since the water is stored in the tank, it has no pressure in the riser as it leaves the tank. As the water leaves the tank and travels down the riser, the water pressure begins to build at a rate of 0.434 psi per ft or 9.8 kPa (without considering friction loss). This means that the highest fixtures, those nearest the tank, will receive the least pressure.

    In this design, the flush tank water closet on the top floor requires a water pressure of 15 psi to operate (Fig. 2-39). So, if 15 psi is required and the water pressure increases 0.434 psi per foot, then 15 psi ÷ 0.434 psi = 34.6 ft of vertical distance would be required between the tank and the top fixture to provide the pressure required. (This distance would be even greater if flush valve water closets were used on the top floors, since they may require more pressure than the flush tank. In addition to the 34.6 ft of distance required to operate the top fixture, a small additional height would be required to overcome the friction loss in the pipes and fittings (an additional 5 ft is usually sufficient unless very small risers are used).

    In this design, the flush tank water closet on the top floor requires a water pressure of 100 kPa to operate (Fig. 2-39). So, if 100 kPa is required and the water pressure increases 9.8 kPa per meter, then 100 kPa ÷ 9.8 kPa = 10.2 m of vertical distance that would be required between the tank and the top of the fixture to provide the pressure required. To overcome the friction loss in the pipes and fittings will require an additional height of about 2 m.

6. The zoning for the downfeed systems is determined next. Since the water pressure is increasing as it goes down the downfeed riser, the vertical distance that a zone can service is limited to the maximum amount of pressure that can be put on the lowest fixture and still have it operate properly. This maximum pressure varies with the type and manufacturer of the fixtures being used; the allowable pressure may range from about 45 to 60 psi (275 to 415 kPa).

    In this design, we will assume that 55 psi is the preferred maximum. This means that the downfeed zone can be 55 psi divided by 0.434 psi per foot or about 126 ft. Since there will be some pressure loss due to friction in the riser pipe, the zone may be slightly longer if required.

    In review then, the upfeed portion of the system can easily service the bottom five floors. This leaves 15 floors to be serviced by the downfeed system. With a floor-to-floor height of 11 ft, this is a vertical distance of about 165 ft, which cannot be served by one downfeed zone. Two downfeed zones will easily provide the water service required to the 15 floors that the upfeed zone (Fig. 2-21) does not serve.

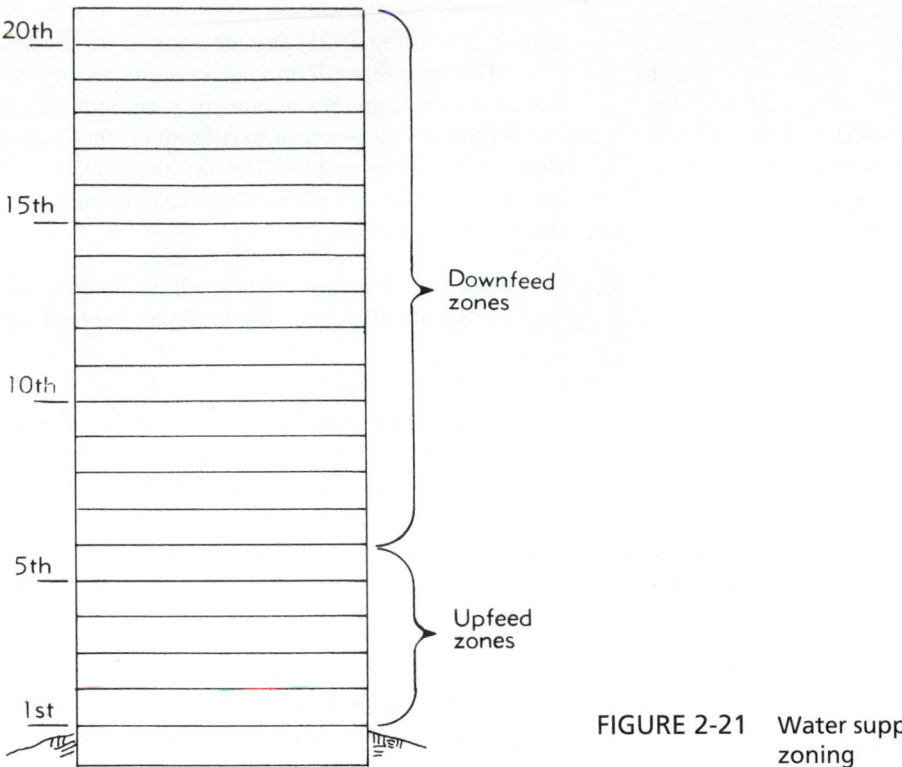

FIGURE 2-21   Water supply zoning

Note: In reviewing the schematic water supply system (Fig. 2-20), note that the house tank used to provide water for zone 2 must be placed high enough above the fixtures it is to supply that there will be sufficient pressure for the fixtures to operate efficiently. In this case, 20 psi will be required to operate the flush valve water closets, calling for a height of 20 psi ÷ 0.434 psi per ft = 46 ft, plus a short distance to overcome friction in the riser (totalling about 50 to 52 ft, depending on the final riser size selected).

In this design we will assume that 380 kPa is the preferred maximum. This means that the downfeed zone can be 380 kPa divided by 9.8 kPa per meter, or about 38.8 m. Since there will be some pressure loss due to friction in the riser pipe, the zone may be slightly longer if required.

In review, then, the upper portion of the system can easily service the bottom five floors. This leaves 15 floors to be serviced by the downfeed system. With a floor-to-floor height of 3.4 m, this is a vertical distance of about 51 m, which cannot be served by one downfeed zone. Two downfeed zones will easily provide the water service required.

Note: In reviewing the schematic water supply system (Fig. 2-20), note that the house tank used to provide water for zone 2 must be placed high enough about the fixtures it is to supply to provide sufficient pressure for the fixtures to operate efficiently. In this design 140 kPa will be required to operate the flush valve water closets, calling for a height of (140 kPa divided by the static head of 9.8 kPa per meter) 14.3 meters plus a short distance to overcome friction in the riser (totaling about 16 m), depending on the final riser size selected.

7. The accumulated information for the upfeed zone (zone 1) is now tabulated (Fig. 2-22), with each floor listed along with its fixture units, the accumulated fixture units

WATER SUPPLY				
UPFEED ZONE, FLOORS 1-5 PROJ. #20–2				
Floor	Accumulated w.s.f.u.	Demand load (gpm)	Friction loss (psi per 100[1])	Pipe size
1	235	97		
2	188	89		
3	141	78		
4	94	66		
4	47	50		

SI UNITS				
WATER SUPPLY				
UPFEED ZONE, FLOORS 1-5 PROJ. #20–2				
Floor	Accumulated w.s.f.u.	Demand load (L/s)	Friction loss (kPa per 100 m)	Pipe size
1	235	6.2		
2	188	5.6		
3	141	4.9		
4	94	4.2		
5	47	3.2		

FIGURE 2-22   Tabulated upfeed data (predominately flush valve)

(remember that the water at the bottom floor must service the floors above it, so the fixture units are the accumulated total of what must be serviced), the demand flow (gpm) for the accumulated fixture units and the pipe size selected.

8. *Static head loss:* From Fig. 2-20, the vertical height of the riser from the top of the main to the top of the upfeed zone is about 55 ft.

$$\text{Static head loss} = 0.434 \text{ psi} \times 55 \text{ ft} = 23.87 \text{ psi}$$

Pressure required to operate fixture at top (5th) floor of the upfeed zone

Flush valve water closet: 20 psi

Pressure left to overcome friction:

$$55 \text{ psi} - (23.87 \text{ psi} + 20 \text{ psi}) = 55 \text{ psi} - 43.87 \text{ psi} = 11.13 \text{ psi}$$

SI From Fig. 2-20, the vertical height of the riser from the top of the main to the top of the upfeed zone is about 16.8 m.

$$\text{Static pressure head loss} = 9.8 \text{ kPa} \times 16.8 \text{ m} = 138 \text{ kPa}$$

Pressure to operate fixture at the top floor of the upfeed zone:

Flush valve water closet: 138 kPa

Pressure remaining to overcome friction:

$$380 - (165 + 138) = 380 - 303 = 77 \text{kPa}$$

9. At this point, the service main should be selected for 902 accumulated water supply fixture units with a demand load of 201 gpm. Referring to Fig. 2-43, the preliminary main pipe selections are:

$$5\text{-in. main} = 0.35 \text{ psi per 100 ft}$$

$$4\text{-in. main} = 1.1 \text{ psi per 100 ft}$$

$$3\tfrac{1}{2}\text{-in. main} = 2.1 \text{ psi per 100 ft}$$

$$3\text{-in main} = 3.9 \text{ psi per } 100 \text{ ft}$$

$$2\tfrac{1}{2}\text{-in. main} = 9.9 \text{ psi per } 100 \text{ ft}$$

SI

$$125 \text{ mm} = \ 7.9 \text{ kPa per } 100 \text{ m}$$

$$100 \text{ mm} = 24.9 \text{ kPa per } 100 \text{ m}$$

$$75 \text{ mm} = 88.1 \text{ kPa per } 100 \text{ m}$$

$$65 \text{ mm} = 224 \text{ kPa per } 100 \text{ m}$$

Since the main distance is given as 40 ft, the friction loss for each pipe would be:

$$5\text{-in. main} = 0.35 \text{ psi per } 100 \text{ ft} = 0.35 \ (0.4) = 0.14 \text{ psi}$$

$$4\text{-in. main} = 1.1 \text{ psi per } 100 \text{ ft} \ = 1.1. \ (0.4) \ = 0.44 \text{ psi}$$

$$3\tfrac{1}{2}\text{-in. main} = 2.1 \text{ psi per } 100 \text{ ft} \ = 2.1 \ (0.4) \ \ = 0.84 \text{ psi}$$

$$3\text{-in. main} = 3.9 \text{ psi per } 100 \text{ ft} \ = 3.9 \ (0.4) \ \ = 1.56 \text{ psi}$$

$$2\tfrac{1}{2}\text{-in. main} = 9.9 \text{ psi per } 100 \text{ ft} \ = 9.9 \ (0.4) \ \ = 3.96 \text{ psi}$$

SI

Since the main distance is given as 12.2 m, the friction loss for each pipe size would be:

$$125 \text{ mm main} = \ 7.9 \text{ kPa per } 100 \text{ m} = \ 7.9 \ (.122) = .96 \text{ kPa}$$

$$100 \text{ mm main} = 24.9 \text{ kPa per } 100 \text{ m} = 24.9 \ (.122) = 3.0 \text{ kPa}$$

$$75 \text{ mm main} = 88.1 \text{ kPa per } 100 \text{ m} = 88.1 \ (.122) = 10.7 \text{ kPa}$$

$$65 \text{ mm main} = 224 \text{ kPa per } 100 \text{ m} \ = 224 \ (.122) \ = 27.3 \text{ kPa}$$

10. The friction loss in the meter (if required) is checked next (Fig. 2-37).

4-in. meter, 201 gpm: 4.1 psi
(a 4-in. meter would be used if 5-in. pipe were used)
3-in. meter, 201 gpm: 10.1 psi
(a 3-in. meter would be used if 4-in. pipe were used)

SI

100 mm meter, 12.68 L/s: 28.3 kPa
(a 100 mm meter would be used if 125 mm pipe were used)
75 mm meter, 12.68 L/s: 69.6 kPa
(a 75 mm meter would be used if 100 mm pipe were used)

11. Obviously, the 3-in. main will not work, and the 4-in. and 5-in. mains are next considered. (It should be noted here that 3½-in pipe is not readily available in many areas, and in such a situation, the 4-in. pipe would be used.)

At this time, a preliminary selection of a 5-in. main and 4-in. meter is made and tabulated.

Pressure available for friction loss	11.13 psi
Friction loss in 5-in. main and 4-in. meter (0.14 psi + 4.1 psi)	− 4.28 psi
Pressure available for riser friction loss	6.85 psi

SI

Obviously the 65 mm main will not work. The 100 mm and 125 mm are considered next.

At this time, a preliminary selection of a 125 mm main and a 100 mm meter is made and tabulated.

Pressure available for friction loss	74 kPa
Friction loss in 125 mm main and 100 mm meter (0.96 + 28.3)	29.3 kPa
Pressure available for riser friction loss	44.7 kPa

12. Before the riser selections can be made, the water supply fixture units and demand load (in gpm) for this upfeed zone must be calculated and put in the tabulated form.

Fixture units, floors 1–5, were previously calculated at 47 w.s.f.u per floor:
47 w.s.f.u. (per floor) × 5 (floors) = 235 (accumulated) w.s.f.u.
Demand load (Fig. 1–28 predominantly flush valve): 97 gpm.

SI

Demand load (Fig. 2-42) predominately flush value: 6.2 L/s

13. This time the available pressure left to overcome friction loss will be converted into loss per 100 ft by using the formula (see Sec. 2-11):

$$\frac{\text{Available psi} \times 100 \text{ (ft)}}{\text{Equivalent length}} = \text{psi loss per 100 ft}$$

$$\frac{6.85 \times 100}{55 + 22.5} = \frac{685}{77.5} = 8.84 \text{ psi loss per 100 ft (maximum)}$$

SI

$$\frac{\text{Available kPa} \times 100 \text{ (m)}}{\text{Equivalent length}} = \text{kPa loss per 100 m}$$

$$\frac{44.7 \times 100}{16.8 + 8.4} = \frac{4470}{25.2} = 177 \text{ kPa}$$

14. Riser pipe selection (from Fig. 2-34).

2-in. pipe = 7.7 psi per 100 ft

2½-in. pipe = 2.9 psi per 100 ft

3-in. pipe = 1.3 psi per 100 ft

SI

50 mm pipe = 173.5 kPa per 100 m

65 mm pipe = 65.5 kPa per 100 m

75 mm pipe = 29.4 kPa per 100 m

15. As previously discussed in regard to upfeed systems, Sec. 2-10, the size of the riser may be decreased in the upper floors, as the gpm decreases, as long as the friction loss for each segment does not increase.

While the 2-in. riser could be used, the 2½-in. pipe is selected to allow some extra pressure at the top fixtures. Now, complete the tabulations for the upfeed zone. (Later, once the design is fully laid out, the actual equivalent length would be rechecked.)

SI

While the 50 mm riser could be used, the 65 mm pipe is selected to allow some extra pressure at the top fixtures.

2-13 HOT WATER

By piping part of the water in the building main (Fig. 2-23) into a heating device, the hot water required on the project can be supplied. The heating device may be a direct or an indirect heater and may operate on oil, gas, electricity, or the sun (solar heat). Direct heaters (Fig. 2-23) are designed solely to provide the hot water required. Indirect heaters (Fig. 2-24) use some type of boiler (heater) to heat the hot water and also to provide heat or steam to the heating system of the project.

Direct heaters come in a variety of sizes and capacities that allow them to be located in the basement, crawl space, or closet; in a cabinet under the counter; or as units which look similar to clothes washers. Many times for projects such as apartments, cold water is run to a direct heater (usually electric) in each apartment, instead of using one large hot water heating unit. This also allows for each individual apartment to be on a separate electric meter and for each resident to pay for the electricity that they use. Residences commonly use a direct heater, and in large homes two units are sometimes used (one near the kitchen-laundry and one near the bathrooms) to cut down the amount of hot water piping required and to provide the almost instant availability of hot water when a faucet is turned on. (The hot water in a pipe will cool off when not used for a while.)

Indirect heaters use the same heating unit to provide hot water or steam to the heating system and to heat the hot water required for use at the fixtures. The same water used in the heating system is not used for the fixtures; instead, a separate compartment or coil containing the water is fed through the unit to be heated. Such units have been used in residences as well as commercial projects. This method is more commonly used in colder climates where the heating system is in operation for more months of the year. Often it is not an economical solution because during the warmer months, the heating system (boiler or furnace) will have to go on to provide hot water when no heating is required. Indirect heaters usually have tanks to store the heated water, or they may have a high-capacity coil capable of providing hot water very quickly.

Some projects use a combination of direct and indirect hot water heaters (Fig. 2-25) whereby the cold water is piped through the indirect heater and then to the direct heater. When the indirect heater is being used for heating, it will provide fully or partially heated water to the direct heater. This means that at times the direct heater will have little or no additional heating to do.

The cold water supply should have a cutoff valve so that the water supply to the heater can be cut off if necessary. This allows for easier repair or replacement of the heater if required. The hot water pipe exits off the top of the tank and should have a relief valve

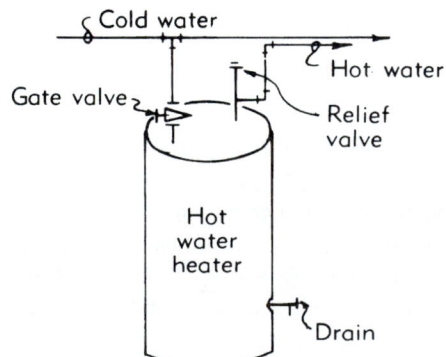

FIGURE 2-23   Hot water heater

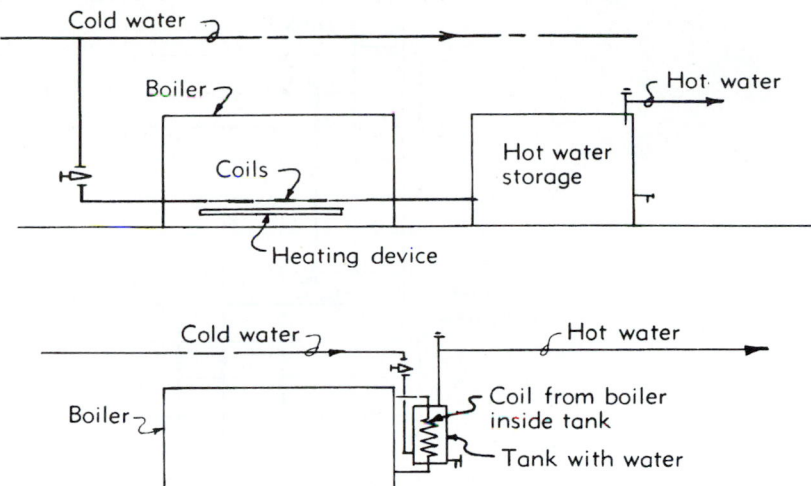

FIGURE 2-24   Indirect heaters

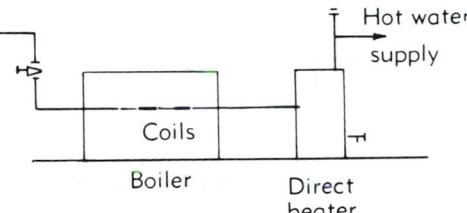

Boiler

Direct
heater

FIGURE 2-25   Combination heater

to allow escape of any excess pressure built up in the system. Heaters which operate on oil or gas will require ventilation to the exterior, usually through a chimney, to get rid of poisonous gases. Electric and solar units do not require venting to the exterior.

2-14 HOT WATER DISTRIBUTION SYSTEMS

Hot water systems make use of upfeed zones to provide the water, under pressure, throughout the project. In taller buildings, the heaters are fed cold water from the tanks supplying water to the downfeed systems (Fig. 2-26). This allows the water from the tank to build up sufficient pressure, as it feeds down to the heater, to create an upfeed zone.

When designing hot water piping systems, the best solution is one which requires the hot water to travel as short a distance as possible from the heater to the point at which it will be used. The longer the supply pipe, the less efficient the system because as the water stays in the pipe, it quickly loses its heat to the surrounding air, even if the pipe is insulated.

For example, review the situation shown in Fig. 2-27, in which the bathroom faucet is about 75 ft (23 m) from the heater. This means that if the water has had time to cool in the pipe (say, during the night), when the faucet is turned on, cool water will come from the

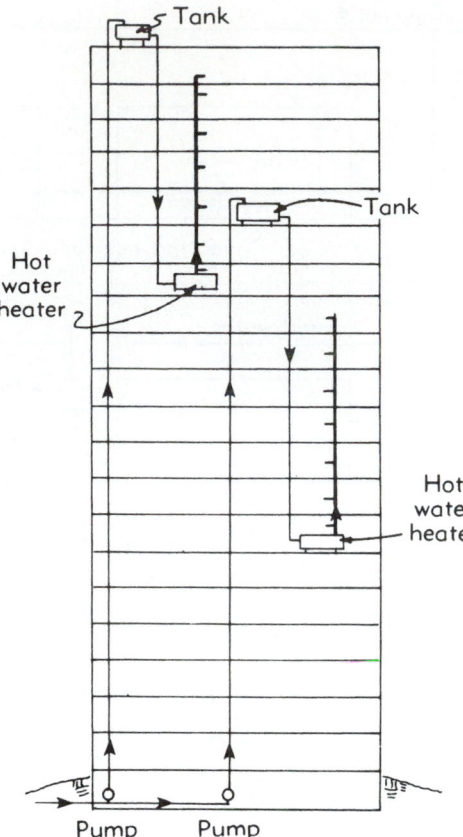

FIGURE 2-26   Multistory hot water zones

faucet until the 75 ft (23 m) of cool water has flowed through the faucet. Only at that time will hot water come out of the faucet. Then, once the faucet has been shut off, there is 75 ft (23 m) of hot water in the pipe which begins to lose its heat. Back at the heater, as the hot water is drawn out, cold water begins to enter and be heated, and enough extra hot water must be heated to fill the 75-ft (23 m) supply pipe again. This process is repeated over and over, several times a day, every day.

One solution to this problem, in projects such as apartment buildings, is to provide each apartment with a direct-type water heater which can be located as conveniently as possible. Typical locations include under the bathroom sink, in a kitchen cabinet, and in a closet.

Most plumbing installations have a noncirculating, direct supply layout, similar to that shown in Fig. 2-28 and discussed in this section. But when the rapid delivery of hot water becomes part of the design requirements and it is not practical to provide small direct heaters throughout the project, a system which circulates the hot water through the supply piping and back to the heater for reheating is used (Fig. 2-29). Circulating systems have a continuous riser going from the tank, through the project, and back to the tank; the fixtures being serviced are fed by a direct branch supply off the riser. The only hot water which stays in the pipe and may cool off is that which is in the branch piping. The hot water will circulate through the system by gravity flow as the hot water rising to the top of the system forces the cooler water back toward the tank.

The circulating system may be upfeed, downfeed, or a combination upfeed-and-downfeed system. A check valve keeps the water flowing in the proper direction. A pump may be required for circulation of the water where the tank is not below the lowest fixture being served.

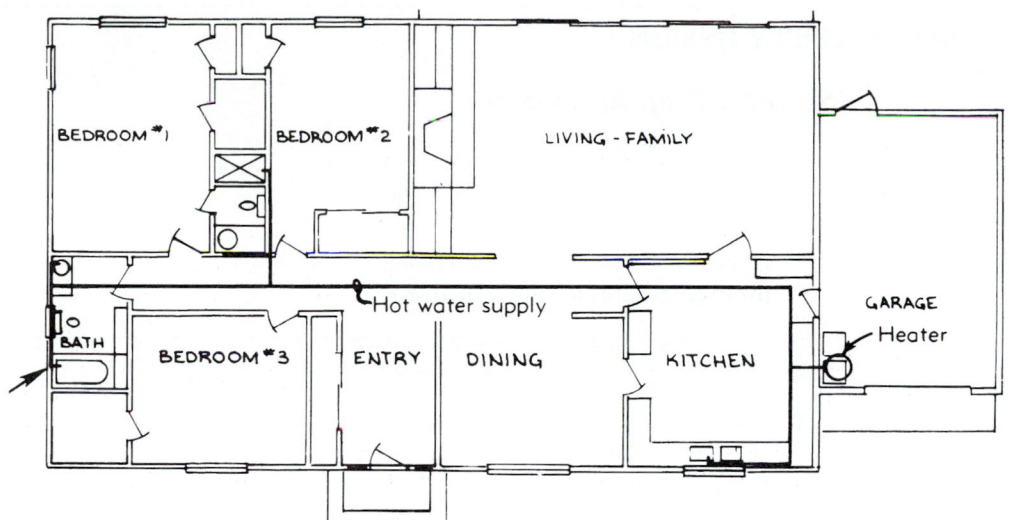

FIGURE 2-27   Hot water supply length

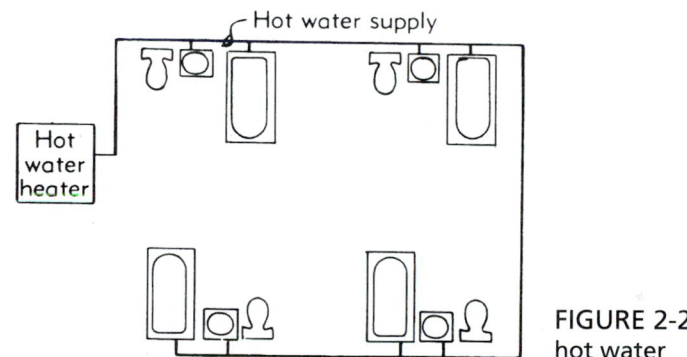

FIGURE 2-28   Noncirculating hot water

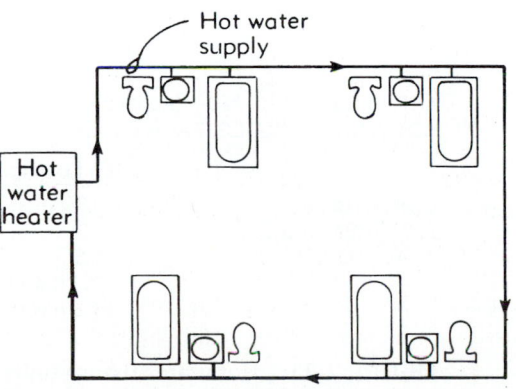

FIGURE 2-29   Circulating hot water

2-15   HOT WATER SUPPLY DESIGN

Step-by-Step Approach

1. In sizing hot water supply piping, the first step is to determine the total demand in fixture units for the project. To calculate the total demand in hot water fixture units for hot water, it is necessary to list the fixtures which require hot water. (The hot water demands for various types of buildings are shown in Fig. 2-40. Use these tables when the exact fixtures are not available.)

To calculate the hot water fixture units for the four-story apartment building, first list the hot water fixtures for each floor (Fig. 2-30).

2. The next step is to list the equivalent fixture units from Fig. 2-40 for each fixture (Fig. 2-31). When added together, this gives the total fixture-unit demand of these fixtures for hot *and* cold water. To determine the fixture units for hot water only, the plumbing code requires that three-fourths of the listed demand be used.

In this case, the hot water demand would be 10 h.w.f.u. $\times$ ¾ = 7.5 h.w.f.u. per floor. So for four floors, the total hot water fixture-unit demand is 7.5 h.w.f.u. (per floor) $\times$ 4 floors = 30 h.w.f.u.

3. The hot water demand (in gallons per minute or gpm) is determined from Fig. 2-42 by finding the total hot water fixture units (in this case, 30) along the left and reading the gpm at the right (30 h.w.f.u., 20 gpm).

SI   30 h.w.f.u., 20gpm (1.26 L/s)

4. Next, the riser pipe size is determined. Remember that when the cold water riser was sized, the total demand in fixture units for hot *and* cold water was used. This means that the friction loss has already been calculated in the cold water supply (Sec. 2-10).

In sizing the cold water riser, a 1¼-in. riser with a friction loss of 3 psi per 100 ft and 7.8 psi for the 60 ft of riser height was used. Since the hot water demand was included in this calculation, the friction loss of 13 psi per 100 ft may be used for the hot water. From Fig. 2-43, using a gpm of 20 and a friction loss of 13 psi, the riser should be a 1¼-in. pipe (copper). In actual design practice, a 1¼-in. pipe would probably be used for the first two floors; then it would be reduced to a 1-in. pipe.

SI In sizing the cold water riser, a 32 mm riser with a friction loss of 293.8 kPa per 100 m and 176.3 kPa for the 20 m of riser height is selected. Since the hot water demand was included in this calculation, the friction loss of 176.3 kPa

APT. BLDG. PROJ. #20-1 (Appendix A)
HOT WATER FIXTURES, EACH FLOOR

h.w.f.u.

2 BATHTUBS
2 KITCHEN SINKS
2 LAVATORIES

FIGURE 2-30   Hot water fixtures

APT. BLDG. PROJ., #20-1 (Appendix A)
HOT WATER FIXUTES, EACH FLOOR

	h.w.f.u. (Fig. 2-40)
2 BATHTUBS	3
2 KITCHEN SINKS	3
2 LAVATORIES	1.5
	7.5 h.w.f.u. per floor
	7.5 h.w.f.u. $\times$ 4 FLOORS = 30 h.w.f.u.

FIGURE 2-31   Hot water fixture units

per 100 m may be used for the hot water. From Fig. 2-43, using 1.26 L/s and a friction loss of 176.3 kPa, the riser could be a 25 mm pipe. In actual practice a 32 mm pipe would probably be used for the first two floors; then it would be reduced to a 25 mm pipe.

5. Selection of the hot water branch pipe serving each floor is based on the demand load in gpm and the permissible friction loss (using the 10 psi which included the distance in the branch).

   The gpm is based on 7.5 h.w.f.u. per floor and, taken from Fig. 2-42, is about 6.5 gpm.

   Using Fig. 2-43, the branch size selected (based on 6.5 gpm and 13 psi per 100 ft) is ¾ in.

   The L/s is based on the 7.5 h.w.f.u. per floor and, taken from Fig. 2-42, is about .41 L/s.

   Using Fig. 2-34, the branch size selected (based on .41 L/s and 293.8 kPa per 100 m) is 20 mm.

6. The minimum size of the pipe from the branch to the fixture (called the supply pipe) is taken from Fig. 2-38. All fixture supply pipe sizes should be checked against the minimum sizes.

2-16 TESTING THE SYSTEM

In many larger cities and municipalities, a separate plumbing permit must be taken out. Oftentimes building departments will even review the drawings to see if the design conforms to the code in force; others may simply require a brief written description of the system.

Also, depending on the locale, once the plumbing is roughed in, the municipal plumbing inspector will check all installed work, particularly the waste portion of the system, for conformance with the code. This inspection should be made before any pipes are covered, and most inspectors will not approve a system unless a complete inspection can be made. It is far better to find a leak in the system before the walls, ceilings, and floor are finished than after.

Tests on a system are made to be certain that it will perform satisfactorily and not solely for inspectors or because it is required. The water supply system is checked by sealing all openings in the system, filling it with potable water, and then pressurizing it (up to the normal operating pressure). Such a check of the system will show any leaks in the fittings.

There are still many areas which do not have inspectors and do not enforce the building codes. The codes are designed to protect human life and to ensure that the completed system will fulfill the function it was designed for. Tests should be made on any system, whether it is inspected or not.

2-17 WATER HEATER TIMERS

Many projects (residential, institutional, and commercial) use time clocks to control the operation of the water heater. Even when the needs of the facility are carefully analyzed to determine peak loading periods, there are many times when the water heater will heat the water, and then the hot water will sit in the tank for hours before it is used. In the second example of hot water demand, the plant runs two shifts. In reviewing the data, the last showers are taken between 12 midnight and 12:30 A.M. The hot water is not used again until 3:30 P.M. when the first shift crew begins to shower. The heating capacity has been calculated to heat all of the hot water required in 8 hours. So during the 15 hours from 12:30 A.M. to 3:30 P.M., the heater will work to heat the water for the first 8 hours (until 8:30 A.M.) and then continue to work to keep the water hot (since there will be a heat loss through the tank) for the next 7 hours (until 3:30 P.M.). In such a situation, a timer would shut the water heater off at 12:30 A.M. and turn it on at about 7 A.M. to begin to heat the water for the 3:30 P.M. shower.

The hot water needs of a residence can also be analyzed. For example, assume that someone will be home all day and prefers to do the clothes washing and to run the dishwasher during the day. Also, assume the family arises at 6:30 A.M. and retires at about 11 P.M. Since a quick recovery water heater will restore hot water in about 45 minutes, after their 11 P.M. showers, all of the hot water will be reheated in the tank by 12 midnight, and then the heater must maintain the temperature of the water until 6:30 A.M. Energy could be saved if the timer were set to shut the water heater off from 11 P.M. until 5:30 A.M. Then at 5:30 A.M., the heater would come on and heat all the water needed before 6:30 A.M.

As people begin to organize their day to conserve energy, the hours that hot water is required will be reduced considerably. The majority of studies show that most of the hot water is required during a 1-hour time period in the morning and a 2-hour time period at night. To provide the hot water required for these periods, the heater will need to work about 3 hours a day. Generally, if it is set to turn on a half hour before rising, to operate for an hour and then to turn back on about suppertime for 2 hours, there will be sufficient hot water to meet all of a family's needs. This will work if clothes are washed in cold water during the day or in hot water in the evening, and if the dishwasher is run only during the evening hours. This type of conservation can reduce the energy used for hot water by as much as 35% to 40%, a significant savings.

2-18 SOLAR HOT WATER

Solar hot water heating systems are no longer a dream of the future. Development continues on systems which capture the heat from the sun to heat water for use in the home or project. The final chapter cannot yet be written on these systems as experimentation continues and as millions of dollars from government and private enterprise are poured into further research and development. Meanwhile, many dependable and cost-effective hot water heating systems are now available for purchase.

The most successful solar hot water heater, in the author's opinion, is the solar flat-plate absorber/collector; the principles of operation are shown in Fig. 2-32. A typical installation is shown in Fig. 2-33. In many areas, the solar heating unit is supplemented with an electric heating coil to provide hot water in case of prolonged cloudiness, which will limit the amount of solar energy available to heat water. This is not to imply that the solar heater cannot be designed to provide all of the hot water required—just that it is most efficient costwise to design the solar system to provide 85% to 95% per year. Whether it is a good investment or not depends on whether it will give a good return on the investment. The economics of the system will vary according to the geographic location, total amount of hot water required, and the current (and future) cost of hot water.

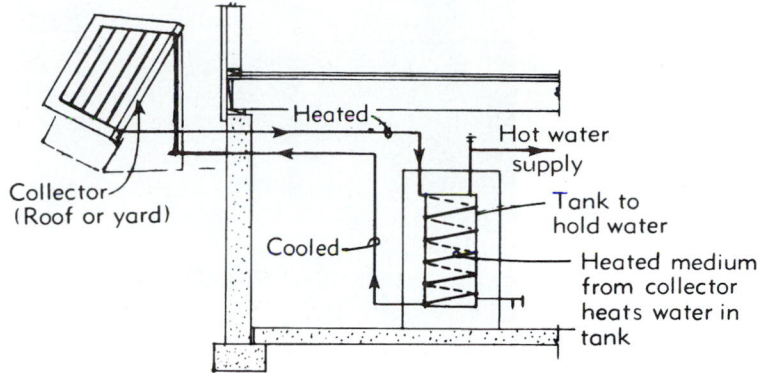

Tank—This tank could be used to just hold water heated by the collector medium or it could be a direct hot water heater used to supplement the collector during times when the collector cannot provide all of the hot water required.

FIGURE 2-32   Solar hot water system

FIGURE 2-33   Solar hot water system installation

2-19   SYSTEM INSTALLATION

The materials and assemblies used in the construction of the project have an important impact on the ease of installation and cost of the mechanicals. In the case of water supply, one of the most important things in any area that has freezing weather is that the pipes be installed in such a manner and location that they will not freeze. Typically, this means do not put the pipes in exterior walls and be certain that the supply coming in is buried deep enough so they will not freeze. The depth varies from inches in the warmer climates to several feet deep in the colder climates.

Whenever poured concrete slabs are used, the plumbing layout will need to be carefully considered. The pipes need to be placed in the ground before the slab is poured, so their accurate placement is crucial. Typically, both the water supply and drainage pipes are laid out next to each other since they go to the same areas of the project. String is usually stretched over the slab area to mark where the pipes should be located (Fig. 2-34). Many times they are planned so they will come up in a wall (Fig. 2-35). The piping must be carefully located and the system checked for leaks before the concrete is poured since any relocation or repairs of pipes would be costly.

FIGURE 2-34   String alignment

FIGURE 2-35   Typical installation
in wall

On larger projects with concrete walls and ceilings, it may be necessary to provide sleeves (holes) in the concrete for the pipes to pass through to get from space to space. It may also be necessary to provide inserts and hangers (Fig. 2-36) to support the pipes.

The open spaces provided in truss-type construction make it easy to run piping through to the desired location. The only points of difficulty would be where it needs to pass by ductwork or some other large pipe that is going in the opposite direction.

In wood frame construction, the holes are usually drilled to allow the passage of the pipes. Typically, these should be at the middle of any load-bearing wood members so a minimum of structural damage is done. There are times when the width of a wall needs to be increased to allow for water pipes running horizontally to pass by drainage pipes (or other pipes) running vertically.

The water system should be tested for leaks before it is covered with other materials, to determine if it is watertight. Tests commonly run on water systems require that it be watertight under a hydrostatic water pressure of 125 psi for a minimum of 1 hour. Any leaks that occur should be repaired with the joint compound originally used.

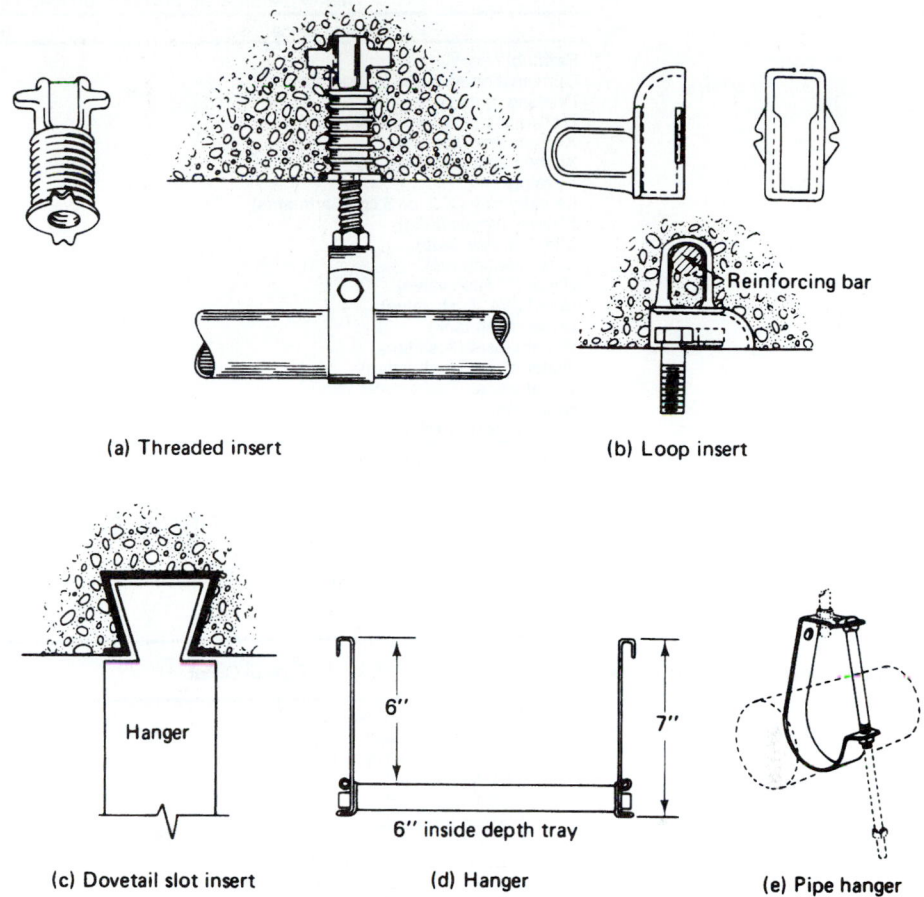

(a) Threaded insert   (b) Loop insert

(c) Dovetail slot insert   (d) Hanger   (e) Pipe hanger

FIGURE 2-36   Typical inserts and hangers

Fixture	Flow Pressure	
	psi	kPa
Ordinary basin faucet	8	55
Self-closing basin faucet	12	80
Sink Faucet - 3/8 in.	10	70
Sink Faucet - 1/2 in.	5	35
Bathtub faucet	5	35
Laundry tub cock - 1/2 in.	5	35
Shower	12	80
Ball-cock for closet	15	100
Flush valve for closet	10-20	70-140
Flush valve for urinal	15	100
Garden hose, 50 ft. and sill cock	30	200

Flow pressure is the pressure in the pipe at the entrance to the particular fixture considered.

Note: Metric added by author

FIGURE 2-37   Fixture pressure and flow

MINIMUM SIZE OF FIXTURE BRANCH PIPING

Fixture Device	Size (in.)	Size (mm)
Bathtub	1/2	15
Combination sink and laundry tray	1/2	15
Drinking fountain	3/8	10
Dishwashing machine (domestic)	1/2	15
Kitchen sink (domestic)	1/2	15
Kitchen sink (commercial)	3/4	20
Lavatory	3/8	10
Laundry tray (1, 2, or 3 compartments)	1/2	15
Shower (single head)	1/2	15
Sink (service, slop)	1/2	15
Sink (flushing rim)	3/4	20
Urinal (1" flush valve)	1	25
Urinal (3/4" flush valve)	3/4	20
Urinal (flush tank)	1/2	15
Water closet (flush tank)	3/8	10
Water closet (flush valve)	1	25
Water closet (flushometer tank)	3/8	10
Hose bibb	1/2	15
Wall hydrant or sill cock	1/2	15

Note: Metric added by author

FIGURE 2-38

Type of Outlet	Demand	
	(gpm)	L/S
Ordinary lavatory faucet	2.0	.13
Self-closing lavatory faucet	2.5	.16
Sink faucet, 3/8" or 1/2"	4.5	.28
Sink faucet, 3/4"	6.0	.38
Bath faucet, 1/2"	5.0	.32
Shower head, 1/2"	5.0	.32
Laundry faucet, 1/2"	5.0	.32
Ballcock in water closet flush tank	3.0	.19
1" flush valve (25 psi flow pressure)	35.0	2.21
1" flush valve (15 psi flow pressure)	27.0	1.7
3/4" flush valve (15 psi flow pressure)	15.0	.95
Drinking fountain jet	0.75	.05
Dishwashing machine (domestic)	4.0	.25
Laundry machine (8 or 16 lbs.)	4.0	.25
Aspirator (operating room or laboratory)	2.5	.16
Hose bibb or sill cock (1/2")	5.0	.32

Note: Metric added by author

FIGURE 2-39   Demand at individual water outlets

WATER SUPPLY FIXTURE UNITS AND FIXTURE BRANCH SIZES

Fixture	Occupancy	Type of Supply Control	Load Values, In Water Supply Fixture Units			Min. Size of Fixture Branch
			Cold	Hot	Total	
Water Closet	Public	Flushometer Valve	10		10	1
Water Closet	Public	Flushometer Tank	5		5	1/2
Water Closet	Public	Flush Tank	5		5	1/2
Urinal	Public	1" Flush Valve	10		10	1
Urinal	Public	3/4" Flush Valve	5		5	3/4
Urinal	Public	Flush Tank	3		3	1/2
Lavatory	Public	Faucet	1.5	1.5	2	1/2
Showerhead	Public	Mixing Valve	3	3	4	1/2
Service Sink	Public	Faucet	2.25	2.25	3	1/2
Kitchen Sink	Public	Faucet	3	3	4	1/2
Drinking Fountain	Public	3/8" Valve	0.25		0.25	3/4
Water Closet	Private	Flush Valve	6		6	1
Water Closet	Private	Flushometer Tank	3		3	1/2
Water Closet	Private	Flush Tank	3		3	1/2
Lavatory	Private	Faucet	0.75	0.75	1	1/2
Bathtub	Private	Faucet	1.5	1.5	2	1/2
Bathroom Group	Private	Flushometer Valve	7.0	2.0	8	1/2
Bathroom Group	Private	Flush Tank	5.0	2.0	6	1/2
Bathroom Group	Private	Flushometer Tank	5.0	2.0	6	1/2
Shower Stall	Private	Mixing Valve	1.5	1.5	2	1/2
Kitchen Sink	Private	Faucet	1.5	1.5	2	1/2
Laundry Trays (1 to 3)	Private	Faucet	2.25	2.25	3	1/2
Combination Fixture	Private	Faucet	2.25	2.25	3	1/2
Dishwashing Machine	Private	Automatic		1	1	1/2
Laundry Machine (8 lbs)	Private	Automatic	1.5	1.5	2	1/2
Laundry Machine (16 lbs)	Public	Automatic	3	3	4	1/2

Note: For fixtures not listed, loads should be assumed by comparing the fixture to one listed using water in similar quantities and at similar rates. The assigned loads for fixtures with both hot and cold water supplies are given for separate hot and cold water loads, and for total load be separate hot and cold water loads being three-fourths of the total load for the fixture in each case.

FIGURE 2-40

Water supply outlets for items not listed shall be computed at their maximum demand, but in no case less than the following valves:

Fixture	Number of Fixture Units	
	Private Use	General Use
3/8 (10 mm)	1	2
1/2 (15 mm)	2	4
3/4 (20 mm)	3	6
1 (25 mm)	6	10

Note: Metric added by author

FIGURE 2-41   Water supply fixture units and fixture branch sizes

TABLE FOR ESTIMATING DEMAND

Supply Systems Predominately For Flush Tanks		Supply Systems Predominately For Flushometer Valves	
Load (Water Supply Fixture Units)	Demand (Gallons per Minute)	Load (Water Supply Fixture Units)	Demand (Gallons per Minute)
6	5		
10	8	10	27
15	11	15	31
20	14	20	35
25	17	25	38
30	20	30	41
40	25	40	47
50	29	50	51
60	33	60	55
80	39	80	62
100	44	100	68
120	49	120	74
140	53	140	78
160	57	160	83
180	61	180	87
200	65	200	91
225	70	225	95
250	75	250	100
300	85	300	110
400	105	400	125
500	125	500	140
750	170	750	175
1000	210	1000	218
1250	240	1250	240
1500	270	1500	270
1750	300	1750	300
2000	325	2000	325
2500	380	2500	380
3000	435	3000	435
4000	525	4000	525
5000	600	5000	600
6000	650	6000	650
7000	700	7000	700
8000	730	8000	730
9000	760	9000	760
10,000	790	10,000	790

Note: Multiply gpm times 0.06308 to obtain L/s.

FIGURE 2-42

FRICTION LOSS IN SMOOTH PIPE

Friction loss in head in kPa Per 100m length

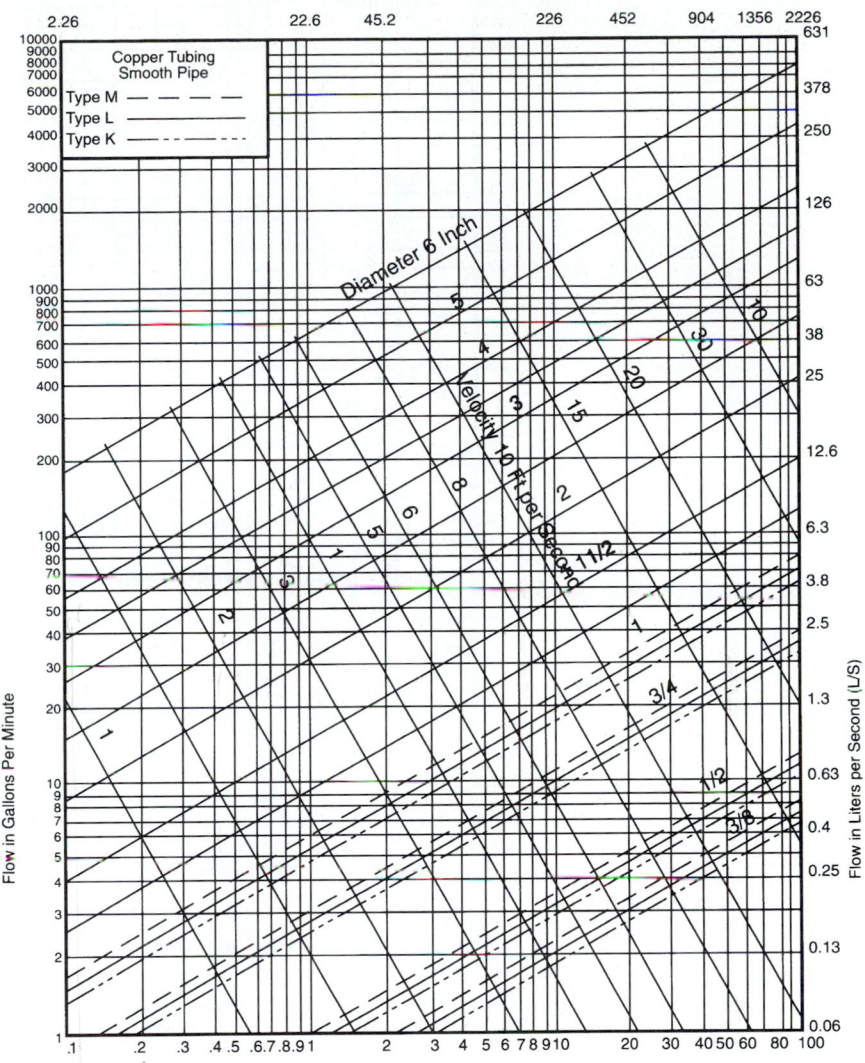

Friction Loss Lbs. per Sq. In. Head per 100 ft. Length

1. This chart applies to smooth new copper tubing with recessed (streamlined) soldered joints and to the actual sizes of types indicated on the diagram.

Note: Metric added by author

FIGURE 2-43   Friction loss (smooth pipe)

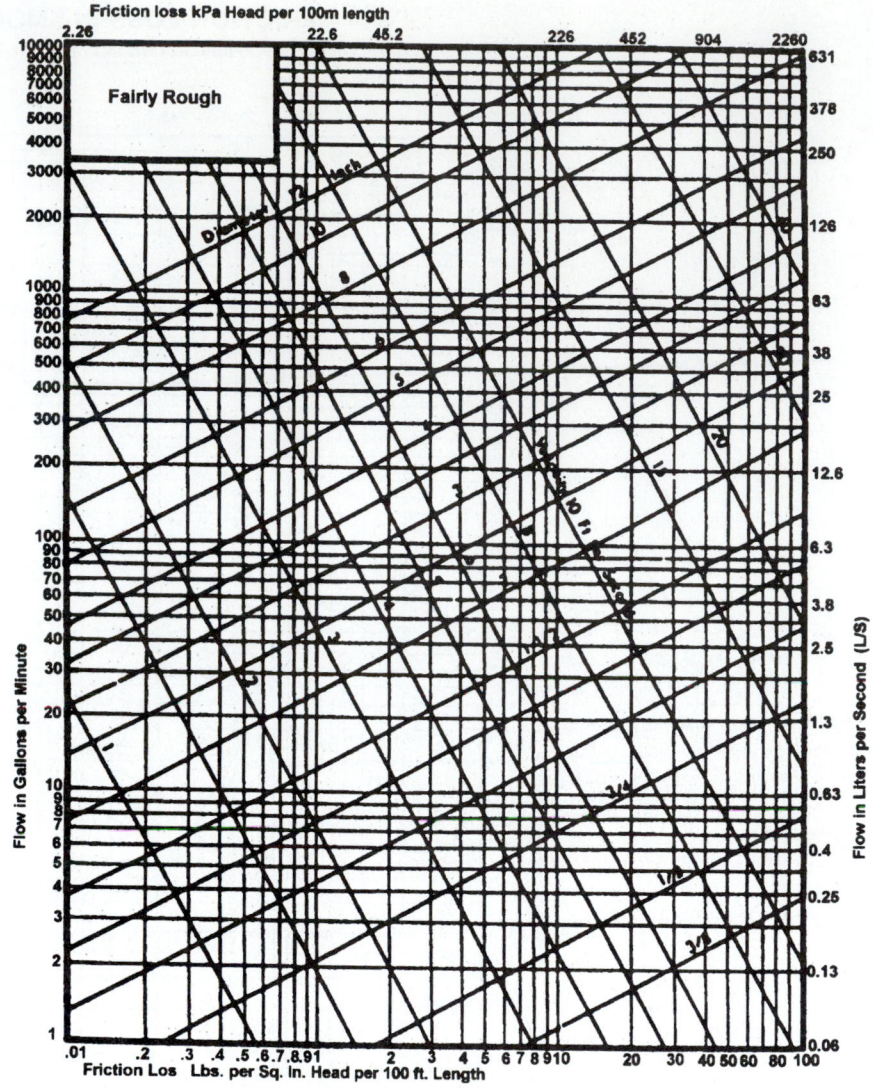

1. This chart applies to fairly rough pipe and to actual diameters which in general will be less than the actual diameters of the new pipe of the same kind.
** Metric added by author.

FIGURE 2-44   Friction loss (rough pipe)

ALLOWANCE IN EQUIVALENT LENGTH OF PIPE FOR FRICTION LOSS IN VALVES AND THREADED FITTINGS

Fitting or valve	Equivalent feet of pipe for various sizes							
	½ in.	¾ in.	1 in.	1¼ in.	1½ in.	2 in.	2½ in.	3 in.
45° elbow	1.2	1.5	1.8	2.4	3.0	4.0	5.0	6.0
90° elbow	2.0	2.5	3.0	4.0	5.0	7.0	8.0	10.0
Tee, run	0.6	0.8	0.9	1.2	1.5	2.0	2.5	3.0
Tee, branch	3.0	4.0	5.0	6.0	7.0	10.0	12.0	15.0
Gate valve	0.4	0.5	0.6	0.8	1.0	1.3	1.6	2.0
Balancing valve	0.8	1.1	1.5	1.9	2.2	3.0	3.7	4.5
Plug-type cock	0.8	1.1	1.5	1.9	2.2	3.0	3.7	4.5
Check valve, swing	5.6	8.4	11.2	14.0	16.8	22.4	28.0	33.6
Globe valve	15.0	20.0	25.0	35.0	45.0	55.0	65.0	80.0
Angle valve	8.0	12.0	15.0	18.0	22.0	28.0	34.0	40.0

ALLOWANCE IN EQUIVALENT LENGTH OF TUBE FOR FRICTION LOSS IN VALVES AND FITTINGS[1]
(Copper Water Tube)

Fitting or valve	Equivalent feet of tube for various sizes							
	½ in.	¾ in.	1 in.	1¼ in.	1½ in.	2 in.	2½ in.	3 in.
45° elbow (wrought)	0.5	0.5	1.0	1.0	2.0	2.0	3.0	4.0
90° elbow (wrought)	0.5	1.0	1.0	2.0	2.0	2.0	2.0	3.0
Tee, run (wrought)	0.5	0.5	0.5	0.5	1.0	1.0	2.0	
Tee, branch (wrought)	1.0	2.0	3.0	4.0	5.0	7.0	9.0	
45° elbow (cast)	0.5	1.0	2.0	2.0	3.0	5.0	8.0	11.0
90° elbow (cast)	1.0	2.0	4.0	5.0	8.0	11.0	14.0	18.0
Tee, run (cast)	0.5	0.5	0.5	1.0	1.0	2.0	2.0	2.0
Tee, branch (cast)	2.0	3.0	5.0	7.0	9.0	12.0	16.0	20.0
Compression Stop	13.0	21.0	30.0	-				
Globe Valve	-	-	-	53.0	66.0	90.0		
Gate Valve	-	-	1.0	1.0	2.0	2.0	2.0	2.0

[1]From "Copper Tube Handbook" 1965, by Copper Development Association, Inc.

ALLOWANCE IN EQUIVALENT LENGTH OF TUBE FOR FRICTION LOSS IN VALVES AND FITTINGS
(Copper Water Tube)

Fitting or Valve	Equivalent meters of pipe for various sizes							
	13 mm	18 mm	25 mm	32 mm	40 mm	50 mm	65 mm	80 mm
45° elbow (wrought)	0.2	0.2	0.3	0.3	0.6	0.6	0.9	1.2
90° elbow (wrought)	0.2	0.3	0.3	0.6	0.6	0.6	0.6	0.9
Tee, run (wrought)	0.2	0.2	0.2	0.2	0.3	0.3	0.6	-
Tee, branch (wrought)	0.3	0.6	0.9	1.2	1.5	2.1	2.8	-
45° elbow (cast)	0.2	0.3	0.6	0.6	0.9	1.5	2.4	3.4
90° elbow (cast)	0.3	0.6	1.2	1.5	2.4	3.4	4.3	5.5
Tee, run (cast)	0.2	0.2	0.2	0.3	0.3	0.6	0.6	0.6
Tee, branch (cast)	0.6	0.9	1.5	2.1	2.8	3.7	4.9	6.1
Compression stop	4.0	6.4	9.2	-	-	-	-	-
Globe valve	-	-	-	16.2	20.1	27.5	-	-
Gate valve	-	-	0.3	0.3	0.6	0.6	0.6	0.6

Note: Metric added by author

ALLOWANCE IN EQUIVALENT LENGTH OF PIPE FOR FRICTION LOSS IN VALVES AND THREADED FITTINGS

Fitting or Valve	Equivalent meters of pipe for various sizes							
	13 mm	18 mm	25 mm	32 mm	40 mm	50 mm	65 mm	80 mm
45° elbow	0.4	0.5	0.6	0.8	0.9	1.2	1.5	1.8
90° elbow	0.6	0.8	0.9	1.2	1.5	2.1	2.4	3.1
Tee, run	0.2	0.2	0.3	0.4	0.5	0.6	0.8	0.9
Tee, branch	0.9	1.2	1.5	1.8	2.1	3.1	3.7	4.6
Gate valve	0.1	0.2	0.2	0.2	0.3	0.4	0.5	0.6
Balancing valve	0.2	0.4	0.5	0.6	0.7	0.9	1.1	1.4
Plug-type cock	0.2	0.4	0.5	0.6	0.7	0.9	1.1	1.4
Check valve, swing	1.7	2.6	3.4	4.3	5.1	6.8	8.6	10.0
Globe valve	4.6	6.1	7.6	10.7	13.7	16.8	19.8	24.4
Angle valve	2.4	3.7	4.6	5.5	6.7	8.6	10.4	12.2

Note: Metric added by author

FIGURE 2-45   Equivalent lengths

Type of Building	Maximum Hour	Maximum Day	Average Day
Men's dormitories	3.8 gal (14.4 L)/student	22.0 gal (83.4 L)/student	13.1 gal (49.7 L)/student
Women's dormitories	5.0 gal (19 L)/student	26.5 gal (100.4 L)/student	12.3 gal (46.6 L)/student
Motels: no. of units[a]			
20 or less	6.0 gal (22.7 L)/unit	35.0 gal (132.6 L)/unit	20.0 gal (75.8 L)/unit
60	5.0 gal (19.7 L)/unit	25.0 gal (94.8 L)/unit	14.0 gal (53.1 L)/unit
100 or more	4.0 gal (15.2 L)/unit	15.0 gal (56.8 L)/unit	10.0 gal (37.9 L)/unit
Nursing homes	4.5 gal (17.1 L)/bed	30.0 (113.7 L)/bed	18.4 gal (69.7 L)/bed
Office buildings	0.4 gal (1.52 L)/person	2.0 gal (7.6 L)/person	1.0 gal (3.79 L)/person
Food service establishments:			
Type A—full meal restaurants and cafeterias	1.5 gal (5.7 L)/max meals/h	11.0 gal (41.7 L)/max meals/h	2.4 gal (9.1 L)/avg meals/day[b]
Type B—drive-ins, grilles, luncheonettes, sandwich and snack shops	0.7 gal (2.6 L)/max meals/h	6.0 gal (22.7 L)/max meals/h	0.7 gal (2.6 L)/avg meals/day[b]
Apartment houses: no. of apartments			
20 or less	12.0 gal (45.5 L)/apt.	80.0 gal (303.2 L)/apt.	42.0 gal (159.2 L)/apt.
50	10.0 gal (37.9 L)/apt.	73.0 gal (276.7 L)/apt.	40.0 gal (151.6 L)/apt.
75	8.5 gal (32.2 L)/apt.	66.0 gal (250 L)/apt.	38.0 gal (144 L)/apt.
100	7.0 gal (26.5 L)/apt.	60.0 gal (227.4 L)/apt.	37.0 gal (140.2 L)/apt.
200 or more	5.0 gal (19 L)	50.0 gal (195 L)/apt.	35.0 gal (132.7 L)/apt.
Elementary schools	0.6 gal (2.3 L)/student	1.5 gal (5.7 L)/student	0.6 gal (2.3 L)/student[b]
Junior and senior high schools	1.0 gal (3.8 L)/student	3.6 gal (13.6 L)/student	1.8 gal (6.8 L)/student[b]

Copyright © by the American Society of Heating, Refrigerating and Air Conditioning Engineers, Inc., Atlanta, GA. Reprinted by permission from the *ASHRAE Systems Handbook,* 1984.

[a] Interpolate for intermediate values.
[b] Per day of operation.

FIGURE 2-46   Hot water demands for various types of buildings

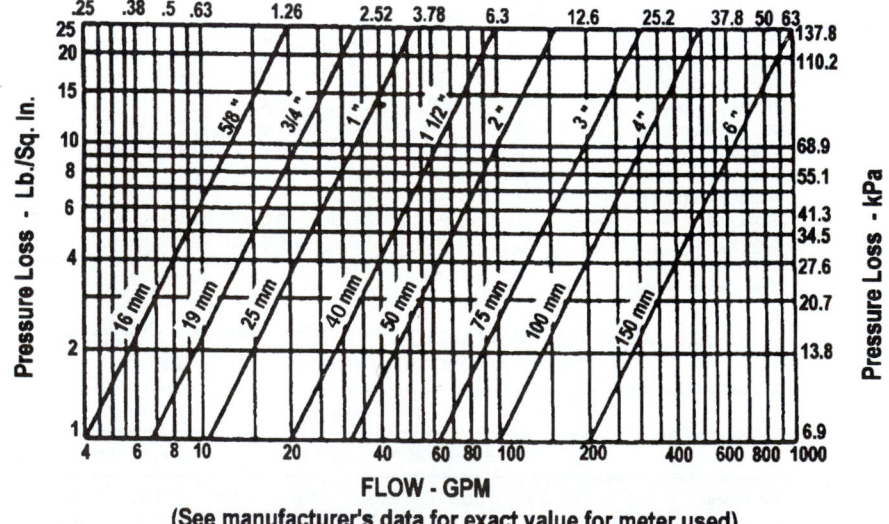

(See manufacturer's data for exact value for meter used)
** Metric added by author

FIGURE 2-47   Meter pressure loss

QUESTIONS

2-1. What are plumbing codes, and why are they used?

2-2. What is meant by *multiple governing codes,* and why must the designer be aware of them?

2-3. Who is most likely to enforce the plumbing code?

2-4. What materials are most commonly used for the pipes and tubing in a water supply system?

2-5. How and where are the following valves used in the system?

    a. Globe   c. Gate

    b. Angle   d. Check

2-6. What is *water shock,* and how can it be reduced in the system?

2-7. How may expansion be allowed for in the system, and what type of pipe expands the most?

2-8. What are the two basic types of water supply systems used, and what determines which will be used?

2-9. What is meant, in water supply systems, by *static head?*

2-10. What is the difference between a direct and an indirect water heater?

2-11. Why is more than one hot water heater sometimes used in larger residences?

2-12. What is the difference between circulating and noncirculating hot water systems, and when would the circulating system most likely be used?

2-13. Briefly, describe and sketch how a solar hot water heater heating system might work.

2-14. What factors affect whether or not it might be economical to use a solar hot water heating system?

2-15. Why are supplemental heating units often installed in solar hot water heating systems?

Design Exercises

2-16. Determine the demand load of water required for the apartment building in Appendix B.

2-17. Design an upfeed water supply system for the apartment building in Appendix B. Use copper pipe and tubing for this design problem. Assume each apartment has its own hot water heater.

2-18. Determine the hot water consumption and the equipment sizes required for the building in Appendix B. Use the information in Fig. 2-46. In this assignment assume one central hot water heater for the building.

2-19. Determine the hot water consumption and the equipment sizes required for a 150-room motel based on the following information:

Hot water required: 30 gal per day per person

Occupancy rate: 2.4 persons per room, 100% occupied

Storage capacity (tank): 60% of total daily use (of which 75% is usable)

Maximum hourly demand: ⅙ of total daily use

Peak demand time: 3 hr.

2-20. Design the upfeed hot water supply system required for the apartment in Appendix B. Use copper pipe and tubing.

CHAPTER 3

Plumbing Drainage

3-1 DRAINAGE PRINCIPLES

In Chapter 2, hot and cold water supply systems were described, with pipes to provide sufficient running water to all of the fixtures throughout a building or project. Following the flow of the water through the system, the next step will be to dispose of the waste matter, both fluid and organic, which is accumulated. The wastes will come from almost all sections of the building—bathrooms, kitchens, and laundry areas, and, in commercial projects, even the equipment being serviced. Because all of the wastes tend to decompose quickly, one of the primary objectives of the plumbing system is to dispose of decaying wastes rapidly, before they cause objectionable odors or become hazardous to health.

Water is used to transport the wastes into the drainage piping and to the point where they will enter a community sewer line leading either to a community sewage treatment plant or to a private sewage treatment system. The sewage from residences, apartments, motels, office buildings, and other similar types of buildings is referred to as *domestic sewage.* Special sewage from laboratories and many industrial plants requires special handling, and such treatment is not discussed here. However, it is important to note that such wastes should not be put into a community sewage system without first getting approval from the governing authorities.

3-2 PLUMBING CODES

The *National Standards Plumbing Code* is used as the basis for the plumbing discussed in this text. Increasingly, all plumbing codes are being modeled after this code, with only minor revisions, to meet the requirements of specific geographic locations. References in this portion of the text relate to the *National Standard Plumbing Code,* and everyone who is learning about plumbing, its design and its requirements, should have a current copy available. In addition, any relevant state or local plumbing codes should be made a part of your library for future reference.

3-3 THE DRAINAGE SYSTEM

In this section the terminology and function of each of the parts of the drainage system are explained. The basic parts of the system are illustrated in Fig. 3-1.

Traps

A *trap* (Fig. 3-2) is a device which catches and holds a quantity of water; this forms a seal which prevents the gases resulting from sewage decomposition from entering the building

71

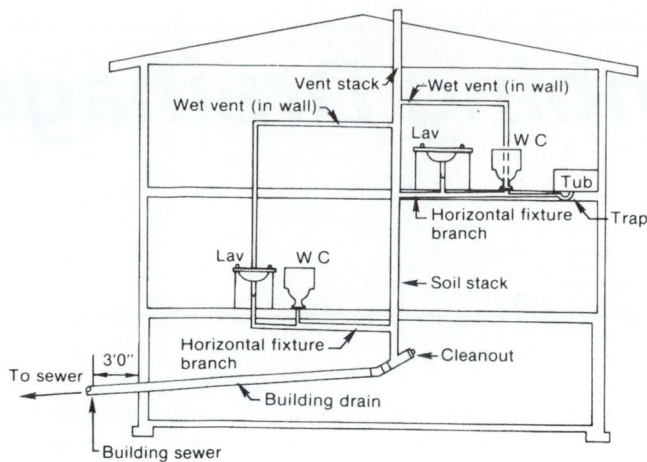

FIGURE 3-1   Drainage system

FIGURE 3-2   Traps

through the pipe. Traps are installed at each fixture as bent pipes unless the fixture is designed with the trap as an integral part of it (as in the case of the water closet in Fig. 3-3).

Traps may be made of copper, plastic, steel, wrought iron, or brass, with plastic most commonly used. Traps in water closets are made of vitreous china and are cast right into the fixture.

The trap is located as close to the fixture as possible, usually within 2 ft (0.6 m) of it. Often, more than one fixture is tied to one trap. Quite often a laundry tray and a kitchen sink, a dishwasher and a kitchen sink, or two kitchen sinks may be connected to a single trap, provided all fixtures are close to one another. There should never be more than three closely located fixtures (such as lavatories) on a single trap, or the trap may not operate properly (it may lose its *water seal,* also called *trap seal*). Since the trap may occasionally need to be cleaned, either there should be a plug in the bottom which may be removed, or the trap should have screwed connections on each end for easy removal.

In locations where the fixtures are infrequently used, care must be taken or the water in the traps may evaporate, and once the water seal is gone, gases may back up from the sewer and drainage pipes through the fixture and into the building. Floor drains (Fig. 3-4),

FIGURE 3-3   Integral trap

FIGURE 3-4   Floor drain

which are used to take away the water after washing floors or which may be used only in case of equipment malfunctions or repairs, present the most serious possibility of losing their water seal. When these floor drains are connected to the drainage system, the possibility of a serious gas problem exists. The designer of the system can avoid such a situation by *not* tying the floor drain into the drainage system. Instead, the floor drains could be tied into a drywell (Fig. 4-2), from which there will be no gases. Many building departments and plumbing codes prohibit the connection of floor drains to the sewage drainage system.

The water seal may also be broken if there is a great deal of air pressure turbulence in the pipes. To reduce the turbulence and to tend to equalize the pressure throughout the system, it is opened to the outside at the top and sufficient air is supplied throughout the system through *vent pipes.*

Vents

Vent pipes allow gases in the sewage drainage system to discharge to the outside and sufficient air to enter the system to reduce the air turbulence in the system. Also, without a vent, once the water discharges from a fixture, the moving waste tends to siphon the water from other fixture traps as it goes through the pipes. This means that the vent piping must serve the various fixtures, or groups of fixtures, as well as the rest of the sewage drainage system. The vent from a fixture or group of fixtures ties in with the main vent stack (Fig. 3-5) or the stack vent (Fig. 3-6), which goes to the exterior. Vent piping may be copper, plastic, cast iron, or steel.

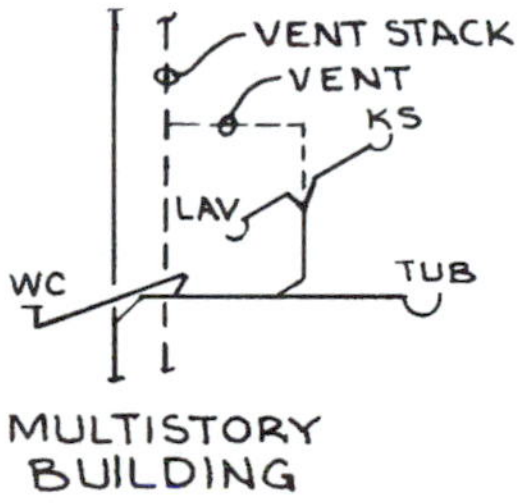

VENT STACK
VENT
KS
LAV
WC
TUB

MULTISTORY
BUILDING

FIGURE 3-5   Vent to vent stack

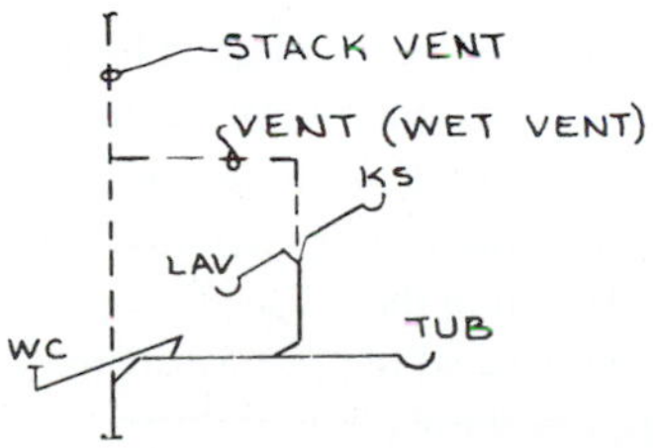

STACK VENT
VENT (WET VENT)
KS
LAV
WC
TUB

FIGURE 3-6   Vent to stack vent

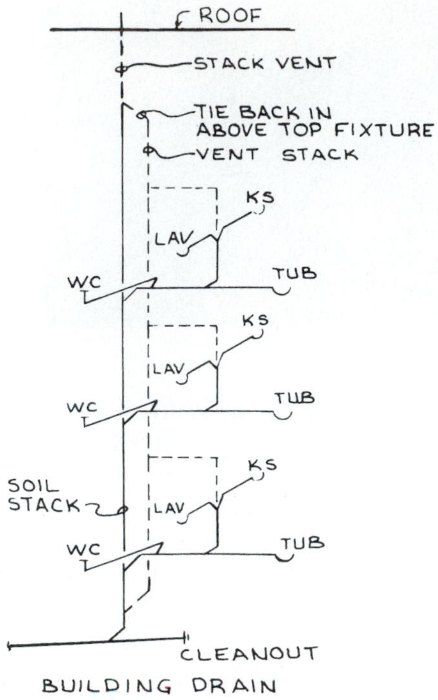

FIGURE 3-7   Vent stack to stack vent

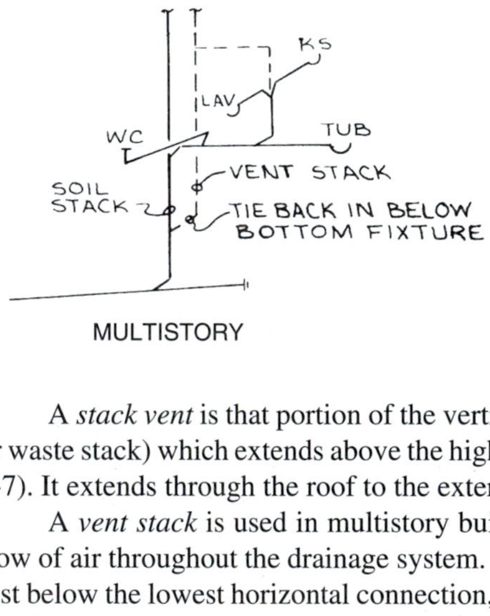

FIGURE 3-8   Vent stack to soil stack

A *stack vent* is that portion of the vertical sewage drainage pipe (which may be a soil or waste stack) which extends above the highest horizontal drain that is connected to it (Fig. 3-7). It extends through the roof to the exterior of the building.

A *vent stack* is used in multistory buildings where a pipe is required to provide the flow of air throughout the drainage system. The vent stack begins at the soil or waste pipe, just below the lowest horizontal connection, and may go through the roof (Fig. 3-8) or connect back into the soil or waste pipe not less than 6 in. (150 mm) above the top of the highest fixture.

Fixture Branches

The fixtures at a floor level are connected horizontally to the stack by a drain called a *fixture branch* (Fig. 3-1). Beginning with the fixture farthest from the stack, the branch must slope ⅛ to ½ in. per ft (10.4 to 41.6 mm per meter) for proper flow of wastes through the branch. Branch piping which serves urinals, water closets, showers, or tubs is usually run in the floor (Fig. 3-9). When these fixtures are not on the branch, the piping may be run in the floor or in the wall behind the fixtures (Fig. 3-10). Branch piping may be copper, plastic, galvanized steel, or cast iron.

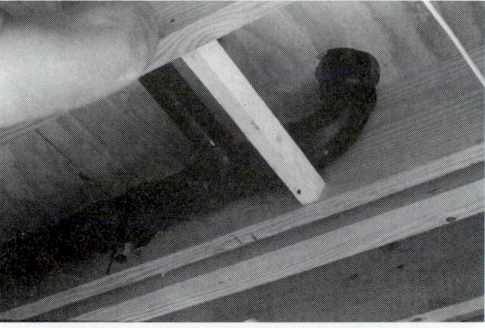

FIGURE 3-9   Branch piping under floor

FIGURE 3-10   Individual fixture drainage

Soil and Waste Stacks

The fixture branches feed into a vertical pipe referred to as a *stack*. When the waste that the stack will carry includes human waste from water closets (or from fixtures which have similar functions), the stack is referred to as a *soil stack*. When the stack will carry all wastes *except* human waste, it is referred to as a *waste stack*. Soil and waste stacks may be copper, plastic, galvanized steel, or cast iron. These stacks service the fixture branches beginning at the top branch and go vertically to the building drain (Fig. 3-11).

In larger buildings, the point where the stack ties into the building drain rests on a masonry pier or steel post so that the downward pressure of the wastes will not cause the piping system to sag. In addition, the stack must be supported at 10-ft (3 m) intervals to limit movement of the pipe. When a stack length is greater than 80 ft (24.4 m), horizontal offsets are used to reduce free fall velocity and air turbulence. Connections to fixture branches and the building drain should be angled 45° or more to allow the smooth flow of wastes.

Most designers try to lay out plumbing fixtures to line up vertically floor after floor so that a minimum number of stacks will be required. Many times, a central core of a

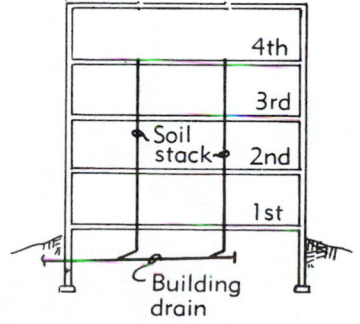

FIGURE 3-11   Stack locations

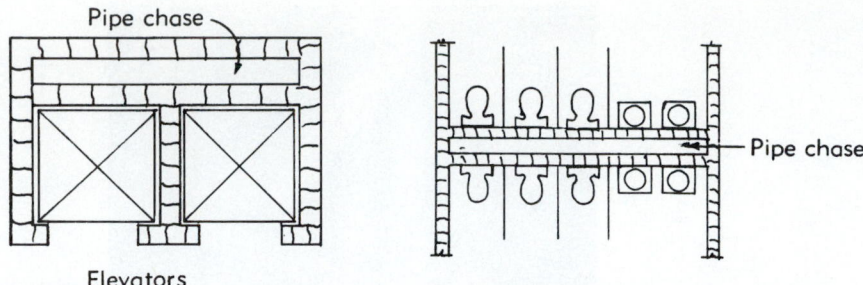

FIGURE 3-12   Pipe chase

multistory building will be used as a plumbing core, and a *pipe chase,* a space which is left to put the pipes in, runs from the first floor to the roof of the building (Fig. 3-12).

Building Drains (Also Called House Drains)

The soil or waste stacks feed into a horizontal pipe referred to as the *building drain.* The building drain slopes ⅛ to ¼ in. per ft (10.4 to 20.8 mm per m) as it feeds the waste into the building sewer outside the building. By definition, the building drain extends to a point 3 ft (1 m) *outside* the wall of the building (Fig. 3-11).

Provision is made to allow cleaning of the building drain by putting a *cleanout* at the end of the drain (Fig. 3-11). Another cleanout is sometimes placed just inside the building wall in case it is necessary to clean the building drain or sewer line. Cleanouts should also be placed no more than 50 ft (15 m) apart in the long building drains.

Location of the building drain in the building depends primarily on the location (elevation) below grade of the community sewer. Ideally, all of the plumbing wastes of the building will flow into the sewer (whether it is a community or a private sewer system) by gravity. Typically, the drain is placed below the first floor (Fig. 3-11) or below the basement floor (Fig. 3-13). If the height of the sewer requires the drain to be placed above the lowest fixtures (Fig. 3-14), it will be necessary for the low fixtures to drain into a sump pit. When the level in the sump pit rises to a certain point, an automatic float or control will activate a pump which raises the waste out of the pit and into the building drain.

Building drains are usually made of plastic, copper (for above the floor), or extra-heavy cast iron (for below the floor) pipe.

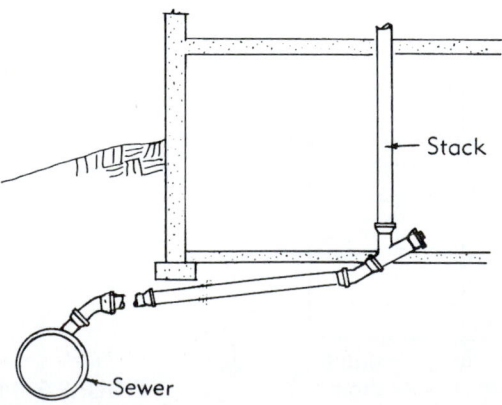

FIGURE 3-13   Underfloor building drain

FIGURE 3-14   Sump pump

Building Traps (Also Called House Traps) and Fresh Air Inlets

Some codes may require a *building trap* on the building drain near the building wall (Fig. 3-15). This trap acts as a seal to keep gases from entering the sewage system from the sewer line. The *National Standard Plumbing Code* and most regional and state codes do not feel a building trap is necessary; instead, it is felt that this trap will impede the flow of wastes in the system. However, when required by local, state, or regional codes, it must be put in the system. When a building trap is used, a *fresh air inlet* (Fig. 3-16) may be required to allow fresh air into the system to be certain that the trap seal is not siphoned through. The fresh air inlet must be a minimum of 4 in. (100 m) or one-half the diameter of the building drain, whichever is larger.

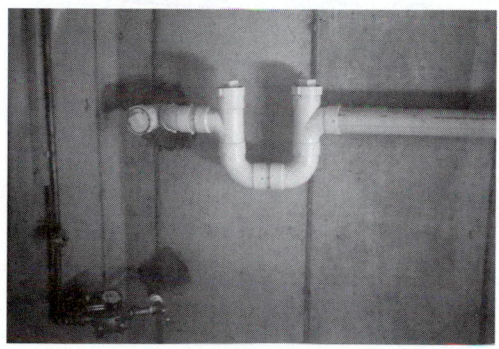

FIGURE 3-15   Building trap

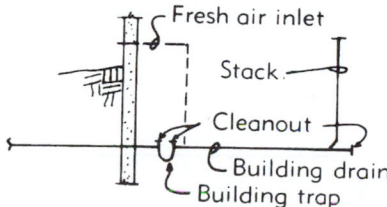

FIGURE 3-16   Fresh air inlet

3-4 PIPES AND FITTINGS

Drainage lines and vents make use of most of the same types of piping used in the water supply system, except that vitrified clay tile may be used in the building sewer line.

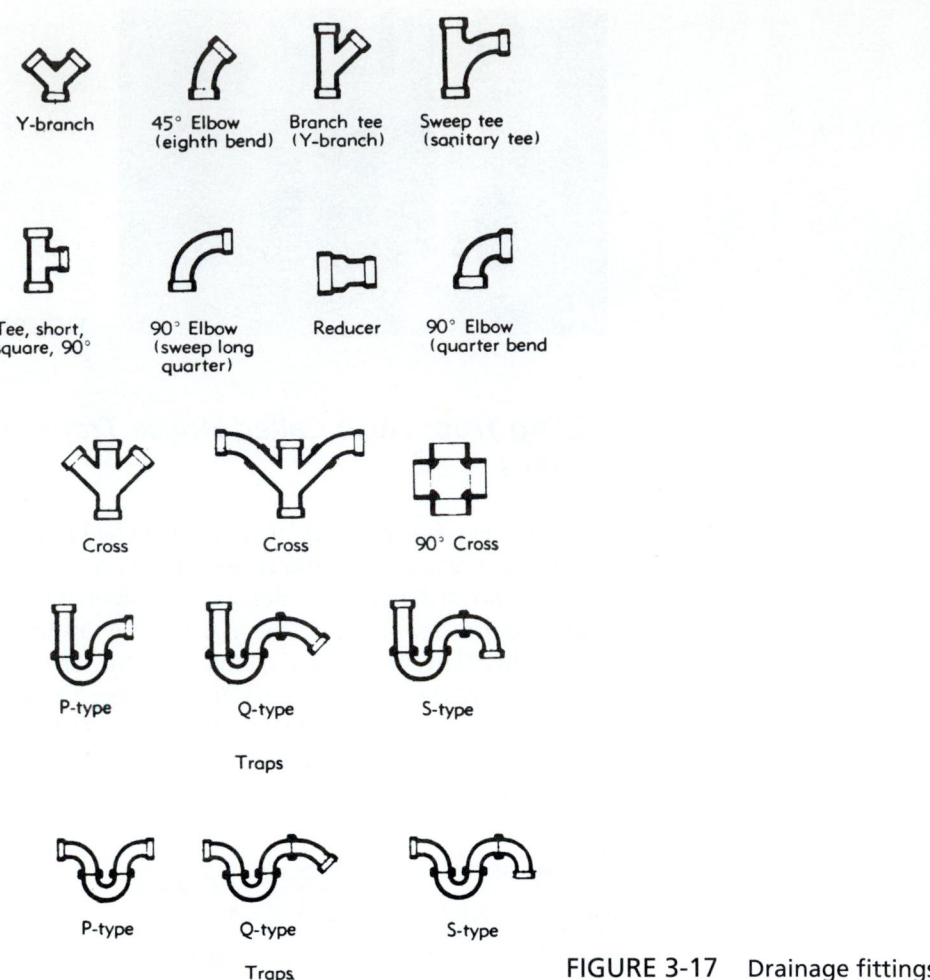

FIGURE 3-17   Drainage fittings

Copper tubing, type DWV, is used in drainage piping but is *not* used in water supply. The *DWV* on the tubing means it can be used for drainage, waste, and venting on the job. Type M copper may be used above grade and type L copper below grade. Plastic piping is used extensively in drainage systems because of its low cost and speedy installation. Other piping sometimes used in the building sewer line includes concrete and bituminous pipes. Fittings used in drainage systems are shown in Fig. 3-17.

3-5 PLUMBING DRAINAGE DESIGN

First, a complete review of the project being designed must be made. In this case, the four-story apartment building used for the water supply problem in Sec. 2-10 is used. A review of the drawings for the apartments shows that each floor contains two bathroom groups and two kitchen sinks. Assume the project will be connected to a community sewer.

In this project, it has been decided that there will be two plumbing stacks, one for each tier of apartments (Fig. 3-18).

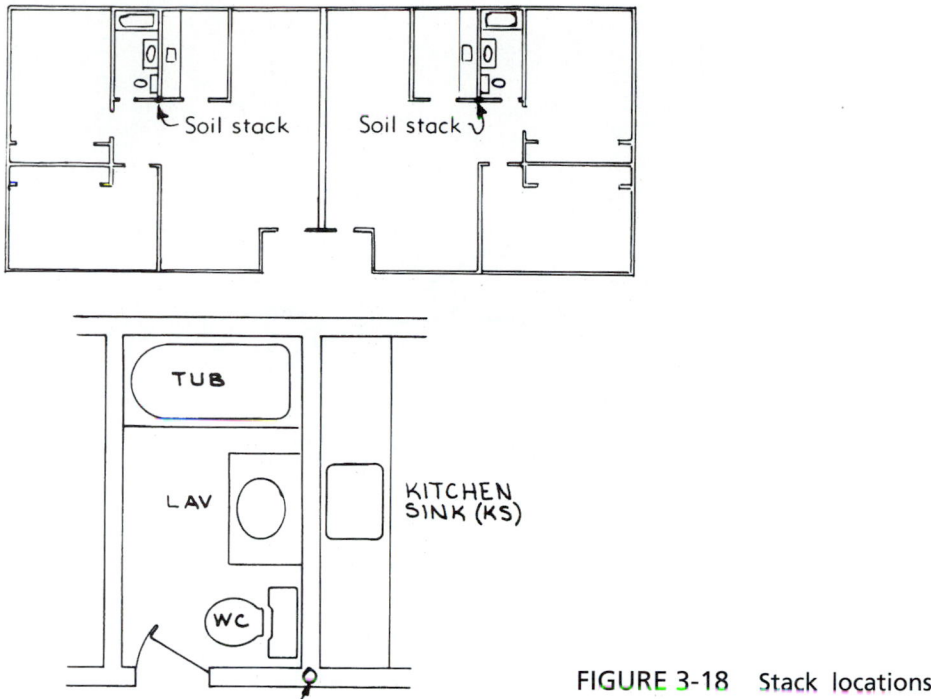

FIGURE 3-18   Stack locations

Step-by-Step Approach

1. The first step will be to sketch an isometric of the drainage piping. It will be easiest to follow if you first locate the stack on the plan (Fig. 3-18). Then sketch the vertical stack (Fig. 3-19), and, beginning at the top floor, sketch the fixture branch and then the connection at each fixture.

   Next, sketch in the other three floors as shown in Fig. 3-20. Now, add the vent stack from a point just below the bottom fixture branch to a point above the top fixture and the building drain (Fig. 3-21).

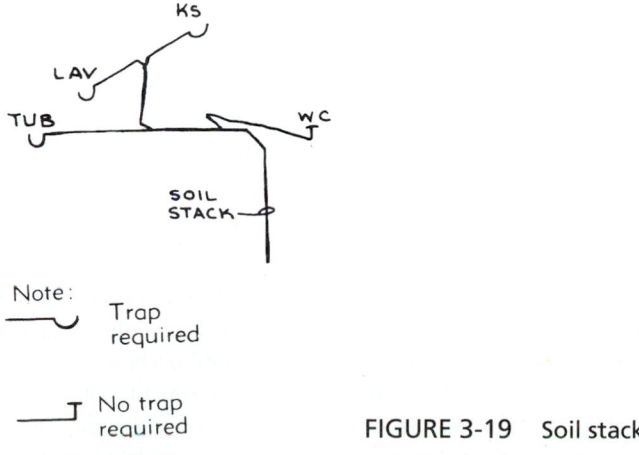

Note:
⎯◡ Trap required

⎯⏉ No trap required

FIGURE 3-19   Soil stack

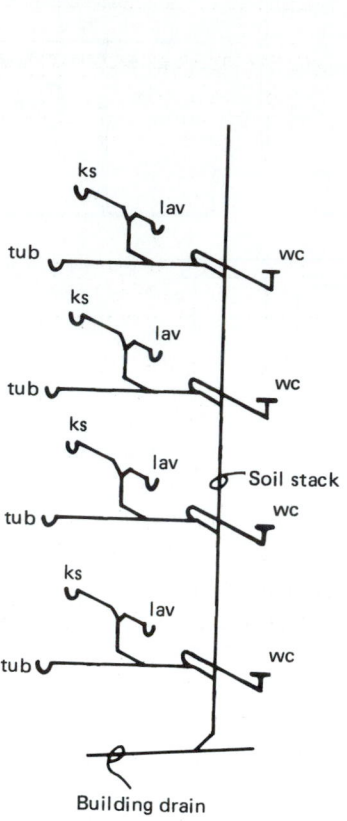

FIGURE 3-20   Add fixture branches

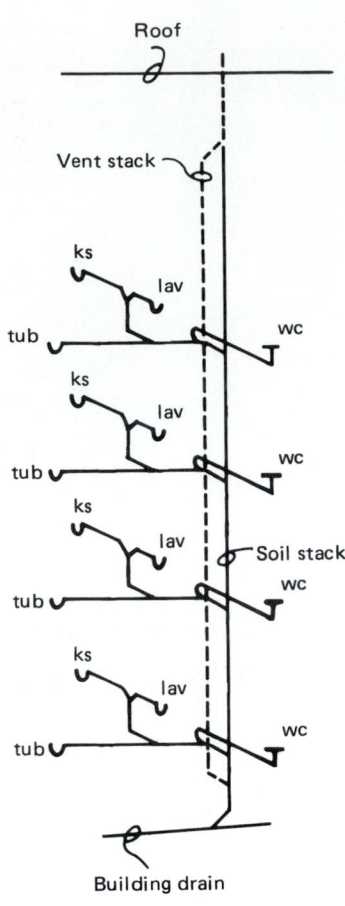

FIGURE 3-21   Add vent stack

2. Next, the minimum trap size for each fixture is selected from Fig. 3-46 and the trap size noted on the schematic pipe layout, as shown in Fig. 3-22. Assume the kitchen sink has a small P.O. (plumbing outlet) plug.

3. The fixture branch is the first drainage pipe to be sized. The first portion of the branch to be sized is from the fixture farthest from the stack to the next fixture, as shown in Fig. 3-23. Begin by determining the fixture units that this short piece of pipe will serve from Fig. 3-23.

> In this problem, the branch serves a bathtub with a fixture-unit value of 2, based on Fig. 3-43.

4. Next, select the branch size to serve the bathtub from the table of sizes in Fig. 3-45. Be sure to check the figures for horizontal fixture branches and stacks each time you use Fig. 3-45. The left row of numbers gives the various diameters of pipe which may be used. The rest of the columns list the maximum number of fixture units which can be connected to a given pipe size. Each of the fixture-unit columns defines the type of piping being selected.

> In this case, the horizontal fixture branch size is being selected, so go to that column (Fig. 3-45). Go down the column until the fixture-unit number is the same as or more than the amount being served—in this case, 3 f.u., which is greater than the 2 f.u. value of the bathtub. Now, move horizontally to the left and select the minimum pipe size required—in this case, 1½ in. or 38 (40) mm. Now, note the information accumulated in tabular form as shown:

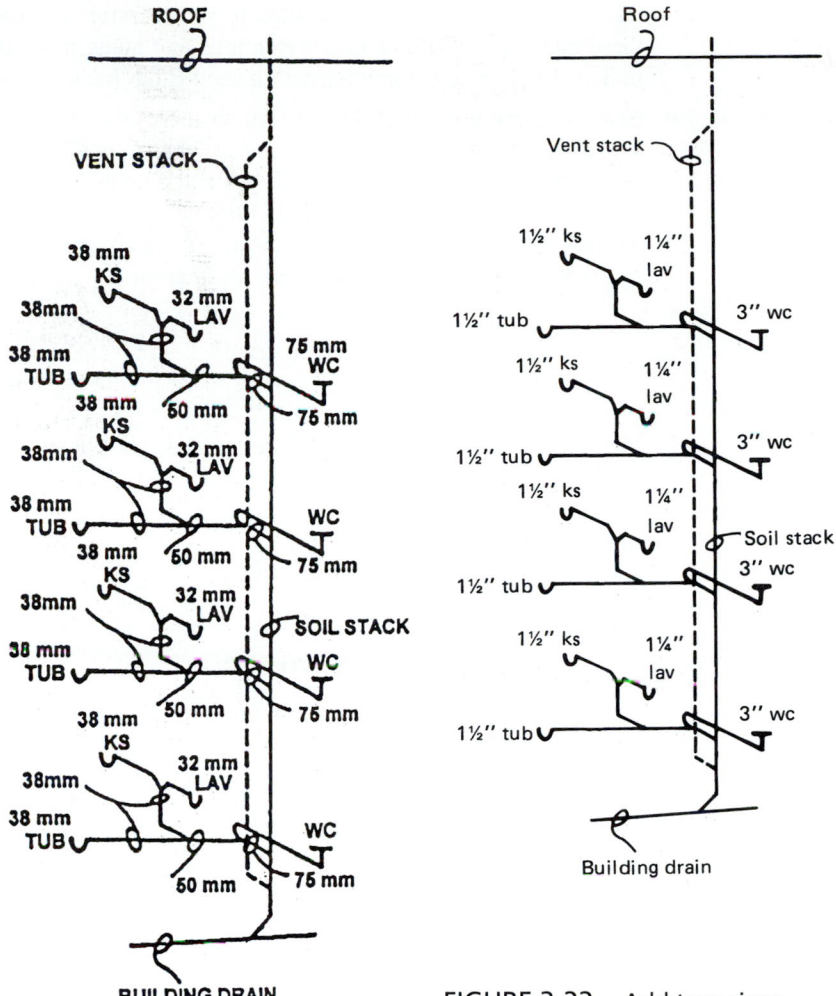

FIGURE 3-22   Add trap sizes

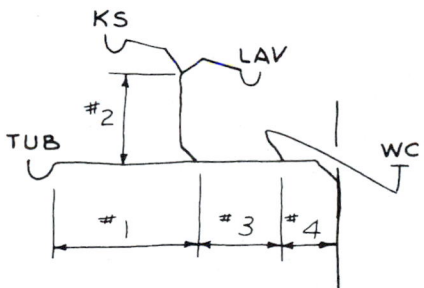

FIGURE 3-23   Typical fixture branch

	f.u.	PIPE SIZE
#1-tub	2	1½ in. (38 or 40 mm)

5. For all fixtures, there is a maximum horizontal distance between the fixture trap and a vent. The distance depends on the pipe size being used and is listed in Fig. 3-48. Add this information to the tabular form:

	f.u.	PIPE SIZE	DISTANCE
#1-tub to lav.	2	1½ in. (38 or 40 mm)	5 ft. (1.5 m)

There are many different ways to vent a fixture or group of fixtures. A few of these methods are shown in this example, and many more are shown and discussed in Sec. 3-6. For now, the information needed has been accumulated.

6. Now, size that portion of the pipe which serves the lavatory and kitchen sink down to the branch. In Fig. 3-23, this length is labeled as #2. From Fig. 3-43, the number of fixture units for the kitchen sink being served is 2 f.u. and for the lavatory 1 f.u. for a total of 3 f.u. Referring to Fig. 3-45, the minimum pipe size would be 1½ in. or 38 mm, the maximum horizontal distance from fixture trap to vent is limited to 5 ft. or 1.5 m. Now, list the accumulated information.

	f.u.	PIPE SIZE	DISTANCE
#2-k.s. & lav. to branch	3	1½ in. (38 or 40 mm)	5 ft (1.5 m)

7. The section of branch marked #3 serves the tub, kitchen sink, and lavatory. From Fig. 3-43:

Bathtub	2 d.f.u.
Kitchen sink	2 d.f.u.
Lavatory	1 d.f.u.
Total	5 d.f.u.

From Fig. 3-45 the pipe size is 2 in. or 50 mm.
Add this information to the tabular form.

	f.u.	PIPE SIZE
#3-k.s., lav., & tub	5	2 in. (50 mm)

8. Now, the last section (#4) is calculated and tabulated. At this point, the branch services all of the fixtures, which are a fixture group and a kitchen sink. From Fig. 3-43 the fixture units are:

1 bathroom group	6 d.f.u.
1 kitchen sink	2 d.f.u.
Total	8 d.f.u.

From Fig. 3-45, the pipe size is 2½ in. or 65 mm, but since the pipe cannot be smaller than the largest fixture trap, a 3-in. or 75 mm pipe is required (water closet, 3-in. or 75 mm trap).

	f.u.	PIPE SIZE
#4-bathroom group & k.s.	8	3 in. (75 mm)

9. Now, all of the branch sizes are tabulated (Fig. 3-24). Note the sizes on the isometric sketch, as shown in Fig. 3-25. Since each floor is exactly the same, all of the branches are the same size, and all floors are noted on the isometric. If any of the branches were different, the procedure just explained would be followed to get the branch pipe sizes.

10. The stack size is selected next. Because the stack handles human waste, it is a *soil* stack. The stack must be selected and sized for the total fixture units that it must handle. As noted on the sketch in Fig. 3-26, each floor has a total of 8 f.u., for an overall total of 32 d.f.u. The soil stack is selected from Fig. 3-45 under the column "More Than Three Branch Intervals" since we are dealing with the entire stack height.

	d.f.u.	Pipe	Dist. to Vent
#1 Tub to Lav	2	1½"	5'-0"
#2 KS, Lav to Tub	3	1½"	5'-0"
#3 KS, Lav, Tub to WC	5	2"	8'-0"
#4 KS, Lav, Tub, WC to Stack	8	3"	10'-0"

SI UNITS

		Pipe	Dist. to Vent
#1 Tub to Lav	2	38 mm	1.5 m
#2 KS, Lav to Tub	3	38 mm	1.5 m
#3 KS, Lav to Tub to WC	5	50 mm	2.4 m
#4 KS, Lav, Tub, WC to Stack	8	75 mm	3.0 m

FIGURE 3-24   Horizontal branch tabulation

SI UNITS

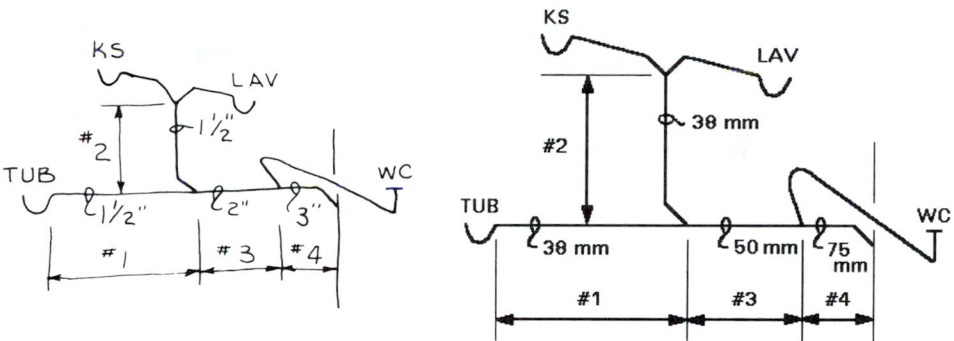

FIGURE 3-25   Horizontal branch

Based on 32 d.f.u. for the entire stack, the minimum stack size from the chart would be 2½ in., except that the stack must be at least as large as the fixture branch—in this design, 3 in. Note the stack size on the sketch as shown in Fig. 3-27 and tabulated in Fig. 3-28.

SI

Based on 32 d.f.u., for the entire stack, the minimum size from the chart would be 65 mm, except that the stack must be at least as large as the fixture branch—in this design, 75 mm. Note the stack size on the sketch as shown in Fig. 3-27, and tabulated in Fig. 3-28.

11. The vent stack is sized next, using the table in Fig. 3-47. To use the table, it is necessary to know the maximum size of the soil stack (left column) and the fixture units connected to the vent stack; the maximum length of vent stack is noted in ft. The soil stack has already been sized, and the drainage fixture units have been totaled as 32 for this stack. The developed length of the vent stack is the lineal ft of pipe required from the lowest point where it connects with the soil stack to the point where it terminates outside the building, as illustrated in Fig. 3-29.

The next step is to determine the size and location of the main (vent) stack. The code requires that every building "shall have at least one main stack." This main stack must be sized to handle the total fixture units on the system.

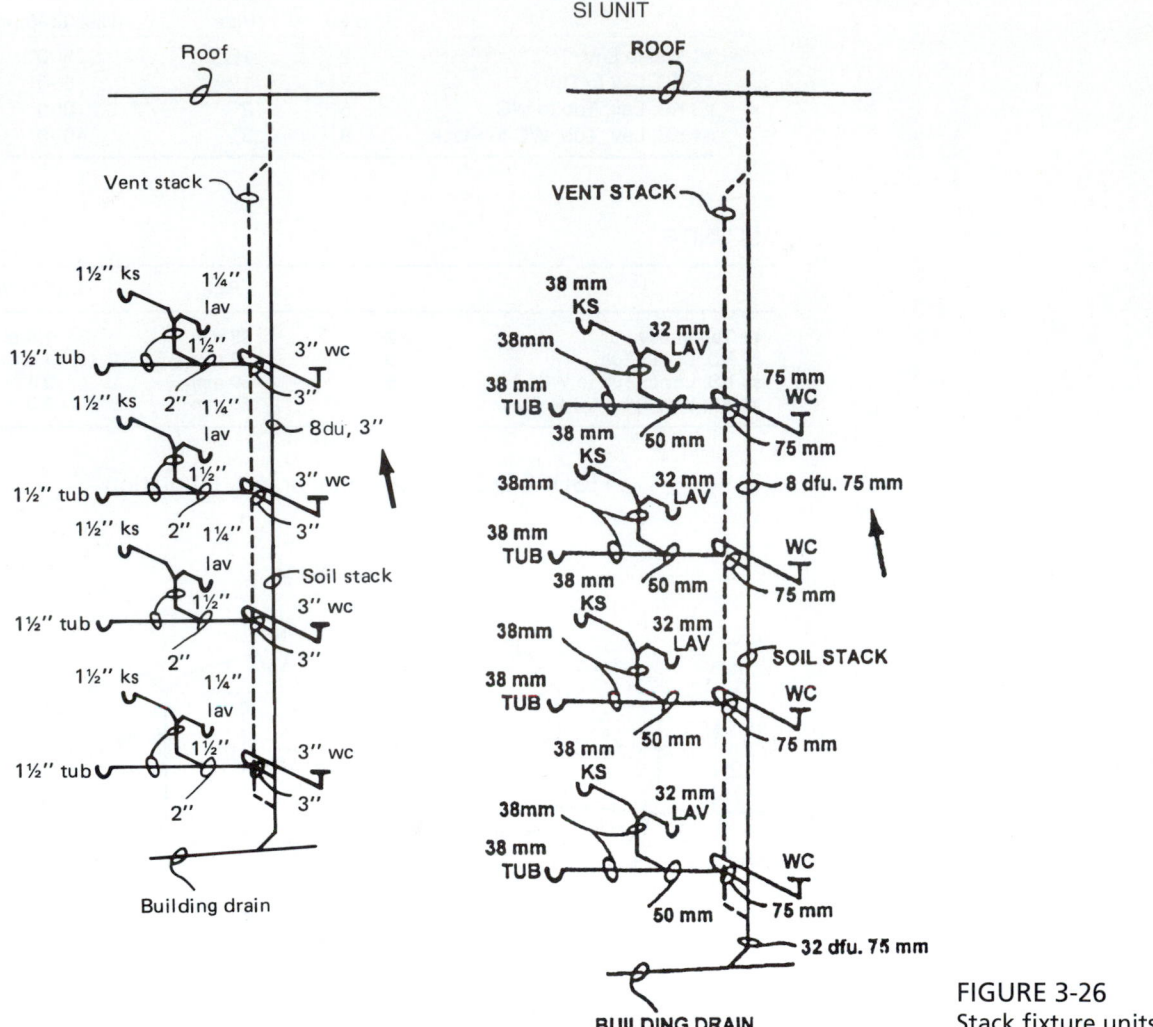

FIGURE 3-26
Stack fixture units

In this problem, the developed length is about 52 ft. To size the stack use the table in Fig. 3-47, find the soil stack size (4 in.) and then move horizontally to the right to check the fixture-unit column. At this point, you should note that there are three listings for a 4-in. soil stack, all with different fixture-unit values. In this case, there are 64 d.f.u., which means the listing of 100 must be used. Now move to the right; the next number represents the developed length of vent in the design—in this case, 52 lineal ft. In the table, it is more than 35, so the 100 column is used. From the 100, move vertically up to read a vent size of 2½ in.

The plumbing code states that the main vent and vent stack shall be sized in accordance with the table "and be not less than 3 in. in diameter." In addition, it states that the size of the vent stack must not be reduced all the way through the roof. So in this design, the minimum vent stack size of 3 in. is used.

In this problem the developed length is about 15.9 m. To size the stack, find the soil stack size (100mm) and move horizontally to the right to check the fixture unit column. At this point, note that there are three listings for a 100 mm soil stack, all with different fixture-unit values. In this case there are 64 d.f.u., which

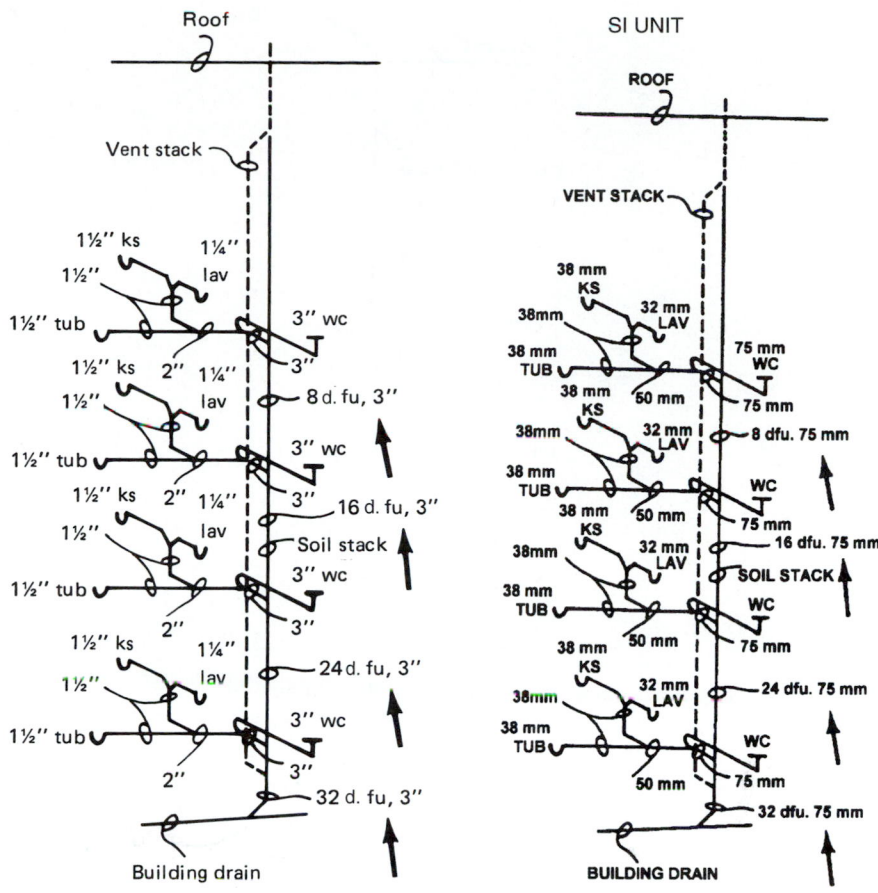

FIGURE 3-27   Stack sizes

STACK SIZE			
Floor	fu	Accum. d.f.u.	Pipe size
4	8	8	3″ (75 mm)
3	8	16	3″ (75 mm)
2	8	24	3″ (75 mm)
1	8	32	3″ (75 mm)

FIGURE 3-28   Tabulated stack size

means the listing of 100 d.f.u. must be used. Now move to the right; the next number represents the developed length of vent in the design—in this case, 15.9 m. In the table it is more than 11 m, so the 30 m column is used. From the 30 m, move vertically up to read a vent size of 65 mm.

The plumbing code states that the main vent and vent stack shall be sized in accordance with the table "and be not less than 75 mm in diameter." In addition, it states that the size of the vent stack must not be reduced all the way through the roof. So in this design, the minimum vent stack size of 75 mm is used.

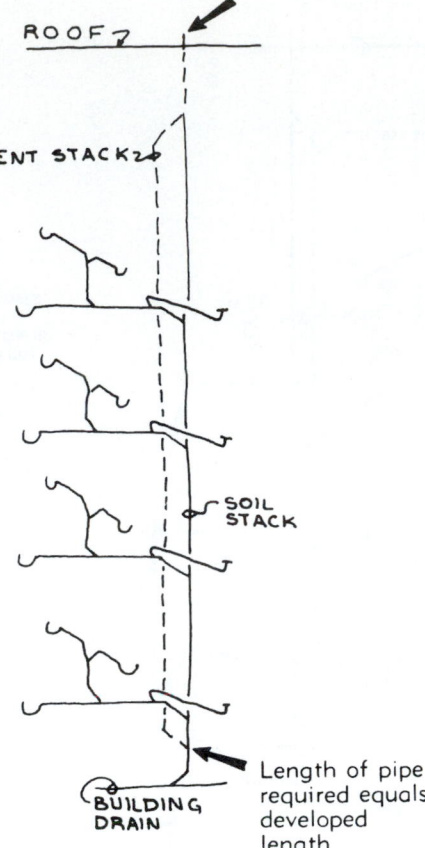

FIGURE 3-29   Vent stack, developed length

Code requirements also stipulate that the vent stack will terminate no less than 6 in. (150 mm) above the roof (Fig. 3-30), and if the roof is to be used for other than weather protection (for example, as a terrace or balcony), the vent stack must run at least 5 ft (1.5 m) above the roof. Most codes require at least a 3-in. (75 mm) vent through the roof.

12. The next step in the design is to size the building drain. At this point, review the basic system being designed.

   The building drain must be sized for the total amount of drainage fixture units connected to the point at which the main vent (step 11) is connected.

   In this problem, there are two stacks serving the building, which means that the two stacks must be connected to the building drain. A sketch similar to Fig. 3-31 helps to prevent any confusion or the possibility that a stack may be forgotten. On the sketch, and in tabular form, list the accumulated fixture units.

   Next, the slope of the drain must be determined and noted on the sketch; in this case, a ⅛-in. slope per ft (10.4 mm per m) is used. Building drain sizes are found in Fig. 3-44. The pipe size is listed on the left, the slope of the pipe (referred to as *fall per ft*) is listed along the top slope and the fixture units are listed below the slope per ft. In this design, the main vent is considered to be the end vent, as noted in Fig. 3-32, so the entire building drain must be sized for 64 d.f.u.

13. The plumbing drainage system has now been designed, with the exception of checking the location of the vent in relation to the fixtures. As the fixture branch sizes were being selected, the maximum distance from the trap to a vent was also tabulated

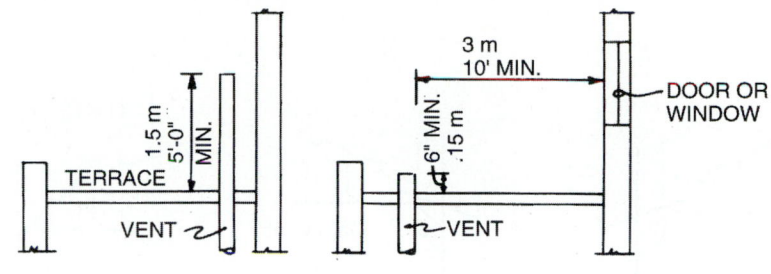

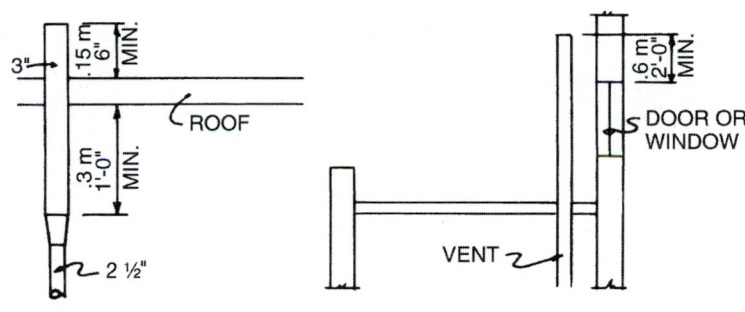

Most codes require at least a 3" vent through the roof beginning at least one foot below the roof.

FIGURE 3-30   Vent termination

STACK	DFU	ACCUM. DFU
#1	32	32
#2	32	64

STACK #2
32 DFU

STACK #1
32 DFU

BUILDING DRAIN
SLOPE 1/8" PER FOOT
(10.4 mm per m)

FIGURE 3-31   Building drain sketch

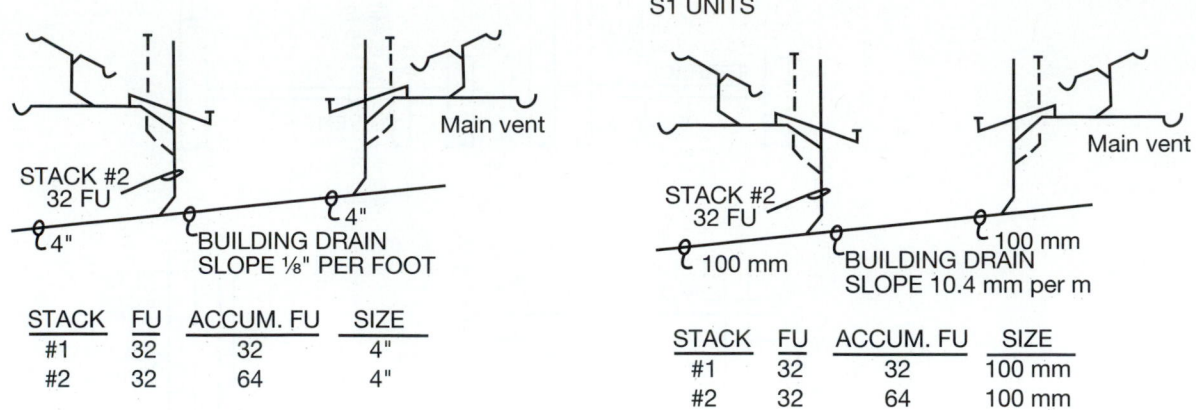

FIGURE 3-32   Building drain size tabulation

(Fig. 3-24). While an experienced designer would make a tentative decision as to vent locations early in the design, it has been left to the end here so that a more thorough explanation of the various methods of venting and possible venting solutions may be given.

In reviewing the fixture branch and trap sizes in Fig. 3-24, each of the tub, lavatory, and sink traps can be no more than 5 ft (1.5 m) from a vent, and the water closet no more than 10 ft (3 m). A check of the floor plan shows that the total width of the bathroom is 7 ft-6 in. (2.25 m). This means that the vent could be located in the wall near the lavatory and be within the limits of all listed horizontal distances.

If fixtures are so spread out that one or more fixtures are beyond the maximum horizontal distance, there are two options.

a. Increase the size of the horizontal fixture branch, which automatically increases the horizontal distance. For example, change length #1 from 1½ in. (38 mm) to 2 in. (50 mm), and it changes the horizontal distance from 5 ft to 8 ft (1.5 m to 2.4 m). Many times this is the most economical solution.

b. Add more vents. Instead of trying to service a group of fixtures with one vent, perhaps a vent should be added to service any fixtures which are beyond the allowable distance. A variety of venting solutions for various groups of fixtures is shown in Figs. 3-33 through 3-36.

In this design, a single vent off the top of the lavatory-kitchen sink fixtures (referred to as a *wet vent* and discussed in the following paragraph) will be used.

The size of the wet vent is selected from Figs. 3-33 through 3-36. But a little explanation is needed. First, a review of the sketch in Fig. 3-33 shows that the portion of the vent referred to as a *wet vent* is actually a part of the drainage system for the lavatory and the kitchen sink. Because it acts as both a *vent* and a *drainage* pipe, it is referred to as a *wet vent*. For single bathroom groups, selection of the wet vent is based on the number of fixture units that the wet vent serves. A careful check of the code requirements for wet venting a multistory bathroom group (Fig. 3-36) indicates that a wet vent and its extension must be 2 in. in diameter and can serve the kitchen sink, lavatory, and bathtub. In Fig. 3-36 it states that "each water closet below the top floor is individually back

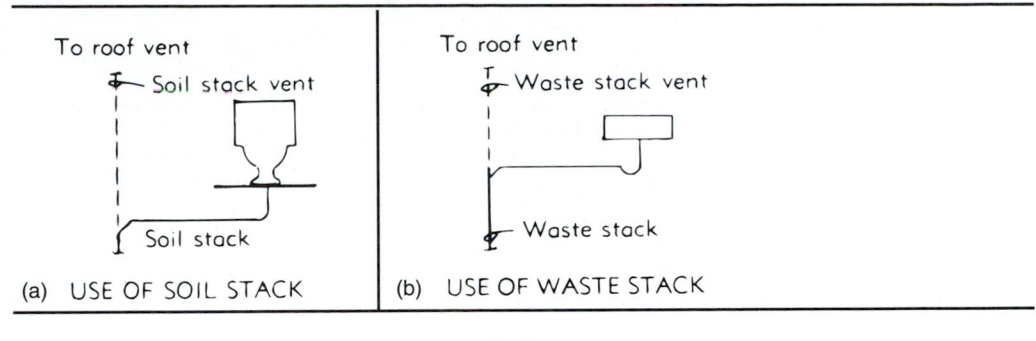

(a) USE OF SOIL STACK

(b) USE OF WASTE STACK

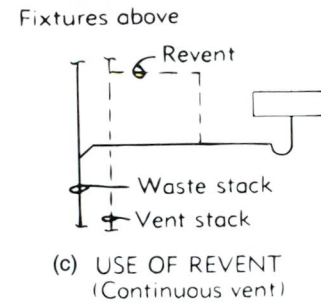

(c) USE OF REVENT
(Continuous vent)

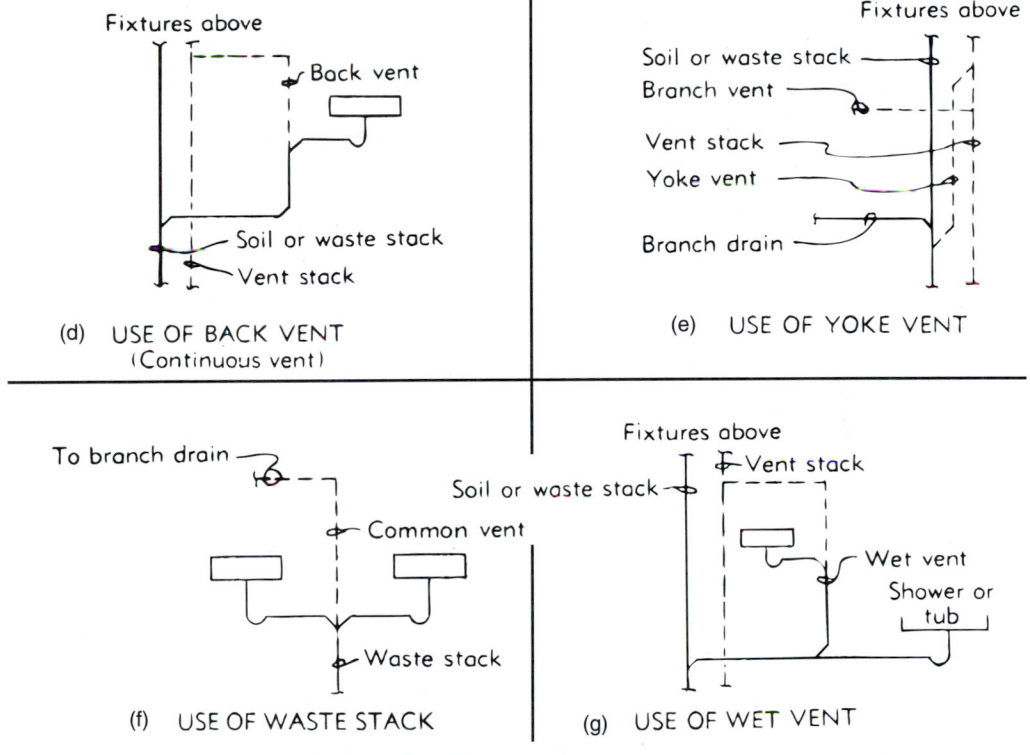

(d) USE OF BACK VENT
(Continuous vent)

(e) USE OF YOKE VENT

(f) USE OF WASTE STACK

(g) USE OF WET VENT

FIGURE 3-33 Vents

vented" unless wet vented as illustrated in Fig. 3-34. When the minimum size of the vent is larger than the waste sizes previously selected, the larger size must be used.

Note the size of the venting required on the schematic sketch, and the plumbing drainage design is completed for this project.

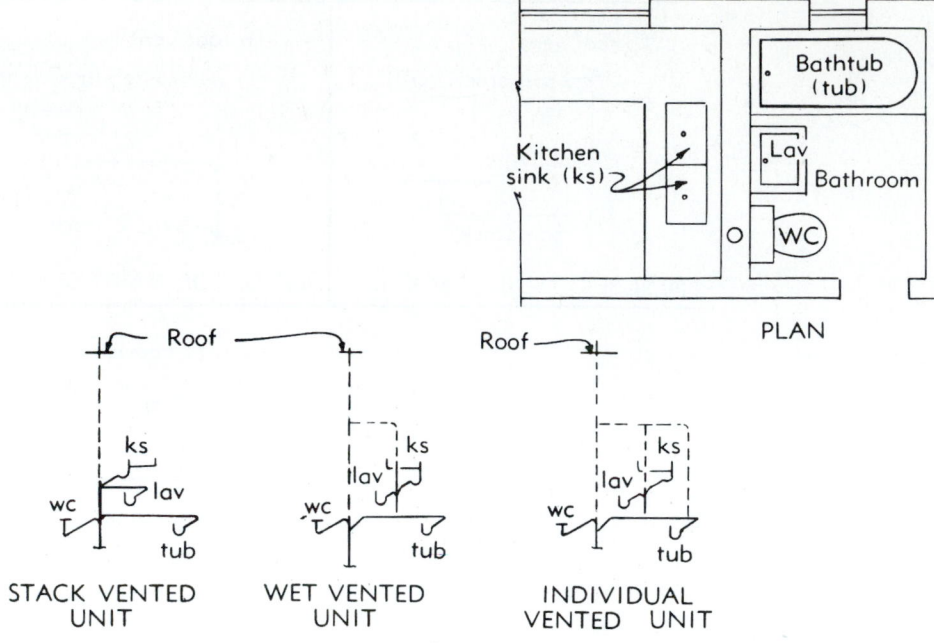

FIGURE 3-34   Vent for a one-family dwelling

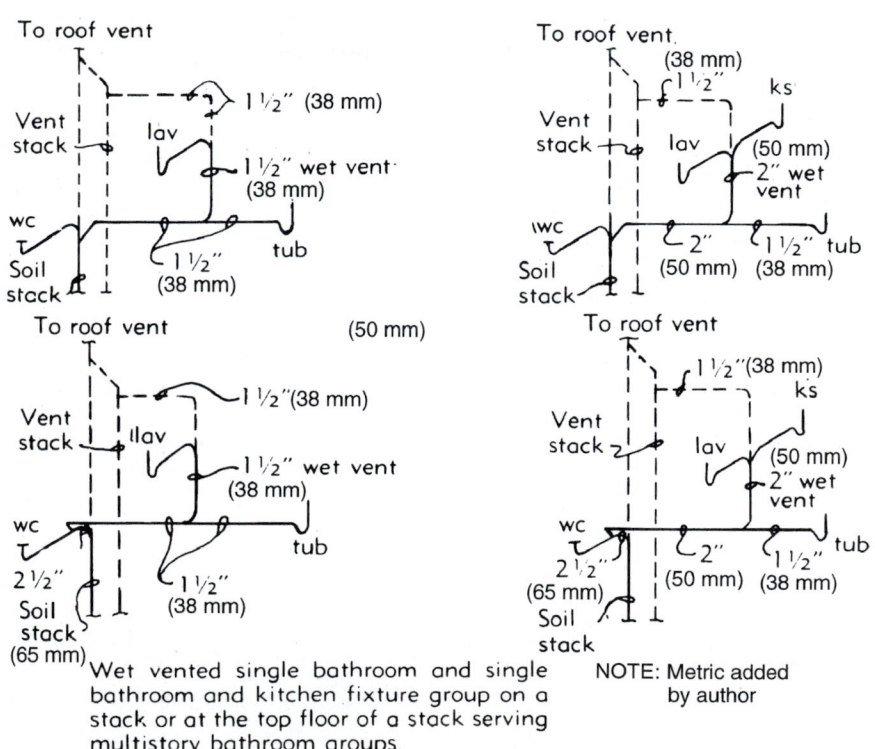

Wet vented single bathroom and single bathroom and kitchen fixture group on a stack or at the top floor of a stack serving multistory bathroom groups.

NOTE: Metric added by author

FIGURE 3-35   Wet venting—top floor

WET VENTING

Single Bathroom Groups—A single bathroom group of fixtures may be installed with the drain from an individually vented lavatory serving as a wet vent for a bathtub or shower stall and for the water closet, provided that: (1) not more than one fixture unit is drained into a 1½-inch diameter wet vent or not more than four fixture units drain into a 2-inch diameter wet vent, and (2) the horizontal branch shall be a minimum of 2 inches and connect to the stack at the same level as the water closet drain or below the water-closet drain when installed on the top floor. It may also connect to the water-closet bend.

Multistory Bathroom Group—On the lower floors of a multistory building, the waste pipe from one or two lavatories may be used as a wet vent for one or two bathtubs or showers provided that: the wet vent and its extension to the vent stack is 2 inches in diameter; each water closet below the top floor is individually back vented.

SIZE OF VENT STACKS

Number of wet-vented fixtures	Diameter of vent stacks	
	in inches	mm
1 or 2 bathtubs or showers	2	50
3 to 5 bathtubs or showers	2½	65
6 to 9 bathtubs or showers	3	75
10 to 16 bathtubs or showers	4	100

In multistory bathroom groups, wet vented in accordance with the paragraph above, water closets below the top floor group need not be individually vented if the 2-inch wet vent connects directly into the water-closet bend at a 45-degree angle to the horizontal portion of the bend and in the direction of flow.

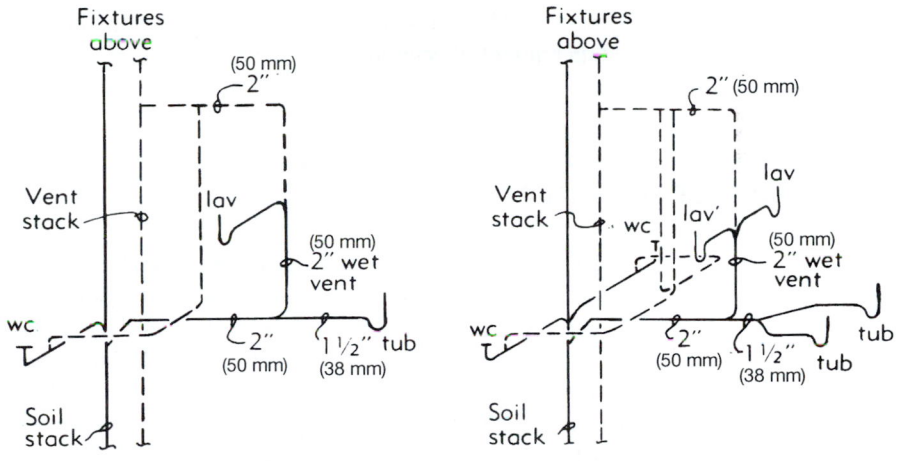

FIGURE 3-36   Wet venting—below top floor

3-6 VENTING

The basics of venting have been outlined in Sec. 3-3 and in the design worked out in Sec. 3-5. A further explanation and several illustrations to show methods of venting are included in Fig. 3-33 through 3-36.

3-7 PLUMBING ECONOMY

Economies in plumbing are possible through the careful planning and location of fixtures in clusters, back-to-back or otherwise grouped to form as few wet-walls (walls in which the plumbing pipes are located) as possible.

In multistory construction, locating fixtures above each other saves considerable money since a minimum amount of piping and the smallest sizes possible may be used for both supply and disposal.

In residences designed with low cost as a primary objective, it is even possible to use the same wet-wall for a back-to-back bathroom and kitchen. For middle-priced and custom-designed residences, the primary concern is the location of fixtures where they will best suit the plan. Most designers find no problem in planning and designing a building so that a certain amount of economy is achieved at no sacrifice to the overall plan.

3-8 SYSTEM INSTALLATION

On a small project, the drainage piping typically varies in size from 1 in. to 4 in. but can be much larger on large hotels, apartments, and office buildings. The size of the pipe often requires special provisions in wall width or furred-out areas (Fig. 3-37) to run them.

Poured concrete slabs will require that the plumbing layout be carefully considered. The pipes need to be placed in the ground before the slab is poured, so their accurate placement is crucial. Typically, both the water supply and drainage pipes are laid out next to each other, since they go to the same areas of the project. String is usually stretched over the slab area to mark where the pipes should be located (Fig. 3-38). Many times, they are planned so they will come up in a wall (Fig. 3-39). However, the tub, shower, and water closet piping will need to be placed in the exact location where the fixture is to go. All piping must be carefully located and the system checked for leaks before the concrete is poured because any relocation or repairs of pipes would be costly.

FIGURE 3-37   Furred out space

FIGURE 3-38   Layout marked with string

FIGURE 3-39   Piping in wall

FIGURE 3-40   Piping in wood frame construction

On larger projects with concrete walls and ceilings, it is usually necessary to provide sleeves (holes) in the concrete for the pipes to pass through to get from space to space. It will also be necessary to provide inserts and hangers to support the pipes.

The open spaces provided in truss-type construction make it easier to run piping through to the desired location. The only points of difficulty would be where it needs to pass by ductwork or some other large pipe that is going in the opposite direction. This will require coordination with the contractor installing any heating, air conditioning, or ventilating ductwork.

In wood frame construction, the holes are sometimes (Fig. 3-40) drilled to allow the passage of the pipes. Typically, these should be at the middle of any load-bearing wood members so that a minimum of structural damage is done. There are times when the width of a wall needs to be increased to allow for pipes running horizontally to pass by drainage pipes (or other pipes) running vertically.

Pipe tunnels (Fig. 3-41) may be used on large projects to provide concealed space for the passage of mechanicals at ground level and from building to building. Hangers from the top or side of the tunnel are used to support the pipes. Access may be from either end of the tunnel, or access floors may be provided.

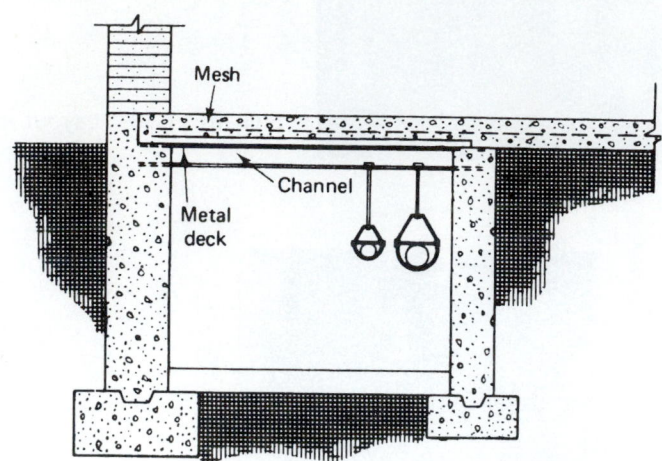

FIGURE 3-41   Pipe tunnel

Piping symbols:

Vent —— —— —— —— —— ——

Cold water —— — · — — · — — · ——

Hot water —— — ·· — — ·· — — ·· ——

Hot water return —— — — — — ——

Gas —— G —— G ——

Soil, waste or leader ————————
  (above grade)

Soil, waste or leader —— —— —— —— ——
  (below grade)

Fixture symbols:

Baths

Water closet (with tank)

Water closet (flush valve)

Shower

Lavatory

DW   Dishwasher

SS   Service sink

HWT   Hot water tank

HWT

DF   Drinking fountain

M   Meter

HB   Hose bib

C/O   CO   Cleanouts

FD   Floor drain

RD   Roof drain

90 Elbow

Gate valve

Reducer

Check valve

Tee (up)

Tee (down)

90 Elbow (up)

90 Elbow (down)

U.T.R. = Vent to Roof.

C.W.R.	cold water return
DEG.	degree
D.F.	drinking fountain
D.H.W.	domestic hot water
DR.	drain
D.W.	dishwasher
F.	fahrenheit
FDR.	feeder
FIXT.	fixture
F.D.	floor drain
F.H.	fire hose
F.E.	fire extinguisher unit
H.W.	hot water
H.W.C.	hot water circulating line
H.W.R.	hot water return

A.F.D.	area floor drain
H.W.S.	hot water supply
H.W.P.	hot water pump
I.D.	inside diameter
LAV.	lavatory
LDR.	leader
O.D.	outside diameter
(R)	roughing only
R.D.	roof drain
S.C.	sill cock
S.S.	service sink
TOIL.	toilet
UR.	urinal
V.	vent
W.C.	water closet
W.H.	wall hydrant

FIGURE 3-42 Plumbing symbols

DRAINAGE FIXTURE UNIT VALUES FOR VARIOUS PLUMBING FIXTURES

Type of Fixture or Group of Fixtures	Drainage Fixture Unit Valve (d.f.u.)
Automatic clothes washer (2" (50 mm) standpipe and trap required - direct connection)	3
Bathtub group consisting of a water closet, lavatory, and bathtub or shower stall:	6
Bathtub (with or without overhead shower)	2
Bidet	1
Clinic Sink	6
Clothes Washer	2
Combination sink-and-tray with food waste grinder	4
Combination sink-and-tray with one 1-1/2" (38 mm) trap	2
Combination sink-and-tray with separate 1-1/2" (38 mm) traps	3
Dental unit or cuspidor	1
Dental lavatory	1
Drinking fountain	1/2
Dishwasher, domestic	2
Floor drains with 2" (50 mm) waste	3
Kitchen sink, domestic, with one 1-1/2" (38 mm) trap	2
Kitchen sink, domestic, with food waste grinder	2
Kitchen sink, domestic, with food waste grinder and dishwasher, 1-1/2" (38 mm) trap	3
Kitchen sink, domestic, with dishwasher, 1-1/2" (38 mm) trap	3
Lavatory with 1-1/4" (32 mm) waste	1
Laundry tray (1 or 2 compartments)	2
Shower stall, domestic	2
Showers (group) per head[1]	2
Sinks:	
Surgeon's	3
Flushing rim (with valve)	6
Service (trap standard)	3
Service (P trap)	2
Pot, Scullery, etc.	4
Urinal, syphon jet blowout	6
Urinal, wall lip	4
Wash sink (circular or multiple) each set of faucets	2
Water closet, private	4
Water closet, general use	6
Fixtures not listed above:	
Trap Size 1-1/4" (32 mm) or less	1
Trap Size 1-1/2" (38 mm)	2
Trap Size 2" (50 mm)	3
Trap Size 2-1/2" (65 mm)	4
Trap Size 3" (75 mm)	5
Trap Size 4" (100 mm)	6

1. A shower head over a bathtub does not increase the fixture unit valve.

Note: Metric added by author

FIGURE 3-43   Drainage fixture unit for various plumbing fixtures values

BUILDING DRAINS AND SEWERS[1]

Diameter of Pipe		Maximum Number of Fixture Units That May Be Connected to Any Portion of the Building Drain or the Building Sewer.			
		Slope			
		1/16 in. per ft. 5.2 mm/m	1/8 in. per ft. 10.4 mm/m	1/4 in. per ft. 20.8 mm/m	1/2 in. per ft. 41.6 mm/m
Inches	mm				
2	50			21	26
2½	65			24	31
3	75			42[2]	50[2]
4	100		180	216	250
5	125		390	480	575
6	150		700	840	1,000
8	200	1,400	1,600	1,920	2,300
10	250	2,500	2,900	3,500	4,200
12	300	2,900	4,600	5,600	6,700
15	400	7,000	8,300	10,000	12,000

1. On site sewers that serve more than one building may be sized according to the current standards and specifications of the Administrative Authority for public sewers.
2. Not over two water closets or two bathroom groups, except that in single family dwellings, not over three water closets or three bathroom groups may be installed.

Note: Metric added by author

FIGURE 3-44   Building drains and sewers

HORIZONTAL FIXTURE BRANCHES AND STACKS

Diameter of Pipe		Maximum Number of Fixture Units That May Be Connected to:			
		Any Horizontal Fixture Branch[1]	One Stack of Three Branch Intervals or Less	Stacks with More Than Three Branch Intervals	
				Total for Stack	Total at One Branch Interval
Inches	mm				
1½	38	3	4	8	2
2	50	6	10	24	6
2½	65	12	20	42	9
3	75	20[2]	48[2]	72[2]	20[2]
4	100	160	240	500	90
5	125	360	540	1,100	200
6	150	620	960	1,900	350
8	200	1,400	2,200	3,600	600
10	250	2,500	3,800	5,600	1,000
12	300	3,900	6,000	8,400	1,500
15	400	7,000			

1. Does not include branches of the building drain.
2. Not more than 2 water closets or bathroom groups within each branch interval nor more than 6 water closets or bathroom groups on the stack.

Stack shall be sized according to the total accumulated connected load at each story or branch interval and may be reduced in size as this load decreases to a minimum diameter of 1/2 of the largest size required.

Note: Metric added by author

FIGURE 3-45   Horizontal fixture branches and stacks

MINIMUM SIZE OF NON-INTEGRAL TRAPS

Plumbing Fixture	Trap Size	
	Inches	mm
Bathtub (with or without overhead shower)	1½	38
Bidet	1¼	32
Clothes Washing Machine Standpipe		
Combination sink and wash (laundry) tray with food waste grinder unit	1½[1]	38
Combination kitchen sink, domestic, dishwasher, and food waste grinder	1½	38
Dental unit or cuspidor	1¼	32
Dental lavatory	1¼	32
Drinking fountain	1¼	32
Dishwasher, commercial	2	50
Dishwasher, domestic (non-integral trap)	1½	38
Floor drain	2	50
Food waste grinder, Commercial Use	2	50
Food waste grinder, Domestic Use	1½	38
Kitchen sink, domestic, with food waste grinder unit	1½	38
Kitchen sink, domestic	1½	38
Kitchen sink, domestic, with dishwasher	1½	38
Lavatory, common	1¼	32
Lavatory (barber shop, beauty parlor or surgeon's)	1½	38
Lavatory, multiple type (wash fountain or wash sink)	1½	38
Laundry tray (1 or 2 compartments)	1½	38
Shower stall or drain	2	50
Sink (surgeon's)	1½	38
Sink (flushing rim type, flush valve supplied)	3	75
Sink (service type with floor outlet trap standard)	3	75
Sink (service trap with P trap)	2	50
Sink, commercial (pot, scullery, or similar type)	2	50
Sink, commercial (with food grinder unit)	2	50

[1]Separate trap required for wash tray and separate trap required for sink compartment with food waste grinder unit.

Note: Metric added by author

FIGURE 3-46   Minimum size of non-integral traps

Size of Soil or Waste Stack	Fixture Units Connected	Diameter of Vent Required - Inches								
		1¼	1½	2	2½	3	4	5	6	8
		Maximum Length of Vent - Feet								
Inches										
1½	8	50	150							
2	12	30	75	200						
2	20	26	50	150						
2½	42		30	100	300					
3	10		30	100	100	600				
3	30			60	200	500				
3	60			50	80	400				
4	100			35	100	260	1000			
4	200			30	90	250	900			
4	500			20	70	180	700			
5	200				35	80	350	1000		
5	500				30	70	300	900		
5	1100				20	50	200	700		
6	350				25	50	200	400	1300	
6	620				15	30	125	300	1100	
6	960					24	100	250	1000	
6	1900					20	70	200	700	
8	600						50	150	500	1300
8	1400						40	100	400	1200
8	2200						30	80	350	1100
8	3600						25	60	250	800
10	1000							75	125	1000
10	2500							50	100	500
10	3800							30	80	350
10	5600							25	60	250

Size of Soil or Waste Stack	Fixture Units Connected	Diameter of Vent Required - mm								
		32	40	50	65	80	100	125	150	200
		Maximum Length of Vent - meters								
mm										
40	8	15	45							
50	12	9	23	61						
50	20	8	15	45						
65	42		9	31	92					
80	10		9	31	31	183				
80	30			18	61	153				
80	60			15	24	122				
100	100			11	31	79	305			
100	200			9	27	76	275			
100	500			6	21	55	214			
125	200				11	24	107	305		
125	500				9	21	92	275		
125	1100				6	15	61	214		
150	350				8	15	61	122	396	
150	620				5	9	38	92	336	
150	960					7	31	76	305	
150	1900					6	21	61	214	
200	600					15	45	153	396	
200	1400					12	31	122	366	
200	2200					9	24	107	336	
200	3600					8	18	76	244	
250	1000						23	38	305	
250	2500						15	31	153	
250	3800						9	24	107	
250	5600						8	18	76	

Note: Metric added by author

FIGURE 3-47   Size and length of vents

MAXIMUM LENGTH OF TRAP ARM

Diameter of Trap Arm (Inches)	Length - Trap to Vent
1¼	3' 6" - 1.1 m
1½	5' - 1.5 m
2	8' - 2.4 m
3	10' - 3.0 m
4	12' - 3.7 m

Note: Metric added by author

FIGURE 3-48   Maximum length of trap arm

QUESTIONS

3-1. What is a *trap,* where is it located, and how does it work?

3-2. Why are vents required on the waste system? Where are they located in reference to the fixture?

3-3. What is a *wet vent,* and how does it differ from other types of vents?

3-4. What is the difference between a *stack vent* and a *vent stack?* Using a sketch, show the location of a stack vent and a vent stack in a multistory design.

3-5. What is the difference between a *soil stack* and a *waste stack?*

3-6. Sketch and locate the house (building) drain and the sewer.

3-7. What provisions must be made to provide drainage for fixtures located below the level of the building drain and the sewer?

3-8. Why are sketches of the drainage piping made by the designer?

3-9. Sketch a typical drainage piping design for a residence with a single bathroom with a kitchen sink on the other side of the wall (Fig. E3-9).

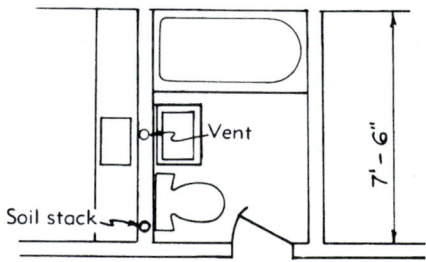

Figure E3-9

3-10. What is the difference between a *flush tank* and *flush valve* water closet?

Design Exercises

3-11. Design the plumbing drainage for the for the apartment building in Appendix B. Use copper pipe and tubing for this design problem.

3-12. Design the plumbing drainage for the residence in Appendix C. Use copper pipe for this design problem.

3-13. Design the plumbing drainage for the residence in Appendix D. Use copper pipe for this design problem.

3-14. Design the plumbing drainage for the bathrooms shown in Fig. E3-14. Assume that the building is 2 floors high and that the average floor-to-floor height is 12 ft (3.6 m).

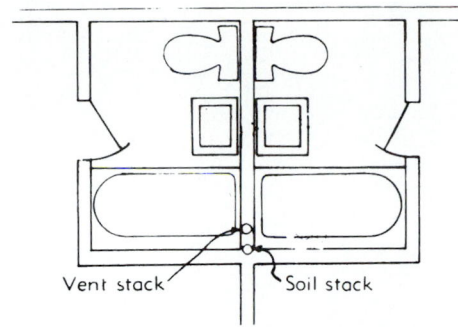

Vent stack   Soil stack

Figure E3-14

3-15. Now resize the building drain in Exercise 3-14 to serve the stack design in Exercise 3-14 plus five other stacks exactly like it plus one stack which serves two service sinks and one drinking fountain on each floor. The building drain slope is ⅛ in. per ft (10.4 mm per m).

3-16. Design the plumbing drainage for the restrooms shown in Fig. E3-16. Assume that the building is 3 stories high and that 9 ft-6 in. (2.9 m) is the average floor-to-floor height.

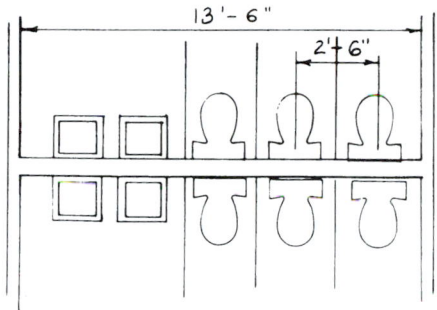

13'-6"

2'-6"

Figure E-3-16

3-17. Size the building drain to serve the stack in Exercise 3-16 plus two other stacks exactly like it. The building drain slope is ½ in. per ft (41.6 mm per m).

CHAPTER 4

Storm Drainage

4-1 TYPES OF SYSTEMS

Whenever it rains, the drainage and runoff from roofs, courtyards, and paved areas (such as parking lots) must be carried away from the building and properly disposed of. This water may be directed to drains in the building roofs, parking areas, courtyards, and the like, and then be directed into:

1. A community storm sewer line.
2. A private storm sewer and drywell, or be run off onto a low portion of the client's land or into a creek, stream, lake, or pond.
3. A community sewer line.

4-2 COMMUNITY STORM SEWER

If the water is directed to a community storm sewer system, the only concern will be that the elevation of the sewer line is low enough that the private storm line can run into it (Fig. 4-1). Many communities have such systems, which are also used to drain rainwater from the streets and safely away.

4-3 PRIVATE STORM SEWER-DRYWELL

If the water is collected into a private storm sewer line, the line can:

1. Be directed into drywells (Fig. 4-2), which allow the water to be absorbed into the ground.
2. Be run so that it will empty into an area of low elevation on the plot.
3. Be run into a nearby creek or stream.
4. Be run into a public or private lake or pond.

If the line serves a large area, the force of the water, after a rain, may cause considerable damage where it runs out the end of the line. This should be carefully considered by an engineer with experience in drainage.

Running the private line into a creek, stream, or lake may require a permit and may not be allowed in some areas. Usually, before approval, the design of the system will be checked to be certain there is no possibility of sewage wastes (chemical, human, or industrial wastes) getting into the storm sewer line and contaminating a lake, stream, or creek.

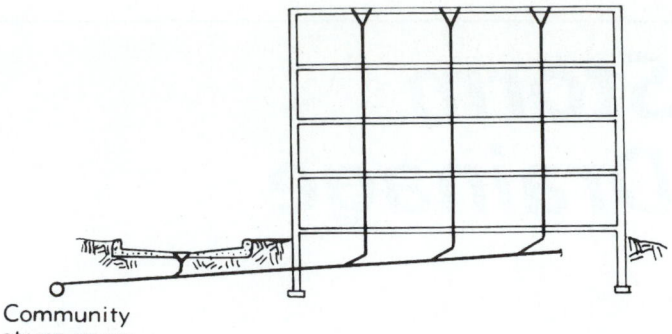

Community
storm sewer

FIGURE 4-1 Community
storm sewer

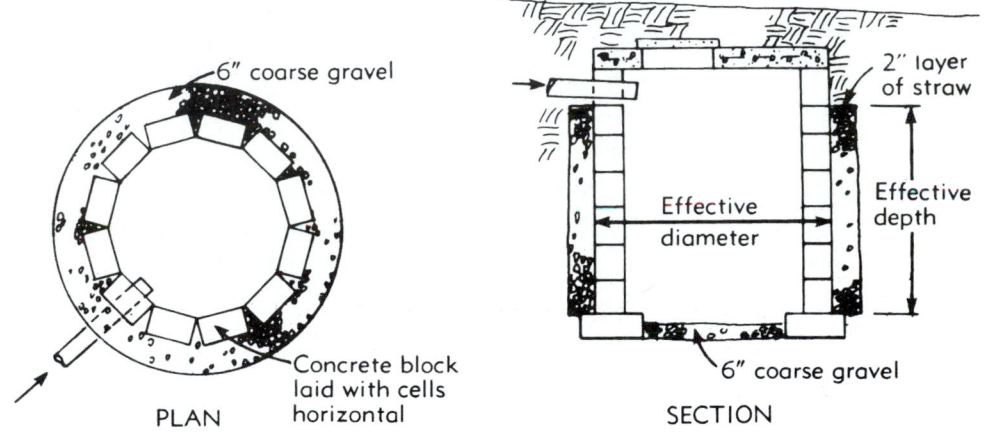

PLAN

SECTION

FIGURE 4-2 Drywell

Many times on large projects, the storm water is run into a pond so that the water will be available for nonpotable uses, such as watering lawns and gardens and circulating in fountains.

4-4 COMMUNITY SEWER LINE

In many cities, especially in the more urban sections, the storm sewers from the city streets and the private buildings, driveways, and parking areas all run into the city sewer line which is used for sewage waste; this is called a *combined sewer.* This should be done only if no other solution is available and if the city allows storm lines to be tied into its sewage lines. This storm water creates a tremendous excess work load for the city (county or municipality) sewage treatment plant. This unnecessary burden often requires that the sewage treatment plants built be much larger than they would be if only sewage were to be treated. In cities with such systems, the cost to separate the storm and sewage lines now would be prohibitive. However, in many cities there are separate storm and sewage lines, and it is illegal to tie storm drainage lines into sewage lines. Cities with only sewage lines may or may not allow storm water to be introduced into the sewage lines, and this should be carefully checked with the local authorities.

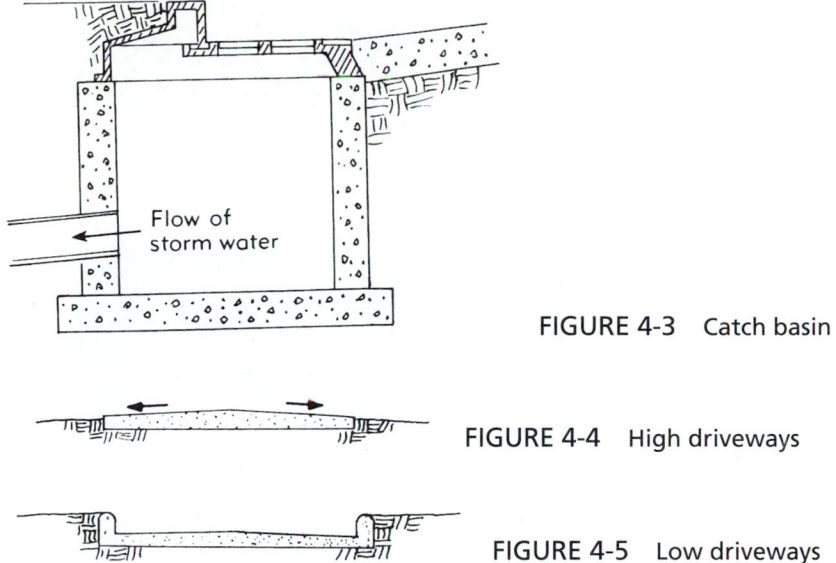

FIGURE 4-3   Catch basin

FIGURE 4-4   High driveways

FIGURE 4-5   Low driveways

When designing drainage for driveways, parking lots, and surrounding ground, the site plan of the project must be checked to determine what effect the existing and revised contours will have on the flow of the surface water after a rain. It is most important that the flow of water be away from the building and not toward it. On large projects (usually not individual residences), the ground should be contoured so water will flow toward the storm sewer system (usually a catch basin for collecting the water, Fig. 4-3). On projects without a storm sewer system, the water should be directed away from buildings, driveways, and parking lots.

The detail for the construction of driveways and parking lots should also be checked. When the water is simply being allowed to run off onto the surrounding ground, the driveway and any curbs should be constructed higher than the surrounding ground (Fig. 4-4) so that the water will run off and onto the ground. When a storm sewer system will be used to collect and carry away the water, the driveway and curbs may be set lower than the surrounding ground (Fig. 4-5) and should be generally pitched toward the catch basins which collect the water. When this detail is used and there is no storm sewer system, the driveways and parking lots become shallow "swimming pools."

4-5   ROOF DRAINAGE DESIGN

The water from the roof may be taken into consideration by any of three methods:

1. Install roof drains.
2. Install gutters.
3. Allow the water to run off without drains or gutters.

Roof drains (Fig. 4-6) are commonly placed in "flat" or built-up roofs to be certain that the water will not stay on the roof after a rain. Because it is generally considered detrimental to the roofing materials to have water left on the roof, the roof should not be flat but have at least a small pitch (slope) to it. When placed in a flat roof, it is very important that the roof surface be pitched toward the drain to be certain that *all* the water is drained off the roof. This slope may be accomplished on a "flat" roof deck by the use of a layer of lightweight concrete or asphalt. When the deck is made of poured gypsum or concrete, the slope is put in as the deck is poured.

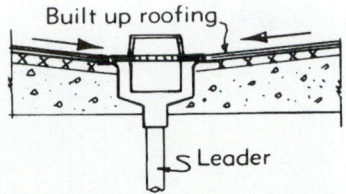

FIGURE 4-6   Roof drain

FIGURE 4-7   Storm leader to drain

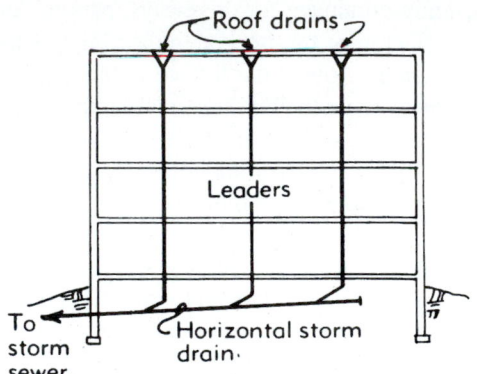

FIGURE 4-8   Leader through ceiling

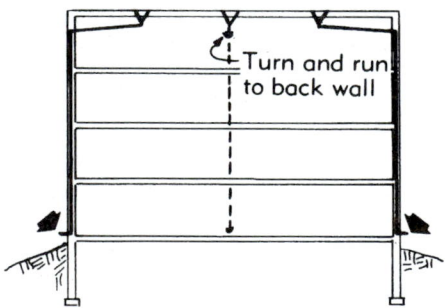

FIGURE 4-9   Leader through wall

Properly installed roof drains are quite effective in draining a roof. The drains connect to pipes (called *leaders*) which carry the water away from the drain and into a horizontal storm drain (Fig. 4-7 and Fig. 4-8) or to the exterior of the building (Fig. 4-9). The leaders may be concealed in the walls or columns (Fig. 4-10) if the sight of an exposed pipe is objectionable. However, once enclosed, it is more expensive to make repairs if necessary. Leaders are usually made of cast iron, galvanized steel, galvanized wrought iron, copper, brass, lead, or plastic pipe.

The method used to size roof drain leaders and horizontal storm drains is described later in this section.

FIGURE 4-10   Concealed leader

If the leader runs to the outside of the building and empties, some precautions must be taken so that the water will be directed away from the building. Immediately adjacent to the building, a concrete pad or splashblock or other similar device is required so that the water coming from the end of the pipe will not hit the soil with such force that it will wash it away, causing soil erosion and permitting the possibility of wet foundation walls. Undermining of the construction is even possible in extreme cases. Minor problems, such as staining the building, may also occur. Once the water is directed away from the building in some manner, be certain that the surrounding contours keep the water moving *away* from the building.

In locales where the leaders can be tied into the sewage system, the system is referred to as a *combined* sewer. Since many locales do not allow the installation of combined sewers, be certain to check with local authorities. Figure 4-11 illustrates how the leader might be tied into the building drain (discussed in Sec. 4-5). In this situation, most codes require a trap on the leader (storm sewer systems seldom require traps, mainly with combined sewers), and they may specify that the trap shall be a minimum of 10 ft (3 m) from any stack.

Sizing of the leader for a combined sewer and its effect on the building drain are discussed in Sec. 4-6.

The water may also be directed off the roof into gutters. The roof surface should be pitched to direct the flow of water toward the gutter. The gutters are tied to leaders (often called *downspouts*) which may be tied to a storm sewer line (either community or private) or which may empty outside the building onto a pad, splashback, or other means to disperse the water (Fig. 4-12).

Another possible solution is to run the leader into a small catch basin (Fig. 4-13) or disposal area filled with gravel (Fig. 4-14). This may be effectively used on smaller buildings, such as residences, but it is important that the catch basin be located at least 10 ft (3 m) from the foundation walls to reduce any chance for wet walls from the water.

Gutters and leaders commonly used may be made of copper, steel, aluminum, or vinyl. Vinyl gutters can be made in one piece (without seams), but if the gutters are prop-

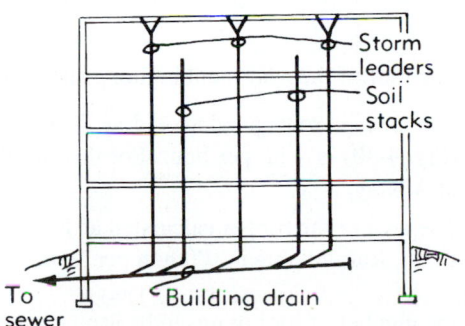

FIGURE 4-11   Leaders and soil stacks to building drain

FIGURE 4-12   Leader and splash-block

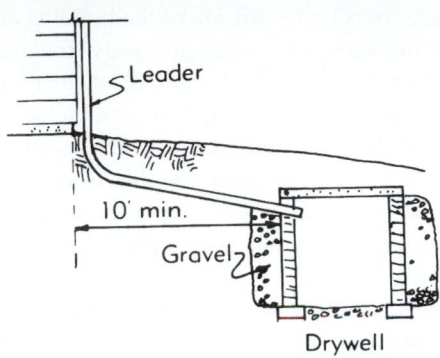

FIGURE 4-13   Leaders to drywell

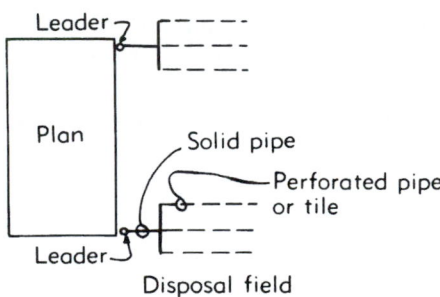

FIGURE 4-14   Leaders to disposal field

erly installed, seams are no problem (too often they are not properly installed). It is most important that the gutters be installed with a definite slope toward the leaders.

Sizing Roof Drains and Leaders— Step-by-Step Approach

1. The first step in the design is to review the building. Rainfall rates are shown in Fig. 4-30.

The roof of the apartment building shown in Appendix A is generally flat.

The area of roof is calculated as 40 ft × 80 ft = 3,200 sq ft. Rainfall rate (Fig. 4-30) is 4 in. per hour. For this problem assume the building is located in Albany, N.Y.

SI   The area of the roof is calculated at 12.2 m × 24.2m = 297.68 sq m (use 300 sq m). Rainfall rate is 100 mm per hour.

2. Next, the number of roof drains to be used must be determined.

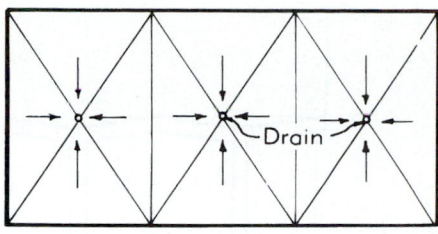

Roof drainage

FIGURE 4-15   Roof drain plan

If two drains are used, each will have to handle 3,200 sq ft ÷ 2 = 1,600 sq ft of roof; if three are used, 1,067 sq ft each; if four are used, 800 sq ft each; and so on.

At this time, a preliminary decision is made to use three roof drains (Fig. 4-15), each serving 1,067 sq ft.

If two drains are used, each will have to handle 150 sq m (300 sq m ÷ 2); if three are used, 100 sq m each; and if four are used, 75 sq m. At this time a preliminary decision is made to use three roof drains (Fig. 4-15), each serving 100 sq m.

3. The leader size required for this situation is selected from Fig. 4-26 and is based on the square footage of roof area the leader must handle.

In this design, a 3-in. leader is required for each roof drain. In reviewing the decision for three drains, if two had been used, a 3-in. leader would have been required, and if four had been used, a 3-in. leader would have been required. In this design, we will continue with the three roof drains (any of the others could also have been used). The horizontal storm drain, if used, is sized later in this section.

In this design, a 75 mm leader is required for each roof drain. If two drains had been used, a 75 mm roof drain would have been required; and if four drains had been used, a 75 mm roof leader would have been required. In this design, we will continue to use the three roof drains. The horizontal storm drain, if used, is sized later in this chapter.

Sizing Roof Gutters and Leaders— Step-by-Step Approach

1. Gutters are also sized according to the square footage of roof they serve.

Using the four-story apartment building as a design problem (Appendix A), assume that the roof is pitched to the back (Fig. 4-16) and that a gutter is installed to collect the water. This means that the gutter will serve the entire roof area.

2. Next, the leaders are selected (in the same manner as discussed above). The leaders' size is based on the number of leaders used and the square footage each leader will serve.

In this design, assume four leaders will be used, each serving 3,200 sq ft ÷ 4 = 800 sq ft. From Fig. 4-26, each leader would be 3 in. in diameter. If the leaders are connected to a horizontal storm drain, the drain is then sized as discussed later in this section.

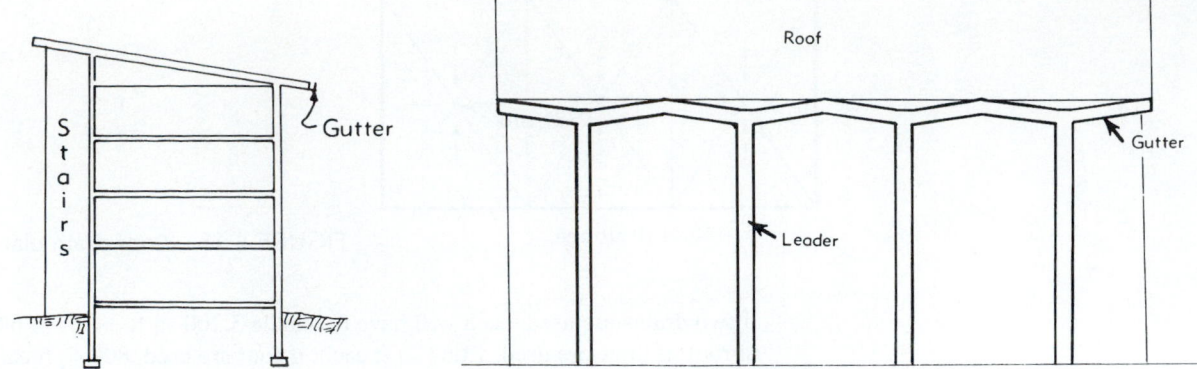

FIGURE 4-16   Sloped roof and gutters and leaders

In this design, assume four leaders will be used, each serving 75 sq m (300 sq m ÷ 4). From Fig. 4-26, each leader would be 75 mm in diameter. If the leaders are connected to a horizontal storm drain, the drain is then sized as discussed later in this chapter.

3. The gutter size depends on the area of the roof which each portion of the gutter serves and the slope of the gutter when it is installed.

In this design, each of the four leaders serves 800 sq ft. The layout would be such that the gutter would feed into the leader from two sides, so that each portion of the gutter would serve 400 sq ft of roof area. Assuming a ⅛-in. slope per foot, from Fig. 4-28 a 6-in. diameter is selected for the gutters.

In this design, each of the four leaders serves 75 sq m. The layout would be such that the gutter would feed into the leader from two sides, so that each portion of the gutter serves 37.5 sq m of roof area. Assuming a 10.4 mm per m slope from Fig. 4-27, a 150 mm diameter is selected for the gutters.

Sizing the Horizontal Storm Drain— Step-by-Step Approach

Once the roof drains or gutters and leaders have been selected and sized, the next step is to size the horizontal storm drain (if one is to be used). The horizontal storm drain data applies to any horizontal storm drain location (such as under the roof slab or below the floor slab).

1. The horizontal storm drain is sized from Fig. 4-27; its size depends on the square footage being served and the slope at which the pipe is installed. The pipe may be increased in size as it collects the leaders, so the first step will be to make a sketch of the system (Fig. 4-17).

   Next, add the square footage that each leader serves and the slope selected for the horizontal drain to the sketch.

   In this design, three leaders are used (Sec. 4-5), serving 1,067 sq ft each, and the slope is ¼ in. per ft.

   In this design three leaders are used serving 100 sq m each, and the slope is 20.9 mm per m.

2. With this information, the first length of drain (labeled "A" on sketch) is sized from Fig. 4-27 as a 3-in. pipe. Be certain that you use the column for a ¼-in. slope, in this case. The next length of drain ("B") services 2,134 sq ft and is a 4-in. pipe, while the

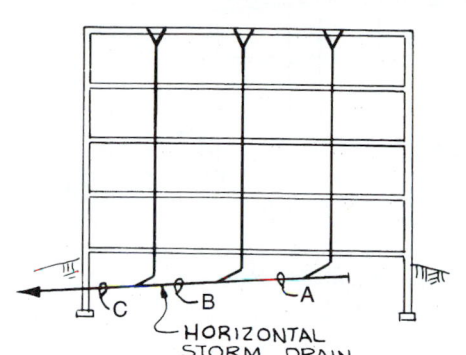

HORIZONTAL STORM DRAIN			
	s.f.	Accum. s.f.	Pipe size
A	1067	1067	
B	1067	2134	
C	1066	3200	

SI UNITS

HORIZONTAL STORM DRAIN			
	sq m	Accum. sq m	Pipe size
A	100	100	
B	100	200	
C	100	300	

FIGURE 4-17 Horizontal storm drain

HORIZONTAL STORM DRAIN			
	s.f.	Accum. s.f.	Pipe size
A	1067	1067	3″
B	1067	2134	4″
C	1066	3200	5″

Slope ¼″ per foot

SI UNITS

HORIZONTAL STORM DRAIN			
	sq m	Accum. sq m	Pipe size
A	100	100	75 mm
B	100	200	100 mm
C	100	300	125 mm

Slope 20.8 mm per m

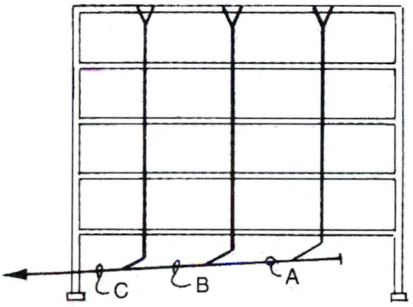

FIGURE 4-18 Horizontal storm drain size

last length ("C") serves 3,200 sq ft and is a 4-in. pipe. Note the pipe sizes on the sketch as shown in Fig. 4-18.

The length of drain labeled *A* in Fig. 4-18 is taken from Fig. 4-27. Using a slope of 20.9 mm per m, a 75 mm drain pipe is selected. The next length of drain, *B,* services 200 sq m and a 100 mm pipe is required. The last length, *C,* serves 300 sq m and a 125 mm pipe is required.

4-6   SIZING A COMBINED SEWER

If the roof leaders (from Sec. 4-5) are to be connected to the building drain (Fig. 4-19), it will be necessary to convert the roof area into an equivalent number of fixture units so that

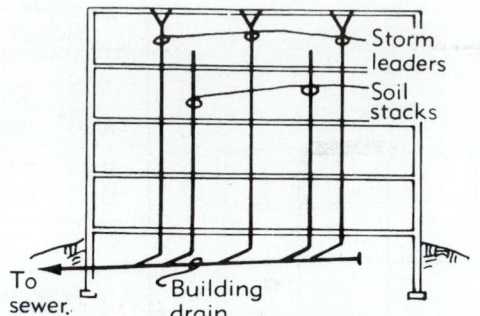

FIGURE 4-19   Combined sewer

the building drain can be sized to reflect the increased load. (Building drains were sized for sewage waste from the stacks in Sec. 3-5.)

Step-by-Step Approach

1. First, a schematic sketch of the stacks, leaders, and building drain should be made so that the relationship of the stacks and the leaders to the building drain can be seen.

   In this design, we will continue with the four-story apartment building. The stacks have also been previously sized (Sec. 3-5), and a schematic of the design would look similar to Fig. 4-20.

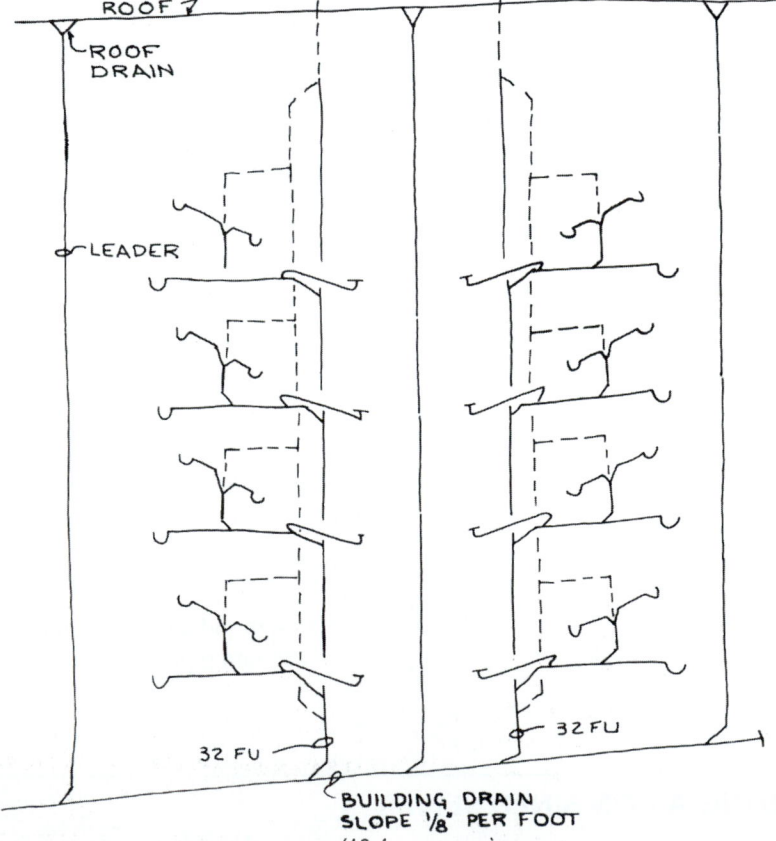

FIGURE 4-20   Combined sewer schematic

The f.u. of each of the stacks is taken from Sec. 3-5 and added to the sketch, as is the slope of the building drain selected (⅛-in. per foot or 10.4 mm per m slope).

2. Next, the fixture units served by each stack must be converted into equivalent square feet or square meters. The code sets up a ratio of f.u. and equivalent square feet or equivalent sq. m. (Fig. 4-29). The equivalent square feet, based on the code, for the first 256 f.u. is 1,000 sq ft.

$$256 \text{ f.u.} = 1,000 \text{ sq ft.}$$

Any additional f.u. are converted into equivalent square feet as the basis that 1 f.u. equals 3.9 sq ft. In this design, the 64 f.u. would be the equivalent of 1,000 sq. ft.

Based on the code, the equivalent square meters for the first 256 f.u. is 93 sq m.

$$256 \text{ f.u.} = 93 \text{ sq m}$$

Any additional f.u. are converted into equivalent square meters on the basis that 1 f.u. equals .36m.

In this design, the 64 f.u. would be the equivalent of 93 sq m

3. Add the equivalent square feet to the roof area being collected for the total area being served by the building drain.

The total area is 3,200 sq ft + 1,000 equivalent sq ft × 2 leaders = 5,200 sq ft. This is noted on the schematic in Fig. 4-21.

From Fig. 4-30, the building drain size is determined. Based on a ⅛ in. slope, the building drain must be 6 in. to the main stack. The complete tabulation is shown in Fig. 4-22.

A review of the sketch in Fig. 4-20 indicates that there is less than 200 ft developed length.

The total area is 300 sq m + 93 equivalent sq m × 2 leaders = 486 sq m

This is noted on the schematic in Fig. 4-21.

From Fig. 4-30, the building drain size is determined. Based on a 10.4 mm per m slope, the building drain must be 150 mm to the main stack. The complete compilation is shown in Fig. 4-22.

A review of the sketch in Fig. 4-20 indicates there is less than 60 m of developed length.

4. The equivalent square feet are based on a rainfall of 4 in. per hr. A check of the local weather service will indicate whether the rainfall in the proposed building location is more or less. If so, then the equivalent square feet must be adjusted proportionately.

Assume the rainfall rate in the proposed location for the apartment is 5 in. per hr. How many equivalent square feet would the 64 f.u. equal? At 4 in. per hour:

$$64 \text{ f.u.} = 1,000 \text{ equivalent sq ft}$$

5 in. per hour:

$$64 \text{ f.u.} = 1,000 \times \frac{5}{4} = 1,250 \text{ equivalent sq ft}$$

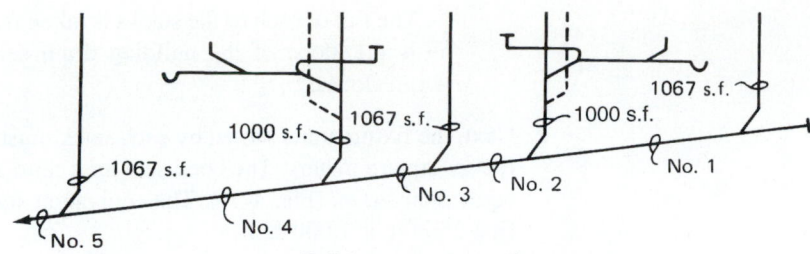

Building drain slope 1/8″ per foot

S1 UNITS

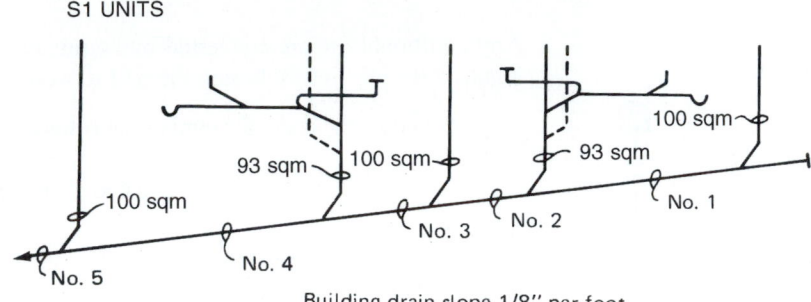

Building drain slope 1/8″ per foot

FIGURE 4-21   Leader and riser fixture sizes

SI UNITS

	BUILDING DRAIN				BUILDING DRAIN		
	s.f.	Accum. s.f.	Pipe size		sq m	Accum. sq m	Pipe size
No. 1	1067	1067	4″	No. 1	100	100	100 mm
No. 2	1000	2067	4″	No. 2	93	193	100 mm
No. 3	1067	3134	5″	No. 3	100	293	125 mm
No. 4	1000	4134	5″	No. 4	93	386	125 mm
No. 5	1086	5200	6″	No. 5	100	486	150 mm

Slope 1/8″ per foot   Slope 10.4 mm per m

FIGURE 4-22   Building drain tabulations

SI

Assume the rainfall rate in the proposed location of the apartment is 125 mm per hour. How many equivalent feet would the 64 f.u. equal?

At 100 mm per hour:

$$64 \text{ f.u.} = 93 \text{ equivalent sq m}$$

At 125 mm per hour:

$$63 \text{ f.u.} = 93 \times \frac{125}{100} = 116 \text{ equivalent sq m}$$

4-7   FOUNDATION DRAINS

Many times drains are placed around the foundation of a building (Fig. 4-23) to direct water away from the building (usually after a rain). These footing drains may be of hard fi-

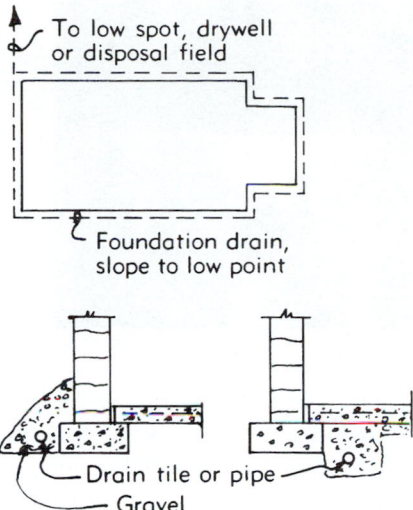

FIGURE 4-23   Foundation drains

brous materials with holes, of plastic pipe with holes, or of clay tile spaced about ¼ in. apart with the upper half of the joint covered. The pipes are laid on a layer of gravel, with the holes toward the bottom, and then gravel is placed over the pipe or tile. The drain must be installed at a slope so the water will run to a low point and then run into a drywell, storm sewer, or low spot on the site. Whenever possible, this drain should not be connected to the sewer system.

4-8   SYSTEM INSTALLATION

On a small project, the interior storm drainage piping typically varies in size from 3 in. to 8 in. (75 mm to 200 mm). The size of exterior piping may range to 24 in. (600 mm) and larger. The size of the pipe often requires special provisions in wall width or furred-out areas to run them. In warehouses, the pipe may run exposed (Fig. 4-8) next to a column or may be enclosed in fireproofing that protects the column (Fig. 4-24).

The storm drains that serve the exterior areas of the project are larger and require extensive planning since they often run under roads and buildings (Fig. 4-25). Poured concrete slabs will require that the interior storm drainage layout be carefully considered. The pipes need to be placed in the ground before the slab is poured, so their accurate placement is crucial. All piping must be carefully located and the system checked for leaks before the concrete is poured because any relocation or repairs of pipes would be costly.

FIGURE 4-24   Storm drainage enclosed in fireproofing

FIGURE 4-25   Storm drainage installed under roadway

The open spaces provided in truss-type construction make it easier to run piping through to the desired location. The only points of difficulty would be where it needs to pass by ductwork or some other large pipe that is going in the opposite direction. This will require coordination with the contractor installing any heating, air conditioning, or ventilating ductwork.

In wood frame construction, there are times when the width of a wall needs to be increased to allow for pipes running horizontally to pass by drainage pipes (or other pipes) running vertically.

Pipe tunnels (Fig. 3-41) may be used on large projects to provide concealed space for the passage of mechanicals at ground level and from building to building. Hangers from the top or side of the tunnel are used to support the pipes. Access may be from either end of the tunnel, or access doors may be provided.

SIZE OF VERTICAL CONDUCTORS AND LEADERS

Quantities are Horizontal Projected Roof Areas in Square Feet

Rainfall in Inches	Size of Drain or Leader in Inches*					
	2	3	4	5	6	8
1	2880	8800	18,400	34,600	54,000	116,000
2	1440	4400	9200	17,300	27,000	58,000
3	960	2930	6130	11,530	17,995	38,660
4	720	2200	4600	8650	13,500	29,000
5	575	1760	3680	6920	10,800	23,200
6	480	1470	3070	5765	9000	19,315
7	410	1260	2630	4945	7715	16,570
8	360	1100	2300	4325	6750	14,500
9	320	980	2045	3845	6000	12,890
10	290	880	1840	3460	5400	11,600
11	260	800	1675	3145	4910	10,545
12	240	730	1530	2880	4500	9660

FIGURE 4-26 (a)   Size of vertical conductors and leaders. (See page 117 for SI Units.)

SI UNITS

Quantities are Horizontal Projected Roof Areas in Square Meters

Rainfall in mm	\multicolumn{6}{c}{Size of Drain or Leader in Millimeters*}					
	50	**75**	**100**	**125**	**150**	**200**
25	267	817	1709	3214	5016	10,776
50	133	408	854	1607	2508	5388
75	89	272	569	1071	1671	3591
100	66	204	427	803	1254	2694
125	53	163	341	642	1003	2155
150	44	136	285	535	836	1794
175	38	117	244	459	716	1539
200	33	102	213	401	627	1347
225	29	91	190	357	557	1197
250	26	81	170	321	501	1077
275	24	74	155	292	456	979
300	22	67	142	267	418	897

*Round, square, or rectangular rainwater pipe may be used and is considered equivalent when enclosing a scribed circle equivalent to the leader diameter.

FIGURE 4-26 (b)   Size of vertical conductors and leaders.

Size of Horizontal Rainwater Piping

Size of Pipe in Inches	\multicolumn{5}{c}{Maximum Rainfall in Inches per Hour}				
1/8"/ft. Slope	**2**	**3**	**4**	**5**	**6**
3	1644	1096	822	657	548
4	3760	2506	1800	1504	1253
5	6680	4453	3340	2672	2227
6	10,700	7133	5350	4280	3566
8	23,000	15,330	11,500	9200	7600
10	41,400	27,600	20,700	16,580	13,800
12	66,600	44,400	33,300	26,650	22,200
15	109,000	72,800	59,500	47,600	39,650

Size of Pipe in Inches	\multicolumn{5}{c}{Maximum Rainfall in Inches per Hour}				
1/4"/ft. Slope	**2**	**3**	**4**	**5**	**6**
3	2320	1546	1160	928	773
4	5300	3533	2650	2120	1766
5	9440	6293	4720	3776	3146
6	15,100	10,066	7550	6040	5033
8	32,600	21,733	16,300	13,040	10,866
10	58,400	38,950	29,200	23,350	19,450
12	94,000	62,600	47,000	37,600	31,350
15	168,000	112,000	84,000	67,250	56,000

Size of Pipe in Inches	\multicolumn{5}{c}{Maximum Rainfall in Inches per Hour}				
1/2"/ft. Slope	**2**	**3**	**4**	**5**	**6**
3	3288	2295	1644	1310	1096
4	7520	5010	3760	3010	2500
5	13,360	8900	6680	5320	4450
6	21,400	13,700	10,700	8580	7140
8	46,000	30,650	23,000	18,400	15,320
10	85,800	55,200	41,400	33,150	27,600
12	133,200	88,800	66,600	53,200	44,400
15	238,000	158,800	119,000	95,300	79,250

FIGURE 4-27 (a)   Roof area for horizontal drains pressure system—head
(See page 118 for SI Units.)

SI UNITS

Size of Horizontal Rainwater Piping

Size of Pipe in mm 10.4 mm/m Slope	Maximum Rainfall in Millimeters per Hour				
	50	75	100	125	150
75	152	101	76	61	50
100	349	232	174	139	116
125	620	413	310	248	206
150	994	662	497	397	331
200	2136	1424	1068	854	706
250	3846	2564	1923	1540	1282
300	6187	4124	3093	2475	2062
375	10,126	6763	5527	4422	3683

Size of Pipe in mm 20.9 mm/m Slope	Maximum Rainfall in Millimeters per Hour				
	50	75	100	125	150
75	215	143	107	86	71
100	492	328	246	197	164
125	877	584	438	350	292
150	1402	935	701	561	467
200	3028	2019	1514	1211	1009
250	5425	3618	2712	2169	1806
300	8732	5815	4366	3493	2912
375	15,607	10,404	7803	6247	5202

Size of Pipe in mm 41.7 mm/m Slope	Maximum Rainfall in Millimeters per Hour				
	50	75	100	125	150
75	305	213	152	121	101
100	698	465	349	279	232
125	1241	826	620	494	413
150	1988	1272	994	797	663
200	4274	2847	2136	1709	1423
250	7692	5128	3846	3079	2564
300	12,374	8249	6187	4942	4124
375	22,110	14,752	11,055	8853	7362

FIGURE 4-27(b)   Roof area for horizontal drains pressure system—head

Size of Gutters

Diameter of Gutter in Inches	Maximum Rainfall in Inches per Hour				
1/16"/ft. Slope	2	3	4	5	6
3	340	226	170	136	113
4	720	480	360	288	240
5	1250	834	625	500	416
6	1920	1280	960	768	640
7	2760	1840	1380	1100	918
8	3980	2655	1990	1590	1325
10	7200	4800	3600	2880	2400

Diameter of Gutter in Inches	Maximum Rainfall in Inches per Hour				
1/8"/ft. Slope	2	3	4	5	6
3	480	320	240	192	160
4	1020	681	510	408	340
5	1760	1172	880	704	587
6	2720	1815	1360	1085	905
7	3900	2600	1950	1560	1300
8	5600	3740	2800	2240	1870
10	10,200	6800	5100	4080	3400

Diameter of Gutter in Inches	Maximum Rainfall in Inches per Hour				
1/4"/ft. Slope	2	3	4	5	6
3	680	454	340	272	226
4	1440	960	720	576	480
5	2500	1668	1250	1000	834
6	3840	2560	1920	1536	1280
7	5520	3680	2760	2205	1840
8	7960	5310	3980	3180	2655
10	14,400	9600	7200	5750	4800

Diameter of Gutter in Inches	Maximum Rainfall in Inches per Hour				
1/2"/ft. Slope	2	3	4	5	6
3	960	640	480	384	320
4	2040	1360	1020	816	680
5	3540	2360	1770	1415	1180
6	5540	3695	2770	2220	1850
7	7800	5200	3900	3120	2600
8	11,200	7460	5600	4480	3730
10	20,000	13,330	10,000	8000	6660

FIGURE 4-28(a)   Size of roof gutters. (See page 120 for SI Units.)

SI UNITS

Size of Gutters

Diameter of Gutter in mm	Maximum Rainfall in Millimeters per Hour				
5.2 mm/m Slope	50	75	100	125	150
75	31	21	15	12	10
100	66	44	33	26	22
125	116	77	58	46	38
150	178	119	89	71	59
175	256	170	128	102	85
200	369	246	184	147	123
250	668	445	334	267	223

Diameter of Gutter in mm	Maximum Rainfall in Millimeters per Hour				
10.4 mm/m Slope	50	75	100	125	150
75	44	29	22	17	14
100	94	63	47	37	31
125	163	108	81	65	54
150	252	168	126	100	84
175	362	241	181	144	120
200	520	347	260	208	173
250	947	631	473	379	315

Diameter of Gutter in mm	Maximum Rainfall in Millimeters per Hour				
20.9 mm/m Slope	50	75	100	125	150
75	63	42	31	25	21
100	133	89	66	53	44
125	232	155	116	92	77
150	356	237	178	142	118
175	512	341	256	204	170
200	739	493	369	295	246
250	133	891	668	534	445

Diameter of Gutter in mm	Maximum Rainfall in Millimeters per Hour				
41.7 mm/m Slope	50	75	100	125	150
75	89	59	44	35	29
100	189	126	94	75	63
125	328	219	164	131	109
150	514	343	257	206	171
175	724	483	362	289	241
200	1040	693	520	416	346
250	1858	1238	929	743	618

FIGURE 4-28 (b)   Size of roof gutters

When the total fixture unit load on the combined drain is less than 256 fixture units, the equivalent drainage area in horizontal projection shall be taken as 1000 square feet.

When the total fixture unit load exceeds 256 fixture units, each fixture unit shall be considered the equivalent of 3.9 square feet of drainage area.

If the rainfall to be provided for is more or less than 4 inches per hour, the 1000 square foot equivalent and the 3.9 shall be adjusted by dividing by 4 and multiplying by the rainfall per hour to be provided for.

Extracted from *1987 National Standard Plumbing Code* with permission of PHCC.

FIGURE 4-29   Roof area to fixture units

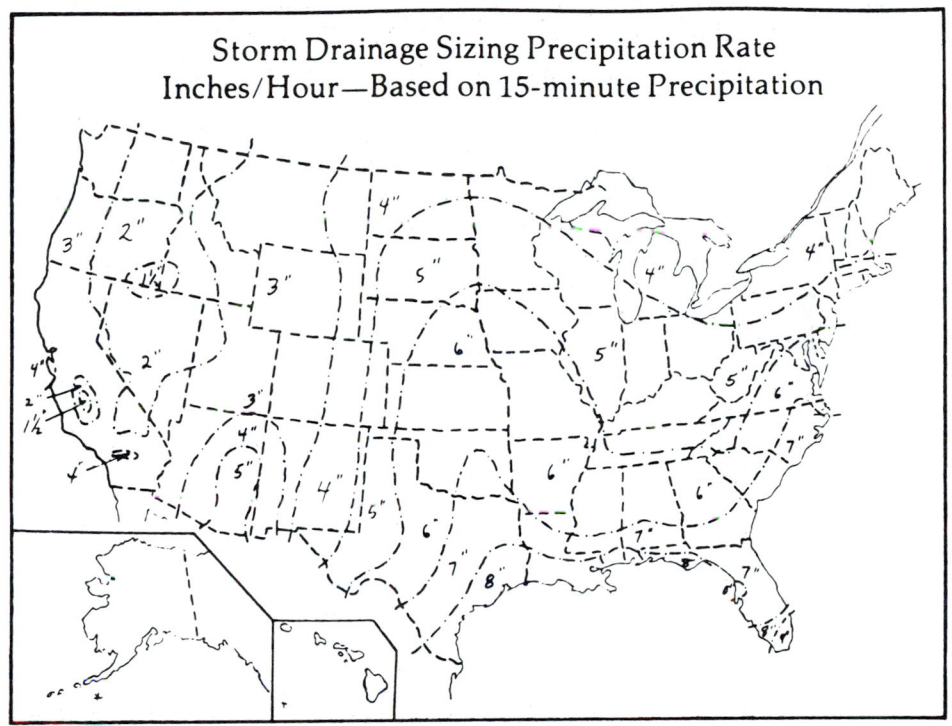

Extracted from the *1987 National Standard Plumbing Code* with permission of PHCC.

Canadian Rainfall Rates, mm per hour based on 15 minute duration

Alberta	British Columbia	Sudbury: 56 mm
Calgary: 36 mm	Kamloops: 10 mm	Toronto: 72 mm
Edmonton: 48 mm	Prince George: 13 mm	Saskatchewan
Grande Prairie: 40 mm	Vancouver: 20 mm	Lloydminister: 48 mm
Jasper: 24 mm	Victoria: 16 mm	Prince Albert: 46 mm
Atlantic Provinces	Manitoba	Regina: 57 mm
Battle Harbour: 24 mm	Brandon: 62 mm	Saskaton: 52 mm
Bonavista: 44 mm	Grand Rapids: 56 mm	Quebec
Cape Race: 44 mm	The Pas: 47 mm	Chibougaman: 52 mm
Chatham: 50 mm	Winnipeg: 68 mm	Montreal: 64 mm
Halifax: 44 mm	Ontario	Rivierie-du-Loup: 40 mm
Schefferville: 30 mm	London: 72 mm	Roberval: 60 mm
Yarmouth: 48 mm	Ottawa: 64 mm	Quebec: 54 mm
	Sault Ste. Marie: 48 mm	

FIGURE 4-30   Storm drainage sizing precipitation rate (inches/hour—based on 15 minute precipitation)

QUESTIONS

4-1. What methods may be used to dispose of the water from roofs, courtyards, and parking lots?

4-2. What is a *combined sewer,* and under what conditions should this system of storm drainage be used?

4-3. What methods are commonly used for roof drainage?

4-4. Why should flat roofs be avoided?

4-5. When the roof drainage water is run into roof drains or gutters, what solutions may be used to disperse the water at the end of the leader?

Design Exercises

4-6. Design the storm drainage system for the apartment building in Appendix B. In this design, use four roof drains on the roof and size the leaders and horizontal storm drainage piping (¼-in. slope or 20.8 mm per m) to a community storm drainage system and a 4-in. (100 mm) per hr rainfall.

4-7. Design the storm drainage system for the apartment building in Appendix B. In this design, assume a sloped roof (Fig. 4-16) with gutters (¼-in. or 20.8 mm per m slope) and leaders to the ground and a 3-in. (75 mm) per hr rainfall.

4-8. Design the storm drainage system for the building in Appendix B. Use three roof drains and size the leaders and horizontal storm drainage piping (½-in. or 41.6 mm per m slope) to a community storm drainage system and a 5-in. (125 mm) per hr rainfall.

4-9. Design the storm drainage system for the residence in Appendix C. Use gutters (⅛-in. or 10.4 mm per m slope) and leaders to the ground, and assume a 4-in. (100 mm) per hr rainfall.

4-10. Design the storm drainage system for the residence in Appendix D. Use gutters (¼-in. or 20.8 mm per m slope) and leaders to the ground, and assume a 3-in. (75 mm) per hr rainfall.

CHAPTER 5

Private Sewage Disposal

5-1 TYPES OF SYSTEMS

Whenever possible, the sanitary drainage system should connect to a community (public) sewer, but when no community sewer is available, a private sewage disposal system must be installed. This is particularly true in surburban and rural areas and is one of the initial items that the designer should check. The information contained in this chapter is based on the *National Standard Plumbing Code,* but the codes in force in the geographic area of construction should always be checked *before* the system is designed. In some areas, the municipality will not only specify exactly where on the site the system will be placed but will also simply state the size of all equipment required and the installation method. In addition, many areas limit the minimum lot size on which a private sewage disposal system might be placed (often one-half acre) and the minimum size if both a sewage disposal system and a well are required (often about one acre). Most of these requirements are established because of the type of soil in the area, and because there is always a concern that the potable water supply might be contaminated by a sewage disposal system (either the system serving this project or a neighbor's sewage system).

5-2 PRIVATE SYSTEMS

Private sewage disposal systems (Fig. 5-1) usually consist of the building sewer which leads from the project into a septic tank and then the line into a distribution box which feeds the fluid (effluent) into the disposal or leach fields. There are variations on this design: as shown in Fig. 5-2, the septic tank may feed into a seepage pit(s), and the packaged unit in Fig. 5-3 is designed to serve larger complexes.

The watertight septic tank is placed underground, where it receives the sewage from the building and holds it for about a day while the suspended solids settle to the bottom and putrefy. The liquids pass out of the tank at the other end into the distribution box. Septic tanks may be concrete and of rectangular shape (Fig. 5-4) plastic or asphalt-protected steel, usually round (Fig. 5-5).

The size of the septic tank will be one of the first considerations in private sewage disposal design. For individual residences, the tank design is based on the number of bedrooms and the maximum number of persons served and assumes a garbage disposal is used (Fig. 5-16); for institutional and commercial projects, it is based on the estimated sewage flow in gallons per day (Fig. 5-17).

A three-bedroom home which will require a septic tank size of 1,000 gal (3,785 liters) minimum (Fig. 5-16). In many locales the tank sizes available may be limited, and it may be necessary to use a larger tank than the minimum. Using a larger tank will not interfere with the operation of the system. If a larger capacity is required than is available in one tank

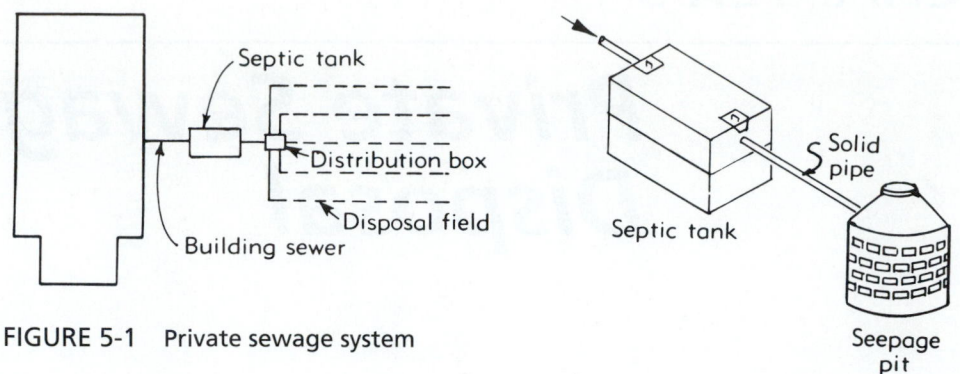

FIGURE 5-1   Private sewage system

FIGURE 5-2   Septic tank to seepage pit

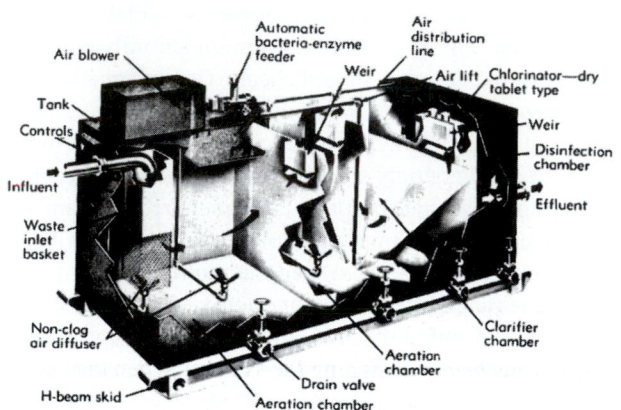

FIGURE 5-3   Package sewage unit

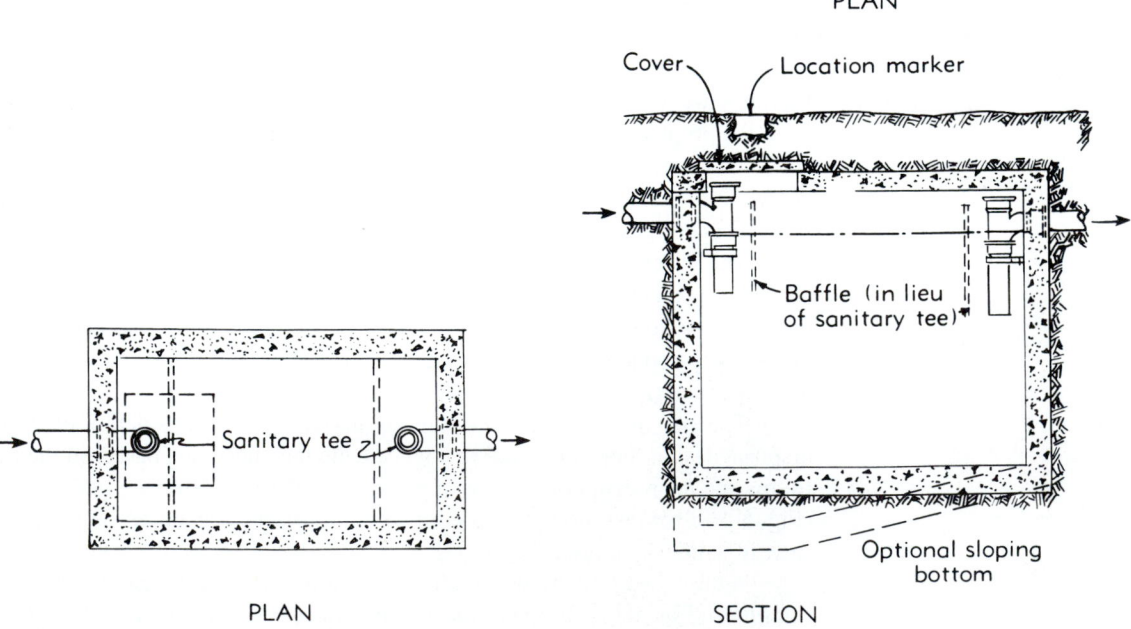

PLAN

PLAN

SECTION

FIGURE 5-4   Rectangular septic tank

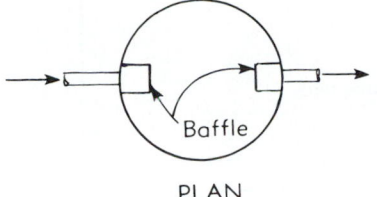

PLAN

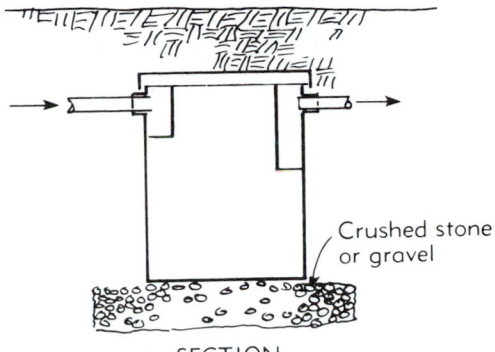

SECTION

FIGURE 5-5 Steel septic tank

(for example, if 1,000 gal is required and only 600-gal tanks are available), two 600-gal tanks may be connected (Fig. 5-6). The tanks should be set close enough together that a single length of pipe (with no joints) can be used to join the outlet of the first tank to the inlet of the second.

Many municipalities require tanks larger than those recommended by the *National Standard Plumbing Code.* It is the designer's responsibility to know or find out local requirements *before* designing the system. When the designer is working in an area where he is not familiar with local codes and requirements, he must visit the local authorities and discuss the design, obtain local requirements in writing, or perhaps call to determine what codes are in effect in the locale.

The code also sets minimum distances for the location of the various parts of the private sewage disposal system (Fig. 5-18); the septic tank must be a minimum of 50 ft (15 m) from any well or suction line (Fig. 5-7), and when possible, the tank is put even farther away. Many local codes require longer distances and therefore must be checked; in general, about 100 ft (30 m) is the preferred distance, but this is not always feasible. These distances greatly reduce the danger of contaminating drinking water if leaks should occur in the tank or pipes (lines).

The depth of the tank will be determined by the depth of the sewer line from the building to the tank. It is important that the sewer line be sloped gradually toward the tank (about ⅛ to ¼ in per ft or 10.4 to 20.8 mm per m). Too much slope and the sewage will flow into

FIGURE 5-6 Septic tanks in series

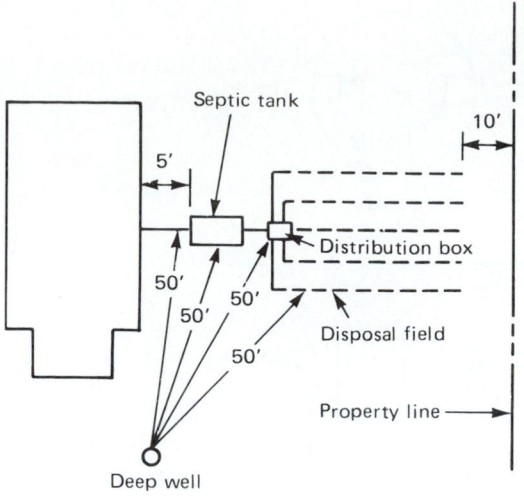

FIGURE 5-7   Private disposal system distances

the tank too rapidly and disturb the natural action in the tank. Too little slope and the sewer line may become clogged. The top of the tank is usually located 12 to 36 in. (.3 to 1 m) below the ground surface so it can be serviced as required.

Institutions and commercial projects base their sewage treatment system sizes on the expected flow of sewage from the project in a day. The table in Fig. 5-17 lists various types of building uses and the gallons per person per day the system must be designed for.

Example   In designing a private sewage system for the four-story apartment building (Appendix A), the first step is to determine the number of apartments. Reviewing the plans, there are two 2-bedroom apartments per floor or a total of 8 apartments with 16 bedrooms.

The required tank capacity may be obtained from Fig. 5-16.

8 bedrooms (one-bedroom units)   = 3,000 gallons (11,355 liters)

8 extra bedrooms at 150 gallons each = 1,200 gallons (4,542 liters)

Minimum septic tank capacity   = 4,200 gallons (15,897 liters)

The *distribution box* receives the effluent from the septic tank and distributes it equally to each individual line of the disposal field (as illustrated in Fig. 5-8). The box is connected to the septic tank with a tight sewer line.

Disposal or tile fields are the preferred method for distributing the effluent. They consist of rows (called *lines*) of pipe through which the effluent passes. The lines may be made

FIGURE 5-8   Distribution box

of clay tile (usually 12 in. or .3 m long), bituminized pipe, or plastic pipe (Fig. 5-9). The clay tiles are laid with about ¼ in. (6 mm) of space between them (Fig. 5-10) to allow the effluent to be absorbed into the gravel fill the tile is placed on. The top portion of the ¼-in. (6 mm) space is covered with a piece of felt so that soil will not fall into the pipe and clog or stop the flow of effluent. Most fields installed today use the bituminized or plastic pipes, which have holes in them. These pipes are installed much faster and do not have any open joints to be covered with felt as the tile does. These pipes are installed with the holes down, and as the effluent flows through the pipes, it is absorbed into the gravel.

The tile or pipe is set into a trench which varies from 18 to 30 in. (.5 to .9 m) in depth and 24 to 36 in. (.6 to 1 m) in width (Fig. 5-20). Trenches must be sloped in the direction of flow; the code limits the maximum slope to 6 in. (150 mm) in 100 ft (30 m) so that the effluent will not simply flow to the end of the line and then back up. The bottom of the trench is filled with a layer of filter material not less than 6 in. (150 mm) deep below the pipe line, extending the full width of the trench and a minimum of 2 in. (50 mm) above the pipe, and covered with a layer of straw. The code also limits any individual line to a length of 100 ft (30 m) and sets the minimum separation between lines at 6 ft (2 m). The disposal field may take any of a number of shapes (Fig. 5-11), depending on the contours (slope) of the ground, the size of the lot and the location of any well or stream on the property. The minimum distance between the field and the building is 10 ft (3 m), between field and property line 10 ft (3 m), between field and stream 25 ft (7.5 m), and between field and well 100 ft. If the well has an outside watertight casing which extends down to a depth of 50 ft (15 m) or more, the minimum separation between field and well may be reduced to 50 ft (15 m).

The length of line required in the disposal field depends on the ability of the soil to absorb sewage. The subsurface conditions are checked first by digging a hole about 5 ft (1.5 m) below final grade to observe the type of soil encountered and whether any groundwater is evident. Next, a soil percolation test is made to measure the ability of the soil to absorb sewage. The procedure for making the soil percolation test is:

FIGURE 5-9   Perforated tile

FIGURE 5-10   Tile

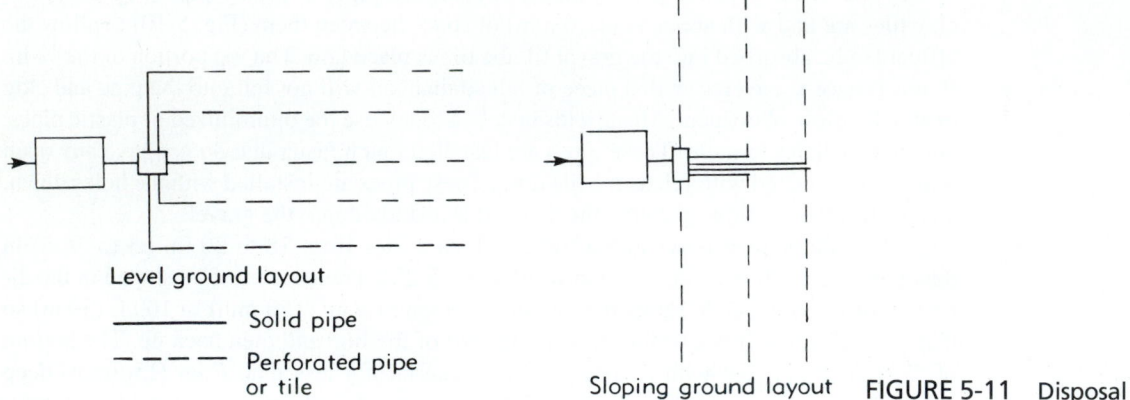

Level ground layout

——————— Solid pipe

— — — — Perforated pipe
             or tile

Sloping ground layout   FIGURE 5-11   Disposal field patterns

1. Dig a hole about 12 in. (300 mm) in diameter and 30 in. (750 mm) deep, keeping the sides of the hole vertical (Fig. 5-12).

2. Presoak the hole by filling it with water and allowing the water to completely seep away. The hole should be presoaked several hours before the test and again at the time of the test.

3. After presoaking, remove any loose soil that might have fallen in from the sides of the hole.

4. Carefully fill the hole to a depth of 6 in. (150 mm) with clean water with as little splashing as possible.

5. Record the time (in minutes) that it takes for the water level to drop 1 in. (25 mm).

6. Now, repeat the test a minimum of three times until the time it takes the water to drop 1 in. (25 mm) for two successive tests is approximately the same. Then take the last test as the stabilized rate of percolation; it is the time recorded for this test which will be used to size the disposal field.

The tile field length is sized from Fig. 5-20, which lists the gallons per day per person, and Fig. 5-19, which lists the time required for the water to drop 1 in. and the square feet of trench required per 100 gallons (378 liters) of sewage per day.

Institutional and commercial projects are sized from Figs. 5-17, 5-19, and 5-20, which list the time required for percolation in minutes and the absorption of the soil for tile fields and seepage pits.

Example   Assuming a soil percolation time of 5 minutes. From Fig. 5-17, the gallons per day per person required is 75. For a three-bedroom residence with four people, 300 gpd of sewage flow

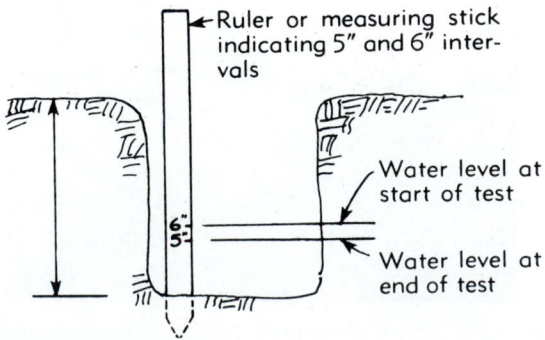

←Ruler or measuring stick indicating 5" and 6" intervals

Water level at start of test

Water level at end of test

FIGURE 5-12   Percolation pit

is generated. Using Fig. 5-19, 42 sq ft of tile length is required per 100 gpd. For 300 gpd, 126 sq ft is required.

 Assuming a soil percolation time of 5 minutes, from Fig. 5-17, the liters per day (L/d) per person is 284. For a three-bedroom residence with four people (284 × 4), 1136 L/d of sewage flow is generated. Using Fig. 5-19, a trench .9m wide, 4.3 m min length is required for every 378 L/d. For 1136 L/d, a trench .9m wide and 14.8 m long is required.

$$1136 \div 378 \times 3.7 = 3 \times 4.3 = 12.9 \text{ m of trench .9m wide}$$

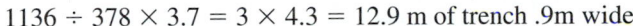

Example Continuing with the four-story apartment building (Appendix A) with 8 apartments, averaging 2.5 persons per apartment. From Fig. 5-17, a rate of 75 gpd per person must be used (the same as individual residences), for 1,500 gpd. Adding 50% for food waste disposals increases it to 2,250 gpd. Assuming a soil percolation time of 3 minutes, from Fig. 5-19, the amount that the soil can absorb is 100 gpd per 35 sq ft. To determine the amount of square footage required, divide the sewage flow by 100 times the sq ft required. In this design, divide 2,250 gpd by 100 gpd times 35 sq ft and 788 sq ft of absorption area are required. Using a trench width of 3 ft, 263 lineal ft of line are required.

 Continuing with the four-story apartment building (Appendix A) with 8 apartments averaging 2.5 persons per apartment, from Fig. 5-17, a rate of 284 L/d per person must be used (the same as individual residences), for 5672 L/d (20 × 284). Adding 50% for food waste disposal increases it to 8508 L/d. Assuming a soil percolation of 3 minutes, from Fig. 5-19, a trench .9m wide and 3.7 m in length is required for each 378 L/d. To determine the total length of trench required, divide the total sewage flow by 378 and multiply it by the length per 378 L/d.

$$(8508 \div 378) \times 3.7 = 22.5 \times 3.7 = 83.25\text{m length required (.9m wide)}$$

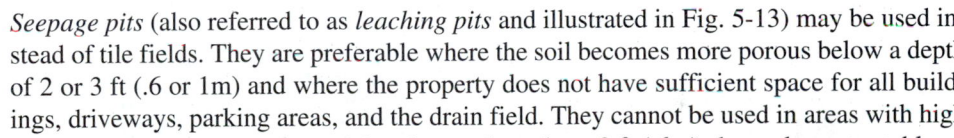

Seepage pits (also referred to as *leaching pits* and illustrated in Fig. 5-13) may be used instead of tile fields. They are preferable where the soil becomes more porous below a depth of 2 or 3 ft (.6 or 1m) and where the property does not have sufficient space for all buildings, driveways, parking areas, and the drain field. They cannot be used in areas with high water tables since the bottom of the pit must be at least 2 ft (.6m) above the water table.

The seepage pit is usually made of concrete block or is a precast concrete unit. Typically, 8-in. (200 mm)-thick blocks are used, and they are laid with the cells (holes) placed horizontally (Fig. 5-13) to allow the effluent to seep into the ground. The tapered cells of the block are set with the widest area to the outside to reduce the amount of loose material behind the lining that might fall into the pit. The typical precast concrete seepage pit shown in Fig. 1–5 is the same unit used for drywells (Sec. 4–3). The bottom of the pit is lined with coarse gravel a minimum of 1 ft (.3 m) deep before the block or concrete is placed. Between the block or concrete and the soil is a minimum of 6 in. (.15 m) of clean crushed stone or gravel. Straw is placed on top of the gravel to keep sand from filtering down and reducing the effectiveness of the gravel. The top of the pit should have an opening with a watertight cover to provide access to the pit if necessary. The construction of the pit above the inlet pipe should be watertight.

When more than one seepage pit is used, the pipe from the settling tank must be laid out so that the effluent will be spread uniformly to the pits. To provide equal distribution, a distribution box with separate laterals (Fig. 5-14)—each lateral feeding no more than two pits—provides the best results. The distance between the outside walls of the pits should be a minimum of 3 pit diameters and not less than 10 ft (3 m).

The size of the seepage pit is based on the outside area of the walls. The areas for pits of various diameters are given in Fig. 5-13. Many designers exclude the bottom area of the

ABSORPTION AREAS					
Diam.	Depth				
	4'	5'	6'	8'	10'
4'	50.2	62.8	75.3	100.4	125.6
5'	62.8	78.5	94.2	125.6	157.0
6'	75.4	94.2	113.0	150.7	188.4
8'	100.4	125.6	150.7	200.9	251.2

SI UNITS

ABSORPTION AREAS (sq. m)					
Diam.	Depth				
m	1.22	1.53	1.83	2.44	3.05
1.22	4.66	5.83	7.00	9.33	11.67
1.53	5.83	7.29	8.75	11.67	14.59
1.83	7.00	8.75	10.50	14.00	17.50
2.44	9.33	12.56	14.00	18.67	23.34

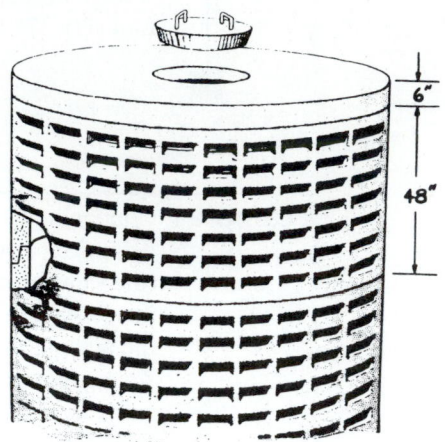

FIGURE 5-13   Typical seepage pit

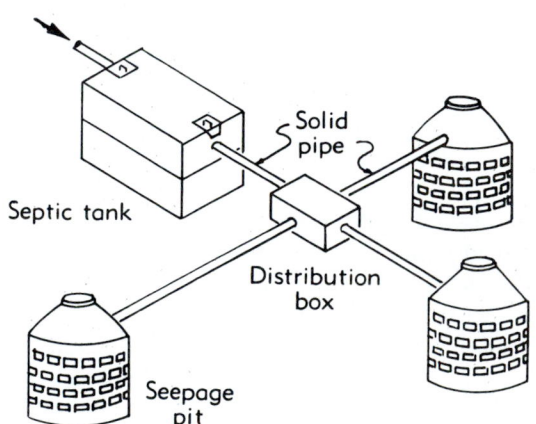

FIGURE 5-14   Distribution box to seepage pits

pit from the absorption area required to allow for a safety factor, while others calculate the total area available and then size the system to allow some safety factor. The latter approach is used in this text.

Seepage pits are sized from Fig. 5-21, which lists the time required for percolation and the absorption area required per 100 gpd. The four-story apartment building has a sewage flow of 2,250 gpd (sized earlier in this section). Assuming a soil percolation of 3 minutes, from Fig. 5-21, the soil can absorb 100 gal per 45 sq ft per day. To determine the amount of absorption area required, divide the sewage flow by 100, times the absorption rate. In this design:

$$\frac{2250}{100} \times 45 = 1013 \text{ sq. ft.}$$

The size of the pits to be used is selected from Fig. 5-13, and 7 seepage pits 6 ft in diameter and 8 ft deep are selected. Other combinations may be used depending on the area required to place the pits, the sizes available locally, whether precast concrete is used, and the designer's own preference.

SI Seepage pits are sized from Fig. 5-21, which lists the time required for percolation and the absorption flow required per 378 L/d. The four-story apartment building has a sewage flow of 8508 L/d (sized earlier in this section). Assuming a soil percolation of 3 minutes, from Fig. 5-21, the soil can absorb 378 L/d per 4.18 sq m. To determine the amount of absorption area required, divide the sewage flow by 378, times the area needed per 378 L/d. In this design:

$$(8508 \div 378) \times 4.18 = 22.5 \times 4.18 = 94 \text{ sq m}$$

The size of the pits to be used is selected from Fig. 5-13, and 7 seepage pits 1.83 m in diameter and 2.44 m deep are selected. Other combinations may be used depending on the area required to place the pits and sizes available locally.

5-3 SEWAGE PROBLEMS

The uncertainty of exactly how much the soil will actually absorb is reflected in the values given in the various tables and charts. Even so, many designers prefer to oversize the system slightly to allow for poor absorption and also to allow for future increased amounts of effluent, either because more people are using the facility than anticipated or because of an addition to the individual residence or an addition of various water-using fixtures which may not have been included in the original design.

One of the most wasteful uses of the private system is the connection of a washing machine to it. Many times, the tying-in of a washing machine to an older system has resulted in more water flow than the ground could handle through the system installed. It is suggested that when a washing machine is installed (especially where there are several children in the family), a drywell should be installed. Problems may also occur if all of the family washing is done on one day. This puts a tremendous additional flow into the system for a brief period, and as a result, the system may back up into the house. One of the simplest solutions to this is to spread the washing out over several days, giving the ground a chance to absorb the water. Other solutions are to connect the washer to a drywell, to increase the size of the disposal field, or to add a seepage pit.

The connection of gutters, storm drainage, and roof drains may also cause periodic overloads on the private sewage system. When these are connected, the designer must increase the size of the system to accommodate the periodic additional flow. Most designers prefer to run such connections into drywells if no storm drain system is available.

5-4 SYSTEM INSTALLATION

Private sewage disposal requires finding an appropriate location on the site to place all of the system components. Typically, a backhoe will be needed to dig the holes for the septic tank or seepage pits and to run any trenches for pipe lines and distribution fields (Fig. 5-15).

FIGURE 5-15   Private sewage trench

To tie into community systems, a trench must be dug from the project to the community sewage line, and the final elevation of the two must be checked to make certain the sewage will flow properly.

Extracted from the 1988 Supplement to the 1987 National Standard Plumbing Code with permission of PHCC.

*Septic tank sizes all for connection of food waste disposable units.

CAPACITY OF SEPTIC TANKS*

Single family dwellings - number of bedrooms	Multiple dwellings units or apartments - one bedroom each	Other uses; maximum fixture units served	Minimum septic tank capacity in:	
	units		gallons	liters
1 - 3		20	1,000	3,785
4	2	25	1,200	4,542
5 or 6	3	33	1,500	5,678
7 or 8	4	45	2,000	7,570
	5	55	2,250	8,516
	6	60	2,500	9,463
	7	70	2,750	10,408
	8	80	3,000	11,355
	9	90	3,250	12,301
	10	100	3,500	13,248

Extra bedroom, 150 gallons (568 L) each.
Extra dwelling units over 10, 250 gallons (946 L) each.
Extra fixture units over 100, 25 gallons (95 L) per fixture unit.

Note: Metric added by author

FIGURE 5-16   Capacity of septic tanks.

SEWERAGE FLOWS ACCORDING TO TYPE OF ESTABLISHMENT

Type of Establishment	
Schools (toilet and lavatories only)	15 Gal. (57 L) per day per person
Schools (with above plus cafeteria)	25 Gal. (95 L) per day per person
Schools (with above plus cafeteria and showers)	35 Gal. (132 L) per day per person
Day workers at schools and offices	15 Gal. (57 L) per day per person
Day camps	25 Gal. (95 L) per day per person
Trailer parks or tourist camps (with built-in bath)	50 Gal. (189 L) per day per person
Trailer parks or tourist camps (with central bathhouse)	35 Gal. (132 L) per day per person
Work or construction camps	50 Gal. (189 L) per day per person
Public picnic parks (toilet wastes only)	5 Gal. (19 L) per day per person
Public picnic parks (bathhouse, showers and flush toilets)	10 Gal. (38 L) per day per person
Swimming pools and beaches	10 Gal. (38 L) per day per person
Country clubs	25 Gal. (95 L) per locker
Luxury residences and estates	150 Gal. (568 L) per day per person
Rooming houses	40 Gal. (151 L) per day per person
Boarding schools	50 Gal. (189 L) per day per person
Hotels (with connecting baths)	50 Gal. (189 L) per day per person
Hotels (with private baths - 2 persons per room)	100 Gal. (378 L) per day per person
Boarding schools	100 Gal. (378 L) per day per person
Factories (gallons per person per shift - exclusive of industrial wastes)	25 Gal. (95 L) per day per person
Nursing homes	75 Gal. (284 L) per day per person
General hospitals	150 Gal. (568 L) per day per person
Public institutions (other than hospitals)	100 Gal. (378 L) per day per person
Restaurants (toilet and kitchen wastes per unit of serving capacity)	25 Gal. (95 L) per day per person
Kitchen wastes from hotels, camps, boarding houses, etc. serving three meals per day	10 Gal. (38 L) per day per person
Motels	50 Gal. (189 L) per bed space
Motels with bath, toilet, and kitchen wastes	60 Gal. (227 L) per bed space
Drive-in theaters	5 Gal. (19 L) per car space
Stores	400 Gal. (1514 L) per toilet room
Service stations	10 Gal. (38 L) per vehicle served
Airports	3-5 Gal. (11-19 L) per passenger
Assembly halls	2 Gal. (8 L) per seat
Bowling alleys	75 Gal. (284 L) per lane
Churches (small)	3-5 Gal. (11-19 L) per sanctuary seat
Churches (large with kitchens)	5.7 Gal. (22 L) per day sanctuary seat
Dance halls	2 Gal. (8 L) per day per person
Laundries (coin operated)	400 Gal. (1514 L) per machine
Service stations	1000 Gal. (3780 L) (First Bay) 500 Gal. (1892 L) (Each add. Bay)
Sub-divisions or individual homes	75 Gal. (284 L) per day per person
Marinas - Flush toilets	36 Gal. (136 L) per fixture per hr
Urinals	10 Gal. (38 L) per fixture per hr
Wash basins	15 Gal. (57 L) per fixture per hr
Showers	150 Gal. (568 L) per fixture per hr

Note: Metric added by author

FIGURE 5-17   Sewerage flows according to type of establishment.

Location of Sewage Disposal System

Minimum Horizontal Distance In Clear Required From:	Building Sewer		Septic Tank		Disposal Field		Seepage Pit or Cesspool	
Buildings or structures[1]	2 feet	(0.6 m)	5 feet	(1.5 m)	8 feet	(2.4 m)	8 feet	(2.4 m)
Property line adjoining private property	Clear[2]		5 feet	(1.5 m)	5 feet	(1.5 m)	8 feet	(2.4 m)
Water supply wells	50 feet[3]	(15.2 m)	50 feet	(15.2 m)	100 feet	(30.5 m)	150 feet	(45.7 m)
Streams	50 feet	(15.2 m)	50 feet	(15.2 m)	50 feet[7]	(15.2 m)[7]	100 feet[7]	(30.5 m)[7]
Trees	–		10 feet	(3.0 m)	–		10 feet	(3.0 m)
Seepage pits or cesspools	–		5 feet	(1.5 m)	5 feet	(1.5 m)	12 feet	(3.7 m)
Disposal field	–		5 feet	(1.5 m)	4 feet[4]	(1.2 m)	5 feet	(1.5 m)
On site domestic water service line	1 foot[5]	(0.3 m)	5 feet	(1.5 m)	5 feet	(1.5 m)	5 feet	(1.5 m)
Distribution box	–		–		5 feet	(1.5 m)	5 feet	(1.5 m)
Pressure public water main	10 feet[6]	(3.0 m)	10 feet	(3.0 m)	10 feet	(3.0 m)	10 feet	(3.0 m)

Note:
When disposal fields and/or seepage pits are installed in sloping ground, the minimum horizontal distance between any part of the leaching system and ground surface shall be fifteen (15) feet (4.6 m).
1. Including porches and steps, whether covered or uncovered, breezeways, roofed porte-cocheres, roofed patios, carports, covered walks, covered driveways and similar structures or appurtenances.
2. See also Section 313.3 of the Uniform Plumbing Code.
3. All drainage piping shall clear domestic water supply wells by at least fifty (50) feet (15.2 m). This distance may be reduced to not less than twenty-five (25) feet (7.6 m) when the drainage piping is constructed of materials approved for use within a building.
4. Plus two (2) feet (0.6 m) for each additional foot (0.3 m) of depth in excess of one (1) foot (0.3 m) below the bottom of the drain line. (See also Section I 6.)
5. See Section 720.0 of the Uniform Plumbing Code.
6. For parallel construction – For crossings, approval by the Health Department shall be required.
7. These minimum clear horizontal distances shall also apply between disposal field, seepage pits, and the ocean mean higher high tide line.

FIGURE 5-18   Minimum distances between components of an individual sewage disposal system (in feet).

The tile lengths for each 100 gallons (378 L) of sewage per day are as follows:

Time in Minutes for 1-inch (25 mm) Drop	The Length for Trench Widths of		
	1-foot (.3 m)	2-feet (.6 m)	3-feet (.9 m)
1	25 (7.6 m)	13 (4.0 m)	9 (2.7 m)
2	30 (9.2 m)	15 (4.6 m)	10 (3 m)
3	35 (10.7 m)	18 (5.5 m)	12 (3.7 m)
5	42 (12.8 m)	21 (6.4 m)	14 (4.3 m)
10	59 (18 m)	30 (9.2 m)	20 (6.1 m)
15	72 (22.6 m)	37 (11.3 m)	25 (7.6 m)
20	91 (27.8 m)	46 (14 m)	31 (9.5 m)
25	105 (32 m)	53 (16.2 m)	35 (10.7 m)
30	125 (38.1 m)	63 (19.2 m)	42 (12.8 m)

Note: Metric added by author

FIGURE 5-19   Tile length

SIZE AND SPACING FOR DISPOSAL FIELDS

Width of trench at bottom		Recommended depth of trench		Spacing tile lines[1]		Effective absorption area per lineal ft. of trench	
(in.)	(m)	(in.)	(m)	(ft.)	(m)	(sq. ft.)	(sq. m)
18	.46	18 to 30	.46 to .76	6.0	1.8	1.5	.14
24	.61	18 to 30	.46 to .76	6.0	1.8	2.0	.19
30	.76	18 to 36	.46 to .92	7.6	2.3	2.5	.23
36	.92	24 to 36	.61 to .92	9.0	2.8	3.0	.28

1. A greater spacing is desirable where available area permits.
2. Maximum length 100 ft. (30.5 m)

Note: Metric added by author

FIGURE 5-20   Disposal fields.

| Time in Minutes for | Effective Absorption Area | |
1-inch (25 mm) Drop	Square Feet	Square Meters
1	32	2.98
2	40	3.72
3	45	4.18
5	56	5.20
10	75	6.97
15	96	8.92
20	108	10.03
25	139	12.91
30	167	15.51

Note: Minimum 125 sq. ft. (11.6 sq. m) required. Bottom of pit shall not be considered part of the absorption area.

Extracted from the National Standard Plumbing Code with permission of PHCC.

Note: Metric added by author

FIGURE 5-21 Effective absorption area in seepage pits for each 100 gallons (378 liters) of sewage per day.

QUESTIONS

5-1. When are private sewage systems generally used?

5-2. Draw a sketch of the major parts of a private sewage system in relation to a residence.

5-3. Why do most codes set minimum distances between the parts of the sewage system and wells and streams?

5-4. What is the function of a distribution box?

5-5. How is the ability of the soil to absorb sewage determined?

5-6. What are *seepage pits,* and when are they used?

5-7. Why should the designer consider not connecting pipes carrying certain waste water (such as water from washing machines) to the private sewage system?

5-8. How would the reuse water system discussed in Sec. 1-9 and illustrated in Fig. 1-18 affect sewage treatment system designs, both private and community?

Design Exercises

5-9. Design a private sewage system (septic tank and disposal field) for the apartment building in Appendix B. In this design, use the sewage flows in Fig. 5-17, an absorption rate of 5 minutes, and use 2 ft (.6m) width of trench.

5-10. Design a private sewage system (seepage pits) for the apartment building in Appendix B. In this design, use the sewage flows in Fig. 5-17 and an absorption rate of 2 minutes. Due to the water table, limit the seepage pits to a 6-ft (2 m) maximum depth.

5-11. Design a private sewage system (septic tank and disposal field) for the building in Appendix C. In this design, use the sewage flows in Fig. 5-16 and 5-17, an absorption rate of 5 minutes, and use 2 ft (.6 m) width of field.

5-12. Redesign Exercise 5-11 with a seepage pit system.

5-13. Design a private sewage system (septic tank and disposal field) for the building in Appendix D. In this design, use an absorption rate of 4 minutes.

CHAPTER 6

Comfort

6-1   COMFORT

Comfort means different conditions to different people. Inside any room it means being able to carry on any desired activity without being either chilly or too hot. Most people don't care what combination of factors cause them to be comfortable as long as they feel comfortable.

6-2   TRANSFER OF BODY HEAT

Since we are concerned here with people, a short study of the body, its heat, and how it is transferred is required. Body heat is transferred to the surrounding air and surfaces by three natural processes which usually occur at the same time. These natural processes are:

1. Convection.
2. Radiation.
3. Evaporation.

Convection
The process of heat transfer by convection is based on two fundamental principles:

1. Heat flows from a hot to a cold surface.
2. Heat rises.

   Applying these principles to the body (Fig. 6-1):

1. As the body gives off heat, it will flow to the surrounding cooler air.
2. As the surrounding air warms, it moves in an upward direction.
3. As the warm air moves upward, cooler air is pulled in behind it; then the cooler air heats and flows upward, and more cool air moves in as the cycle of convection continues.

Radiation
The process of heat transfer by radiation is based on one fundamental principle (Fig. 6-2):

Heat flows from a hot to a cold surface.

But radiation is different from convection in one respect: no air movement is required to transfer the heat. In the process of radiation, *heat rays* transfer the heat from the heat source, just as the sun's rays heat any surface they touch. The only difference is that, in this case, people are the heat source.

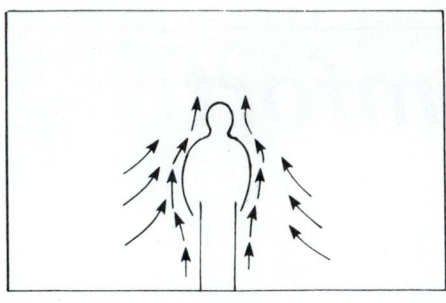

FIGURE 6-1   Convection

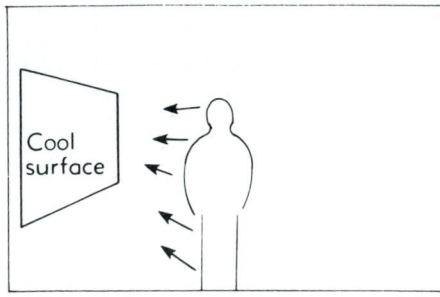

FIGURE 6-2   Radiation

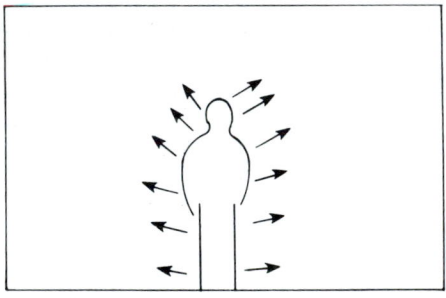

FIGURE 6-3   Evaporation

Evaporation

The process of heat transfer by evaporation is based on one fundamental principle (Fig. 6-3):

> Moisture evaporates from a warm surface.

The evaporation of moisture from a warm surface removes heat from the surface and cools it. As moisture is given off through the pores of the skin, it evaporates, removing heat from the body.

Air Temperature and Body Heat

The amount of body heat given off is affected by the air temperature surrounding it in the following ways:

1. *Convection.* Cool air increases the rate of heat loss by convection, while warm air slows the rate.
2. *Radiation.* Cool air will lower the surface temperature of the surroundings, increasing the rate of heat loss by radiation; warm air raises the temperature of the surfaces, lowering the rate of heat loss due to radiation.

3. *Evaporation.* Generally, cool air increases the rate of evaporation while warm air reduces the rate; however, the rate of evaporation also depends on the relative humidity (the amount of moisture in the air) and the amount of air movement.

6-3   SPACE CONDITIONS AFFECTING COMFORT

The conditions within a space (room) which affect the comfort of the occupant, because they affect the rate of heat loss from the body, are:

1. Room air temperature (also called dry-bulb temperature).

2. Humidity or moisture content of the room.

3. Surface temperature of surrounding surfaces in the room (also called mean radiant temperature or MRT).

4. Rate of air motion.

The occupant will feel comfortable when each of these is within a certain range.

The room air temperature affects the rate of convective and evaporative body heat losses. Since the room temperature will be below the body temperature of 98.6°F (37°C), there will be convective heat loss from the body. Generally, the body feels most comfortable in a temperature range of 72° to 78°F (22.2° to 25.5°C) in the winter and 72° to 76°F (22.2° to 24.4°C) in the summer. Of course, the range of comfortable temperatures varies with each individual and also is affected by what the person is wearing and how active he or she is.

The humidity (moisture content) of the air affects the rate of evaporative heat loss from the body. A high humidity (often occurring in the summer) will cause the surrounding air to absorb less heat from the body, making the occupant feel warmer. A low humidity (often occurring in the winter) allows the air to absorb greater quantities of body heat, making the occupant feel cooler. As discussed later in the text, part of the design problem may be to lower the summer humidity and increase the winter humidity.

The surface temperatures of surrounding surfaces affect the radiant heat loss from the body. The surface temperatures in the room may vary widely. Windows and exterior walls will probably be cooler in the winter than interior walls and the furniture. The temperatures of the floors and ceilings will depend on the air temperature on the other side of the construction. Comfort can be increased (in the winter) by directing the flow of warm air over the colder surfaces. This is why most heating elements (registers, baseboard heaters) are located under a window.

The rate of air motion affects the transfer of body heat by convection and evaporation. In the summer, increased air motion increases the evaporation rate of heat from the body to help keep it cool. This is why buildings without air conditioners use fans to increase the movement of the air on hot days. In the winter, a slower rate of air motion is desired, or the occupants will feel cool due to evaporative heat loss. But even in the winter it is desirable to have a flow of air to keep the air from becoming stagnant.

6-4   COMFORTABLE ENVIRONMENT

The building occupant will feel comfortable within certain ranges of air temperature, humidity, surface temperatures of surrounding surfaces, and air motion.

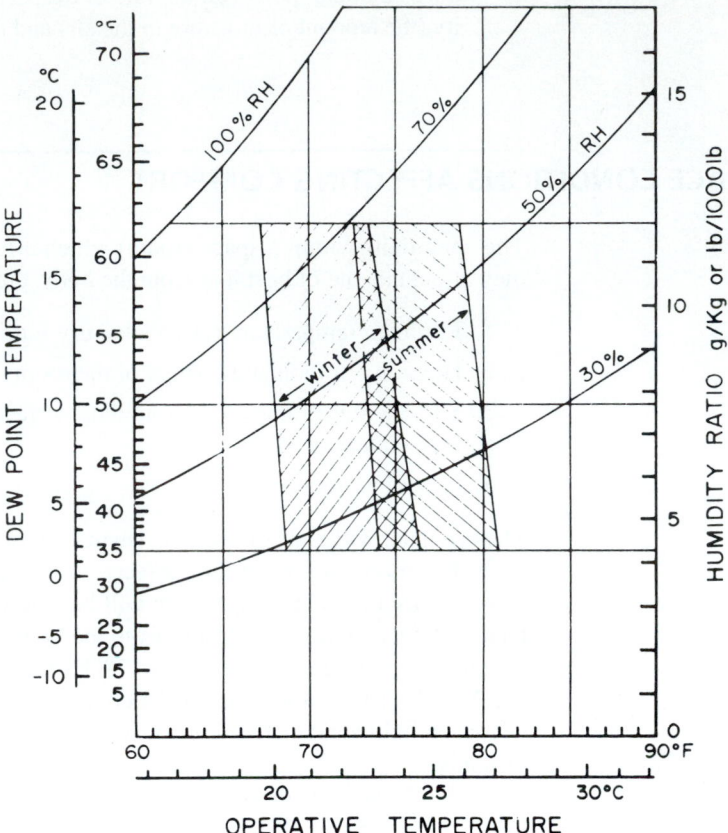

FIGURE 6-4   Comfort ranges

Experiments were made by ASHRAE (The American Society of Heating, Refrigerating and Air Conditioning Engineers) to determine the ranges of these variable conditions which feel comfortable to the occupants. The results were then tabulated and made into the "comfort chart" shown in Fig. 6-4. Generally, the results show that there is a definite link between humidity and air temperature when both air movement and surface temperatures remain constant. This air temperature to humidity relationship shows that to produce comfort, as the temperature increases, the humidity should decrease. This means that in the winter, it would be most economical to provide comfort with lower air temperature and higher humidity.

While the experiments and chart were done for central heating systems and have other design limitations noted by ASHRAE, they do show the important humidity/air temperature relationships. But, first, an explanation of how to read the chart will be helpful. The chart has five basic parts listed here and identified in Fig. 6-4:

1. Dry-bulb temperature (along the bottom).

2. Dew point temperature (along the left side).

3. Relative humidity in percent (across the chart from the lower left to the upper right, with percentages in the upper right).

4. The shaded areas show the range of conditions in which most people will feel comfortable.

6-5   WINTER COMFORT

Use of the comfort chart to determine winter comfort is shown in Fig. 6-4.

1. Determine the most likely dry-bulb temperature for the space along the bottom of the chart.
2. From the dry-bulb temperature, move up into the comfort areas on the chart.
3. Determine the range of relative humidity which will provide comfort to the most occupants.

This will be used to determine the various combinations of dry-bulb temperature and relative humidity which will provide comfort and still provide an effective temperature of about 70° to 77°F (21.1° to 25°C).

Using a dry-bulb (room air) temperature of 70°F (21.1°C) what relative humidity would be required? Use of the chart to determine various dry-bulb to humidity relationships is shown in Fig. 6-4.

1. Find 76°F (24.4°C) dry-bulb temperature along the bottom of the chart (point 1).
2. Move vertically up to the point where the 76°F (24.4°C) dry-bulb line intersects the shaded area.
3. Read the relative humidity off as about 35% to 65% lines, where individuals will be most comfortable.

Homes with no humidity control will tend to have low relative humidity in the winter. So, with a winter heating condition and a 70°F (21.1°C) dry-bulb reading, it is necessary to have about 60% relative humidity for the highest percentage of occupants to feel comfortable.

The approximate humidity inside a building can be found by putting a hygrometer (Fig. 6-5) in the space; it will show the percentage of humidity just as a thermometer shows temperature. Also, the chart in Fig. 6-6 shows the relationship of outside air temperature and relative humidity to heated interior air. For example, with an outside temperature of 45°F (7.2°C) and a relative humidity of 80%, the indoor relative humidity will be only 30%.

FIGURE 6-5   Hygrometer (humidity meter)

INDOOR RELATIVE HUMIDITY CONVERSION CHART

OUTDOOR RELATIVE HUMIDITY	−20°	−10°	−5°	0°	+5°	+10°	+15°	+20°	+25°	+30°	+35°	+40°	+45°	+50°	°F
100%	2%	3%	4%	6%	7%	9%	11%	14%	17%	21%	26%	31%	38%	46%	
95%	2%	3%	4%	5%	7%	8%	10%	13%	16%	20%	24%	30%	36%	44%	
90%	2%	2%	4%	5%	6%	8%	10%	12%	15%	19%	23%	28%	34%	41%	
85%	2%	2%	4%	5%	6%	8%	9%	12%	15%	18%	22%	27%	32%	39%	
80%	2%	2%	4%	5%	6%	7%	9%	11%	14%	17%	20%	25%	30%	37%	
75%	2%	2%	3%	4%	5%	7%	8%	10%	13%	16%	19%	23%	28%	36%	
70%	1%	2%	3%	4%	5%	6%	8%	10%	12%	15%	18%	22%	26%	32%	
65%	1%	2%	3%	4%	5%	6%	7%	8%	11%	14%	17%	20%	25%	30%	
60%	1%	2%	3%	3%	4%	5%	7%	8%	10%	13%	15%	19%	23%	28%	
55%	1%	1%	2%	3%	4%	5%	6%	8%	9%	12%	14%	17%	21%	25%	
50%	1%	1%	2%	3%	4%	4%	6%	7%	9%	10%	13%	16%	19%	23%	
45%	1%	1%	2%	3%	3%	4%	5%	6%	8%	9%	12%	14%	17%	21%	
40%	1%	1%	2%	2%	3%	4%	4%	6%	7%	8%	10%	12%	15%	18%	
35%	1%	1%	2%	2%	3%	3%	4%	5%	6%	7%	9%	11%	13%	16%	
30%	1%	1%	1%	2%	2%	3%	3%	4%	5%	6%	8%	9%	11%	14%	
25%	1%	1%	1%	1%	2%	2%	3%	3%	4%	5%	6%	8%	10%	12%	
20%	+%	1%	1%	1%	1%	2%	2%	3%	3%	4%	5%	6%	8%	10%	
15%	+%	+%	1%	1%	1%	1%	2%	2%	3%	3%	4%	5%	6%	7%	
10%	+%	+%	+%	1%	1%	1%	1%	1%	2%	2%	3%	3%	4%	5%	
5%	+%	+%	+%	+%	+%	+%	1%	1%	1%	1%	1%	1%	2%	2%	
0%	0%	0%	0%	0%	0%	0%	0%	0%	0%	0%	0%	0%	0%	0%	
	−29°	−23°	−21°	−18°	−15°	−12°	−9°	−7°	−4°	−1	+2°	+5°	+7°	+10°	°C

OUTDOOR TEMPERATURE

To find indoor relative humidity when outside air is heated to 70 degrees, find outdoor RH at left and outdoor temperature along the bottom. Indoor RH that results in where lines cross.

FIGURE 6-6   Humidity conversion

It becomes obvious that it is often necessary to add moisture to the air in the winter. Humidity can be introduced into the space by installing humidifiers in the ducts of forced air systems and by placing large portable humidifiers in spaces with water, steam, or electric baseboard or panel heat.

6-6   SUMMER COMFORT

Summer comfort ranges can also be found on the comfort chart (Fig. 6-4).

1. Determine the most likely dry-bulb temperature for the space along the bottom of the chart.

2. Follow the dry-bulb line from that point on the chart where it enters the comfort areas.

This is the effective temperature at which the largest percentage of occupants feel comfortable, and it is used to determine the various comfortable combinations of dry-bulb temperature and relative humidity.

A range of about 35% to 65% relative humidity (Fig. 6-6) and a dry-bulb range of 74° to 80°F (23.3° to 26.7°C) is required for the highest percentage of occupants to feel comfortable.

During the summer, homes with no humidity control will tend to have high relative humidity. Yet during the summer, a low humidity will make higher temperatures feel comfortable and save cooling costs (where used). However, humidity control is often more expensive than cooling the air.

6-7 OVERALL COMFORT

The overall comfort will still vary from person to person, as well as with air velocity, types of heating and cooling systems, and types and materials of construction.

Women will generally prefer higher effective temperatures than men. Similarly, older people prefer higher effective temperatures, as do people who live in southern climates. Of course, whether a person will be comfortable or not also depends on the level of activity and the type of clothing the individual wears. A person with a sweater on who is cleaning the house will need a lower effective temperature than someone with a short-sleeved shirt or blouse on who is watching television.

The suggested inside winter dry-bulb temperatures for a wide variety of spaces are shown in Fig. 6-7. Those spaces in which people will be relatively inactive, such as classrooms and hospital rooms, will require higher temperatures than spaces in which people will be active.

In the summer, comfort is also greatly affected by the relationship of indoor-outdoor temperatures. This is especially true in spaces which people occupy for short periods of time. People who are in residences or offices for a long period of time become used to a 75°F (23.8°C) temperature and are comfortable in it. But people occupy a space for short durations, generally up to 1 hr., they will be most comfortable if there is just a 10° to 15°F (6° to 9°C) difference between the outside and the inside temperatures. So, if it is 95°F (35°C) outside, it should be about 80°F inside stores and shops for a person to feel com-

Type of Building	F	C
SCHOOLS-		
Classrooms	72-74	22-23
Assembly rooms	68-72	20-22
Gymnasiums	55-65	13-18
Toilets and baths	70	21
Wardrobe and locker rooms	65-68	18-20
Kitchens	66	19
Dining and lunch rooms	65-70	18-21
Playrooms	60-65	15-19
Natatoriums	75	24
HOSPITALS-		
Private rooms	72-74	22-23
Private rooms (surgical)	70-80	21-27
Operating rooms	70-95	21-35
Wards	72-74	22-23
Kitchens and laundries	66	19
Toilets	68	20
Bathrooms	70-80	21-27
THEATERS-		
Seating space	68-72	20-22
Lounge rooms	68-72	20-22
Toilets	68	20
HOTELS-		
Bedrooms and baths	75	24
Dining rooms	72	22
Kitchens and laundries	66	19
Bathrooms	65-68	18-20
Toilets and service rooms	68	20
HOMES	73-75	23-24
STORES	65-68	18-20

Note: Metric added by author

FIGURE 6-7   Inside winter dry-bulb temperatures

fortable. The common practice in stores, shops, grocery markets, etc., of setting their cooling at a 68° to 70°F (20° to 21.1°C) inside temperature does not make for comfortable conditions. This is obvious when you see the number of people who must take sweaters into grocery stores, fast food restaurants, and other stores and shops during the summer. Also, notice that the customers who didn't bring sweaters are almost always shivering. Yet, it is comfortable to some of those working there since it is likely that there is a low temperature/high humidity relationship.

Also, no matter how much heat is introduced into a space, it is very difficult for the occupant to feel comfortable sitting next to a very cold surface, such as a window. The combination of the body heat loss on the side of the body nearest the window and the cold air coming through any cracks around the window makes a difficult situation. In one case, the temperature in a person's office was turned up to 76°F and he still felt uncomfortable. By rearranging the office so that there was an interior wall at his back instead of a window, he felt comfortable at 72°F (22.2°C).

6-8   AIR CONDITIONING

Whenever the term *air conditioning* is used, almost everyone automatically thinks of a cool building on a hot day. However, in a technical sense, air conditioning means exactly what it implies—conditioning the air inside a space; this means heating, cooling, humidity con-

Functional Area	Air Changes per Hour	Cfm/ Person	Functional Area	Air Changes per Hour	Cfm/ Person
Anesthesia, hospital	8-12	—	Kitchens	10-30	—
Animal room	12-16	—	Laundries	10-60	—
Auditorium	10-20	10	Libraries	15-25	10
Autopsy, hospital	8-12	10	Locker room	2-15	—
Bakery	20-60	—	Machine shop	8-12	—
Bowling alley	15-30	30	Mechanical equipment	8-12	—
Churches	15-25	5	Media room, hospital	6-10	—
Cystoscopy, hospital	8-10	20	Nursery	10-15	—
Classroom	10-30	40	Offices	6-20	10
Conference	25-35	—	Operating room, hospital	10-15	—
Corridors	3-10	—	Ozalid room	8-12	—
Delivery room, hospital	8-12	—	Paint finishing	18-22	—
Dairies	5-15	—	Radiology	6-10	—
Dishwashing	30-60	—	Restaurant (dining room)	6-20	10
Drycleaning	20-40	—	Retail stores	18-22	10
Foundries	5-20	—	Residences	5-20	—
Gymnasiums	—	1½ sq ft	Telephone equipment	6-10	—
Garages	6-30	—	Traffic and flight cont.	18-22	10
Hydrotherapy, hospital	6-10	—	Toilets	8-20	—
Isolation ward	6-10	—	Transmitter, receiver		
Janitor and cleaning	8-12	—	and electronic	10° rise	—
			Welding	18-22	—

NOTE—Individual design conditions may cause variation in these values of considerable magnitude.

FIGURE 6-8   Ventilation design criteria

trol, ventilation, filtering, and any other conditioning that may be required. While realizing that manufacturers will continue to sell air coolers as air conditioners and that when clients say they want (or don't want) air conditioning, they mean cooling, the technician and the engineer must be aware of the real meaning of the term.

Modern building design has increased the need for air conditioning since there is an increased tendency to build glass buildings which don't have a window that opens (no operable sash). Even in northern states, which have only short periods of warm weather, total air conditioning is often required since there are no windows to open for fresh air and ventilation.

6-9 VENTILATION

Many times it is necessary to ventilate (introduce outside air into a space) in order for the occupants to be comfortable. This is particularly true in any spaces where smoking is permitted and where dust or fumes are present.

Ventilation requirements, as set by ASHRAE, are shown in Fig. 6-8, and many times the codes in force also have ventilation requirements. Rooms with exterior exposures, windows, and exterior doors and with low occupancy may get all the ventilation they need through air leakage around the windows and doors. Rooms without exterior exposures commonly need ventilation so they will not become stale with cigarette smoke and other odors.

Many buildings with forced air systems leave the fan on even when the heating and cooling system is not operating. This provides ventilation throughout the building and reduces the possibility of stagnant air layers in a room.

QUESTIONS

6-1. Sketch and label the three natural processes by which body heat is transferred.

6-2. Briefly describe the three natural processes by which body heat is transferred and how they affect comfort.

6-3. In what ways does the surrounding air temperature affect the amount of body heat given off?

6-4. What are the three space conditions which affect the comfort of the occupant of the space?

6-5. What effect does humidity have on the comfort of a body?

6-6. What is meant by *effective temperature* in relation to comfort?

CHAPTER 7

Heating and Air-Conditioning Systems

7-1 TYPES OF SYSTEMS

The selection of a heating system and the possible inclusion of a cooling system will depend on local climate conditions, degree of comfort desired, client's budget, and fuel costs. An increasing number of clients want a system which incorporates total year-round air conditioning.

There are three basic methods of delivering heat to a space: forced air, hot water, and radiant electric. Of these three, only one, forced air, is used for central air conditioning in residences. Large projects often use chilled water systems to provide air conditioning.

In *forced air systems,* the air is heated or cooled in a central unit and then delivered to the room through supply ducts. Air is returned to the central unit for treatment (to heat, cool, add humidity, or purify) and then recirculated through the rooms. (Chapter 9 contains a complete discussion and description of forced air systems for heating and cooling.)

Hot water heating systems heat the water in a central unit and pass it through pipes to a heating device in the room. As the water goes through the pipes and devices, it cools, and then it goes back through the central unit to be reheated. Cooling with water follows the same principle. Hot water heating is discussed in Chapter 10.

Radiant electric heat delivers the heat by electricity running through a cable, and the resistance produced gives off the heat. The radiant heating devices are actually in the room and may be ceiling, floor, or baseboard units. There is a complete discussion on radiant electric systems in Chapter 11.

In addition, *infrared heaters* use a lamp (bulb) which transfers heat by radiation to any people or objects which its heat rays come in contact with. It is effectively used when it is necessary to warm people, yet the surrounding air need not be heated or, perhaps, cannot be effectively heated. They are commonly used in bathrooms where for a short period of time (such as when a person steps out of a hot shower in the winter) extra heat is needed for the person to feel comfortable. Other locations include covered walkways and entries to commercial and industrial buildings.

Water may be delivered either hot or chilled, and chilled water systems are used for air conditioning in many larger projects. *Chilled water systems* cool the water at a central point, in a condenser, and then distribute chilled water throughout the building—either to convectors in the space or through pipes embedded in the floor or ceiling.

7-2 HEATING SYSTEM COMBINATIONS

It is not unusual for industrial and commercial buildings to use different types of systems to heat and cool different areas of a building. The systems may vary in use of hot water heat, with finned tube units being used in offices and convectors units being used in storage or warehouse areas. Similarly, completely different systems may be used, with radiant hot water heat in the floor of an office area and electric unit heaters in the warehouse or storage areas. Any combination may be used, and the selection may vary, depending on the type of heat required, the type of fuels available, and, to a great extent, how much heat is required, how often it is required, and what method of supplying the heat will provide the best results in the designer's opinion.

Residential designs may use combinations of systems such as hot water for heat and forced air for cooling, hot water or forced air heat with electric supplements (particularly in bathrooms or kitchens), and finned tube units with supplemental hot water unit heaters in areas such as a kitchen or basement.

As mentioned previously, chilled water may be used to provide cooling in a building by running it through the same pipes used for hot water heating. This means that once the system is changed over from one function to the other—say, from heating to cooling as summer approaches—it is not easily reversed if cold weather comes for several days. This type of system has little flexibility in that it provides either heating or cooling and cannot provide both at one time.

However, different uses, activities, type of exterior wall construction, and location in the building (especially larger buildings) may make it desirable to have heating in some spaces while cooling is required in others. This may be accomplished by putting in separate pipes for hot and chilled water. In this manner, areas which require cooling can get chilled water while those requiring heat can get hot water. Of course, there is the extra cost of piping and controls to regulate the flow of the water to the desired location.

The actual design of any particular system will involve the size of the building, amount of insulation, doors, windows, climate in the area, and fuel to be used.

7-3 FUELS

The most commonly used sources of building heat are the sun, electricity, gas, oil, and coal. Use of the sun as an energy source is discussed in Chapter 14. The decision on which of the other four fuels to use is based on availability and cost of operation. In this section, we will review the advantages and disadvantages of each fuel and the method used to determine the cost of operation.

Electricity

Electricity is used as a fuel for a variety of heating systems, including baseboard radiant heat and electric coils in the ceiling and/or walls (refer to Chapter 11). In addition, there are electric furnaces for forced hot air systems, and electricity is used with heat pumps both to operate the system and to provide supplemental electric resistance heat for the system. Electricity has as its advantage its simplicity. It requires no chimney to remove toxic gases, and when baseboard strips and ceilings and wall coils are used, the system has individual room controls, providing a high degree of flexibility and comfort. Electric systems cost significantly less to install than other systems (including electric furnaces for forced hot air), and all electric systems (except heat pumps) require much less upkeep and maintenance than those using the other fuels.

The primary disadvantage of electric heat is its yearly cost of operation when compared with those of other fuels. Electricity rates vary tremendously throughout the country, and the rates are continuing to climb. It is necessary to determine the rates in the geographical area of construction by doing a cost analysis. Electric heat is quite popular, and generally most economical, in the southern regions since winter is shorter and heat bills are generally lower. It is also used extensively in apartments, offices, and similar buildings where the developer is primarily interested in building the units as inexpensively as possible and where the cost of heating is usually paid by the person renting the space. Its minimal need for maintenance and repairs is also a factor in such construction. Builders of some homes, especially those which they may want to be able to sell at the lowest price, may use electric baseboard heat.

Gas

Gas, also a popular fuel for heating, is used to heat the water in hot water systems and the air in forced air systems. Gas fuels available include natural gas, which is piped to the residence or building, and propane gas, which is delivered in pressurized cylinders in trucks and tanks and stored in tanks at or near the building. Since natural gas is simply available as needed, with no storage or individual delivery required, it is considered simpler to use. However, it is not available in many areas, and the more surburban the area, the less likely it is that natural gas will be available. So the designer must first determine if natural gas is available. Secondly, periodically there is a shortage of natural gas. While this shortage may be alleviated in the future, many areas do not permit any new natural gas customers. In periods of shortages, residential customers can be reasonably assured that they will have sufficient natural gas for use, but industrial and commercial customers cannot be so assured. Since 1975, hundreds of businesses have been faced with the option of converting to another fuel or closing, and many did close for the winter months. While the reasons for such shortages may be debated as to whether they are real or contrived—caused by government, industry, or both—the designer is concerned with one thing: Is it available, and will it continue to be available, or not?

The primary advantages of natural gas have been its relatively low cost and its simple and clean burning which reduces the maintenance required on the heating unit. As with all of the fuels, as costs go up in the future, it is difficult to say which will be the most economical.

Propane gas is equally as clean as natural gas, and its availability is not limited to areas where supply pipes have been installed. To date, the cost of propane gas has generally been higher than that of natural gas.

Oil

Oil is one of the most popular fuels, and it is used extensively in the Northeast. The primary reason for its use has been its availability and historic low cost. It does require delivery by trucks to storage tanks located in or near the building, and the heating unit will generally require more maintenance than a gas heating unit. The selection of oil as a heating fuel has diminished somewhat since the oil embargo in the early 1970s. In addition, the cost of this fuel has risen dramatically since that time. Costs of other fuels have also risen, but it is the fear of not having oil if another embargo is imposed that is one of the most important concerns.

Oil is available in various weights (Fig. 7-1) with various heating values and at different costs. Generally, the lower the number, the more refined it is and the higher the cost. Number 2 oil is commonly used in residences, while numbers 4, 5, and 6 are commonly used in commercial and industrial projects.

Commercial standard number	Weight (lb./gal.)	Btu per gallon	Btu average	kJ/L
1	6.675–7.076	132,900–138,800	136,000	16,340
2	6.870–7.481	135,800–144,300	140,000	16,820
4	7.529–8.212	145,000–153,000	149,000	17,900
5	7.627–8.328	146,200–154,600	150,000	18,020
6	7.909–8.448	149,700–156,000	154,000	18,500

FIGURE 7-1   Fuel oil characteristics

Coal

Coal is rarely used for residential heating in new construction, and its use in industrial and commercial construction fluctuates. Its primary advantage is that it is available, and there are ample supplies so that a shortage seems unlikely at this time. Generally, its cost is competitive with those of other fuels. Its disadvantages lie in the amount of space required for storage and the fact that it does not burn as completely as oil or gas, thus producing more pollution. Government regulations for clean air have also limited the use of some more polluting coals for industrial purposes.

Coal's decreased use for heating residences is due to the handling required in delivery and the inconvenience of having a coal bin as part of a basement. Also, originally the coal had to be shoveled into the furnace by the occupant of the house, which is sufficient reason to change to oil, gas, or electricity. Now, coal heating units are fed by automatic stokers which require no hand shoveling.

Since coal is used so little as a residential and small commercial heating fuel, it will not be considered further.

7-4   COOLING PRINCIPLES

While the principle of providing heat to a source (water or air), which is then used to heat the space, is easily understood, the principles of providing cooling should be discussed. The fundamental principles on which the cooling process is based are derived from physics:

1. As a gas is compressed, it will liquefy at a given point, and as it *liquefies, is* will *release* a large amount of latent *heat* from within the gas/liquid.

2. As the pressure on the liquid is lowered, it vaporizes back to a gas, and as it boils through the vaporizing process, it *absorbs* a large amount of latent heat into the liquid/gas.

The refrigerant medium in the cooling system is cycled through three components (Fig. 7-2):

1. A *compressor* which will compress the refrigerant, causing it to liquefy.

2. A heat transfer surface which will distribute the *heat released* to a surrounding medium such as water or air. This heat transfer surface is called a *condenser*.

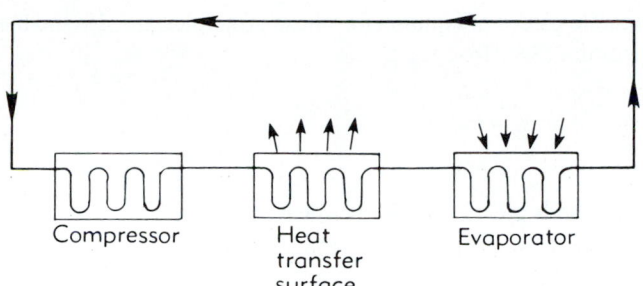

FIGURE 7-2   Refrigerant cycle

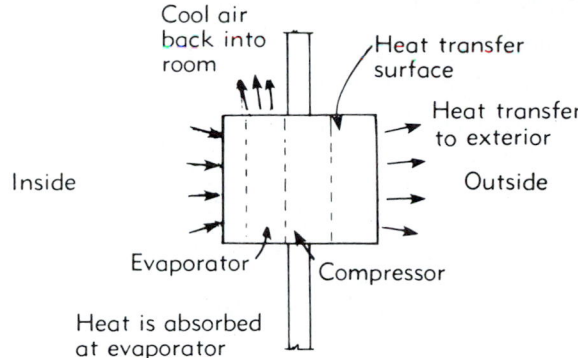

FIGURE 7-3   Single unit cooling

3. A second heat transfer surface which will *extract heat* from the surrounding medium, such as water or air, as it is *absorbed* into the refrigerant. This heat transfer surface is called an *evaporator*.

The refrigerant is run continuously through the cycle while the system is in operation:

1. The refrigerant is compressed to a liquid in the compressor, generally located in or near the condenser.

2. The liquid passes through the condenser (heat transfer surface), which allows the latent heat to be released. The condenser is often located on the exterior of the building. For most residences and small commercial projects, it is located on the ground; however, it may be located on the roof, often the case on larger projects. The heat is released through the condenser to a surrounding medium. This surrounding medium is the outside air for most installations, but water, such as a pond, can be used. In the typical installation, the condenser unit has a fan which pushes the air past the refrigerant to take as much heat away from it as possible.

3. The refrigerant then passes out of the condenser to the second heat transfer surface (the evaporator), which will extract heat from the surrounding medium. So, as the liquid vaporizes to a gas, it draws heat out of the surrounding medium as it passes through the evaporator. The surrounding medium may be air or water. In a forced air system, air would be the medium, and this drawing of heat from the air and into the refrigerant causes the air to cool. As the air is forced back through the system, it is cool air. When water is used as the medium, the heat is drawn from the water, making it cool or chilled; the water is then circulated through the system to cool the space.

These basic principles apply for all types of cooling systems. Some equipment, such as a room air conditioner, may be designed to combine all the components in one unit (Fig.

7-3), or they may be separate pieces of equipment. These principles are also the basic principles of heat pump operation (Sec. 14–8).

QUESTIONS

7-1. What three basic methods are used to deliver heat to a space?

7-2. Why might combinations of systems be used in a building?

7-3. What type of system is most commonly used for residential cooling?

7-4. When might a chilled water system be used?

7-5. Describe the advantages and disadvantages of electricity as a fuel.

7-6. Why is electric heat often used in projects such as apartments?

7-7. List the fuels most commonly used for heating in a residence.

7-8. What are the two fundamental principles on which the cooling process is based?

7-9. Describe the three components used in the refrigeration process.

7-10. What types of surrounding mediums are generally used to take the heat from the refrigerant?

CHAPTER 8

Heat Loss

8-1   HEAT LOSS

The amount of heat lost from the building to the surrounding air will determine the size of the heating plant, the size of the heat convectors (radiators, ducts, etc.), and the heating cost per year. The amount of heat loss is measured in BTU (British Thermal Units); a BTU is defined as *the amount of heat required to raise 1 pound of water 1 degree Fahrenheit.* The SI equivalent is the joule (J). A joule is 1/4.184 the amount of heat required to raise a gram of water 1 degree Celsius (C). The kilojoule is also commonly used, and it equals 1000 joules and 0.9478 Btu. Watts (W) is the power required to produce energy at the rate of one joule per second. Watts is commonly used to specify heat loss and gain, as well as sizes for furnaces, air conditioning and other heat producers including lightbulbs (1W = 3.412 Btuh). The temperature used is Celsius (°C) but several formulas refer to Kelvin (K) when they indicate a change in temperature. This is not a problem since a change of one degree Kelvin equals one degree Celsius. Heat loss will depend on the type of construction assembly, the types of doors and windows used, and the climate in the geographical area where the building is being constructed.

8-2   FACTORS IN CALCULATING HEAT LOSS

Heat is lost from a space by transmission, infiltration, and ventilation.

1. Before the heat loss for a building can be calculated, the places at which the losses occur must be identified. Since the heat inside will flow to colder areas, the heat will pass through walls, floors, ceilings, doors, and windows.

2. The heat loss is calculated separately for each room. Therefore, it is necessary to determine each type of assembly of materials used.

3. For each room, the area of each type of material assembly which has one side exposed to the inside of the room and the other side exposed to a lower temperature space—such as the exterior, attic, basement, or crawl space or a colder room such as a garage or enclosed porch which is not heated—must be calculated. This heat loss is referred to as *heat transmission.* Calculations for heat transmission loss are shown in this chapter.

4. In every building, no matter how well constructed, there is a certain amount of cold air which leaks into the building, referred to as *infiltration heat loss,* and an equal amount of hot air which leaks out. Most commonly, infiltration will occur around doors and windows. The tight construction of the building will save the building owner a considerable amount of money over the life of the building.

5. *Ventilation* is the introduction of fresh air into the building, or parts of the building, at a controlled rate. Many times the air used to ventilate the building is heated before it is introduced into the building, and this must be considered in the design of the system.

8-3   RESISTANCE RATINGS

Since heat passes through different materials at different rates, each material is considered to have a certain resistance to the flow of heat through it. In general, the heavier and denser the material, the less resistance it has to the flow of heat through it and the lower its resistance (R) value is; for lighter, less dense materials, the higher the R value, the better its insulating value.

EXAMPLE   A 3″ clay tile has an R value of about 0.80 while a 3″ thickness of mineral fiber batt insulation has an R value of about 11.0. Increasing the thickness of the material also affects the insulating value of the fiber batt insulation material in proportion to its original value. For example, doubling the 3″ clay tile to 6″ will double the R value from 0.80 to 1.60 (unless the design varies). Doubling the mineral fiber thickness from 3″ to 6″ will double its R value from 11 to 22. To determine the heat flow of an assembly of materials (U value), it is first necessary to find the R value for each material and total them.

 A 76-mm-thick clay tile has an R value of about 0.14, while a 75-mm thickness of mineral fiber batt insulation has an R value of about 1.94. Increasing the thickness of the material also affects the insulation value of the material in proportion to its original value. For example, doubling the 76 mm thickness of the clay tile to 152 mm doubles the R value from 0.14 to 0.28. Doubling the mineral fiber thickness from 75 mm to 150 mm doubles its R value from 1.94 to 3.88. To determine the heat flow of an assembly of materials, it is necessary first to find the R value of each material and then to total the R values of all component materials.

The resistance listings for most materials can be found in Fig. 8-45. When materials not listed here are encountered, the heat flow characteristics will have to be obtained from the manufacturer.

There are four values used to measure the heat resistance of a material; they are the $U, R, C,$ and k values. Each of the values is related, and the designer must understand the relationships:

1. The k value of the material used in the construction measures heat flow conductance in terms of how effective it is *per inch of thickness* or *per meter of thickness* (1000 mm).

2. The C (conductance) value of the material measures the material's ability to conduct heat in terms of how effective it is for the *thickness being used*. The more effective the material, the lower the C value ($C = k \div t$).

3. The R (resistance) value of the material measures heat flow resistance in terms of how effective it is for the *thickness being used*. (This term is inversely related to the C value.) The more effective the material, the higher the R value ($R = 1 \div C$).

4. The U value of the entire assembly of materials measures heat flow resistance in terms of how effective it is for the type and thickness of the materials used in the total assembly ($U = 1 \div R$).

EXAMPLE   To show the relationship of the values, we will use a 2-in.-thick piece of expanded polystyrene, density 1.0 lb/ft^3 foam insulation (Fig. 8-1). Polystyrene foam has a k value of 0.26 per inch of thickness. Since it is 2 in. thick, its insulating value is given as a C value

$$k = 0.26 \text{ per inch}$$
$$C = \frac{k}{t} = \frac{0.26}{2} = 0.13$$
$$R = \frac{1}{c} = \frac{1}{0.13} = 7.7$$
$$R = \frac{t}{k} = \frac{2}{0.26} = 7.7$$

SI UNITS
$$k = 0.037 \text{ (per 1000 mm)}$$
$$C = \frac{k}{t} = \frac{0.037 (1000)}{50} = 0.74$$
$$R = \frac{1}{C} = \frac{1}{0.74} = 1.35$$
$$R = \frac{t}{k} = \frac{50}{0.037 (1000)} = 1.35$$

FIGURE 8-1   Heat flow terms

as 0.13 (or one-half the *k* value). These values are directly proportional to each other, based on the thickness of material used. The *lower* these values, the better the material is as a heat insulator. The *R* value is equal to 1 divided by the *C* value. So, for the 2-in.-thick polystyrene foam, $R = 1/C = 1 \div 0.13 = 7.7$. This indicates the resistance value of that particular material.

It is quite unlikely that a wall would be built only of 2-in. polystyrene foam; this insulating material would be used with other materials, such as metal facings, masonry, and concrete. For this example, assume that the wall is made of 6-in. poured concrete, limestone aggregate (140 lb/ft³), 2-in. polystyrene foam (1.0 lb/ft³) and ⅜-in. gypsum board (Fig. 8-2). From Fig. 8-45, the resistance values listed are:

⅜-in. gypsum board	0.32
6-in. concrete, limestone (0.09 per in.)	0.54
2-in. polystyrene foam (3.85 per in.)	7.70
Outside air film (see Sec. 8-4)	0.17
*Inside air film (see Sec. 8-4)	0.68
Total resistance *R*	9.41

SI To show the relationship in values, we will use a 50 mm thick piece of expanded polystrene, density 16 kg/m2, foam insulation (Fig 8-1). Polystrene foam has a *k* value of 0.037 per meter (1000 mm) of thickness. Since it is 50 mm thick, its insulating value calculated as a *C* value is 0.74 (or *k* times 1000 divided by the thickness, Fig 8-1). These values are proportional to each other, based on the thickness of the material used. The lower these values, the better the material as a heat insulator. The *R* value (Fig. 8-1)

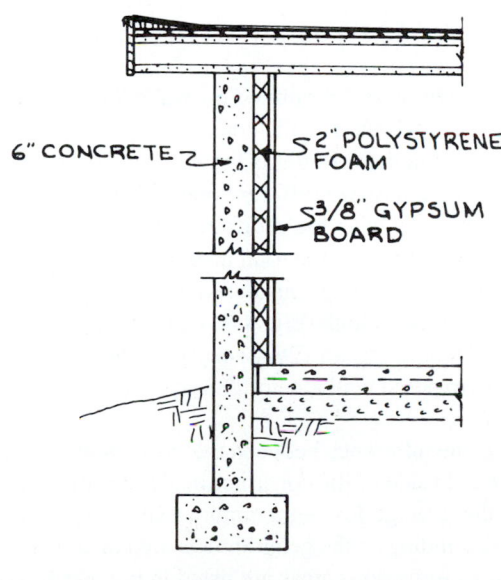

6" CONCRETE

2" POLYSTYRENE FOAM

3/8" GYPSUM BOARD

SECTION   **FIGURE 8-2**   Wall section

is equal to 1 divided by the *C* value. So, for the 50 mm thick expanded polystyrene, $R = 1/C = 1 \div 0.74 = 1.35$.

It is unlikely that a wall would be built out of only 50 mm expanded polystrene; this insulating material would be used with other materials, such as metal facings, masonry, and concrete. For this example, assume the wall will be made of 150 mm thick poured concrete (limestone aggregate, .1920 kg/m^3), 50 mm expanded polystyrene, and 9.5 mm gypsum board (Fig. 8-2). From Fig. 8-45, the resistance figures are:

9.5 mm gypsum board	0.06
150 mm concrete	0.09
50 mm polystyrene	1.35
*Inside air film (Sec. 8.4)	0.02
Outside air film (Sec. 8.4)	0.12
Total resistance *R*	1.64

*Nonreflective, vertical surface, horizontal heat flow.

Heat flow from the interior of the building to the cooler exterior is determined by multiplying the heat flow characteristics of a given assembly of materials (referred to as the *U value*) times the area (*A*) of the material involved times the difference between the inside temperature desired and the outside design temperature on the other side of the construction (referred to as ΔT and given in °F). In equation form:

$$U \times A \times \Delta T = \text{W (Watts per square meter per degree Kelvin)}$$

SI $$U \times A \times T = W \times m^2 \times K \text{ (Watts per square meter per degree Kelvin)}$$

In the above equation, the *U* value is the most important item since it can vary so much, depending on the material assembly used. With any given *A* and ΔT, the number of Btuh required will vary in direct relationship with the *U* value for the assembly of materials used. The *U* value is determined by adding the *R* values of each material plus the resistance to heat flow caused by the flow of air inside and outside the building and then dividing 1 by the total *R* ($U = 1 \div R = 1/R$).

$$U = 1/R = 1 \div 9.41 = 0.106 \text{ (use 0.11)}$$

SI $$U = 1/R = 1 \div 1.64 = 0.6097 \text{ (use 0.61)}$$

The *U* value must be determined for walls, floors, and ceilings and for each material or assembly of materials used in the construction.

The area used in the formula is the *net exposed area* of the material assembly being considered. If a room has two different material assemblies (Fig. 8-3), each must be considered separately since each has a different *U* value. Likewise, the area of windows and doors must be deducted from the total area to obtain the net exposed wall area of the particular material assembly being considered. Assuming that the room being designed has two exposed walls 8 ft (2.4 m) high (Fig. 8-4) with a total of 36 lineal ft (11 m), this would give a gross square footage of 288 (26.4 sq m). Deduct the two windows 3 ft (.9 m) × 4 ft (1.2 m) for a net exposed wall area of 288 sq ft − 24 sq ft = 264 sq ft (26.4 sq m − 2.2 sq m = 24.2 sq m).

The ΔT is the difference between the inside temperature desired and the design temperature on the cold side of the construction. If the cold side of the construction is the outside air, then the average low temperature expected during the year is used. This varies considerably, depending on the geographical area of construction. Some typical outside design temperatures for various areas are listed in Fig. 8-53. Assuming an indoor design temperature of 70°F (21°C), if the building is being built in Rutland, Vermont, with an outside

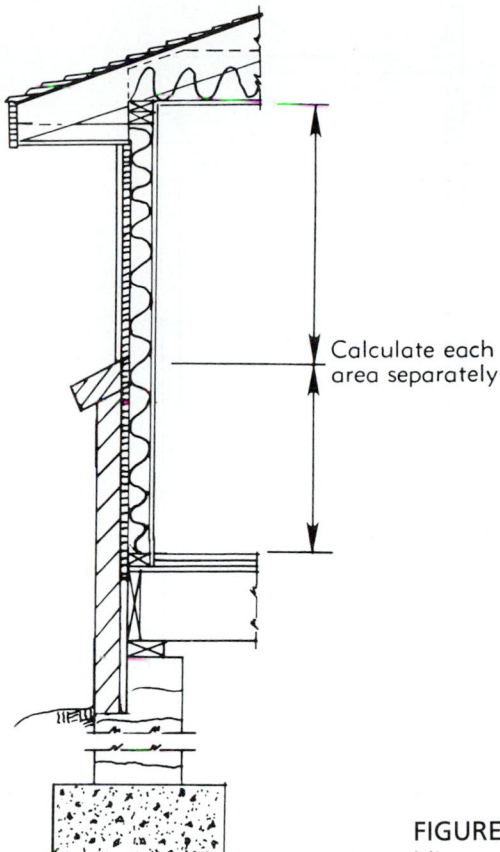

Calculate each area separately

FIGURE 8-3 Different wall assemblies

design temperature of −8°F (−22°C), the ΔT will be 78°F (43°C) (Fig. 8-5). If the same building is constructed in Raleigh, North Carolina, the outside design temperature is 20°F (−7°C) and the ΔT will be 50°F (28°C). Assuming that similar buildings are being built in Raleigh and Rutland, the heat loss for the walls in that one room would be:

Raleigh $U \times A \times \Delta T = 0.11 \times 264 \times 50 = 1,452$ Btuh

Rutland $U \times A \times \Delta T = 0.11 \times 264 \times 78 = 2,265$ Btuh

SI Raleigh $U \times A \times \Delta T = 0.62 \times 24.2 \times 28 = 420$ W

■ Rutland $U \times A \times \Delta T = 0.62 \times 24.2 \times 43 = 645$ W

It is important that the designer become proficient in calculating U values for various assemblies of materials. The flow of heat through any barrier—in this case, the construction of the walls—is resisted by the materials used for the construction, the inside and outside air films, and any air spaces left. It is, therefore, important that the designer carefully review the drawings, especially wall sections and details, to determine what materials will be used on the project. In addition, the designer should check to be certain exactly what the materials being specified actually *are*. For example, the wall section (Fig. 8-6) may show 2 in. (50 mm) of rigid insulation. The designer must also know exactly what type of rigid insulation is going to be used and must be certain that the material the calculations were based on is actually specified on the project and, even further, that no material substitution is allowed. This is because if the rigid insulation in Fig. 8-6 is polyurethane (unfaced), it is about 40% more effective than polystyrene foam insulation.

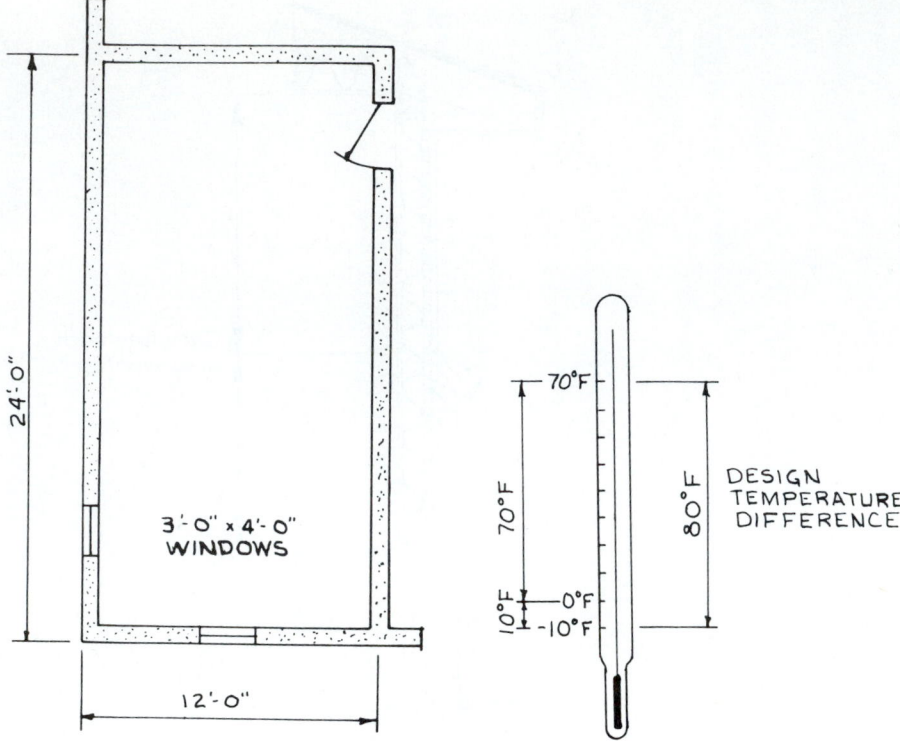

FIGURE 8-4   Room plan   FIGURE 8-5   Calculating ΔT

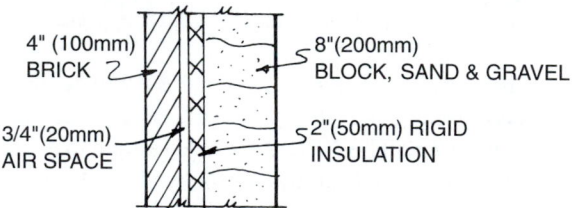

FIGURE 8-6   Wall section

EXAMPLE   *Problem:* Determine the resistance values (Fig. 8-45) for each of the materials in the wall section shown in Fig. 8-7. In cases where a range of *R* values are given, such as for brick, check with the manufacturer. If the manufacturer is unknown, use the lowest value.

First, list all of the materials of construction; then list their resistance values next to them. When necessary, convert the *C* and *k* values to *R* values.

$$R = 1/C \text{ and } R = t/k$$
$$(t = \text{thickness of the material, in inches})$$

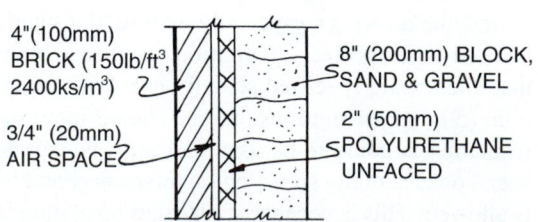

FIGURE 8-7   Wall section

SI UNITS:			
4″ Brick	0.40	100 mm Brick	0.07
8″ Block, Gravel	0.97	200 mm Block	0.17
2″ Polyurethane	11.12	50 mm Polyurethane	1.93

FIGURE 8-8   Material *R* values

The materials and *R* values for the wall section shown in Fig. 8-7 are indicated in Fig. 8-8. *Note:* When a range of *R* values are given, use the lowest.

8-4   AIR FILMS

Both inside a room and outside the building, there is a movement of air over the surfaces, creating a film of air which causes the air and wall to exchange convective heat. This air film helps resist the flow of heat through the construction. The air film's resistance value depends first on whether it is an *inside* air film (inside the room) or an *outside* air film (on the outside of the wall).

The resistance values of inside air films are shown in Fig. 8-46. These values are based on still air (no wind), and proper selection from the table is based on:

1. Surface position or orientation. (Is it a vertical wall, horizontal ceiling, or floor?)

2. Direction of heat flow—for a wall, horizontal; for a ceiling, upward; for a floor over a basement, downward. (In what direction is the heat flowing in the room in which the heat loss is being calculated?)

3. Surface emissivity. (Is the surface reflective or not? Most surfaces used in buildings, such as gypsum board painted with flat paint and concrete block, are not reflective. Use the nonreflective or reflective value, depending on the wall finish to be used in the building.) (Fig. 8-48)

EXAMPLE   *Given:* For the wall section in Fig. 8-7, the interior wall is concrete block which will be nonreflective (from Fig. 8-48, an emissivity value of 0.90).
Problem: Using Fig. 8-46, find the resistance value of the inside air film.

The resistance value of this inside air film—which is for a vertical, nonreflective wall, with horizontal heat flow—is taken from Fig. 8-46 as *R* = 0.68 (R = 0.12 SI). This value is included with the tabulation of material resistance values from Fig. 8-8 and combined here (Fig. 8-9):

The resistance values of exterior air films are shown in Fig. 8-46. Proper selection of the *R* value is based on:

1. Winter: 15 mph air velocity; use *R* = 0.17.

2. Summer: 7½ mph air velocity; use *R* = 0.25.

SI

1. Winter: 24 km/h air velocity; use *R* = 0.030

2. Summer: 12 km/h air velocity; use *R* = 0.044

This value is now included with the tabulation of resistance values for materials and inside air film (Fig. 8-9) and combined here (Fig. 8-10):

		SI UNITS:	
4″ Brick	0.40	100 mm Brick	0.07
8″ Block, Gravel	0.97	200 mm Block	0.17
2″ Polyurethane	11.12	50 mm Polyurethane	1.93
→ Inside Air Film	0.68	Inside Air Film	0.12

FIGURE 8-9   Inside air film resistance

		SI UNITS:	
4″ Brick	0.40	100 mm Brick	0.07
8″ Block, Gravel	0.97	200 mm Block	0.17
2″ Polyurethane	11.12	50 mm Polyurethane	1.93
Inside Air Film	0.68	Inside Air Film	0.12
→ Outside Air Film	0.17	Outside Air Film	0.03

FIGURE 8-10   Outside air film resistance

8-5   AIR SPACES

Many times an assembly of material will have an air space left between the materials. For example, there is usually an air space between the brick veneer on a building and the material behind it. A review of the wall section in Fig. 8-7 shows that it calls for a ¾-in. air space. Only unvented air spaces can be claimed as having an insulating value. Be certain to review the drawing details and determine whether or not the air space is vented. These air spaces also resist the flow of heat and have an R or a C value.

The resistance values of air spaces are shown in Fig. 8-47. Proper selection of an R value from the table is based on:

1. Position of air space (horizontal, 45° slope, vertical—for this wall, it is vertical).
2. Direction of heat flow (up, horizontal, down—for this wall, the heat flow is horizontal).
3. Air space thickness (thicknesses other than those listed may be interpolated from the values given: for thickness greater than 3.5 in., use the 3.5-in. value).
4. Value of E (emissivity) (Are the surfaces on each side of the wall space reflective or not? For nonreflective materials, use $E = 0.82$; for reflective materials, such as aluminum foil, use $E = 0.05$. In this case, there are only nonreflective surfaces.)
5. Air space temperatures. You will need to estimate both the mean teamperature and the temperature difference of the air space. You may interpolate for other values.

EXAMPLE   For the ¾-in. air space in the wall section (Fig. 8-7), with a vertical air space, horizontal flow of heat, nonreflective surfaces, E at 0.82, and 50°F mean temperature, 30°F temperature difference, the resistance value R is 0.94. This value is included in the tabulation of material resistance values from Fig. 8-10 and combined (Fig. 8-11).

 For the 20 mm air space in the wall section (Fig. 8-7), with a vertical air space, horizontal flow of heat, nonreflective surfaces, E at 0.82, 10°C mean temperature, 16.7 temperature difference, the resistance value is 0.17. This value is included in the tabulation of material resistance values from Fig. 8-10 and combined (Fig. 8-11).

		SI UNITS:	
4″ Brick	0.40	100 mm Brick	0.07
8″ Block, Gravel	0.97	200 mm Block	0.17
2″ Polyurethane	11.12	50 mm Polyurethane	1.93
Inside Air Film	0.68	Inside Air Film	0.12
Outside Air Film	0.17	Outside Air Film	0.03
→ ¾″ Air Space	0.94	20 mm Air Space	0.17

FIGURE 8-11   Air space resistance

8-6   TOTAL RESISTANCE VALUES

With all of the materials, air films, and air spaces listed with their appropriate R values, the next step is to total the R values of the wall and convert this total to the equivalent U value.

The total resistance value of the wall with a U value is calculated and shown in Fig. 8-12.

		SI UNITS:	
4″ Brick	0.40	100 mm Brick	0.07
8″ Block, Gravel	0.97	200 mm Block	0.17
2″ Polyurethane	11.12	50 mm Polyurethane	1.93
Inside Air Film	0.68	Inside Air Film	0.12
Outside Air Film	0.17	Outside Air Film	0.03
¾″ Air Space	0.94	20 mm Air Space	0.17
Total R	14.28	Total R	2.49

$$U = \frac{1}{R} = \frac{1}{14.28} = 0.07 \qquad\qquad U = \frac{1}{R} = \frac{1}{2.49} = 0.4016 \text{ (use 0.40)}$$

FIGURE 8-12   Calculate total R and U values

8-7   CALCULATING R AND U VALUES

The method described in Secs. 8-3 through 8-6 of combining the R values for each separate material, for the inside and outside air films, and for any air spaces gives the total R value of the assembly. It is this total R of the assembly which is used to calculate the U value for the particular assembly and is used in the basic heat loss formula:

$$U \times A \times \Delta T = \text{Btuh}$$

SI
$$U \times A \times \Delta T = W$$

To become proficient in the use of the tables and to be certain that the basic relationships of the various R values are understood, examples for floors and ceilings are given next. Keep in mind that:

1. The higher the R value, the more effective the material at resisting the flow of heat.

2. The higher the R value, the lower the heat loss. The lower the heat loss, the less expensive the entire installation in terms of the yearly fuel bills.

EXAMPLE   *Problem:* Calculate the R and U values for the wall assembly shown in Fig. 8-13. The tabulations are shown in Fig. 8-14.

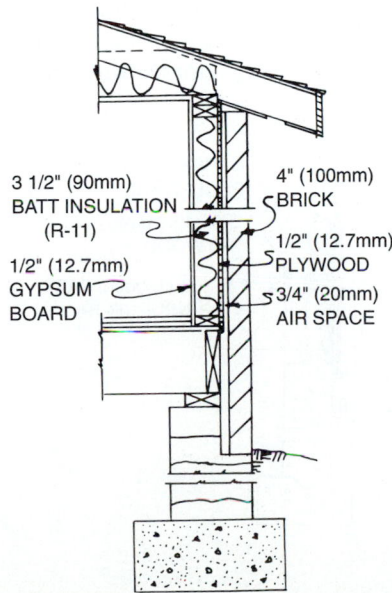

3 1/2" (90mm)
BATT INSULATION
(R-11)

1/2" (12.7mm)
GYPSUM
BOARD

4" (100mm)
BRICK

1/2" (12.7mm)
PLYWOOD

3/4" (20mm)
AIR SPACE

FIGURE 8-13   Wall assembly

½″ Plywood	0.62		SI UNITS:	
4″ Brick	0.40		12.7 mm Plywood	0.11
3½″ Batt. Insul.	11.00		100 mm Brick	0.07
½″ Gypsum Board	0.45		90 mm Batt. Insul.	1.94
Inside Air Film	0.68		12.7 mm Gypsum Board	0.08
Outside Air Film	0.17		Inside Air Film	0.12
¾″ Air Space	0.94		Outside Air Film	0.03
Total *R*	14.26		20 mm Air Space	0.17
			Total R	2.52

$$U = \frac{1}{R} = \frac{1}{14.26} = 0.07$$

$$U = \frac{1}{R} = \frac{1}{2.52} = 0.3968 \text{ (use 0.40)}$$

FIGURE 8-14   *R* and *U* calculations

EXAMPLE   *Problem:* Calculate the *R* and *U* values for the roof assembly shown in Fig. 8-15. The tabulations are also shown in Fig. 8-16. Note that the *R* values for the inside and outside air films are different from those used for the wall assembly and that there is no air space.

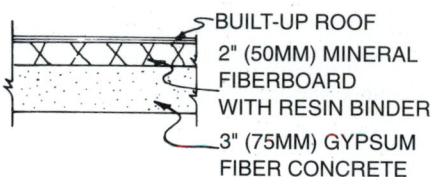

BUILT-UP ROOF
2″ (50MM) MINERAL FIBERBOARD WITH RESIN BINDER
3″ (75MM) GYPSUM FIBER CONCRETE

FIGURE 8-15   Ceiling assemblies

3″ Gyp. Conc.	1.8		SI UNITS:	
2″ Fiberboard Insul.	6.9		75 mm Gyp. Conc.	0.31
Built-up Roof	0.33		50 mm Fiberboard Insul.	1.79
Outside Air Film	0.17		Built-up Roof	0.06
Inside Air Film	0.61		Outside Air Film	0.03
Total *R*	9.81		Inside Air Film	0.11
			Total R	2.30

$$U = \frac{1}{R} = \frac{1}{9.81} = 0.10$$

$$U = \frac{1}{R} = \frac{1}{2.30} = 0.44$$

FIGURE 8-16   Ceiling *R* and *U* calculations

EXAMPLE   *Problem:* Calculate the *R* and *U* values for the floor assembly shown in Fig. 8-17. Note that the tabulation (Fig. 8-18) shows *R* values for outside and inside air films (using still air for both) different from those used for the wall and roof assemblies.

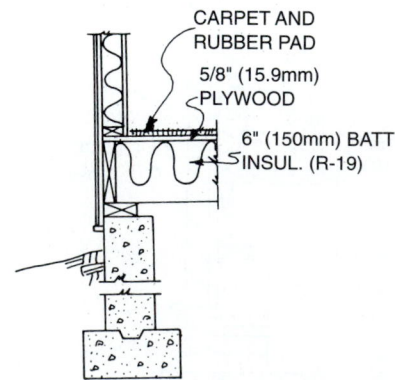

CARPET AND RUBBER PAD
5/8″ (15.9mm) PLYWOOD
6″ (150mm) BATT INSUL. (R-19)

FIGURE 8-17   Floor assembly

		SI UNITS:	
Carpet, Rubber Pad	1.23	Carpet, Rubber Pad	0.22
⅝" Plywood	0.77	15.9 mm Plywood	0.14
6" Batt. Insul.	19.00	150 mm Batt. Insul.	3.32
Inside Air Film	0.92	Inside Air Film	0.16
Outside Air Film	0.92	Outside Air Film	0.16
Total R	22.84	Total R	4.00

$$U = \frac{1}{R} = \frac{1}{22.84} = 0.0437 \text{ (use 0.044)} \qquad U = \frac{1}{R} = \frac{1}{4.00} = 0.25$$

FIGURE 8-18   Floor R and U calculations

In reviewing the various assemblies, it becomes obvious that as more and better insulation is used, the R values of the other materials in the assembly become relatively less important.

EXAMPLE   In Fig. 8-12, the total R value is 14.28 with a U value of 0.07. If the concrete block were changed to lightweight block, the R value of the block would change from 0.97 to 1.90. This would cause the total R to change from 14.28 to 15.21 and the U value to change from 0.07 to 0.0657. In terms of real savings, this is about 6.1%.

   In Fig. 8-12, the total R value is 2.49 with a U value of 0.40. If the concrete block were changed to lightweight (low mass) block, the R value of the block would change from 0.17 to 0.33. This would cause the U value to change from 0.40 to 0.38. In terms of real saving, this is about 5%. Note: this varies from the above example (6.1%) due to rounding off numbers.

The designer must be able to see that small changes in the R value will have little effect on overall R value, on U value, and on subsequent heat loss of the building. This is particularly important when reading technical "sales sheets" and talking to materials salespeople.

As an example of what is meant, on one residential project, it was proposed to use ½-in.-thick asphalt-impregnated fiberboard instead of ½-in.-thick plywood over the 2 × 4 studs because the fiberboard has a higher insulation value. A review of Fig. 8-46 quickly shows that:

½-in. sheathing (vegetable fiberboard) $R = 1.32$
½-in. plywood $R = 0.62$

Now, while the point could be made that the impregnated sheathing is twice as resistant as the plywood, it must be judged in relation to the entire wall.

A review of the tabulations (Fig. 8-14) indicates that the wall with ½-in. plywood has an R of 14.26; if ½-in. impregnated sheathing were used, the R value would increase by 0.70 (1.32 − 0.62) to 14.96 (Fig. 8-19) —about a 5% increase in the total R.

SI

12.7 mm sheathing (vegetable fiberboard)	R = 0.23
12.7 mm plywood	R = 0.11

Now, while the point could be made that the sheathing is twice as resistant as the plywood, it must be judged in relation to the entire wall.

A review of the tabulations (Fig. 8-14) indicates the wall with the 12.7 mm plywood has an R of 2.49; if 12.7 mm impregnated sheathing were used, the R value would increase by 0.12 (0.23 − 0.11) from 2.49 to 2.61 (Fig. 8-19) about a 5% increase in the total R.

There may be other reasons to use the impregnated sheathing (because it is less expensive, for example), but not because it is a better insulation. The designer (as well as the architect, engineer, and owner) is subjected to this type of logic much of the time, and newspapers, magazines, television, and direct mailings send this type of logic into every household and business.

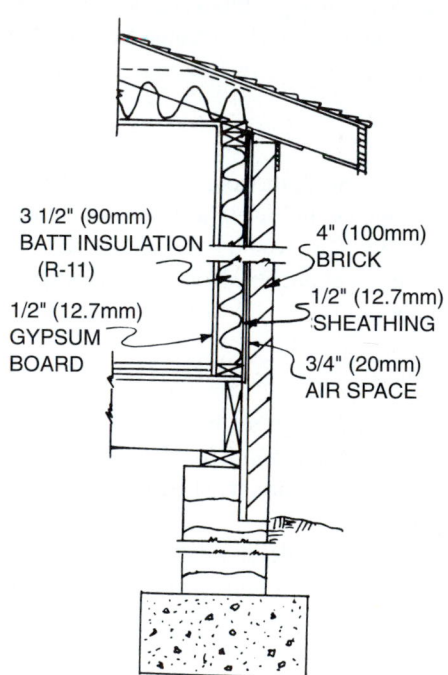

½" Sheathing	1.32	12.7 mm Sheathing	0.23
4" Brick	0.40	100 mm Brick	0.07
3½" Batt. Insul.	11.00	90 mm Batt Insul.	1.94
½" Gypsum Bd.	0.45	12.7 Gypsum Bd.	0.08
Inside Air Film	0.68	Inside Air Film	0.12
Outside Air Film	0.17	Outside Air Film	0.03
¾" Air Space	0.94	20 mm Air Space	0.17
Total *R*	14.96	Total *R*	2.64

SI UNITS:

FIGURE 8-19   Material comparisons

There is one other type of *R* value information that the designer should be aware of. To illustrate the type of analysis required, the same example has been prepared in another way (this is a fictitious example and not taken from any manufacturer's brochure or data sheet):

Using sheathing in this design gives an R value of 2.26.
Heat flow: Horizontal
R value:

¾-in. air space + ½-in. impregnated fiber sheathing board = 2.26

Now, this in itself is not misrepresentation since the *R* value would be 2.26. But the *R* value of the board is 1.32 and the air space is about 0.94, so any comparison of materials should be based on the *R* of the 1.32, not 2.26.

20 mm air space + 12.7 impregnated fiber sheathing board = 0.40

Now, this in itself is not a misrepresentation since the *R* value would be 0.39. But the *R* value of the board is 0.23 and the air space is about 0.17, so comparision of materials should be based on the *R* of 0.23, not 0.40.

Also, the designer who is familiar with construction would realize that in the brick veneer construction shown in Fig. 8-20, the air space would easily be available; but the wood frame construction, with plywood panel facing, shown in Fig. 8-21 would not have an air space at the sheathing unless special construction methods were used. This is because *R*-11 insulation is 3½ in. (90 mm) thick, filling the entire space of the 2 × 4 (50 × 100 mm) studs; the sheathing is attached to the studs and the exterior finish material attached to the sheathing and studs. The only way there would be an air space is to use 2¼-in. (56 mm)-thick batt insulation which has a lower *R* rating.

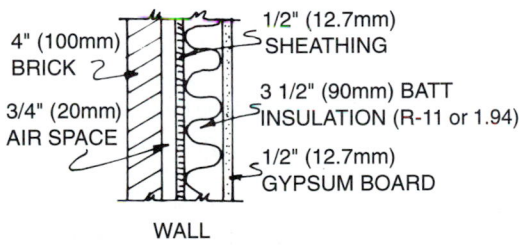

FIGURE 8-20   Air space

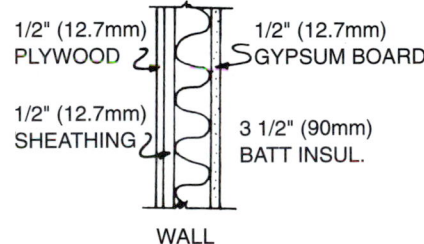

FIGURE 8-21   No air space

8-8   DOORS AND GLASS

Door heat loss depends on the thickness of doors and the type of material. The values in Fig. 8-50 show the U value for typical wood frame doors without any glazing (glass) in the door itself. It also has values for the exterior door when there is also a glass/wood or metal-framed storm door. The values of doors that are similar to those listed but that vary in thickness may be extrapolated from the given information. Doors with glazing would be analyzed as a window (Fig. 8-49).

Window U values are listed in Fig. 8-49, and they are categorized according to the type of glazing; whether there are gas-filled air spaces; the type of frame and whether the frames are insulated or not; and whether the window is glass only, operable or fixed, a double door, or a sloped skylight. While the glazing U value will vary depending on the wind speeds, these values are based on a winter wind speed of 15 miles (24 km) per hour.

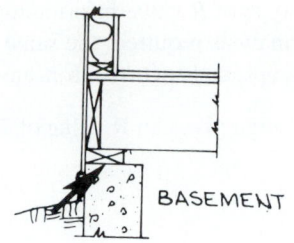

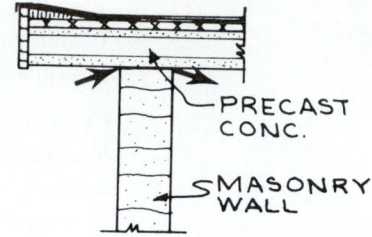

FIGURE 8-22   Construction leakage   FIGURE 8-23   Construction leakage

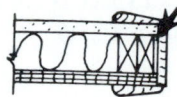 

FIGURE 8-24   Caulk frames and trim

8-9   INFILTRATION

Air leaks into a building through cracks around windows and doors, between the foundation wall and the sill plate, and at many of the connections between walls made of different materials (Fig. 8-22 through 8-24). This air leakage is caused by a combination of the wind pressure on the building and the difference in temperature between the inside and outside of the building. As the wind velocity increases, the amount of air leakage also increases. The close proximity of any large trees, fences, and other buildings close by can also affect the amount of infiltration.

The infiltration will be calculated based on the number of air changes it is estimated there will be per hour. This will give the number of cubic feet or cubic meters per hour (cfh or cmh). The estimated room air changes per hour is taken from Fig 8-52 and depends on the floor area, fit of the construction (envelope evaluation), and whether there are fireplaces in the building. It is also based on building proportions (floor plan aspect) of between 1:1 and 3:1.

The infiltration heat loss per room is calculated by multiplying the volume per hour times the amount of heat required to raise the air temperature times the ΔT (the design temperature difference).

$$\text{Btuh} = \text{Volume} \times 0.018 \times \Delta T$$

$$W = \text{Volume} \times 0.35 \times \Delta T$$

EXAMPLE   A room of 12 ft $\times$ 24 ft $\times$ 8 ft high, total building area over 2100 square feet, with best construction, fireplace in building, would have an air change of 0.3 per hour (Fig. 8-52) plus 0.1 for the fireplace for a total of 0.4. This would amount to 0.4 air changes per hour times the area of the room (12 $\times$ 24 $\times$ 8) $\times$ 0.4 = 921.6 cubic feet per hour. Assuming a design temperature difference of 60°, the heat loss would be calculated as

$$922 \times 0.018 \times 60 = 995.76 \text{ (use 996 Btuh)}$$

A room 3.6 m $\times$ 7.3 m $\times$ 2.4 m total volume, building over 195 sq m (2100 sq ft), with best construction, fireplace in building, would have an air change of 0.3 per hour (Fig.

8-52), plus 0.1 for the fireplace, for a total air change of 0.4 per hour. This would amount to 0.4 air changes per hour times the volume of the room $(3.6 \times 7.3 \times 2.4) \times 0.4 = 25.2$ cubic meters (m^3). Assuming a design temperature difference of 33°C, the heat loss would be calculated as:

$$25.2 \times 0.35 \times 33 = 291 \text{ W}$$

8-10   TEMPERATURES IN UNHEATED SPACES

To determine the heat loss from a heated room to an unheated room or space, it is first necessary to determine the temperature in the unheated space, after which the ΔT can be calculated. The temperature in the unheated space will fall somewhere between the outside and inside temperatures and will vary depending on the amount of surface area adjacent to heated rooms compared to the amount of surface area adjacent to the exterior.

Temperatures in Spaces other than Attics

All unheated spaces, except attics, use the equation:

$$t_u = \frac{t_i(A_1 U_1 + A_2 U_2 + \ldots) + t_o(A_a U_a + A_b U_b + \ldots)}{(A_1 U_1 + A_2 U_2 + \ldots) + (A_a U_a + A_b U_b + \ldots)}$$

where

t_u = temperature in unheated space (°F or °C)
t_i = indoor design temperature of heated room (°F or °C)
t_o = outdoor design temperature (°F or °C)
$A_1, A_2, A_3, \ldots$ = areas of the surfaces which are exposed to the heated space (sq ft or sq m)
$A_a, A_b, A_c, \ldots$ = areas of the surfaces which are exposed to the exterior (sq ft or sq m)
$U_1, U_2, U_3, \ldots$ = coefficients of heat transmission for the material assemblies in A_1, A_2, $A_3, \ldots$
$U_a, U_b, U_c, \ldots$ = coefficients of heat transmission for the material assemblies in A_a, A_b, $A_c \ldots$

Each of the areas involved, and their U values, must be considered in the formula.

EXAMPLE   *Given:* The residence shows in Appendix C.
Problem: Calculate the temperature in the unheated garage adjoining the kitchen.

1. Determine the area of the cold wall.

From the floor plan, the cold wall has a total length of 22 ft 0 in. (6.7 m) (the same as the depth of the garage) and is 8 ft (2.4 m) high—for a gross wall area of:

$$22 \text{ ft} \times 8 \text{ ft} = 176 \text{ sq ft}$$

Now, the net wall area is determined by deducting any door or window openings.

In this design, there is one 3 ft 0 in. × 6 ft 8 in. door with an area of 20 sq ft.

Gross wall area	176 sq ft
Opening area	−20 sq ft
Net wall area	156 sq ft

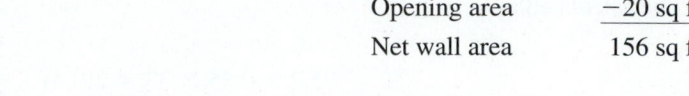

In this design there is one 0.91m × 2.03m door with an area of 1.84 sq m (use 1.8 sq m).

Gross wall area	16.1 sq m
Opening area	−1.8 sq m
Net wall area	14.3 sq m

2. Determine the U value of the cold wall.

From Fig. 8-34, the U value of the cold wall is 0.076.

SI From Fig. 8-34, the U value of the cold wall is 0.43.

3. The door area has been calculated as 18 sq ft in step 1 of this example, and the hollow core flush door U value is taken as 0.47 (Fig. 8-50).

SI The door area has been calculated as 1.7 sq m in step 1 of this example, and the hollow core flush door U value is taken as 2.67 (Fig. 8-50).

4. At this point, all of the areas and U values of the cold wall have been calculated and accumulated. Put this cold wall information in the equation.

$$t_u = \frac{70[(156 \times 0.076) + (20 \times 0.47) + t_o(A_aU_a + A_bU_b + \ldots]}{(156 \times 0.076) + (20 \times 0.47) + (A_aU_a + A_bU_b + \ldots)}$$

SI
$$t_u = \frac{21[(14.3 \times 0.431) + (1.8 \times 2.67) + t_o(A_aU_a + A_bU_b + \ldots]}{(14.3 \times 0.431) + (1.8 \times 2.67) + (A_aU_a + A_bU_b + \ldots)}$$

5. Next, the areas and U values of the surfaces which are exposed to the outside are determined. Begin by listing each of the material assemblies which must be considered.
Ceiling
Wall
Floor (slab on grade, neglected)
Doors

6. Determine the area of the exterior wall.

In this design, the exterior walls referred to are the exterior walls of the garage. From the floor plan in Appendix C, the garage has a total exterior wall length of 14 ft 6 in. + 22 ft 0 in. + 14 ft 6 in., or 51 lineal feet. Using an 8-ft-high ceiling:

Gross exterior wall area = 51 ft × 8 ft = 408 sq ft

Now, the net exterior wall area is determined by deducting the areas of any door or window openings.

In this design, there are two doors, one 3 ft 0 in. × 6 ft 8 in. and one 8 ft 0 in. × 7 ft 0 in. Thus, the door opening area is:

(3.0 ft × 6.7 ft) + (8.0 ft × 7.0 ft) = 20 sq ft + 56 sq ft = 76 sq ft

Gross wall area	408 sq ft
Opening area	−76 sq ft
Net wall area	332 sq ft

SI

In this design, the exterior walls referred to are the exterior walls of the garage. From the floor plan in Appendix C, the garage has a total exterior wall length of 4.4 m + 6.7 m + 4.4 m = 15.5 lineal meters. Using a 2.4 m high ceiling:

Gross exterior wall = 15.5 m × 2.4 m = 37.2 sq m

Now, the net exterior wall area is determined by deducting the areas of any door or window openings.

In this design, there are two doors, one .9m × 2.03 m and one 2.4 m × 2.1 m. Thus the door opening area is:

(0.9 m × 2.03 m) + (2.4 m × 2.1 m) = 6.9 sq m

Gross wall area	37.2 sq m
Opening area	−6.9 sq m
Net wall area	30.3 sq m

7. Determine the U value of the exterior wall. Many times, this information is not as readily available to the designer as is the information for the material assembly for the heated portion of the building. Check the drawings and specifications carefully to determine if insulation is required in these walls, and, if so, how thick it must be. Quite often, insulation is not required in these exterior walls.

In this example, the exterior garage walls have the same material assembly as the rest of the exterior walls, and the U value of 0.07 is taken from Fig. 8-34.

8. Determine the area of the ceiling.

In this design, the ceiling referred to is the ceiling in the garage. This ceiling has an area of:

14 ft 6 in. × 22 ft 0 in. = 14.5 ft × 22 ft = 319 sq ft

SI   4.4 m × 6.7 m = 29.48 sq m (use 29.5 sq m)

9. Determine the U value of the ceiling.

In this design, the ceiling is insulated and has the same U value as the ceiling assembly used when determining the residence heat loss, 0.048 from Fig. 8-34.

10. The door area has been calculated as 76 sq ft (6.9 sq m) in Step 6 of this example, and the door U value is taken as 0.47 (2.67), the same as that used in determining the residence heat loss (Fig. 8-34).

11. Determine the temperature of the unheated space. Gather all of the accumulated information and put it into the equation:

$$t_u = \frac{70[(156 \times 0.076) + (20 \times 0.47)] + 10[(332 \times 0.07) + (319 \times 0.048) + (76 \times 0.47)]}{(156 \times 0.076) + (20 \times 0.47)] + [(332 \times 0.07) + (319 \times 0.048) + (76 \times 0.47)]}$$

$$= \frac{70(12 + 8.5) + 10(23 + 15 + 36)}{12 + 8.5 + 23 + 15 + 36} = \frac{1435 + 740}{94.5}$$

$$= 23°F$$

$$tu = \frac{21[(14.3 \times 0.431) + (1.8 \times 2.67) + (-12)[(32.2 \times .40) + (29.5 \times .27) + (6.9 \times 2.67)]}{(14.3 \times 0.431) + (1.8 \times 2.67) + (32.2 \times .40) + (29.5 \times .27) + (6.9 \times 2.67)}$$

$$= \frac{21(10.97 + (-12))(39.3)}{6.2 + 4.54 + 12.9 + 7.97 + 18.42} = \frac{-2.41}{50.03} = -4.92 \ (-5.0°C)$$

Temperatures in Attics

To calculate the heat loss of the ceiling area of a building which has an uninsulated attic, it is necessary to estimate the temperature in the attic.

The attic temperature will fall somewhere between the outside and inside temperatures and will vary depending on the amount of surface area adjacent to heated rooms compared to the amount of surface area adjacent to the exterior.

To estimate the attic temperature, use the equation:

$$ta = \frac{A_c U_c t_i + t_o(A_r U_r + A_w U_w + A_g U_g)}{A_c U_c + A_r U_r + A_w U_w + A_g U_g}$$

where

t_a = attic temperature (°F)
t_i = indoor temperature near top floor ceiling (°F)
t_o = outdoor design temperature (°F)
A_c = area of ceiling (sq ft)
A_r = area of roof (sq ft)
A_w = net area of vertical attic wall surface (sq ft)
A_g = area of attic glass (sq ft)
U_c = coefficient of heat transmission of ceiling
U_r = coefficient of heat transmission of roof
U_w = coefficient of heat transmission of vertical wall surface
U_g = coefficient of heat transmission of glass

EXAMPLE   *Given:* The residence shown in Appendix C.
Problem: Calculate the temperature in the attic.

1. Determine the area of the ceiling.

    From the floor plan, the total area of the ceiling in heated spaces is:

    36 ft 0 in. × 61 ft 6 in. = 2,214 sq ft

    SI   11 m × 18.8 m = 207 sq m

2. Determine the *U* value for the ceiling.

    From Fig. 8-34, the *U* value of the ceiling is 0.048.

    SI   From Fig. 8-34, the *U* value of the ceiling us 0.28.

3. Determine the area of the roof.

    To determine the area of the roof, it will be necessary to check the drawings in Appendix C. The quickest approach is to simply measure the length of the rafter from the outside edge of the wall to the ridge (highest point). This will give one-half the width of the roof. In this design, the roof measures about 20.2 ft, so the total roof width would be about 40.4 ft. The length of the roof is taken as the length of the building, or 61.5 ft.

$$\text{Roof area} = 40.4 \text{ ft} \times 61.5 \text{ ft} = 2{,}485 \text{ sq ft}$$

SI

In this design, the roof measures about 6.15 m, so the total roof width would be about 12.3 m. The length of the roof is taken as the length of the building, or 18.8 m.

$$\text{Roof area} = 12.3 \text{ m} \times 18.8 \text{ m} = 231.2 \text{ sq m}$$

4. Determine the U value for the roof construction.

In this example, the construction of the roof is shown in Appendix C. The U value is taken from Fig. 8-34 as 0.48 (3.03).

5. Determine the area of the vertical wall surface (at the gable ends).

The area of the vertical wall surface may also be measured from the drawings. In this design, the triangular wall surface has a base dimension of 36 ft 0 in. and a height of about 9 ft 0 in.; there are two such surfaces (one at each end of the house). The area of vertical wall surface is:

$$\frac{bh}{2} \times 2 = \frac{36 \text{ ft} \times 9 \text{ ft}}{2} \times 2 = 324 \text{ sq ft}$$

SI

In this design, the triangular wall surface has a base dimension of 11 m and a height of 2.75 m.; there are two such surfaces (one at each end of the house). The area of vertical wall surface is:

$$\frac{bh}{2} \times 2 = \frac{11 \times 2.75}{2} \times 2 = 30.25 \text{ sq m}$$

6. Determine the U value of the vertical walls.
Many times, this information is not as readily available to the designer as is the information for the material assembly for the heated portion of the building. Check the drawings and specifications carefully to determine if insulation is required in these walls and, if so, how thick it must be. Quite often, insulation is not required in these exterior walls.

In this example, the vertical walls are constructed of the same basic materials as the rest of the walls, but without the insulation in the walls. To calculate the U value of the construction, it is necessary to take the R value of the wall with insulation, from Fig. 8-34; deduct the R value of the insulation; and add the R value of the air space left where the insulation was.

Making use of the R values of the materials to calculate the R value of the wall in Fig. 8-34:

½-in. plywood	0.63
4-in. brick	0.40
Outside air	0.17
Inside air	0.68
	$R = 1.00$

$$U = 1 \div R = 0.53$$

SI

12.7 mm Plywood	0.11
100 mm Brick	0.07
Outside air	0.03
Inside air	0.12
Total R	0.33

$$U = 1/R = 1/0.33 = 3.03$$

7. Determine the temperature of the unheated attic. Gather all of the accumulated information and put it into the equation:

$$T_a = \frac{(2{,}214 \times 0.047 \times 70) + 10[(2{,}485 \times 0.48) + (324 \times 0.53)]}{(2{,}214 \times 0.047) + (2{,}485 \times 0.48) + (324 \times 0.53)}$$

$$= \frac{7{,}284 + 10(1{,}193 + 169)}{104 + 1{,}193 + 169} = \frac{7{,}284 + 13{,}620}{1{,}466}$$

$$= 14.26°F$$

SI

$$T_a = \frac{(205.7 \times 0.267 \times 21) + (-12)[(230.9 \times 2.72) + (30.1 \times 2.95)]}{(205.7 \times 0.267) + (230.9 \times 2.72) + (30.1 \times 2.95)}$$

$$= \frac{1{,}153 + (-12)(628 + 88.8)}{54.9 + 628 + 88.8} = \frac{1{,}153 + (-8602)}{771.7} = \frac{-7449}{771.7} =$$

$$= -9.65 \text{ (use } -10°C)$$

What we find in determining unheated attic temperatures is that when the ceiling of the building is very well insulated and when there is no insulation in the roof and vertical walls, the temperature will be quite close to the outside design temperature.

The added flow of air through attic vents or louvers will bring the temperature down further.

The only times that the unheated attic temperature is significantly above the outside design temperature are:

1. When the ceiling of the building is not well insulated.

2. When the ceiling is well insulated and the roof is insulated also.

Temperatures Below Floors

To calculate the heat loss through the floor area of a building, it is necessary to estimate the temperature in the basement or crawl space below the floor. Since the temperature cannot be readily calculated, the designer should use the following guidelines:

1. Basements and crawl spaces located almost entirely below ground will normally have a temperature midway between the inside and outside design temperatures being used.

2. If the basement has windows, the temperature will be lower than if it does not have windows.

3. If the heating unit (boiler or furnace) is located in the basement, the temperature will be higher than if the heater were located elsewhere.

4. The crawl space temperature will be affected by the size and number of vents; the presence of heating unit, hot water heater, piping, and ductwork; and the amount of insulation used.

In short, the designer must use their judgment in determining the basement or crawl space temperature to use in the heat loss calculation.

8-11 DESIGNER RESPONSIBILITIES

The designer of the heating system for a building will have to review carefully all drawings and specifications for the construction. It is important that the designer not simply accept the word of the builder, general contractor, architect, or owner that certain materials, assemblies of materials, details, windows, and glass will be used. Whenever possible, all these things should be properly shown and specified for the designer's review.

In addition, the designer should clearly note on the design proposal the assemblies of materials, insulation, window types, and any other items on which the design is based. This is done because, many times, changes which may greatly affect the design are made, and the designer is never told. Common changes include the type of windows; number of windows; type or thickness of insulation; ceiling height; materials in walls, ceilings, and floors; and type of window glass. Sometimes, even room sizes are changed without letting the designer know. For these reasons, it is important that the designer make every attempt to keep in close contact with the others involved in the construction.

The designer has a further responsibility to the others in the sense that the designer should make recommendations on methods of construction and changes in materials which can reduce heat loss and save money.

8-12 OBTAINING BUILDING DIMENSIONS

When calculating the heat loss of a proposed building, the dimensions used are taken from the written dimensions on the drawings of the building. If written dimensions are incomplete, check the scale at which the drawings are made, and then check at least one dimension in each direction to be certain that it is drawn fairly close to scale. Be very careful since in many architectural and engineering offices, once the building has been drawn, if there are any small changes in room or building sizes, the written dimensions are simply changed, and the scale drawing is not revised. The written dimension takes precedence over any scaled dimension and should be used. Of course, if large discrepancies are found, check with whomever drew the plans to find out what is correct.

Room Measurements

For heat loss calculations, the inside dimensions of a room are used when available. When working from the floor plan, the dimensions given are used. They should be accurate to the nearest foot for walls; for example, a wall 12 ft 8 in. (3.86 m) long is considered 13 ft (3.9 m). Ceiling heights should be accurate to the nearest one-half foot; a 7 ft 10 in. (2.39 m) ceiling height is considered 8 ft (2.4 m).

When calculating the areas, the following guidelines are used:

1. Any closet is figured as a part of the room that it opens into. Any exposed closet wall, ceiling, or floor is included as part of the heat loss of the room. Dressing rooms and walk-in closets are generally considered separate rooms with their own heat loss and will require heat to be supplied to them.

2. Any cabinets, bookshelves, or other types of pantry or closet areas are ignored when calculating the heat loss of a room. The calculations will be made as if they are not on the walls.

3. When two or more types of wall assemblies are used in one room (Figs. 8-3 and 8-25), the wall areas must be calculated separately since each assembly will have a different heat transmission value.

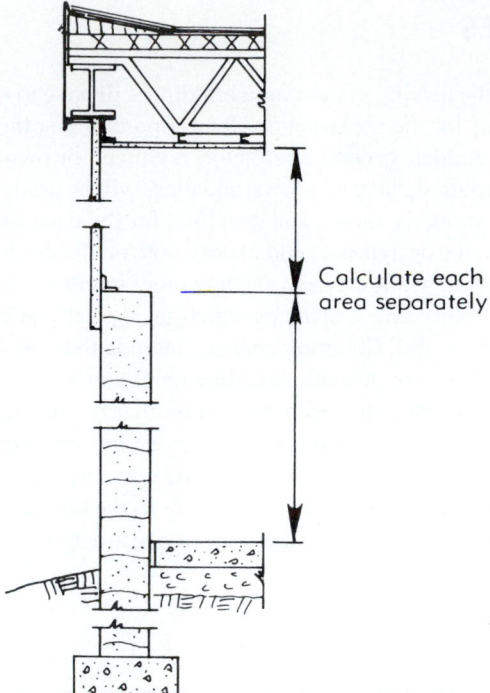

FIGURE 8-25   Wall assemblies

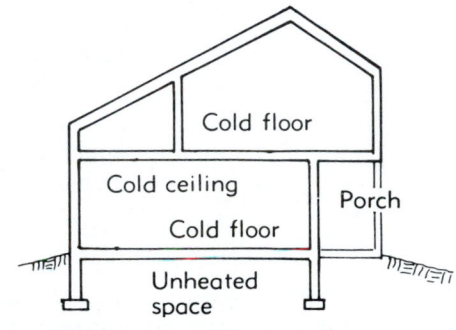

FIGURE 8-26   Cold floor and ceilings

Many times, special care must be taken in calculating the area of a wall, floor, or ceiling because part of the area may be a cold partition, floor, or ceiling. Typical *cold partitions* are found between a heated room and an adjoining unheated space, such as any unheated garage, enclosed porch, attic or cellar stairway, crawl space, basement, or storage area. The area of cold partitions must be carefully separated from the areas of partitions which are exposed to the exterior since they will have different ΔT's.

Cold ceilings are found between a heated room and an unheated attic, storage area, or other unheated space. In a one-story house, all of the rooms have cold ceilings. In some two-story designs (such as Fig. 8-26), the second-story rooms have cold ceilings while some of the first-floor rooms may have full or partial cold ceilings. That area of the ceiling located below an unheated space must be calculated by multiplying the length times the width and then recorded. The ceiling portions of any first-floor rooms that are under heated second-floor rooms have no heat loss since the temperature would be balanced on each side of the ceiling (Fig. 8-27).

Cold floors are found between a first-floor room and an unheated basement or crawl space, and the area of these floors must be calculated. Floors over heated basements are not

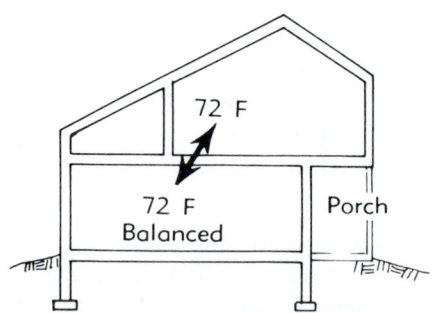

FIGURE 8-27   Temperature balance

considered cold floors. If the basement is heated, check to be certain that it is completely heated since, quite often, only a portion is heated and part is left unheated. Any floor over the unheated portion is considered a cold floor. Cold floors also occur when a second-floor heated room is fully or partially over an unheated, enclosed first-floor space, such as an enclosed porch or a garage. The area of the floor exposed to the cold must be calculated also.

When the floor construction is a concrete slab at or near the grade, calculate and record the lineal feet of exposed edge of slab around the building. In addition, the square footage of floor area is calculated and recorded. The heat loss factor for slab construction is given in Fig. 8-51.

Window and Door Measurements

Windows and doors are measured inside the casings (Fig. 8-28), and while they are measured to the nearest inch, they are usually recorded to the nearest 0.1 ft (0.1 m), since it will be necessary to determine their area in square feet. The values used in Fig. 8-29 may be used in place of inches as the dimensions are recorded. When totalling up the door and window areas of the room, these measurements should be rounded off to the nearest square foot or 0.1 m. The door and window areas should be kept separate. Many windows and doors come in stock (standardized) sizes; the commonly used sizes, with the areas listed, are given in Fig. 8-30.

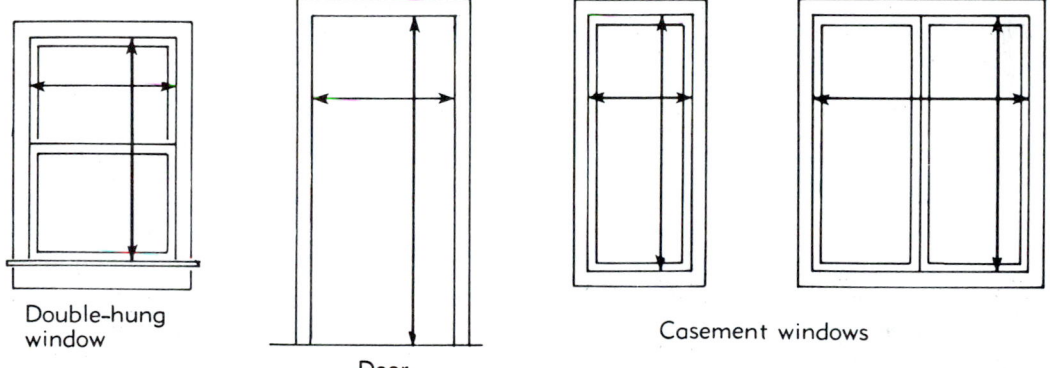

Double-hung window

Door

Casement windows

FIGURE 8-28   Door and window dimensions

Actual dimension (in.)	Recorded dimension (ft.)
1	0.1
2	0.2
3	0.3
4	0.3
5	0.4
6	0.5
7	0.6
8	0.7
9	0.8
10	0.8
11	0.9

FIGURE 8-29   Dimension conversion

Window Type	Size	Sq.ft	Size (m)	Sq.m
Double-hung	2'-0" × 3'-10"	7.6	0.61 × 1.16	0.71
	2'-0" × 4'-2"	8.4	0.61 × 1.28	0.78
	2'-4" × 3'-10"	8.7	0.73 × 1.16	0.85
	2'-4" × 4'-2"	9.7	0.73 × 1.28	0.93
	2'-4" × 4'-6"	10.4	0.73 × 1.37	1.00
	2'-8" × 3'-10"	10.3	0.81 × 1.16	0.94
	2'-8" × 4'-2"	11.3	0.81 × 1.28	1.04
	2'-8" × 4'-6"	12.2	0.81 × 1.37	1.11
	3'-0" × 4'-2"	12.6	0.92 × 1.28	1.18
	3'-0" × 4'-6"	13.5	0.92 × 1.37	1.26
	3'-0" × 5'-2"	15.6	0.92 × 1.59	1.46
	3'-4" × 5'-2"	17.2	1.02 × 1.59	1.62
	3'-4" × 5'-6"	18.2	1.02 × 1.68	1.71
Casement	2'-0" × 3'-0"	6	0.61 × 0.92	0.56
	2'-0" × 3'-6"	7	0.61 × 1.07	0.65
	2'-0" × 4'-0"	8	0.61 × 122	0.74
	2'-0" × 5'-0"	10	0.61 × 1.53	0.93
	2'-0" × 6'-0"	12	0.61 × 1.83	1.12
	2' 4" × 3'-6"	8.1	0.73 × 1.07	0.78
	2'-4" × 4'-0"	9.2	0.73 × 1.22	0.89
Awning	3'-0" × 2'-0"	6	0.92 × 0.61	0.56
	3'-0" × 3'-6"	10.5	0.92 × 1.07	0.98
	4'-0" × 2'-0"	8	1.22 × 0.61	0.74

FIGURE 8-30   Window areas

2-3'-0" × 4'0" Double Hung Wood
Windows, Insulating Glass
or
2-3/0 × 4/0 D.H., Wood, I.G.   FIGURE 8-31   Window notation

When recording the area of windows, note whether a storm window (SW) or insulating glass (IG) is specified (Fig. 8-31) since this will greatly affect the heat transmission value of the window. Similarly, note also the type of exterior door used and whether or not there is a storm door.

8-13   HEAT LOSS CALCULATIONS

The actual procedure in estimating heat loss is best illustrated by doing the calculations for a typical building. For this example, the single-story house shown in Appendix C is used. Throughout these calculations, keep in mind that accuracy should be limited to the first three figures since this is an *estimate* of the heat loss, *not* an exact measurement. Similarly, those using calculators should not use such accuracy that the heat loss might be listed as 587.236; instead, 587 or 588, even 590, would be a logical answer.

Step-by-Step Approach

To record the heat loss, some type of form is required; the one shown in Fig. 8-32 is typical. Whatever form is used should be set up to list the rooms separately. Begin the heat loss estimate by putting the name of the project, the date, and the designer's initials on the form and on the sheet of paper which will be used for calculations. Now, the actual heat loss estimate may begin:

Room	Portion considered	Dimensions	Height	Gross area or l.f.	Openings	Opening area	Net area or c.f.	U value	ΔT	Heat loss (Btuh)	Room heat loss (Btuh)

FIGURE 8-32   Typical heat loss form

1. Select the outdoor design temperature for the geographical area of construction (Fig. 8-53).

   Izn this case, the geographical area is 35° latitude, and the outdoor design temperature is 10°F (−12°C).

2. Select the indoor design temperature for the design.

   Typically, this is about 70°F (21°C) for all residences or living quarters; occasionally it may vary slightly in an office design.

3. Calculate the design temperature difference. This is the difference between the outdoor and indoor design temperatures.

   In this design, the difference is 70°F minus 10°F, or 60°F.

   Note: Take care; if the outdoor design temperature was −10°F, then the difference would be 80°F (Fig. 8-33).

   SI   21°C minus −12°C, or 33°C

   SI   If the outdoor design temperature was −23°C, the difference would be 45°C

4. Locate and identify the construction of all walls, ceilings, floors, and windows from the drawings.

   From Appendix C: *Exterior walls* are wood frame, with brick, plywood sheathing, 2 × 4 studs, 3½-in. blanket or batt insulation with a vapor barrier, and ½-in. gypsum board finish. The *cold partition* is a frame partition, with ½-in. gypsum board on each side and 3½-in. batt-type insulation. *Ceilings* are ½-in. gypsum board, 6-in. batt-type insulation, and ½-in. plywood. The *attic* is vented and unheated. *Floors* are carpet and pad on ½-in. plywood on wood joists, with 3½-in. batt-type insulation with a vapor barrier and no ceiling below. The *crawl space* is vented and unheated. *Windows* are wood, double-hung, with insulating glass (³⁄₁₆-in. air space) and weatherstripped. *Exterior doors* are weatherstripped and have storm doors.

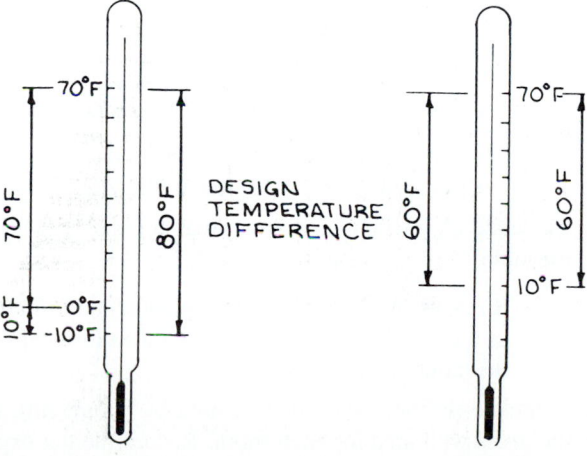

FIGURE 8-33   Calculating ΔT

Design Temperatures	Inside	Outside	ΔT
Exterior Walls	70°F	10°F	60°F
Cold Walls	70°F	23°F	47°F
Ceiling	70°F	14°F	56°F
Crawl Space	70°F	35°F	35°F

U Values:

Exterior Wall

½″ Plywood	0.62
4″ Brick	0.40
3½″ Batt Insul.	11.00
½″ Gypsum Bd.	0.45
Inside Air Film	0.68
Outside Air Film	0.17
¾″ Air Space	0.94
Total R =	14.26

$$U = \frac{1}{R} = \frac{1}{14.26} = 0.07$$

Floor

Carpet (fiborous pad)	2.08
½″ Plywood	0.62
3½″ Batt Insul.	11.00
Inside Air	0.92
Outside Air	0.17
Total R =	14.79

$$U = \frac{1}{R} = \frac{1}{14.79} = 0.0676 \text{ (use 0.068)}$$

Roof

Shingles (Asphalt)	0.44
Bldg. Paper (felt)	0.06
⅝″ Plywood	0.77
Inside Air	0.62
Outside Air	0.17
Total R =	2.06

$$U = \frac{1}{R} = \frac{1}{2.06} = 0.48$$

Ceiling

½″ Gypsum Bd.	0.45
6″ Batt Insul.	19.00
Inside Air	0.61
Outside Air	0.61
Total R =	20.67

$$U = \frac{1}{R} = \frac{1}{20.67} = 0.048$$

Cold Wall

½″ Gypsum Bd.	0.45
3½″ Batt Insul.	11.00
½″ Gypsum Bd.	0.45
Inside Air	0.68
Outside Air	0.68
	13.26

$$U = \frac{1}{R} = \frac{1}{13.26} = 0.076$$

Glass, I.G., 0.25 Air Space
$$U = 0.65$$

Windows: 0.59

Doors:
  D1, U = 0.60
  D4, U = 0.47
  D9 & D10 = 0.59

FIGURE 8-34   *R* and *U* values calculated

5. The *U* values of the construction must be found next by totalling up the resistances of the assembly of materials. Calculations should be made and then the *U* value recorded in the table under the appropriate column. The *U* values are calculated in Fig. 8.34.

6. Next, the infiltration factors for the rooms should be determined and then recorded on the table.

   From Fig. 8-52, the air change values are:

   Best construction, one fireplace, over 2100 sq. ft, 0.4 (AC) air changes per hour.

7. Next, the gross wall area of each room is calculated and recorded in the table.

   The dimensions are taken from the floor plan(s) of the building (Appendix C) as discussed in Sec. 8-12. Record the gross wall area of each room in the table.

8. Calculate the window and door areas for each room separately, and record them in the table.

   Sizes are taken from the floor plan(s) and from the window and door schedules of the building. Record the gross area of windows and doors for each room in the table.

9. Calculate the net exposed wall area by subtracting the windows and doors from the gross wall area for each room. Record the net exposed wall area for each room in the table.

SI UNITS

Design Temperatures	Inside	Outside	ΔT
Exterior Walls	21°C	−12°C	33°C
Cold Walls	21°C	−5°C	26°C
Ceiling	21°C	−10°C	31°C
Crawl Space	21°C	2°C	19°C

U Values:

Exterior Wall

12.7 mm Plywood	0.11
100 mm Brick	0.07
90 mm Batt Insul.	1.94
12.7 mm Gypsum Bd.	0.08
Inside Air Film	0.12
Outside Air Film	0.03
20 mm Air Space	0.17
Total R =	2.52

$$U = \frac{1}{R} = \frac{1}{2.52} = 0.3968 \text{ (use 0.40)}$$

Floor

Carpet (fiborous pad)	0.37
12.7 mm Plywood	0.11
90 mm Batt Insul.	1.94
Inside Air	.16
Outside Air	.03
Total R =	2.61

$$U = \frac{1}{R} = \frac{1}{2.61} = 0.3831 \text{ (use 0.38)}$$

Roof

Shingles (Asphalt)	0.08
Bldg. Paper (felt)	0.01
15.9 mm Plywood	0.14
Inside Air	0.11
Outside Air	0.03
Total R =	0.37

$$U = \frac{1}{R} = \frac{1}{0.37} = 2.7$$

Ceiling

12.7 mm Gypsum Bd.	0.08
150 mm Batt Insul.	3.32
Inside Air	0.11
Outside Air	0.11
Total R =	3.62

$$U = \frac{1}{R} = \frac{1}{3.62} = 0.2762 \text{ (use 0.28)}$$

Cold Wall

12.7 mm Gypsum Bd.	0.08
90 mm Batt Insul.	1.94
12.7 mm Gypsum Bd.	0.08
Inside Air	0.12
Outside Air	0.12
Total R =	2.34

$$U = \frac{1}{R} = \frac{1}{2.34} = 0.4274 \text{ (use 0.43)}$$

Glass, I.G., 6.4 mm Air Space

$$U = 3.71$$

Windows: 3.34

Doors:

D1, U = 3.41
D4, U = 2.67
D9 & D10 = 3.34

FIGURE 8-34   (*continued*)

10. Calculate the area of any cold partitions, ceiling, and floors (discussed in Sec. 8-12), and record the net area for any such rooms.

11. Calculate the design heat loss for each of the areas of heat loss in the room. For the transmission heat loss, multiply the *U* value times the area times the design temperature difference:

$$U \times A \times \Delta T = \text{Btuh}$$

$$U \times A \times \Delta T = W$$

SI

12. For infiltration heat loss, multiply the infiltration air change volume times heat per cfh (cmh) times the design temperature difference:

$$\text{cfh} \times 0.018 \times \Delta T = \text{Btuh}$$

SI

$$\text{cmh} \times 0.35 \times \Delta T = W$$

The design heat losses calculated are recorded in the table.

Room	Portion considered	Dimensions	Height	Gross area or l.f.	Openings	Opening area	Net area or c.f.	U value	ΔT	Heat loss (Btuh)	Room heat loss (Btuh)
Living-Family	Ext.Wall	32.25 + 7.0	8	314	2-6/0 × 6/8	80	234	0.07	60	983	
	Cold Wall	12.7	8	104	—	—	104	0.076	47	372	
	Floor	32.25 × 18.8	—	603	—	—	603	0.068	35	1435	
	Ceiling	32.25 × 18.8		603	—	—	603	0.048	56	1621	
	Window	2-6/0 × 6/8		80			80	0.59	60	2832	
	Infiltration		8	603	—	—	4824	0.4 × 0.018	60 —	2084 —	9327
Kitchen	Ext. Wall	15.8 + 7.0	8	183	W2 4.0 × 3.0	12	171	0.07	60	718	
	Cold Wall	10.3	8	82	3.0 × 6.8	21	61	0.076	47	218	
	Floor	15.8 × 17.3	—	274	—	—	274	0.068	35	652	
	Ceiling	—		274	—	—	274	0.048	56	736	
	Door	—	—	21	—	—	21	0.47	47	464	
	Window	4.0 × 3.0		—	—		12	0.59	60	425	
	Infiltration		8	274			2192	0.4 × 0.18	60 —	947 —	4160
Dining Room	Ext. Wall	14.5	8	116	6.0 × 5.0	30	86	0.07	60	360	
	Floor	14.5 × 13.3	—	193	—	—	193	0.068	35	459	
	Ceiling	14.5 × 13.3	—	193	—	—	193	0.048	56	508	
	Window	6.0 × 5.0	—	30	—	—	30	0.59	60	1062	
	Infiltration		8	193			1544	.4 × 0.018	60 —	667 —	3056
Entry and Hall	Ext. Wall	7.5 + 2.5	8	80	6.0 × 6.8	40	40	0.07	60	168	
	Floor	7.5 + 2.5 × 13.3 + 3.0 × 39	—	289 —	— —	— —	289 —	0.068 —	35 —	688 —	
	Ceiling	—	—	289	—	—	289	0.048	56	777	
	Window	2.8 × 6.7	—	—	—	—	19	0.59	60	673	
	Door	—	—	21	—	—	21	0.60	60	756	
	Infiltration	—	8	289	—	—	2312	0.4 × 0.018	60 —	999 —	4061
Bedroom 3	Ext. Wall	14.5 + 6.8 + 6.8	8	225	4.0 × 4.0	16	209	0.07	60	878	
	Floor	14.5 × 13.3 + 6.8 × 6.8	—	239 —	— —	— —	239 —	0.068 —	35 —	569 —	
	Ceiling	—	—	239	—	—	239	0.048	56	643	
	Window	4.0 × 4.0	—	16	—	—	16	0.59	60	567	
	Infiltration	—	8	239	—	—	1912	0.4 × 0.018	60 —	826 —	3483
Bedroom 1 (inc. 2 closets)	Ext. wall	12.7 + 18.7 + 2.5	8	272 —	4.0 × 4.0 + 3.0 × 4.0	28 —	244 —	0.07 —	60 —	1025 —	
	Floor	12.7 × 18.7 + 2.5 × 4.0 + 4.0 × 6.0		271 — —	— — —	— — —	271 — —	0.068 — —	35 — —	645 — —	
	Ceiling	—	—	271	—	—	271	0.048	56	728	
	Window	—	—	28	—	—	28	0.59	60	991	
	Infiltration	—	8	271	—	—	2168	.04 × 0.018	60 —	937 —	4326 + Bath
Bedroom 2	Ext. Wall	11.5 + 2.5	8	112	3.0 × 4.0	12	100	0.07	60	420	
	Floor	11.5 × 18.8 + 2.5 × 4.0		226 —	— —	— —	226 —	.068 —	35 —	538 —	
	Ceiling	—	—	226	—	—	226	0.048	56	608	
	Window	—	—	12	—	—	12	0.59	60	425	
	Infiltration	—	8	226	—	—	1808	0.4 × 0.018	60 —	781 —	2772
Bath	Ext. Wall	10.5	8	84	2.0 × 2.0	4	80	0.07	60	336	
	Floor	10.5 × 6.8	—	72	—	—	72	.068	35	171	
	Ceiling	—	—	72	—	—	72	.048	56	194	
	Window	—	—	4	—	—	4	0.59	60	142	
	Infiltration	—	8	72	—	—	576	0.4 × 0.018	60 —	249 —	1092
Interior Bath (May add to bedroom 1)	Floor	8.0 × 5.0		40			40	0.68	35	95	
	Ceiling			40			40	0.48	56	108	203

Total Heat Loss 32,480
Bedroom 1 + Interior Bath = 4529 BTU

FIGURE 8-35 Heat loss calculations

Room	Portion considered	Dimensions	Height	Gross area or Volume	Openings	Opening area	Net area or Volume	U value	ΔT	Heat loss	Room heat loss
Living-Family	Ext. Wall	9.8 + 2.1	2.4	28.6	2-1.8 × 2	7.2	21.4	0.40	33	283	
	Cold Wall	3.9	2.4	9.4	—	—	9.4	0.43	26	105	
	Floor	9.8 × 5.7	—	55.9	—	—	55.9	0.38	19	404	
	Ceiling	9.8 × 5.7	—	55.9	—	—	55.9	0.28	31	485	
D9 & D10	Glass	2-1.8 × 2	—	7.2	—	—	7.2	3.34	33	794	
	Infiltration	9.8 × 5.7	2.4	134	—	—	134	.4 ×	33	619	
	—	—	—	—	—	—	—	0.35	—	—	2690
Kitchen	Ext. Wall	4.8 + 2.1	2.4	16.6	w2 1.22 × 0.92	1.1	15.5	0.40	33	205	
	—	—	—	—	—	—	—	—	—	—	
	Cold Wall	3.1	2.4	7.4	0.92 × 2.1	1.9	5.5	0.43	26	62	
	Floor	4.8 × 5.3	—	25.4	—	—	25.4	0.38	19	183	
	Ceiling	—	—	25.4	—	—	25.4	0.28	31	220	
	Door	—	—	1.9	—	—	1.9	2.67	26	132	
	Glass	1.22 × 0.92	—	—	—	—	1.1	3.34	33	121	
	Infiltration	4.8 × 5.3	2.4	25.4	—	—	61	.4 ×	33	282	
	—	—	—	—	—	—	—	0.35	—	—	1205
Dininig Room	Ext. Wall	4.4	2.4	10.6	1.83 × 1.53	2.8	7.8	0.40	33	103	
	Floor	4.4 × 4.1	—	18	—	—	18	0.38	19	130	
	Ceiling	4.4 × 4.1	—	18	—	—	18	0.28	31	156	
	Glass	1.83 × 1.53	—	2.8	—	—	2.8	3.34	33	309	
	Infiltration		2.4	18	—	—	43.2	.4 ×	33	200	
	—	—	—	—	—	—	—	0.35	—	—	898
Entry and Hall	Ext. Wall	2.3 + 0.76	2.4	7.3	1.8 × 2	3.6	3.7	0.40	33	49	
	Floor	2.3 + 0.76 ×	—	30.7	—	—	30.7	0.38	19	222	
	—	4.1 + 0.92 × 11.9	—	—	—	—	—	—	—		
	Ceiling	—	—	30.7	—	—	30.7	0.28	31	267	
	Glass	0.85 × 2	—	—	—	—	1.7	3.34	33	187	
	Door	—	—	2	—	—	2	3.41	33	225	
	Infiltration	—	2.4	30.7	—	—	73	.4 ×	33	337	
	—	—	—	—	—	—	—	0.35	—	—	1287
Bedroom 3	Ext. Wall	4.4 + 2.1 × 2.1	2.4	20.6	1.2 × 1.2	1.4	19.2	0.40	33	253	
	Floor	4.4 × 4.1 +	—	22.4	—	—	22.4	0.38	19	162	
	—	2.1 × 2.1	—	—	—	—	—	—	—	—	
	Ceiling	—	—	22.4	—	—	22.4	0.28	31	195	
	Glass	1.2 × 1.2	—	1.4	—	—	1.4	3.34	33	154	
	Infiltration	—	2.4	22.4	—	—	53.8	.4 ×	33	249	
	—	—	—	—	—	—	—	0.35	—	—	1013
Bedroom 1 (Inc. 2 Closets)	Ext. Wall	3.9 + 5.7 + 0.8	2.4	25	1.2 × 1.2 + 0.92 × 1.2	2.6	22.4	0.40	33	296	
	Floor	3.9 × 5.7 +		25.35	—	—	25.35	0.38	19	183	
	—	0.8 × 1.2 +	—	—	—	—	—	—			
	—	1.2 × 1.8	—	—	—	—	—	—			
	Ceiling	—	—	25.35	—	—	25.35	0.28	31	220	
	Glass	—	—	2.6	—	—	2.6	3.34	33	287	
	Infiltration	—	2.4	25.35	—	—	60.8	0.4 ×	33	281	
	—	—	—	—	—	—	—	0.35	—	—	1267
Bedroom 2	Ext. Wall	3.5 + 0.8	2.4	10.3	0.92 × 1.2	1.1	9.2	0.40	33	122	
	Floor	3.5 × 5.7 +	—	20.9	—	—	20.9	0.38	19	151	
	—	0.8 × 1.2	—	—	—	—	—	—			
	Ceiling	—	—	20.9	—	—	20.9	0.28	31	182	
	Glass	—	—	1.1	—	—	1.1	3.34	33	121	
	Infiltration	—	2.4	20.9	—	—	50	0.4 ×	33	231	
	—	—	—	—	—	—	—	0.35	—	—	807
Bath	Ext. Wall	3.2	2.4	7.7	0.61 × 0.61	0.4	7.3	0.40	33	96	
	Floor	3.2 × 2.1	—	6.7	—	—	6.7	0.38	19	49	
	Ceiling	—	—	6.7	—	—	6.7	0.28	31	58	
	Glass	—	—	0.4	—	—	0.4	3.34	33	44	
	Infiltration	—	2.4	6.7	—	—	16.1	0.4 ×	33	75	
	—	—	—	—	—	—	—	0.35	—	—	322
Interior Bath (May add to Bedroom 1)	Floor	2.4 × 1.5	—	3.6	—	—	3.6	0.38	19	26	
	Ceiling	—	—	3.6	—	—	3.6	0.28	31	31	57

Total Heat Loss 9546
Bedroom 1 and Interior Bath = 1324

FIGURE 8-35   Continued (SI Units)

13. Calculate the total heat loss for each room by adding the design heat losses obtained for each room. Each room is totalled separately and recorded in the table.

14. Calculate the total heat loss for the entire building by adding the heat losses of each room together, and record this total in the table. This total heat loss will be used later when the heating unit (furnace) size is being selected; however, this value is *not* the heating unit's size (sizing of heating units is discussed in Sec. 9-6).

The total heat loss for the building in Appendix C is shown in Figs. 8-34 and 8-35.

Actual sizing of the system is discussed separately for the various types of heating systems commonly used. Water heating is discussed in Chapter 9; forced air heating in Chapter 10; forced air cooling in Chapter 13; electric baseboard, ceiling, and wall heating in Chapter 11; and solar heating in Chapter 14.

The total heat loss calculated for the residence in Appendix C may seem quite low in comparison to calculations in various other reference books and manuals. This large difference is attributed to the thicknesses of insulation used in this text and to the limited amount of window area used.

8-14   SEALING THE CRACKS

It is most important that the space (cracks) between different materials be properly sealed so that air will not infiltrate through them. This is a common problem in commercial buildings after about five years. One of the most common points of this air infiltration tends to be around the outside frame of a window or door where it rests against the surrounding wall (Fig. 8-36). Whether the window is operable (opens) or fixed has no effect on this infiltration since it occurs around the edges of the frame. This tendency can be reduced by any one method or a combination of several methods:

1. Detail the window and door to rest against a protruding portion of the structure or wall, as illustrated in Fig. 8-37. This reduces the chance of frame and wall separation which commonly occurs when a detail such as Fig. 8-36 is used and when the building undergoes expansion and contraction.

2. When using detail such as Fig. 8-38, place a compressible filler between the wall and the window frame. This filler will reduce the chance of air flow as materials contract.

3. All exterior cracks should be filled with a sealant designed to remain flexible so that it will hold to both the frame and the wall during periods of material contraction. Sealants should be called for in details, as shown in Fig. 8-38, and, when possible, they should be used in combination with compressible fillers.

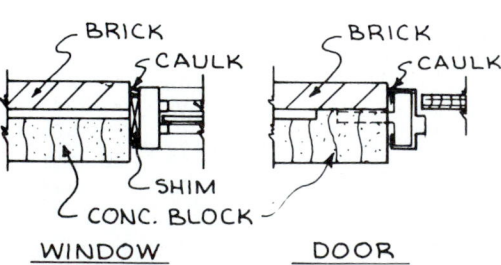

FIGURE 8-36   Frame details

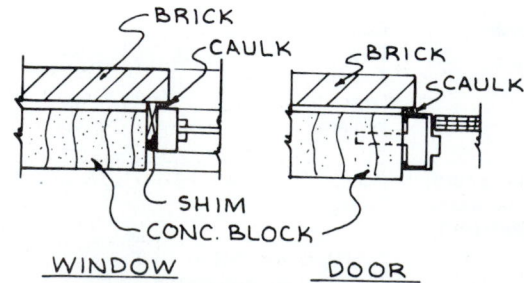

FIGURE 8-37   Frame details

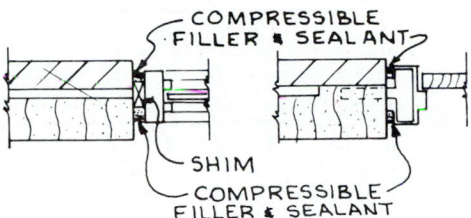

FIGURE 8-38   Frame details

These suggestions should be followed whenever different materials meet and at junctions in the walls, walls and floors, and walls and roof.

The importance of such air flow can be verified in any older building. On a cold day, especially if the wind is blowing, the air infiltrating between the frame and wall can be felt by placing your hand at the juncture of materials. This cool flow of air is usually not considered in the design calculations, yet there it is. In some buildings, with large amounts of glass and glass frame, the air infiltration becomes so great on a cold, windy day that it is impossible to heat the space to the design temperature. This can be solved in new construction by the use of the installation methods outlined. In existing buildings, it can be corrected by removing the dried caulking, resealing around the frames with compressible filler, and using a sealant inside and out.

It is important to realize that the designer of the heating system *must* review the drawings, details, and specifications to determine what is being called for on the project. But just as important, the designer has a responsibility to make suggestions for the architect to consider which will improve the heat flow characteristics.

8-15   FIREPLACES

Since heat rises, a fireplace in a room has a tendency to allow the warm inside air to flow up the chimney. Once that begins, a natural flow of air will occur, continuously drawing warm air up the chimney and pulling in cooler air from the exterior through infiltration (Fig. 8-39). This natural flow of air can be greatly reduced if the fireplace is equipped with a tight-fitting damper that is kept *closed* when the fireplace is not in use. This flow of air increases greatly when there is a fire in the fireplace, and the constant draft of air up the chimney may result in warm air in adjoining rooms being pulled toward the fireplace; these rooms then become cool as air is pulled into them through infiltration cracks. At the same time, the room with the fireplace will become quite warm, even overheated.

When the fuels used to provide heat were relatively inexpensive, little attention was paid to the heat loss through the fireplace. But the tremendous increases in fuel costs to date, along with even higher costs projected for the future, make it necessary to reduce heat loss as much as possible. This does not mean that fireplaces should no longer be built or used. It does mean that a few precautions must be taken to prevent some of the heat from escaping up the chimney.

1. Be certain that the fireplace has a tight-fitting damper and that the damper is closed when the fireplace is not being used.

2. Consider the use of a glass enclosure in front of the fireplace. This reduces the flow of air up the chimney, thus reducing the heat loss.

3. Consider a "snuff box" to put out the fire instead of leaving the fire to go out by itself. Even if the fire is almost out when you go to bed at night, the fire in the fireplace will probably still have hot ashes or coals in the morning. This means that all night long the air from inside the house flows up the chimney. The fire can be extinguished

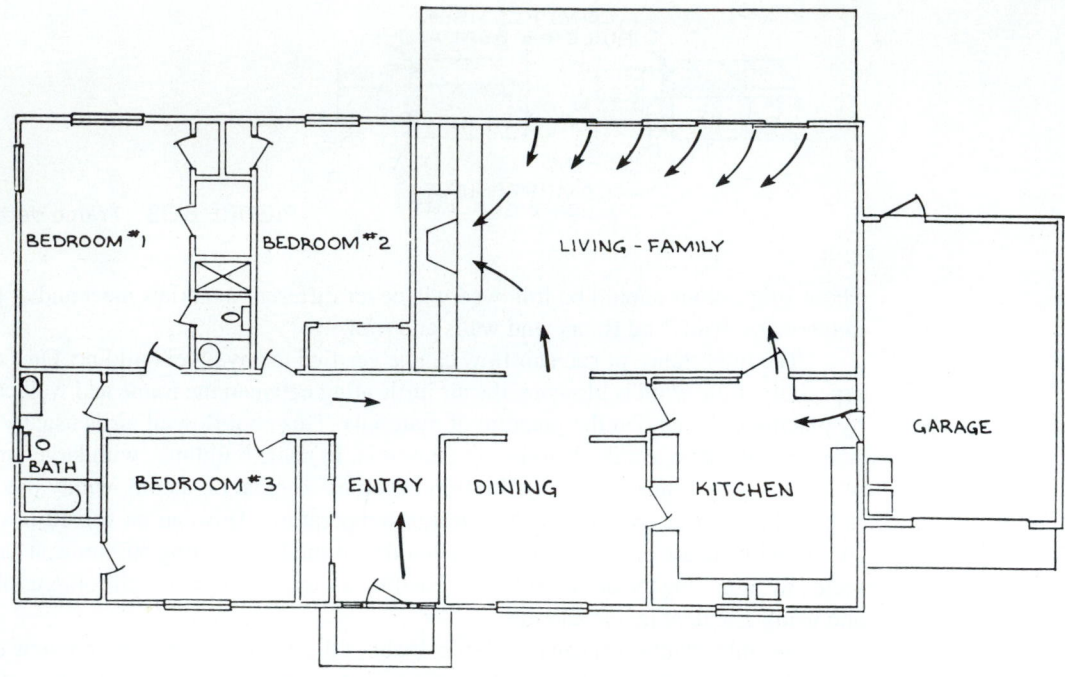

FIGURE 8-39 Flow of warm air toward fireplace

with a metal box that is placed over the fire to smother it. Then the damper can be closed and the flow of air reduced.

There are few calculations that a designer can make for fireplace heat loss. The designer must simply take this into consideration while designing the system by providing slightly more heat in surrounding rooms and by keeping the thermostat away from the fireplace, preferably in another room, to reduce the effect of fireplace heat on it.

8-16 MISCELLANEOUS HEAT LOSS

There are several other places where heat loss may occur that the designer should be aware of, even though the designer will not calculate the loss in designing the system.

  1. Electrical boxes in exterior (or cold) walls and ceilings should be insulated around and in back (Fig. 8-40) to reduce the flow of cold air.

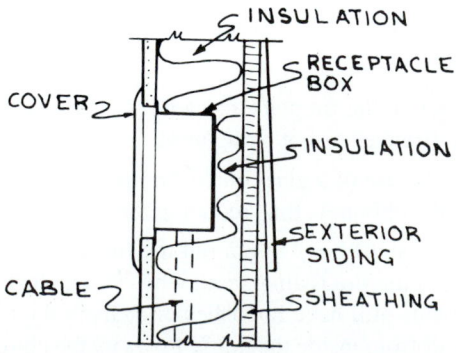

FIGURE 8-40 Insulated electrical box

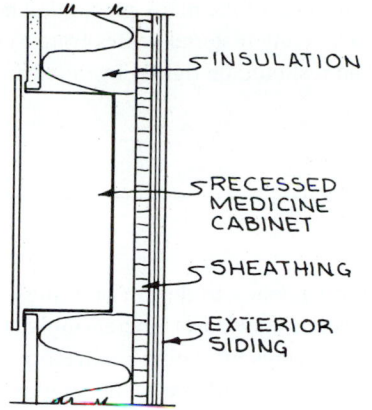

FIGURE 8-41   Recessed medicine cabinet

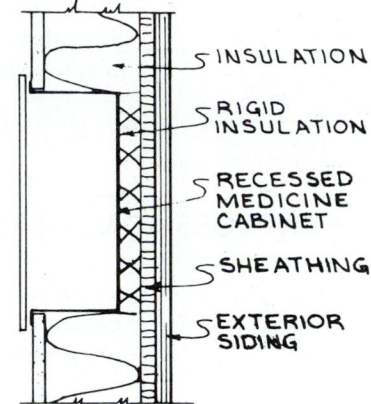

FIGURE 8-42   Insulated recessed medicine cabinet

2. Medicine cabinets recessed into exterior (or cold) walls leave little or no space for insulation between the cabinet and the exterior wall sheathing (Fig. 8-41). A recessed medicine cabinet should be located in an interior wall when possible. If located in an exterior wall, consider placing a thin sheet of rigid insulation, such as polystyrene, between the cabinet and the sheathing (Fig. 8-42).

3. The exhaust fan for a cooking area that is vented to the exterior will draw heat out of the space. Consider using a charcoal filter exhaust which cleans the air but blows it back into the room. This way heat from the space is not exhausted to the exterior, and the fact that the heat from the stove and range is put into the room instead of exhausted to the exterior is an extra bonus.

4. Exhaust fans in bathrooms do an excellent job of reducing humidity in the space and are also effective in eliminating odors. Since the air being exhausted is heated, it will increase the heat loss in the building. However, this can be greatly reduced if the fan is on a switch separate from the light switch (Fig. 8-43) so that it can be shut off when no longer needed.

5. All doorways to attics and cellars or other cold areas should have weatherstripping to reduce infiltration.

6. If the attic is reached through an access door or "hatch" in the ceiling of a heated space, be certain that the access door is insulated and weatherstripped to reduce heat loss.

7. Folding stairways into attics from heated spaces should also be insulated and weatherstripped to reduce heat loss.

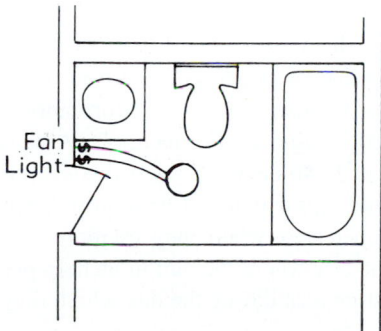

FIGURE 8-43   Separate switches

These are just several of the many places that miscellaneous heat loss can occur. Any place that a pipe, duct, or any material passes from a cold to a warm space is a possible location for heat loss, and it should be tightly installed and sealed and compressible filler used.

8-17 HEAT LOSS CONSIDERATIONS

There is an increased tendency to design buildings with as low a heat loss as possible to conserve fuel and reduce fuel costs. Increasingly, the R values are approaching 20 and often higher. Similar savings are being realized by increased use of insulating glass and weatherstripping and use of smaller amounts of glass. While taking all of these factors into account in the sizing of the system, the designer must be particularly careful since the U values for material assemblies used will result in a total system which is much smaller than in the past. Care must be taken so that the system will have sufficient capacity to provide any extra heat (above that calculated) that may be required due to:

1. The possibility that the materials will not be carefully installed. For example, insulation may not be packed in the smaller areas around windows and doors. Similar problems may occur when windows and doors are not installed snugly into the construction, allowing more infiltration than had been calculated.

2. The possibility that the builder (contractor) may use different materials from those on which the designer based the calculations (this may even be done with the architect's approval).

3. The probability that over the years the construction may deteriorate, allowing increased infiltration loss. This is a particular problem in buildings with window frames caulked into the wall similar to the detail shown in Fig. 8-36. If the sealant separates from the frame or the wall as the building settles, or if the sealant dries, cracks, and separates for any reason, the increased infiltration will require substantial extra heat. Obviously, the owner should have the windows resealed, but the initial separation occurs slowly, and the heat loss will slowly increase.

8-18 DEGREE DAYS

Degree days are a measure of approximately how much the ΔT is per year, and is determined by finding the *daily* difference between a base 65°F (19°C) inside temperature and the average outside temperature during a 24-hour period (not the low temperature but the *average* over the 24 hours). For example, if the average outside temperature is 40°F, then the ΔT for that day is 65°F − 40°F = 25°F. Using SI units, if the average outside temperature was 4°C then the ΔT for that day is 19° − 4° = 15°C. The ΔT for each day is added for a yearly total, which is given as degree days.

Figure 8-53 lists degree days for various cities; if one near your locale is not listed, contact the closest weather bureau for information. From Fig. 8-53, note the variations in degree days based on climate. For example, Albany, N.Y., has 6,900 degree days, while Raleigh, N.C., has 3,440 degree days at the airport. These degree days show the relative amount of heat and fuel which will be required. Since airports are located in the suburbs, and generally in open areas where they are more exposed to the climate, they have more degree days than the city where the buildings help protect one another and where the buildings and streets store heat during the day which they radiate at night.

8-19   FUEL COST COMPARISONS

The amount of fuel required for one year to heat a residence is determined by the following formula:

$$\frac{\text{Heat loss} \times \text{Degree days} \times 24 \text{ hours (a constant)}}{\text{Overall efficiency of fuel} \times \Delta T \times \text{Heating value of fuel}} = \begin{array}{c} \text{Amount of fuel required} \\ \text{in a year} \end{array}$$

For this example, assume 62,500 Btuh for baseboard, ceiling, or partition electric resistance heat and 75,000 Btuh for all other types of systems. The higher amount allows for an additional heat loss in the heating unit and in the pipes or ducts used to distribute the heat to the rooms.

Using the heating values and efficiency from Fig. 8-44, the amount of fuel required each year for a given residence would be:

Oil

$$\frac{(75,000)(3,440)(24)}{(0.80)(60)(140,000)} = 922 \text{ gal}$$

Natural gas

$$\frac{(75,000)(3,440)(24)}{(0.85)(60)(100,000)} = 1,215 \text{ ccf (hundred cubic feet)}$$

Propane

$$\frac{(75,000)(3,440)(24)}{(0.85)(60)(2,300)} = 52,788 \text{ c.f. or } 528 \text{ ccf}$$

Electric forced air

$$\frac{(75,000)(3,440)(24)}{(1.00)(60)(3,413)} = 30,237 \text{ kWh}$$

Electric baseboard, ceiling, partition

$$\frac{(62,500)(3,440)(24)}{(1.00)(60)(3,413)} = 25,198 \text{ kWh}$$

Now, the fuel costs per year can be estimated by multiplying the amount of fuel used times its cost. This is when the designer *must check the fuel costs in the locale where the*

Fuel	Btu	Unit	Efficiency (%)
Oil	140,000	Gal.	65-85
Natural gas	100,000	ccf	70-90
Propane	2,300	cf	70-90
Electricity	3,413	kw	100

cf = cubic feet
ccf = 100 cubic feet

FIGURE 8-44   Heating values and efficiencies

building will be built since it is these costs which determine which is the most economical system. Fuel costs vary considerably and are constantly changing, so call and get local prices for comparison purposes.

Oil	$1.10 per gallon
Natural Gas	$0.60 per ccf
Propane Gas	$2.40 per ccf
Electricity	$0.068 per kWh

The fuel costs per year are:

Oil

$$922 \text{ gal} \times \$1.10 \text{ per gal} = \$1,014.20 \text{ per year}$$

Natural gas

$$1,215 \text{ ccf} \times \$0.60 \text{ per ccf} = \$729 \text{ per year}$$

Propane gas

$$528 \text{ ccf} \times \$2.40 \text{ per ccf} = \$1,267.20 \text{ per year}$$

Electric forced air

$$30,237 \text{ kWh} \times \$0.068 \text{ per kwh} = \$2,056.12 \text{ per year}$$

Electric baseboard, ceiling, partition

$$25,198 \text{ kWh} \times \$0.068 \text{ per kwh} = \$1,713.47 \text{ per year}$$

A comparison of the figures makes it obvious that oil and natural gas are the preferred methods of heating a building. Electricity does not become cost-competitive until systems such as a heat pump are used (Sec. 14-8). In the future, costs will continue to vary, but the chances are that they will remain, proportionally, about the same. When one fuel price goes up, all others do also. After the oil embargo, the retail price of oil went up about 100% due to increased prices paid for the oil; it is significant that the prices of coal, gas and electricity all went up about the same percentage. This is because manufacturers and suppliers price their product based on what the alternatives are. The coal used is mined here near its point of use; the only reason the price of coal doubled is that the price of oil doubled. It is likely that this trend will continue in the future. (This is true in all business, not simply fuels, and it is mentioned here to explain how businesses operate, disregarding who may be involved.) This increasing cost of fuel is the primary reason that such an emphasis is being placed on heating with solar energy. It has greatly stimulated research and development in the solar energy field and has made solar energy systems a practical investment now.

8-20   CONTROLLING HEAT LOSS

Basically, all residences should be insulated in all portions of the building which are exposed to colder temperatures. By observing several considerations, the total heat loss of the building is reduced by one-half (and more) in the typical home built before 1976. Basically, the heat loss is reduced by:

1. 6 in. (150 mm) or more of insulation in the crawl space or basement.
2. 6 in. (150 mm) of insulation in the exterior walls ($R = 19$). (Use 2-in. $\times$ 6-in. studs for framing the walls instead of 2×4.)
3. 12 in. (300 mm) of insulation in the ceiling ($R = 38$).
4. Insulated glass windows.
5. A window area limited to 8% of the floor area. (This does away with the large glass windows and sliding glass doors which have such high heat losses. In the past, it is estimated that the average home had windows which totalled 12% of the floor area.)
6. An insulated metal entry door with 1¾-in. (45 mm) polyurethane insulation.
7. Humidity controlled by a humidifier and a dehumidifier.
8. Locating forced air heating ducts in the heated area. (Drop the ceilings in corridors and install ducts along the upper perimeters of interior walls to reduce the heat loss from the ducts.)

Comparative studies of actual homes built in Arkansas show that heat loss can be reduced by much more than one-half. The real significance of this study is that for very little additional money actually spent on a house, significant savings can be realized. The type of construction, together with a solar heating system or with solar-assisted heat pumps, provides a major basis for future energy savings. A full report on "The Arkansas Story," *Report No. 1, Energy Conservation Ideas to Build On* is available through Owens-Corning Fiberglass Corp., Insulation Operating Division, Fiberglass Tower, Toledo, Ohio 43659. This report belongs in your library.

Another interesting investigation of energy conservation ideas, including the use of solar energy, was conducted by NASA (National Aeronautics and Space Administration) with similar results. However, the use of solar energy *and* water conservation techniques makes it doubly interesting. Copies of this study, "Technical support package for Tech. Brief LAR—12134," NASA Technology Utilization House, are available from Technology Utilization Office, NASA Langley Research Center, Hampton, Virginia 23665.

Typical Thermal Properties of Common Building and Insulating Materials—Design Values[a]

Description	Density, lb/ft³	Conductivity[b] (k), $\dfrac{\text{Btu} \cdot \text{in}}{\text{h} \cdot \text{ft}^2 \cdot {}^\circ\text{F}}$	Conductance (C), $\dfrac{\text{Btu}}{\text{h} \cdot \text{ft}^2 \cdot {}^\circ\text{F}}$	Resistance[c] (R) Per Inch Thickness $(1/k)$, $\dfrac{{}^\circ\text{F} \cdot \text{ft}^2 \cdot \text{h}}{\text{Btu} \cdot \text{in}}$	Resistance[c] (R) For Thickness Listed $(1/C)$, $\dfrac{{}^\circ\text{F} \cdot \text{ft}^2 \cdot \text{h}}{\text{Btu}}$	Specific Heat, $\dfrac{\text{Btu}}{\text{lb} \cdot {}^\circ\text{F}}$
BUILDING BOARD						
Asbestos-cement board	120	4.0	—	0.25	—	0.24
Asbestos-cement board ....0.125 in.	120	—	33.00	—	0.03	
Asbestos-cement board ....0.25 in.	120	—	16.50	—	0.06	
Gypsum or plaster board........0.375 in.	50	—	3.10	—	0.32	0.26
Gypsum or plaster board.........0.5 in.	50	—	2.22	—	0.45	
Gypsum or plaster board.......0.625 in.	50	—	1.78	—	0.56	
Plywood (Douglas Fir)[d]	34	0.80	—	1.25	—	0.29
Plywood (Douglas Fir) ....0.25 in.	34	—	3.20	—	0.31	
Plywood (Douglas Fir) ....0.375 in.	34	—	2.13	—	0.47	
Plywood (Douglas Fir) ....0.5 in.	34	—	1.60	—	0.62	
Plywood (Douglas Fir) ....0.625 in.	34	—	1.29	—	0.77	
Plywood or wood panels........0.75 in.	34	—	1.07	—	0.93	0.29
Vegetable fiber board						
Sheathing, regular density[e] ......0.5 in.	18	—	0.76	—	1.32	0.31
..............0.78125 in.	18	—	0.49	—	2.06	
Sheathing intermediate density[e] ....0.5 in.	22	—	0.92	—	1.09	0.31
Nail-base sheathing[e].........0.5 in.	25	—	0.94	—	1.06	0.31
Shingle backer ....0.375 in.	18	—	1.06	—	0.94	0.31
Shingle backer ....0.3125 in.	18	—	1.28	—	0.78	
Sound deadening board..........0.5 in.	15	—	0.74	—	1.35	0.30
Tile and lay-in panels, plain or acoustic	18	0.40	—	2.50	—	0.14
.....0.5 in.	18	—	0.80	—	1.25	
.....0.75 in.	18	—	0.53	—	1.89	
Laminated paperboard	30	0.50	—	2.00	—	0.33
Homogeneous board from repulped paper	30	0.50	—	2.00	—	0.28
Hardboard[e]						
Medium density	50	0.73	—	1.37	—	0.31
High density, service-tempered grade and service grade	55	0.82	—	1.22	—	0.32
High density, standard-tempered grade	63	1.00	—	1.00	—	0.32
Particleboard[e]						
Low density	37	0.71	—	1.41	—	0.31
Medium density	50	0.94	—	1.06	—	0.31
High density	62.5	1.18	—	0.85	—	0.31
Underlayment.........0.625 in.	40	—	1.22	—	0.82	0.29
Waferboard	37	0.63	—	1.59	—	
Wood subfloor ....0.75 in.	—	—	1.06	—	0.94	0.33
BUILDING MEMBRANE						
Vapor—permeable felt	—	—	16.70	—	0.06	
Vapor—seal, 2 layers of mopped 15-lb felt	—	—	8.35	—	0.12	
Vapor—seal, plastic film	—	—	—	—	Negl.	
FINISH FLOORING MATERIALS						
Carpet and fibrous pad	—	—	0.48	—	2.08	0.34
Carpet and rubber pad	—	—	0.81	—	1.23	0.33
Cork tile ....0.125 in.	—	—	3.60	—	0.28	0.48
Terrazzo ....1 in.	—	—	12.50	—	0.08	0.19
Tile—asphalt, linoleum, vinyl, rubber	—	—	20.00	—	0.05	0.30
vinyl asbestos						0.24
ceramic						0.19
Wood, hardwood finish ....0.75 in.	—	—	1.47	—	0.68	
INSULATING MATERIALS						
Blanket and Batt[f,g]						
Mineral fiber, fibrous form processed from rock, slag, or glass						
approx. 3–4 in.	0.4–2.0	—	0.091	—	11	
approx. 3.5 in.	0.4–2.0	—	0.077	—	13	
approx. 3.5 in.	1.2–1.6	—	0.067	—	15	
approx. 5.5–6.5 in.	0.4–2.0	—	0.053	—	19	
approx. 5.5 in.	0.6–1.0	—	0.048	—	21	
approx. 6–7.5 in.	0.4–2.0	—	0.045	—	22	
approx. 8.25–10 in.	0.4–2.0	—	0.033	—	30	
approx. 10–13 in.	0.4–2.0	—	0.026	—	38	
Board and Slabs						
Cellular glass	8.0	0.33	—	3.03	—	0.18
Glass fiber, organic bonded	4.0–9.0	0.25	—	4.00	—	0.23
Expanded perlite, organic bonded	1.0	0.36	—	2.78	—	0.30
Expanded rubber (rigid)	4.5	0.22	—	4.55	—	0.40
Expanded polystyrene, extruded (smooth skin surface) (CFC-12 exp.)	1.8–3.5	0.20	—	5.00	—	0.29
Expanded polystyrene, extruded (smooth skin surface) (HCFC-142b exp.)[h]	1.8–3.5	0.20	—	5.00	—	0.29

FIGURE 8-45 Typical thermal properties of common building and insulating materials—design values

Typical Thermal Properties of Common Building and Insulating Materials—Design Values[a] (Continued)

Description	Density, lb/ft³	Conductivity[b] (k), Btu·in / h·ft²·°F	Conductance (C), Btu / h·ft²·°F	Resistance[c] (R) Per Inch Thickness (1/k), °F·ft²·h / Btu·in	For Thickness Listed (1/C), °F·ft²·h / Btu	Specific Heat, Btu / lb·°F
Expanded polystyrene, molded beads	1.0	0.26	—	3.85	—	—
	1.25	0.25	—	4.00	—	—
	1.5	0.24	—	4.17	—	—
	1.75	0.24	—	4.17	—	—
	2.0	0.23	—	4.35	—	—
Cellular polyurethane/polyisocyanurate[i] (CFC-11 exp.) (unfaced)	1.5	0.16–0.18	—	6.25–5.56	—	0.38
Cellular polyisocyanurate[i] (CFC-11 exp.)(gas-permeable facers)	1.5–2.5	0.16–0.18	—	6.25–5.56	—	0.22
Cellular polyisocyanurate[j] (CFC-11 exp.)(gas-impermeable facers)...........	2.0	0.14	—	7.04	—	0.22
Cellular phenolic (closed cell)(CFC-11, CFC-113 exp.)	3.0	0.12	—	8.20	—	—
Cellular phenolic (open cell)	1.8–2.2	0.23	—	4.40	—	—
Mineral fiber with resin binder	15.0	0.29	—	3.45	—	0.17
Mineral fiberboard, wet felted						
Core or roof insulation........................	16–17	0.34	—	2.94	—	—
Acoustical tile	18.0	0.35	—	2.86	—	0.19
Acoustical tile	21.0	0.37	—	2.70	—	—
Mineral fiberboard, wet molded						
Acoustical tile[k]	23.0	0.42	—	2.38	—	0.14
Wood or cane fiberboard						
Acoustical tile,[k] ....0.5 in.	—	—	0.80	—	1.25	0.31
Acoustical tile[k] ....0.75 in.	—	—	0.53	—	1.89	—
Interior finish (plank, tile)	15.0	0.35	—	2.86	—	0.32
Cement fiber slabs (shredded wood with Portland cement binder)	25–27.0	0.50–0.53	—	2.0–1.89	—	—
Cement fiber slabs (shredded wood with magnesia oxysulfide binder)	22.0	0.57	—	1.75	—	0.31
Loose Fill						
Cellulosic insulation (milled paper or wood pulp)	2.3–3.2	0.27–0.32	—	3.70–3.13	—	0.33
Perlite, expanded	2.0–4.1	0.27–0.31	—	3.7–3.3	—	0.26
	4.1–7.4	0.31–0.36	—	3.3–2.8	—	—
	7.4–11.0	0.36–0.42	—	2.8–2.4	—	—
Mineral fiber (rock, slag, or glass)[g]						
approx. 3.75–5 in.	0.6–2.0	—	—	—	11.0	0.17
approx. 6.5–8.75 in.	0.6–2.0	—	—	—	19.0	—
approx. 7.5–10 in.	0.6–2.0	—	—	—	22.0	—
approx. 10.25–13.75 in.	0.6–2.0	—	—	—	30.0	—
Mineral fiber (rock, slag, or glass)[g]						
approx. 3.5 in. (closed sidewall application)	2.0–3.5	—	—	—	12.0–14.0	—
Vermiculite, exfoliated	7.0–8.2	0.47	—	2.13	—	0.32
	4.0–6.0	0.44	—	2.27	—	—
Spray Applied						
Polyurethane foam	1.5–2.5	0.16–0.18	—	6.25–5.56	—	—
Ureaformaldehyde foam	0.7–1.6	0.22–0.28	—	4.55–3.57	—	—
Cellulosic fiber	3.5–6.0	0.29–0.34	—	3.45–2.94	—	—
Glass fiber	3.5–4.5	0.26–0.27	—	3.85–3.70	—	—

METALS
(See Chapter 36, Table 3)

ROOFING

Description	Density, lb/ft³	Conductivity (k)	Conductance (C)	1/k	1/C	Specific Heat
Asbestos-cement shingles	120	—	4.76	—	0.21	0.24
Asphalt roll roofing	70	—	6.50	—	0.15	0.36
Asphalt shingles	70	—	2.27	—	0.44	0.30
Built-up roofing ....0.375 in.	70	—	3.00	—	0.33	0.35
Slate...................................0.5 in.	—	—	20.00	—	0.05	0.30
Wood shingles, plain and plastic film faced	—	—	1.06	—	0.94	0.31

PLASTERING MATERIALS

Description	Density, lb/ft³	Conductivity (k)	Conductance (C)	1/k	1/C	Specific Heat
Cement plaster, sand aggregate....................	116	5.0	—	0.20	—	0.20
Sand aggregate ....0.375 in.	—	—	13.3	—	0.08	0.20
Sand aggregate ....0.75 in.	—	—	6.66	—	0.15	0.20
Gypsum plaster:						
Lightweight aggregate ....0.5 in.	45	—	3.12	—	0.32	—
Lightweight aggregate ....0.625 in.	45	—	2.67	—	0.39	—
Lightweight aggregate on metal lath ....0.75 in.	—	—	2.13	—	0.47	—
Perlite aggregate	45	1.5	—	0.67	—	0.32
Sand aggregate	105	5.6	—	0.18	—	0.20
Sand aggregate ....0.5 in.	105	—	11.10	—	0.09	—
Sand aggregate ....0.625 in.	105	—	9.10	—	0.11	—
Sand aggregate on metal lath ....0.75 in.	—	—	7.70	—	0.13	—
Vermiculite aggregate	45	1.7	—	0.59	—	—

MASONRY MATERIALS
Masonry Units

Description	Density, lb/ft³	Conductivity (k)	Conductance (C)	1/k	1/C	Specific Heat
Brick, fired clay................................	150	8.4–10.2	—	0.12–0.10	—	—
	140	7.4–9.0	—	0.14–0.11	—	—
	130	6.4–7.8	—	0.16–0.12	—	—
	120	5.6–6.8	—	0.18–0.15	—	0.19
	110	4.9–5.9	—	0.20–0.17	—	—

FIGURE 8-45 (continued)

Typical Thermal Properties of Common Building and Insulating Materials—Design Values[a] (Continued)

Description	Density, lb/ft³	Conductivity[b] (k), Btu·in / h·ft²·°F	Conductance (C), Btu / h·ft²·°F	Resistance[c] (R) Per Inch Thickness (1/k), °F·ft²·h / Btu·in	For Thickness Listed (1/C), °F·ft²·h / Btu	Specific Heat, Btu / lb·°F
Brick, fired clay *continued*	100	4.2–5.1	—	0.24–0.20	—	—
	90	3.6–4.3	—	0.28–0.24	—	—
	80	3.0–3.7	—	0.33–0.27	—	—
	70	2.5–3.1	—	0.40–0.33	—	—
Clay tile, hollow						
1 cell deep .... 3 in.	—	—	1.25	—	0.80	0.21
1 cell deep .... 4 in.	—	—	0.90	—	1.11	—
2 cells deep .... 6 in.	—	—	0.66	—	1.52	—
2 cells deep .... 8 in.	—	—	0.54	—	1.85	—
2 cells deep .... 10 in.	—	—	0.45	—	2.22	—
3 cells deep .... 12 in.	—	—	0.40	—	2.50	—
Concrete blocks[i]						
Limestone aggregate						
8 in., 36 lb, 138 lb/ft³ concrete, 2 cores	—	—	—	—	—	—
Same with perlite filled cores	—	—	0.48	—	2.1	—
12 in., 55 lb, 138 lb/ft³ concrete, 2 cores	—	—	—	—	—	—
Same with perlite filled cores	—	—	0.27	—	3.7	—
Normal weight aggregate (sand and gravel)						
8 in., 33-36 lb, 126-136 lb/ft³ concrete, 2 or 3 cores	—	—	0.90–1.03	—	1.11–0.97	0.22
Same with perlite filled cores	—	—	0.50	—	2.0	—
Same with verm. filled cores	—	—	0.52–0.73	—	1.92–1.37	—
12 in., 50 lb, 125 lb/ft³ concrete, 2 cores	—	—	0.81	—	1.23	0.22
Medium weight aggregate (combinations of normal weight and lightweight aggregate)						
8 in., 26-29 lb, 97-112 lb/ft³ concrete, 2 or 3 cores..	—	—	0.58–0.78	—	1.71–1.28	—
Same with perlite filled cores	—	—	0.27–0.44	—	3.7–2.3	—
Same with verm. filled cores	—	—	0.30	—	3.3	—
Same with molded EPS (beads) filled cores	—	—	0.32	—	3.2	—
Same with molded EPS inserts in cores	—	—	0.37	—	2.7	—
Lightweight aggregate (expanded shale, clay, slate or slag, pumice)						
6 in., 16-17 lb 85-87 lb/ft³ concrete, 2 or 3 cores...	—	—	0.52–0.61	—	1.93–1.65	—
Same with perlite filled cores	—	—	0.24	—	4.2	—
Same with verm. filled cores	—	—	0.33	—	3.0	—
8 in., 19-22 lb, 72-86 lb/ft³ concrete,	—	—	0.32–0.54	—	3.2–1.90	0.21
Same with perlite filled cores	—	—	0.15–0.23	—	6.8–4.4	—
Same with verm. filled cores	—	—	0.19–0.26	—	5.3–3.9	—
Same with molded EPS (beads) filled cores	—	—	0.21	—	4.8	—
Same with UF foam filled cores	—	—	0.22	—	4.5	—
Same with molded EPS inserts in cores	—	—	0.29	—	3.5	—
12 in., 32-36 lb, 80–90 lb/ft³ concrete, 2 or 3 cores...	—	—	0.38–0.44	—	2.6–2.3	—
Same with perlite filled cores	—	—	0.11–0.16	—	9.2–6.3	—
Same with verm. filled cores	—	—	0.17	—	5.8	—
Stone, lime, or sand						
Quartzitic and sandstone	180	72	—	0.01	—	—
	160	43	—	0.02	—	—
	140	24	—	0.04	—	—
	120	13	—	0.08	—	0.19
Calcitic, dolomitic, limestone, marble, and granite..	180	30	—	0.03	—	—
	160	22	—	0.05	—	—
	140	16	—	0.06	—	—
	120	11	—	0.09	—	0.19
	100	8	—	0.13	—	—
Gypsum partition tile						
3 by 12 by 30 in., solid	—	—	0.79	—	1.26	0.19
3 by 12 by 30 in., 4 cells	—	—	0.74	—	1.35	—
4 by 12 by 30 in., 3 cells	—	—	0.60	—	1.67	—
Concretes						
Sand and gravel or stone aggregate concretes (concretes with more than 50% quartz or quartzite sand have conductivities in the higher end of the range) ..	150	10.0–20.0	—	0.10–0.05	—	—
	140	9.0–18.0	—	0.11–0.06	—	0.19–0.24
	130	7.0–13.0	—	0.14–0.08	—	—
Limestone concretes	140	11.1	—	0.09	—	—
	120	7.9	—	0.13	—	—
	100	5.5	—	0.18	—	—
Gypsum-fiber concrete (87.5% gypsum, 12.5% wood chips)	51	1.66	—	0.60	—	0.21
Cement/lime, mortar, and stucco	120	9.7	—	0.10	—	—
	100	6.7	—	0.15	—	—
	80	4.5	—	0.22	—	—
Lightweight aggregate concretes						
Expanded shale, clay, or slate; expanded slags; cinders; pumice (with density up to 100 lb/ft³); and scoria (sanded concretes have conductivities in the higher end of the range)	120	6.4–9.1	—	0.16–0.11	—	—
	100	4.7–6.2	—	0.21–0.16	—	0.20
	80	3.3–4.1	—	0.30–0.24	—	0.20
	60	2.1–2.5	—	0.48–0.40	—	—
	40	1.3	—	0.78	—	—

FIGURE 8-45   (*continued*)

Typical Thermal Properties of Common Building and Insulating Materials—Design Values[a] (Concluded)

Description	Density, lb/ft^3	Conductivity[b] (k), $\frac{\text{Btu} \cdot \text{in}}{\text{h} \cdot \text{ft}^2 \cdot {}^\circ\text{F}}$	Conductance (C), $\frac{\text{Btu}}{\text{h} \cdot \text{ft}^2 \cdot {}^\circ\text{F}}$	Resistance[c] (R) Per Inch Thickness $(1/k)$, $\frac{{}^\circ\text{F} \cdot \text{ft}^2 \cdot \text{h}}{\text{Btu} \cdot \text{in}}$	Resistance[c] (R) For Thickness Listed $(1/C)$, $\frac{{}^\circ\text{F} \cdot \text{ft}^2 \cdot \text{h}}{\text{Btu}}$	Specific Heat, $\frac{\text{Btu}}{\text{lb} \cdot {}^\circ\text{F}}$
Perlite, vermiculite, and polystyrene beads	50	1.8–1.9	—	0.55–0.53	—	—
	40	1.4–1.5	—	0.71–0.67	—	0.15–0.23
	30	1.1	—	0.91	—	—
	20	0.8	—	1.25	—	—
Foam concretes	120	5.4	—	0.19	—	—
	100	4.1	—	0.24	—	—
	80	3.0	—	0.33	—	—
	70	2.5	—	0.40	—	—
Foam concretes and cellular concretes	60	2.1	—	0.48	—	—
	40	1.4	—	0.71	—	—
	20	0.8	—	1.25	—	—
SIDING MATERIALS (on flat surface)						
Shingles						
Asbestos-cement	120	—	4.75	—	0.21	—
Wood, 16 in., 7.5 exposure	—	—	1.15	—	0.87	0.31
Wood, double, 16-in., 12-in. exposure	—	—	0.84	—	1.19	0.28
Wood, plus insul. backer board, 0.3125 in.	—	—	0.71	—	1.40	0.31
Siding						
Asbestos-cement, 0.25 in., lapped	—	—	4.76	—	0.21	0.24
Asphalt roll siding	—	—	6.50	—	0.15	0.35
Asphalt insulating siding (0.5 in. bed.)	—	—	0.69	—	1.46	0.35
Hardboard siding, 0.4375 in.	—	—	1.49	—	0.67	0.28
Wood, drop, 1 by 8 in.	—	—	1.27	—	0.79	0.28
Wood, bevel, 0.5 by 8 in., lapped	—	—	1.23	—	0.81	0.28
Wood, bevel, 0.75 by 10 in., lapped	—	—	0.95	—	1.05	0.28
Wood, plywood, 0.375 in., lapped	—	—	1.59	—	0.59	0.29
Aluminum or Steel[m], over sheathing						
Hollow-backed	—	—	1.61	—	0.61	0.29
Insulating-board backed nominal 0.375 in.	—	—	0.55	—	1.82	0.32
Insulating-board backed nominal 0.375 in., foil backed	—	—	0.34	—	2.96	—
Architectural (soda-lime float) glass	158	6.9	—	—	—	0.21
WOODS (12% moisture content)[e,n]						
Hardwoods						0.39[o]
Oak	41.2–46.8	1.12–1.25	—	0.89–0.80	—	
Birch	42.6–45.4	1.16–1.22	—	0.87–0.82	—	
Maple	39.8–44.0	1.09–1.19	—	0.92–0.84	—	
Ash	38.4–41.9	1.06–1.14	—	0.94–0.88	—	
Softwoods						0.39[o]
Southern Pine	35.6–41.2	1.00–1.12	—	1.00–0.89	—	
Douglas Fir-Larch	33.5–36.3	0.95–1.01	—	1.06–0.99	—	
Southern Cypress	31.4–32.1	0.90–0.92	—	1.11–1.09	—	
Hem-Fir, Spruce-Pine-Fir	24.5–31.4	0.74–0.90	—	1.35–1.11	—	
West Coast Woods, Cedars	21.7–31.4	0.68–0.90	—	1.48–1.11	—	
California Redwood	24.5–28.0	0.74–0.82	—	1.35–1.22	—	

[a] Values are for a mean temperature of 75 °F. Representative values for dry materials are intended as design (not specification) values for materials in normal use. Thermal values of insulating materials may differ from design values depending on their in-situ properties (e.g., density and moisture content, orientation, etc.) and variability experienced during manufacture. For properties of a particular product, use the value supplied by the manufacturer or by unbiased tests.

[b] To obtain thermal conductivities in Btu/h·ft·°F, divide the k-factor by 12 in./ft.

[c] Resistance values are the reciprocals of C before rounding off C to two decimal places.

[d] Lewis (1967).

[e] U.S. Department of Agriculture (1974).

[f] Does not include paper backing and facing, if any. Where insulation forms a boundary (reflective or otherwise) of an airspace, see Tables 2 and 3 for the insulating value of an airspace with the appropriate effective emittance and temperature conditions of the space.

[g] Conductivity varies with fiber diameter. (See Chapter 20, Factors Affecting Thermal Performance.) Batt, blanket, and loose-fill mineral fiber insulations are manufactured to achieve specified R-values, the most common of which are listed in the table. Due to differences in manufacturing processes and materials, the product thicknesses, densities, and thermal conductivities vary over considerable ranges for a specified R-value.

[h] This material is relatively new and data are based on limited testing.

[i] For additional information, see Society of Plastics Engineers (SPI) *Bulletin* U108. Values are for aged, unfaced board stock. For change in conductivity with age of expanded polyurethane/polyisocyanurate, see Chapter 20, Factors Affecting Thermal Performance.

[j] Values are for aged products with gas-impermeable facers on the two major surfaces. An aluminum foil facer of 0.001 in. thickness or greater is generally considered impermeable to gases. For change in conductivity with age of expanded polyisocyanurate, see Chapter 20, Factors Affecting Thermal Performance, and SPI *Bulletin* U108.

[k] Insulating values of acoustical tile vary, depending on density of the board and on type, size, and depth of perforations.

[l] Values for fully grouted block may be approximated using values for concrete with a similar unit weight.

[m] Values for metal siding applied over flat surfaces vary widely, depending on amount of ventilation of airspace beneath the siding; whether airspace is reflective of non-reflective; and on thickness, type, and application of insulating backing-board used. Values given are averages for use as design guides, and were obained from several guarded hot box tests (ASTM C236) or calibrated hot box (ASTM C976) on hollow-backed types and types made using backing-boards of wood fiber, foamed plastic, and glass fiber. Departures of ±50% or more from the values given may occur.

[n] See Adams (1971), MacLean (1941), and Wilkes (1979). The conductivity values listed are for heat transfer across the grain. The thermal conductivity of wood varies linearly with the density, and the density ranges listed are those normally found for the wood species given. If the density of the wood species is not known, use the mean conductivity value. For extrapolation to other moisture contents, the following empirical equation developed by Wilkes (1979) may be used:

$$k = 0.1791 + \frac{(1.874 \times 10^{-2} + 5.753 \times 10^{-4}M)\rho}{1 + 0.01M}$$

where ρ is density of the moist wood in lb/ft^3, and M is the moisture content in percent.

[o] From Wilkes (1979), an empirical equation for the specific heat of moist wood at 75 °F is as follows:

$$c_p = \frac{(0.299 + 0.01M)}{(1 + 0.01M)} + \Delta c_p$$

where Δc_p accounts for the heat of sorption and is denoted by

$$\Delta c_p = M(1.921 \times 10^{-3} - 3.168 \times 10^{-5}M)$$

where M is the moisture content in percent by mass.

FIGURE 8-45 *(continued)*

Typical Thermal Properties of Common Building and Insulating Materials—Design Values[a]

Description	Density, kg/m³	Conductivity[b] (k), W/(m·K)	Conductance (C), W/(m²·K)	Resistance [c] (R)		Specific Heat, kJ/(kg·K)
				(1/k), K·m/W	For Thickness Listed (1/C), K·m²/W	
BUILDING BOARD						
Asbestos-cement board	1900	0.58	—	1.73	—	1.00
Asbestos-cement board ....3.2 mm	1900	—	187.4	—	0.005	—
Asbestos-cement board ....6.4 mm	1900	—	93.7	—	0.011	—
Gypsum or plaster board....................9.5 mm	800	—	17.6	—	0.056	1.09
Gypsum or plaster board...................12.7 mm	800	—	12.6	—	0.079	
Gypsum or plaster board...................15.9 mm	800	—	10.1	—	0.099	
Plywood (Douglas Fir)[d]	540	0.12	—	8.66	—	1.21
Plywood (Douglas Fir) ....6.4 mm	540	—	18.2	—	0.055	—
Plywood (Douglas Fir) ....9.5 mm	540	—	12.1	—	0.083	—
Plywood (Douglas Fir) ....12.7 mm	540	—	9.1	—	0.11	—
Plywood (Douglas Fir) ....15.9 mm	540	—	7.3	—	0.14	—
Plywood or wood panels ....19.0 mm	540	—	6.1	—	0.16	1.21
Vegetable fiber board						
Sheathing, regular density...............12.7 mm	290	—	4.3	—	0.23	1.30
.................19.8 mm	290	—	2.8	—	0.36	—
Sheathing intermediate density.........12.7 mm	350	—	5.2	—	0.19	1.30
Nail-base sheathing ....12.7 mm	400	—	5.3	—	0.19	1.30
Shingle backer ....9.5 mm	290	—	6.0	—	0.17	1.30
Shingle backer ....7.9 mm	290	—	7.3	—	0.14	—
Sound deadening board..................12.7 mm	240	—	4.2	—	0.24	1.26
Tile and lay-in panels, plain or acoustic	290	0.058	—	17.3	—	0.59
....12.7 mm	290	—	4.5	—	0.22	—
....19.0 mm	290	—	3.0	—	0.33	—
Laminated paperboard......................	480	0.072	—	13.9	—	1.38
Homogeneous board from repulped paper	480	0.072	—	13.9	—	1.17
Hardboard						
Medium density..........................	800	0.105	—	9.50	—	1.30
High density, service-tempered grade and service grade..................................	880	0.82	—	8.46	—	1.34
High density, standard-tempered grade	1010	0.144	—	6.93	—	1.34
Particleboard						
Low density.............................	590	0.102	—	9.77	—	1.30
Medium density..........................	800	0.135	—	7.35	—	1.30
High density............................	1000	0.170	—	5.90	—	1.30
Underlayment.........................15.9 mm	640	—	6.9	—	0.14	1.21
Waferboard	590	0.01	—	11.0	—	—
Wood subfloor ....19.0 mm	—	—	6.0	—	0.17	1.38
BUILDING MEMBRANE						
Vapor—permeable felt........................	—	—	94.9	—	0.011	
Vapor—seal, 2 layers of mopped 0.73 kg/m² felt	—	—	47.4	—	0.21	
Vapor—seal, plastic film.....................	—	—	—	—	Negl.	
FINISH FLOORING MATERIALS						
Carpet and fibrous pad	—	—	2.73	—	0.37	1.42
Carpet and rubber pad	—	—	4.60	—	0.22	1.38
Cork tile ....3.2 mm	—	—	20.4	—	0.049	2.01
Terrazzo ....25 mm	—	—	71.0	—	0.014	0.80
Tile—asphalt, linoleum, vinyl, rubber	—	—	113.6	—	0.009	1.26
vinyl asbestos................................						1.01
ceramic..........................						0.80
Wood, hardwood finish ....19 mm			8.35	—	0.12	
INSULATING MATERIALS						
Blanket and Batt [f,g]						
Mineral fiber, fibrous form processed from rock, slag, or glass						
approx. 75–100 mm	6.4–32	—	0.52	—	1.94	
approx. 90 mm	6.4–32	—	0.44	—	2.29	
approx. 90 mm	19–26	—	0.38	—	2.63	
approx. 140–165 mm	6.4–32	—	0.30	—	3.32	
approx. 140 mm	10–16	—	0.27	—	3.67	
approx. 150–190 mm	6.4–32	—	0.26	—	3.91	
approx. 210–250 mm	6.4–32	—	0.19	—	5.34	
approx. 250–330 mm	6.4–32	—	0.15	—	6.77	
Board and Slabs						
Cellular glass	136	0.050	—	19.8	—	0.75
Glass fiber, organic bonded	64–140	0.036	—	27.7	—	0.96
Expanded perlite, organic bonded	16	0.052	—	19.3	—	1.26
Expanded rubber (rigid)	72	0.032	—	31.6	—	1.68
Expanded polystyrene extruded (smooth skin surface) (CFC-12 exp.)	29–56	0.029	—	34.7	—	1.22
Expanded polystyrene, extruded (smooth skin surface) (HCFC-142b exp.)[h]..........................	29–56	0.029	—	34.7	—	1.21

FIGURE 8-45   (*continued*)

Typical Thermal Properties of Common Building and Insulating Materials—Design Values[a] (*Continued*)

Description	Density, kg/m³	Conductivity[b] (k), W/(m·K)	Conductance (C), W/(m²·K)	Resistance [c](R) (1/k), K·m/W	For Thickness Listed (1/C), K·m²/W	Specific Heat, kJ/(kg·K)
Expanded polystyrene, molded beads	16	0.037	—	26.7	—	—
	20	0.036	—	27.7	—	—
	24	0.035	—	28.9	—	—
	28	0.035	—	28.9	—	—
	32	0.033	—	30.2	—	—
Cellular polyurethane/polyisocyanurate[i] (CFC-11 exp.) (unfaced).........................	24	0.023–0.026	—	43.3–38.5	—	1.59
Cellular polyisocyanurate[i] (CFC-11 exp.) (gas-permeable facers)	24–40	0.023–0.026	—	43.3–38.5	—	0.92
Cellular polyisocyanurate[j] (CFC-11 exp.) (gas-impermeable facers)..........	32	0.020	—	48.8	—	0.92
Cellular phenolic (closed cell) (CFC-11, CFC-113 exp.)	32	0.017	—	56.8	—	—
Cellular phenolic (open cell)......................	29–35	0.033	—	30.5	—	—
Mineral fiber with resin binder	240	0.042	—	23.9	—	0.71
Mineral fiberboard, wet felted						
Core or roof insulation........................	260–270	0.049	—	20.4	—	
Acoustical tile	290	0.050	—	19.8	—	0.80
Acoustical tile	340	0.053	—	18.7	—	
Mineral fiberboard, wet molded						
Acoustical tile[k]	370	0.060	—	16.5	—	0.59
Wood or cane fiberboard						
Acoustical tile[k] ....12.7 mm	—	—	0.80	—	1.25	1.30
Acoustical tile[k] ....19.0 mm	—	—	0.53	—	1.89	
Interior finish (plank, tile)	240	0.050	—	19.8	—	1.34
Cement fiber slabs (shredded wood with Portland cement binder)	400–430	0.072–0.076	—	13.9–13.1	—	—
Cement fiber slabs (shredded wood with magnesia oxysulfide binder)	350	0.082	—	12.1	—	1.30
Loose Fill						
Cellulosic insulation (milled paper or wood pulp)	37–51	0.039–0.046	—	25.6–21.7	—	1.38
Perlite, expanded	32–66	0.039–0.045	—	25.6–22.9	—	1.09
	66–120	0.045–0.052	—	22.9–19.4	—	—
	120–180	0.052–0.060	—	19.4–16.6	—	—
Mineral fiber (rock, slag, or glass)[g]						
approx. 95–130 mm	9.6–32	—	—	—	1.94	0.71
approx. 170–220 mm	9.6–32	—	—	—	3.35	—
approx. 190–250 mm	9.6–32	—	—	—	3.87	—
approx. 260–350 mm	9.6–32	—	—	—	5.28	—
Mineral fiber (rock, slag, or glass)[g]						
approx. 90 mm (closed sidewall application)	32–56	—	—	—	2.1–2.5	—
Vermiculite, exfoliated...........................	110–130	0.068	—	14.8	—	1.34
	64–96	0.063	—	15.7	—	—
Spray Applied						
Polyurethane foam	24–40	0.023–0.026	—	43.3–38.5	—	—
Ureaformaldehyde foam	11–26	0.032–0.040	—	31.5–24.7	—	—
Cellulosic fiber	56–96	0.042–0.049	—	23.9–20.4	—	—
Glass fiber	56–72	0.038–0.039	—	26.7–25.6	—	—

METALS
(See Chapter 36, Table 3)

ROOFING

Description	Density, kg/m³	Conductivity (k), W/(m·K)	Conductance (C), W/(m²·K)	(1/k), K·m/W	For Thickness Listed (1/C), K·m²/W	Specific Heat, kJ/(kg·K)
Asbestos-cement shingles	1900	—	27.0	—	0.037	1.00
Asphalt roll roofing	1100	—	36.9	—	0.026	1.51
Asphalt shingles	1100	—	12.9	—	0.077	1.26
Built-up roofing.......................10 mm	1100	—	17.0	—	0.058	1.46
Slate ....13 mm	—	—	114	—	0.009	1.26
Wood shingles, plain and plastic film faced	—	—	6.0	—	0.166	1.30

PLASTERING MATERIALS

Description	Density, kg/m³	Conductivity (k), W/(m·K)	Conductance (C), W/(m²·K)	(1/k), K·m/W	For Thickness Listed (1/C), K·m²/W	Specific Heat, kJ/(kg·K)
Cement plaster, sand aggregate.....................	1860	0.72	—	1.39	—	0.84
Sand aggregate.......................95 mm	—	—	75.5	—	0.08	0.84
Sand aggregate.......................19 mm	—	—	37.8	—	0.15	0.84
Gypsum plaster:						
Low density aggregate ....127 mm	720	—	17.7	—	0.32	—
Low density aggregate ....16 mm	720	—	15.2	—	0.39	—
Low density agg. on metal lath ....19 mm	—	—	12.1	—	0.47	—
Perlite aggregate	720	0.22	—	4.64	—	1.34
Sand aggregate	1680	0.81	—	1.25	—	0.84
Sand aggregate ....127 mm	1680	—	63.0	—	0.09	—
Sand aggregate ....16 mm	1680	—	51.7	—	0.11	—
Sand aggregate on metal lath ....19 mm	—	—	43.7	—	0.13	—
Vermiculite aggregate	720	0.24	—	4.09	—	—

MASONRY MATERIALS
Masonry Units

Description	Density, kg/m³	Conductivity (k), W/(m·K)	Conductance (C), W/(m²·K)	(1/k), K·m/W	For Thickness Listed (1/C), K·m²/W	Specific Heat, kJ/(kg·K)
Brick, fired clay...............................	2400	1.21–1.47	—	0.83–0.68	—	—
	2240	1.07–1.30	—	0.94–0.77	—	—
	2080	0.92–1.12	—	1.08–0.89	—	—
	1920	0.81–0.98	—	1.24–1.02	—	0.79
	1760	0.71–0.85	—	1.42–1.18	—	—

FIGURE 8-45   (*continued*)

Typical Thermal Properties of Common Building and Insulating Materials—Design Values[a] (*Continued*)

Description	Density, kg/m³	Conductivity[b] (*k*), W/(m·K)	Conductance (*C*), W/(m²·K)	Resistance [c](*R*) (1/*k*), K·m/W	Resistance [c](*R*) For Thickness Listed (1/*C*), K·m²/W	Specific Heat, kJ/(kg·K)
Brick, fired clay *continued*	1600	0.61–0.74	—	1.65–1.36	—	—
	1440	0.52–0.62	—	1.93–1.61	—	—
	1280	0.43–0.53	—	2.31–1.87	—	—
	1120	0.36–0.45	—	2.77–2.23	—	—
Clay tile, hollow						
1 cell deep ....76 mm	—	—	7.10	—	0.14	0.88
1 cell deep ....102 mm	—	—	5.11	—	0.20	—
2 cells deep ....152 mm	—	—	3.75	—	0.27	—
2 cells deep ....203 mm	—	—	3.07	—	0.33	—
2 cells deep ....254 mm	—	—	2.56	—	0.39	—
3 cells deep ....305 mm	—	—	2.27	—	0.44	—
Concrete blocks[i]						
Limestone aggregate						
200 mm, 16.3 kg, 2210 kg/m³ concrete, 2 cores ...	—	—	—	—	—	—
Same with perlite filled cores	—	—	2.73	—	0.37	—
300 mm, 25 kg, 2210 kg/m³ concrete, 2 cores	—	—	—	—	—	—
Same with perlite filled cores	—	—	1.53	—	0.65	—
Normal mass aggregate (sand and gravel) 200 mm,						
15-16 kg, 2020-2180 kg/m³ concrete, 2 or 3 cores ..	—	—	5.1–5.8	—	0.20–0.17	0.92
Same with perlite filled cores	—	—	2.84	—	0.35	—
Same with vermiculite filled cores	—	—	3.0–4.1	—	0.34–0.24	—
300 mm, 22.7 kg, 2000 kg/m³ concrete, 2 cores ...	—	—	4.60	—	0.217	0.92
Medium mass aggregate (combinations of normal and low mass aggregate) 200 mm, 12-13 kg,						
1550-1790 kg/m³ concrete, 2 or 3 cores	—	—	3.3–4.4	—	0.30–0.22	—
Same with perlite filled cores	—	—	1.5–2.5	—	0.65–0.41	—
Same with vermiculite filled cores	—	—	1.70	—	0.58	—
Same with molded EPS (beads) filled cores	—	—	1.82	—	0.56	—
Same with molded EPS inserts in cores	—	—	2.10	—	0.47	—
Low mass aggregate (expanded shale, clay, slate or slag, pumice) 150 mm,						
7.3-7.7 kg, 1360-1390 kg/m³ concrete, 2 or 3 cores	—	—	3.0–3.5	—	0.34–0.29	—
Same with perlite filled cores	—	—	1.36	—	0.74	—
Same with vermiculite filled cores	—	—	1.87	—	0.53	—
200 mm, 8.6-10.0 mm, 1150-1380 kg/m³ concrete,	—	—	1.8–3.1	—	0.56–0.33	0.88
Same with perlite filled cores	—	—	0.9–1.3	—	1.20–0.77	—
Same with vermiculite filled cores	—	—	1.1–1.5	—	0.93–0.69	—
Same with molded EPS (beads) filled cores	—	—	1.19	—	0.85	—
Same with ureaformaldehyde foam filled cores .	—	—	1.25	—	0.79	—
Same with molded EPS inserts in cores	—	—	1.65	—	0.62	—
300 mm, 14.5-16.3 kg, 1280-1440 kg/m³ concrete, 2 or 3 cores	—	—	2.2–2.5	—	0.46–0.40	—
Same with perlite filled cores	—	—	0.6–0.9	—	1.6–1.1	—
Same with vermiculite filled cores	—	—	0.97	—	1.0	—
Stone, lime, or sand						
Quartzitic and sandstone	2880	10.4	—	0.10	—	—
	2560	6.2	—	0.16	—	—
	2240	3.5	—	0.29	—	—
	1920	1.9	—	0.53	—	0.79
Calcitic, dolomitic, limestone, marble, and granite..	2880	4.3	—	0.23	—	—
	2560	3.2	—	0.32	—	—
	2240	2.3	—	0.43	—	—
	1920	1.6	—	0.63	—	0.79
	1600	1.1	—	0.90	—	—
Gypsum partition tile						
76 by 305 by 760, solid	—	—	4.50	—	0.222	0.79
76 by 305 by 760, 4 cells	—	—	4.20	—	0.238	—
102 by 305 by 760, 3 cells	—	—	3.40	—	0.294	—
Concretes						
Sand and gravel or stone aggregate concretes (concretes with more than 50% quartz or quartzite sand have conductivities in the higher end of the range) ..	2400	1.4–2.9	—	0.69–0.35	—	—
	2240	1.3–2.6	—	0.77–0.39	—	0.8–1.0
	2080	1.0–1.9	—	0.99–0.53	—	—
Limestone concretes	2240	1.60	—	0.62	—	—
	1920	1.14	—	0.88	—	—
	1600	0.79	—	1.26	—	—
Gypsum-fiber concrete (87.5% gypsum, 12.5% wood chips)	816	0.24	—	4.18	—	0.88
Cement/lime, mortar, and stucco	1920	1.40	—	0.71	—	—
	1600	0.97	—	1.04	—	—
	1280	0.65	—	1.54	—	—
Low density aggregate concretes						
Expanded shale, clay, or slate; expanded slags; cinders; pumice (with density up to 1600 kg/m³); and scoria (sanded concretes have conductivities in the higher end of the range)	1920	0.9–1.3	—	1.08–0.76	—	—
	1600	0.68–0.89	—	1.48–1.12	—	0.84
	1280	0.48–0.59	—	2.10–1.69	—	0.84
	960	0.30–0.36	—	3.30–2.77	—	—
	640	0.18	—	5.40	—	—

FIGURE 8-45 (*continued*)

Typical Thermal Properties of Common Building and Insulating Materials—Design Values[a] (*Concluded*)

Description	Density, kg/m³	Conductivity[b] (k), W/(m·K)	Conductance (C), W/(m²·K)	Resistance[c] (R) (1/k), K·m/W	Resistance[c] (R) For Thickness Listed (1/C), K·m²/W	Specific Heat, kJ/(kg·K)
Perlite, vermiculite, and polystyrene beads	800	0.26–0.27	—	3.81–3.68	—	—
	640	0.20–0.22	—	4.92–4.65	—	0.63–0.96
	480	0.16	—	6.31	—	—
	320	0.12	—	8.67	—	—
Foam concretes	1920	0.75	—	1.32	—	—
	1600	0.60	—	1.66	—	—
	1280	0.44	—	2.29	—	—
	1120	0.36	—	2.77	—	—
Foam concretes and cellular concretes	960	0.30	—	3.33	—	—
	640	0.20	—	4.92	—	—
	320	0.12	—	8.67	—	—

SIDING MATERIALS (on flat surface)
Shingles

Description	Density, kg/m³	Conductivity (k), W/(m·K)	Conductance (C), W/(m²·K)	(1/k), K·m/W	For Thickness Listed (1/C), K·m²/W	Specific Heat, kJ/(kg·K)
Asbestos-cement	1900	—	27.0	—	0.037	—
Wood, 400 mm, 190-mm exposure	—	—	6.53	—	0.15	1.30
Wood, double, 400 mm, 300-mm exposure	—	—	4.77	—	0.21	1.17
Wood, plus insul. backer board, 8 mm	—	—	4.03	—	0.25	1.30

Siding

Description	Density	k	Conductance (C)	(1/k)	(1/C)	Specific Heat
Asbestos-cement, 6.4 mm, lapped	—	—	27.0	—	0.037	1.01
Asphalt roll siding..........................	—	—	36.9	—	0.026	1.47
Asphalt insulating siding (12.7 mm bed.)...........	—	—	3.92	—	0.26	1.47
Hardboard siding, 11 mm	—	—	8.46	—	0.12	1.17
Wood, drop, 20 by 200 mm	—	—	7.21	—	0.14	1.17
Wood, bevel, 13 by 200 mm, lapped..............	—	—	6.98	—	0.14	1.17
Wood, bevel, 19 by 250 mm, lapped..............	—	—	5.40	—	0.18	1.17
Wood, plywood, 9.5 mm, lapped	—	—	9.03	—	0.10	1.22
Aluminum or Steel[m], over sheathing						
Hollow-backed	—	—	9.14	—	0.11	1.22
Insulating-board backed						
9.5 mm nominal	—	—	3.12	—	0.32	1.34
9.5 mm nominal, foil backed.................	—	—	1.93	—	0.52	—
Architectural (soda lime float) glass	—	—	56.8	—	0.018	0.84

WOODS (12% moisture content)[e,n]
Hardwoods   1.63[o]

Description	Density	Conductivity (k)	Conductance (C)	(1/k)	(1/C)	Specific Heat
Oak	659–749	0.16–0.18	—	6.2–5.5	—	
Birch	682–726	0.167–0.176	—	6.0–5.7	—	
Maple	637–704	0.157–0.171	—	6.4–5.8	—	
Ash	614–670	0.153–0.164	—	6.5–6.1	—	

Softwoods   1.63[o]

Description	Density	Conductivity (k)	Conductance (C)	(1/k)	(1/C)	Specific Heat
Southern Pine	570–659	0.144–0.161	—	6.9–6.2	—	
Douglas Fir-Larch.............................	536–581	0.137–0.145	—	7.3–6.9	—	
Southern Cypress	502–514	0.130–0.132	—	7.7–7.6	—	
Hem-Fir, Spruce-Pine-Fir	392–502	0.107–0.130	—	9.3–7.7	—	
West Coast Woods, Cedars	347–502	0.098–0.130	—	10.3–7.7	—	
California Redwood	392–448	0.107–0.118	—	9.4–8.5	—	

[a] Values are for a mean temperature of 24 °C. Representative values for dry materials are intended as design (not specification) values for materials in normal use. Thermal values of insulating materials may differ from design values depending on their in-situ properties (*e.g.*, density and moisture content, orientation, etc.) and variability experienced during manufacture. For properties of a particular product, use the value supplied by the manufacturer or by unbiased tests.

[b] The symbol λ is also used to represent thermal conductivity.

[c] Resistance values are the reciprocals of C before rounding off C.

[d] Lewis (1967).

[e] U.S. Department of Agriculture (1974).

[f] Does not include paper backing and facing, if any. Where insulation forms a boundary (reflective or otherwise) of an air space, see Tables 2 and 3 for the insulating value of an air space with the appropriate effective emittance and temperature conditions of the space.

[g] Conductivity varies with fiber diameter. (See Chapter 20, Factors Affecting Thermal Performance.) Batt, blanket, and loose-fill mineral fiber insulations are manufactured to achieve specified R-values, the most common of which are listed in the table. Due to differences in manufacturing processes and materials, the product thicknesses, densities, and thermal conductivities vary over considerable ranges for a specified R-value.

[h] This material is relatively new and data are based on limited testing.

[i] For additional information, see Society of Plastics Engineers (SPI) *Bulletin* U108. Values are for aged, unfaced board stock. For change in conductivity with age of expanded polyurethane/polyisocyanurate, see Chapter 20, Factors Affecting Thermal Performance.

[j] Values are for aged products with gas-impermeable facers on the two major surfaces. An aluminum foil facer of 25 μm thickness or greater is generally considered impermeable to gases. For change in conductivity with age of expanded polyisocyanurate, see Chapter 20, Factors Affecting Thermal Performance, and SPI *Bulletin* U108.

[k] Insulating values of acoustical tile vary, depending on density of the board and on type, size, and depth of perforations.

[l] Values for fully grouted block may be approximated using values for concrete with a similar density.

[m] Values for metal siding applied over flat surfaces vary widely, depending on amount of ventilation of air space beneath the siding; whether air space is reflective or non-reflective; and on thickness, type, and application of insulating backing-board used. Values given are averages for use as design guides, and were obained from several guarded hot box tests (ASTM C236) or calibrated hot box (ASTM C976) on hollow-backed types and types made using backing-boards of wood fiber, foamed plastic, and glass fiber. Departures of ±50% or more from the values given may occur.

[n] See Adams (1971), MacLean (1941), and Wilkes (1979). The conductivity values listed are for heat transfer across the grain. The thermal conductivity of wood varies linearly with the density, and the density ranges listed are those normally found for the wood species given. If the density of the wood species is not known, use the mean conductivity value. For extrapolation to other moisture contents, the following empirical equation developed by Wilkes (1979) may be used:

$$k = 0.7494 + \frac{(4.895 \times 10^{-3} + 1.503 \times 10^{-4}M)\rho}{1 + 0.01M}$$

where ρ is density of the moist wood in kg/m³, and M is the moisture content in percent.

[o] From Wilkes (1979), an empirical equation for the specific heat of moist wood at 24 °C is as follows:

$$c_p = 0.1442 \left[\frac{(0.299 + 0.01M)}{(1 + 0.01M)} \right] + \Delta c_p$$

where Δc_p accounts for the heat of sorption and is denoted by

$$\Delta c_p = M(0.008037 - 1.325 \times 10^{-4}M)$$

where M is the moisture content in percent by mass.

FIGURE 8-45 (*continued*)

Surface Conductances and Resistances for Air

Position of Surface	Direction of Heat Flow	Surface Emittance, ϵ					
		Non-reflective $\epsilon = 0.90$		Reflective $\epsilon = 0.20$		$\epsilon = 0.05$	
		h_i	R	h_i	R	h_i	R
STILL AIR							
Horizontal	Upward	1.63	0.61	0.91	1.10	0.76	1.32
Sloping—45°	Upward	1.60	0.62	0.88	1.14	0.73	1.37
Vertical	Horizontal	1.46	0.68	0.74	1.35	0.59	1.70
Sloping—45°	Downward	1.32	0.76	0.60	1.67	0.45	2.22
Horizontal	Downward	1.08	0.92	0.37	2.70	0.22	4.55
MOVING AIR (Any position)		h_o	R	h_o	R	h_o	R
15-mph Wind (for winter)	Any	6.00	0.17	—	—	—	—
7.5-mph Wind (for summer)	Any	4.00	0.25	—	—	—	—

SI Units **Surface Conductances and Resistances for Air**

Position of Surface	Direction of Heat Flow	Surface Emittance, ϵ					
		Non-reflective $\epsilon = 0.90$		Reflective $\epsilon = 0.20$		$\epsilon = 0.05$	
		h_i	R	h_i	R	h_i	R
STILL AIR							
Horizontal	Upward	9.26	0.11	5.17	0.19	4.32	0.23
Sloping—45°	Upward	9.09	0.11	5.00	0.20	4.15	0.24
Vertical	Horizontal	8.29	0.12	4.20	0.24	3.35	0.30
Sloping—45°	Downward	7.50	0.13	3.41	0.29	2.56	0.39
Horizontal	Downward	6.13	0.16	2.10	0.48	1.25	0.80
MOVING AIR (Any position)		h_o	R	h_o	R	h_o	R
Wind (for winter) 6.7 m/s (24 km/h)	Any	34.0	0.030	—	—	—	—
Wind (for summer) 3.4 m/s (12 km/h)	Any	22.7	0.044	—	—	—	—

FIGURE 8-46   Surface conductances and resistances for air

Thermal Resistances of Plane Air Spaces[a,b,c], °F·ft²·h/Btu

Position of Air Space	Direction of Heat Flow	Air Space Mean Temp.[d], °F	Temp. Diff.[d], °F	0.5-in. Air Space[c] Effective Emittance ϵ_{eff}[d,e]					0.75-in. Air Space[c] Effective Emittance ϵ_{eff}[d,e]				
				0.03	0.05	0.2	0.5	0.82	0.03	0.05	0.2	0.5	0.82
Horiz. Up		90	10	2.13	2.03	1.51	0.99	0.73	2.34	2.22	1.61	1.04	0.75
		50	30	1.62	1.57	1.29	0.96	0.75	1.71	1.66	1.35	0.99	0.77
		50	10	2.13	2.05	1.60	1.11	0.84	2.30	2.21	1.70	1.16	0.87
		0	20	1.73	1.70	1.45	1.12	0.91	1.83	1.79	1.52	1.16	0.93
		0	10	2.10	2.04	1.70	1.27	1.00	2.23	2.16	1.78	1.31	1.02
		−50	20	1.69	1.66	1.49	1.23	1.04	1.77	1.74	1.55	1.27	1.07
		−50	10	2.04	2.00	1.75	1.40	1.16	2.16	2.11	1.84	1.46	1.20
45° Slope Up		90	10	2.44	2.31	1.65	1.06	0.76	2.96	2.78	1.88	1.15	0.81
		50	30	2.06	1.98	1.56	1.10	0.83	1.99	1.92	1.52	1.08	0.82
		50	10	2.55	2.44	1.83	1.22	0.90	2.90	2.75	2.00	1.29	0.94
		0	20	2.20	2.14	1.76	1.30	1.02	2.13	2.07	1.72	1.28	1.00
		0	10	2.63	2.54	2.03	1.44	1.10	2.72	2.62	2.08	1.47	1.12
		−50	20	2.08	2.04	1.78	1.42	1.17	2.05	2.01	1.76	1.41	1.16
		−50	10	2.62	2.56	2.17	1.66	1.33	2.53	2.47	2.10	1.62	1.30
Vertical Horiz.		90	10	2.47	2.34	1.67	1.06	0.77	3.50	3.24	2.08	1.22	0.84
		50	30	2.57	2.46	1.84	1.23	0.90	2.91	2.77	2.01	1.30	0.94
		50	10	2.66	2.54	1.88	1.24	0.91	3.70	3.46	2.35	1.43	1.01
		0	20	2.82	2.72	2.14	1.50	1.13	3.14	3.02	2.32	1.58	1.18
		0	10	2.93	2.82	2.20	1.53	1.15	3.77	3.59	2.64	1.73	1.26
		−50	20	2.90	2.82	2.35	1.76	1.39	2.90	2.83	2.36	1.77	1.39
		−50	10	3.20	3.10	2.54	1.87	1.46	3.72	3.60	2.87	2.04	1.56
45° Slope Down		90	10	2.48	2.34	1.67	1.06	0.77	3.53	3.27	2.10	1.22	0.84
		50	30	2.64	2.52	1.87	1.24	0.91	3.43	3.23	2.24	1.39	0.99
		50	10	2.67	2.55	1.89	1.25	0.92	3.81	3.57	2.40	1.45	1.02
		0	20	2.91	2.80	2.19	1.52	1.15	3.75	3.57	2.63	1.72	1.26
		0	10	2.94	2.83	2.21	1.53	1.15	4.12	3.91	2.81	1.80	1.30
		−50	20	3.16	3.07	2.52	1.86	1.45	3.78	3.65	2.90	2.05	1.57
		−50	10	3.26	3.16	2.58	1.89	1.47	4.35	4.18	3.22	2.21	1.66
Horiz. Down		90	10	2.48	2.34	1.67	1.06	0.77	3.55	3.29	2.10	1.22	0.85
		50	30	2.66	2.54	1.88	1.24	0.91	3.77	3.52	2.38	1.44	1.02
		50	10	2.67	2.55	1.89	1.25	0.92	3.84	3.59	2.41	1.45	1.02
		0	20	2.94	2.83	2.20	1.53	1.15	4.18	3.96	2.83	1.81	1.30
		0	10	2.96	2.85	2.22	1.53	1.16	4.25	4.02	2.87	1.82	1.31
		−50	20	3.25	3.15	2.58	1.89	1.47	4.60	4.41	3.36	2.28	1.69
		−50	10	3.28	3.18	2.60	1.90	1.47	4.71	4.51	3.42	2.30	1.71

Position of Air Space	Direction of Heat Flow	Air Space Mean Temp., °F	Temp. Diff., °F	1.5-in. Air Space[c]					3.5-in. Air Space[c]				
Horiz. Up		90	10	2.55	2.41	1.71	1.08	0.77	2.84	2.66	1.83	1.13	0.80
		50	30	1.87	1.81	1.45	1.04	0.80	2.09	2.01	1.58	1.10	0.84
		50	10	2.50	2.40	1.81	1.21	0.89	2.80	2.66	1.95	1.28	0.93
		0	20	2.01	1.95	1.63	1.23	0.97	2.25	2.18	1.79	1.32	1.03
		0	10	2.43	2.35	1.90	1.38	1.06	2.71	2.62	2.07	1.47	1.12
		−50	20	1.94	1.91	1.68	1.36	1.13	2.19	2.14	1.86	1.47	1.20
		−50	10	2.37	2.31	1.99	1.55	1.26	2.65	2.58	2.18	1.67	1.33
45° Slope Up		90	10	2.92	2.73	1.86	1.14	0.80	3.18	2.96	1.97	1.18	0.82
		50	30	2.14	2.06	1.61	1.12	0.84	2.26	2.17	1.67	1.15	0.86
		50	10	2.88	2.74	1.99	1.29	0.94	3.12	2.95	2.10	1.34	0.96
		0	20	2.30	2.23	1.82	1.34	1.04	2.42	2.35	1.90	1.38	1.06
		0	10	2.79	2.69	2.12	1.49	1.13	2.98	2.87	2.23	1.54	1.16
		−50	20	2.22	2.17	1.88	1.49	1.21	2.34	2.29	1.97	1.54	1.25
		−50	10	2.71	2.64	2.23	1.69	1.35	2.87	2.79	2.33	1.75	1.39
Vertical Horiz.		90	10	3.99	3.66	2.25	1.27	0.87	3.69	3.40	2.15	1.24	0.85
		50	30	2.58	2.46	1.84	1.23	0.90	2.67	2.55	1.89	1.25	0.91
		50	10	3.79	3.55	2.39	1.45	1.02	3.63	3.40	2.32	1.42	1.01
		0	20	2.76	2.66	2.10	1.48	1.12	2.88	2.78	2.17	1.51	1.14
		0	10	3.51	3.35	2.51	1.67	1.23	3.49	3.33	2.50	1.67	1.23
		−50	20	2.64	2.58	2.18	1.66	1.33	2.82	2.75	2.30	1.73	1.37
		−50	10	3.31	3.21	2.62	1.91	1.48	3.40	3.30	2.67	1.94	1.50
45° Slope Down		90	10	5.07	4.55	2.56	1.36	0.91	4.81	4.33	2.49	1.34	0.90
		50	30	3.58	3.36	2.31	1.42	1.00	3.51	3.30	2.28	1.40	1.00
		50	10	5.10	4.66	2.85	1.60	1.09	4.74	4.36	2.73	1.57	1.08
		0	20	3.85	3.66	2.68	1.74	1.27	3.81	3.63	2.66	1.74	1.27
		0	10	4.92	4.62	3.16	1.94	1.37	4.59	4.32	3.02	1.88	1.34
		−50	20	3.62	3.50	2.80	2.01	1.54	3.77	3.64	2.90	2.05	1.57
		−50	10	4.67	4.47	3.40	2.29	1.70	4.50	4.32	3.31	2.25	1.68
Horiz. Down		90	10	6.09	5.35	2.79	1.43	0.94	10.07	8.19	3.41	1.57	1.00
		50	30	6.27	5.63	3.18	1.70	1.14	9.60	8.17	3.86	1.88	1.22
		50	10	6.61	5.90	3.27	1.73	1.15	11.15	9.27	4.09	1.93	1.24
		0	20	7.03	6.43	3.91	2.19	1.49	10.90	9.52	4.87	2.47	1.62
		0	10	7.31	6.66	4.00	2.22	1.51	11.97	10.32	5.08	2.52	1.64
		−50	20	7.73	7.20	4.77	2.85	1.99	11.64	10.49	6.02	3.25	2.18
		−50	10	8.09	7.52	4.91	2.89	2.01	12.98	11.56	6.36	3.34	2.22

[a] See Chapter 20, section Factors Affecting Heat Transfer across Air Spaces. Thermal resistance values were determined from the relation, $R = 1/C$, where $C = h_c + \epsilon_{eff} h_r$, h_c is the conduction-convection coefficient, $\epsilon_{eff} h_r$ is the radiation coefficient $\cong$ $0.00686 \epsilon_{eff} [(t_m + 460)/100]^3$, and t_m is the mean temperature of the air space. Values for h_c were determined from data developed by Robinson et al. (1954). Equations (5) through (7) in Yarbrough (1983) show the data in Table 2 in analytic form. For extrapolation from Table 2 to air spaces less than 0.5 in. (as in insulating window glass), assume $h_c = 0.159(1 + 0.0016 t_m)/l$ where l is the air space thickness in inches, and h_c is heat transfer through the air space only.

[b] Values are based on data presented by Robinson et al. (1954). (Also see Chapter 3, Tables 3 and 4, and Chapter 39). Values apply for ideal conditions, i.e., air spaces of uniform thickness bounded by plane, smooth, parallel surfaces with no air leakage to or from the space. When accurate values are required, use overall U-factors determined through calibrated hot box (ASTM C 976) or guarded hot box (ASTM C 236) testing. Thermal resistance values for multiple air spaces must be based on careful estimates of mean temperature differences for each air space.

[c] A single resistance value cannot account for multiple air spaces; each air space requires a separate resistance calculation that applies only for the established boundary conditions. Resistances of horizontal spaces with heat flow downward are substantially independent of temperature difference.

[d] Interpolation is permissible for other values of mean temperature, temperature difference, and effective emittance ϵ_{eff}. Interpolation and moderate extrapolation for air spaces greater than 3.5 in. are also permissible.

[e] Effective emittance ϵ_{eff} of the air space is given by $1/\epsilon_{eff} = 1/\epsilon_1 + 1/\epsilon_2 - 1$, where ϵ_1 and ϵ_2 are the emittances of the surfaces of the air space (see Table 3).

FIGURE 8-47 Thermal resistances of plane air spaces

SI UNITS   Thermal Resistances of Plane Air Spaces[a,b,c], K · m²/W

Position of Air Space	Direction of Heat Flow	Air Space Mean Temp.[d], °C	Air Space Temp. Diff.[d], °C	13-mm Air Space[c] Effective Emittance ϵ_{eff}[d,e] 0.03	0.05	0.2	0.5	0.82	20-mm Air Space[c] Effective Emittance ϵ_{eff}[d,e] 0.03	0.05	0.2	0.5	0.82
Horiz. Up ↑		32.2	5.6	0.37	0.36	0.27	0.17	0.13	0.41	0.39	0.28	0.18	0.13
		10.0	16.7	0.29	0.28	0.23	0.17	0.13	0.30	0.29	0.24	0.17	0.14
		10.0	5.6	0.37	0.36	0.28	0.20	0.15	0.40	0.39	0.30	0.20	0.15
		−17.8	11.1	0.30	0.30	0.26	0.20	0.16	0.32	0.32	0.27	0.20	0.16
		−17.8	5.6	0.37	0.36	0.30	0.22	0.18	0.39	0.38	0.31	0.23	0.18
		−45.6	11.1	0.30	0.29	0.26	0.22	0.18	0.31	0.31	0.27	0.22	0.19
		−45.6	5.6	0.36	0.35	0.31	0.25	0.20	0.38	0.37	0.32	0.26	0.21
45° Slope Up ↗		32.2	5.6	0.43	0.41	0.29	0.19	0.13	0.52	0.49	0.33	0.20	0.14
		10.0	16.7	0.36	0.35	0.27	0.19	0.15	0.35	0.34	0.27	0.19	0.14
		10.0	5.6	0.45	0.43	0.32	0.21	0.16	0.51	0.48	0.35	0.23	0.17
		−17.8	11.1	0.39	0.38	0.31	0.23	0.18	0.37	0.36	0.30	0.23	0.18
		−17.8	5.6	0.46	0.45	0.36	0.25	0.19	0.48	0.46	0.37	0.26	0.20
		−45.6	11.1	0.37	0.36	0.31	0.25	0.21	0.36	0.35	0.31	0.25	0.20
		−45.6	5.6	0.46	0.45	0.38	0.29	0.23	0.45	0.43	0.37	0.29	0.23
Vertical Horiz. →		32.2	5.6	0.43	0.41	0.29	0.19	0.14	0.62	0.57	0.37	0.21	0.15
		10.0	16.7	0.45	0.43	0.32	0.22	0.16	0.51	0.49	0.35	0.23	0.17
		10.0	5.6	0.47	0.45	0.33	0.22	0.16	0.65	0.61	0.41	0.25	0.18
		−17.8	11.1	0.50	0.48	0.38	0.26	0.20	0.55	0.53	0.41	0.28	0.21
		−17.8	5.6	0.52	0.50	0.39	0.27	0.20	0.66	0.63	0.46	0.30	0.22
		−45.6	11.1	0.51	0.50	0.41	0.31	0.24	0.51	0.50	0.42	0.31	0.24
		−45.6	5.6	0.56	0.55	0.45	0.33	0.26	0.65	0.63	0.51	0.36	0.27
45° Slope Down ↘		32.2	5.6	0.44	0.41	0.29	0.19	0.14	0.62	0.58	0.37	0.21	0.15
		10.0	16.7	0.46	0.44	0.33	0.22	0.16	0.60	0.57	0.39	0.24	0.17
		10.0	5.6	0.47	0.45	0.33	0.22	0.16	0.67	0.63	0.42	0.26	0.18
		−17.8	11.1	0.51	0.49	0.39	0.27	0.20	0.66	0.63	0.46	0.30	0.22
		−17.8	5.6	0.52	0.50	0.39	0.27	0.20	0.73	0.69	0.49	0.32	0.23
		−45.6	11.1	0.56	0.54	0.44	0.33	0.25	0.67	0.64	0.51	0.36	0.28
		−45.6	5.6	0.57	0.56	0.45	0.33	0.26	0.77	0.74	0.57	0.39	0.29
Horiz. Down ↓		32.2	5.6	0.44	0.41	0.29	0.19	0.14	0.62	0.58	0.37	0.21	0.15
		10.0	16.7	0.47	0.45	0.33	0.22	0.16	0.66	0.62	0.42	0.25	0.18
		10.0	5.6	0.47	0.45	0.33	0.22	0.16	0.68	0.63	0.42	0.26	0.18
		−17.8	11.1	0.52	0.50	0.39	0.27	0.20	0.74	0.70	0.50	0.32	0.23
		−17.8	5.6	0.52	0.50	0.39	0.27	0.20	0.75	0.71	0.51	0.32	0.23
		−45.6	11.1	0.57	0.55	0.45	0.33	0.26	0.81	0.78	0.59	0.40	0.30
		−45.6	5.6	0.58	0.56	0.46	0.33	0.26	0.83	0.79	0.60	0.40	0.30

Position of Air Space	Direction of Heat Flow	Air Space Mean Temp., °C	Air Space Temp. Diff., °C	40-mm Air Space[c] 0.03	0.05	0.2	0.5	0.82	90-mm Air Space[c] 0.03	0.05	0.2	0.5	0.82
Horiz Up ↑		32.2	5.6	0.45	0.42	0.30	0.19	0.14	0.50	0.47	0.32	0.20	0.14
		10.0	16.7	0.33	0.32	0.26	0.18	0.14	0.27	0.35	0.28	0.19	0.15
		10.0	5.6	0.44	0.42	0.32	0.21	0.16	0.49	0.47	0.34	0.23	0.16
		−17.8	11.1	0.35	0.34	0.29	0.22	0.17	0.40	0.38	0.32	0.23	0.18
		−17.8	5.6	0.43	0.41	0.33	0.24	0.19	0.48	0.46	0.36	0.26	0.20
		−45.6	11.1	0.34	0.34	0.30	0.24	0.20	0.39	0.38	0.33	0.26	0.21
		−45.6	5.6	0.42	0.41	0.35	0.27	0.22	0.47	0.45	0.38	0.29	0.23
45° Slope Up ↗		32.2	5.6	0.51	0.48	0.33	0.20	0.14	0.56	0.52	0.35	0.21	0.14
		10.0	16.7	0.38	0.36	0.28	0.20	0.15	0.40	0.38	0.29	0.20	0.15
		10.0	5.6	0.51	0.48	0.35	0.23	0.17	0.55	0.52	0.37	0.24	0.17
		−17.8	11.1	0.40	0.39	0.32	0.24	0.18	0.43	0.41	0.33	0.24	0.19
		−17.8	5.6	0.49	0.47	0.37	0.26	0.20	0.52	0.51	0.39	0.27	0.20
		−45.6	11.1	0.39	0.38	0.33	0.26	0.21	0.41	0.40	0.35	0.27	0.22
		−45.6	5.6	0.48	0.46	0.39	0.30	0.24	0.51	0.49	0.41	0.31	0.24
Vertical Horiz. →		32.2	5.6	0.70	0.64	0.40	0.22	0.15	0.65	0.60	0.38	0.22	0.15
		10.0	16.7	0.45	0.43	0.32	0.22	0.16	0.47	0.45	0.33	0.22	0.16
		10.0	5.6	0.67	0.62	0.42	0.26	0.18	0.64	0.60	0.41	0.25	0.18
		−17.8	11.1	0.49	0.47	0.37	0.26	0.20	0.51	0.49	0.38	0.27	0.20
		−17.8	5.6	0.62	0.59	0.44	0.29	0.22	0.61	0.59	0.44	0.29	0.22
		−45.6	11.1	0.46	0.45	0.38	0.29	0.23	0.50	0.48	0.40	0.30	0.24
		−45.6	5.6	0.58	0.56	0.46	0.34	0.26	0.60	0.58	0.47	0.34	0.26
45° Slope Down ↘		32.2	5.6	0.89	0.80	0.45	0.24	0.16	0.85	0.76	0.44	0.24	0.16
		10.0	16.7	0.63	0.59	0.41	0.25	0.18	0.62	0.58	0.40	0.25	0.18
		10.0	5.6	0.90	0.82	0.50	0.28	0.19	0.83	0.77	0.48	0.28	0.19
		−17.8	11.1	0.68	0.64	0.47	0.31	0.22	0.67	0.64	0.47	0.31	0.22
		−17.8	5.6	0.87	0.81	0.56	0.34	0.24	0.81	0.76	0.53	0.33	0.24
		−45.6	11.1	0.64	0.62	0.49	0.35	0.27	0.66	0.64	0.51	0.36	0.28
		−45.6	5.6	0.82	0.79	0.60	0.40	0.30	0.79	0.76	0.58	0.40	0.30
Horiz. Down ↓		32.2	5.6	1.07	0.94	0.49	0.25	0.17	1.77	1.44	0.60	0.28	0.18
		10.0	16.7	1.10	0.99	0.56	0.30	0.20	1.69	1.44	0.68	0.33	0.21
		10.0	5.6	1.16	1.04	0.58	0.30	0.20	1.96	1.63	0.72	0.34	0.22
		−17.8	11.1	1.24	1.13	0.69	0.39	0.26	1.92	1.68	0.86	0.43	0.29
		−17.8	5.6	1.29	1.17	0.70	0.39	0.27	2.11	1.82	0.89	0.44	0.29
		−45.6	11.1	1.36	1.27	0.84	0.50	0.35	2.05	1.85	1.06	0.57	0.38
		−45.6	5.6	1.42	1.32	0.86	0.51	0.35	2.28	2.03	1.12	0.59	0.39

[a]See Chapter 20, section Factors Affecting Heat Transfer across Air Spaces. Thermal resistance values were determined from the relation, $R = 1/C$, where $C = h_c + \epsilon_{eff} h_r$, h_c is the conduction-convection coefficient, $\epsilon_{eff} h_r$ is the radiation coefficient $\cong 0.227 \epsilon_{eff} [(t_m + 273)/100]^3$, and t_m is the mean temperature of the air space. Values for h_c were determined from data developed by Robinson et al. (1954). Equations (5) through (7) in Yarbrough (1983) show the data in Table 2 in analytic form. For extrapolation from Table 2 to air spaces less than 13 mm (as in insulating window glass), assume $h_c = 21.8(1 + 0.00274\, t_m)/l$ where l is the air space thickness in mm, and h_c is heat transfer in W/(m²·K) through the air space only.

[b]Values are based on data presented by Robinson et al. (1954). (Also see Chapter 3, Tables 3 and 4, and Chapter 39). Values apply for ideal conditions, i.e., air spaces of uniform thickness bounded by plane, smooth, parallel surfaces with no air leakage to or from the space. When accurate values are required, use overall U-factors determined through calibrated hot box (ASTM C 976) or guarded hot box (ASTM C 236) testing. Thermal resistance values for multiple air spaces must be based on careful estimates of mean temperature differences for each air space.

[c]A single resistance value cannot account for multiple air spaces; each air space requires a separate resistance calculation that applies only for the established boundary conditions. Resistances of horizontal spaces with heat flow downward are substantially independent of temperature difference.

[d]Interpolation is permissible for other values of mean temperature, temperature difference, and effective emittance ϵ_{eff}. Interpolation and moderate extrapolation for air spaces greater than 90 mm are also permissible.

[e]Effective emittance ϵ_{eff} of the air space is given by $1/\epsilon_{eff} = 1/\epsilon_1 + 1/\epsilon_2 - 1$, where ϵ_1 and ϵ_2 are the emittances of the surfaces of the air space (see Table 3).

FIGURE 8-47   (continued)

**Emittance Values of Various Surfaces and
Effective Emittances of Airspaces[a]**

Surface	Average Emittance ϵ	Effective Emittance, E of Airspace	
		One surface emittance ϵ; the other 0.9	Both surfaces emittance ϵ
Aluminum foil, bright	0.05	0.05	0.03
Aluminum foil, with condensate just visible (> 0.7gr/ft^2)	0.30[b]	0.29	—
Aluminum foil, with condensate clearly visible (> 2.9 gr/ft^2)	0.70[b]	0.65	—
Aluminum sheet	0.12	0.12	0.06
Aluminum coated paper, polished	0.20	0.20	0.11
Steel, galvanized, bright	0.25	0.24	0.15
Aluminum paint	0.50	0.47	0.35
Building materials: wood, paper, masonry, nonmetallic paints	0.90	0.82	0.82
Regular glass	0.84	0.77	0.72

[a]These values apply in the 4 to 40 μm range of the electromagnetic spectrum.
[b]Values are based on data presented by Bassett and Trethowen (1984).

FIGURE 8-48 Surface values of reflectivity and emissivity

Overall Coefficients of Heat Transmission of Various Fenestration Products (Btu/h·ft²·°F)

Operator Type	Glass Only			Aluminum w/o Thermal Break				Aluminum with Thermal Break							
	Center Glass	Edge of Glass		Operable	Fixed	Double Door	Sloped Skylight	Operable		Fixed		Double Door		Sloped Skylight	
Spacer Type		Metal	Insul.	All	All	All	All	Metal	Insul.	Metal	Insul.	Metal	Insul.	Metal	Insul.
Glazing ID and Type															
Single Glazing															
1 1/8 in. glass	1.11	1.11	—	1.30	1.17	1.26	1.92	1.07	—	1.11	—	1.10	—	1.93	—
2 1/4 in. acrylic/polycarbonate	0.93	0.93	—	1.15	1.00	1.10	1.72	0.93	—	0.95	—	0.95	—	1.73	—
3 1/8 in. acrylic/polycarbonate	1.02	1.02	—	1.22	1.08	1.18	1.93	1.00	—	1.03	—	1.02	—	1.94	—
4 glass block	0.60														
Double Glazing															
5 1/4 in. air space	0.57	0.65	0.59	0.87	0.69	0.80	1.30	0.67	0.64	0.63	0.60	0.66	0.64	1.13	1.07
6 1/2 in. air space	0.49	0.60	0.51	0.81	0.62	0.74	1.29	0.62	0.58	0.56	0.53	0.59	0.57	1.12	1.06
7 1/4 in. argon space	0.52	0.62	0.54	0.83	0.64	0.76	1.25	0.64	0.60	0.59	0.56	0.62	0.60	1.08	1.02
8 1/2 in. argon space	0.46	0.58	0.49	0.78	0.59	0.71	1.25	0.59	0.56	0.54	0.50	0.57	0.55	1.08	1.02
Double Glazing, $E = 0.40$ on surface 2 or 3															
9 1/4 in. air space	0.50	0.61	0.52	0.81	0.63	0.74	1.24	0.62	0.59	0.57	0.54	0.60	0.58	1.06	1.00
10 1/2 in. air space	0.40	0.54	0.43	0.74	0.54	0.66	1.23	0.55	0.51	0.49	0.45	0.52	0.50	1.06	0.99
10 1/4 in. argon space	0.43	0.56	0.46	0.76	0.57	0.69	1.17	0.57	0.54	0.51	0.48	0.54	0.52	1.00	0.93
12 1/2 in. argon space	0.36	0.51	0.40	0.71	0.50	0.63	1.18	0.52	0.48	0.45	0.41	0.49	0.46	1.01	0.94
Double Glazing, $E = 0.20$ on surface 2 or 3															
13 1/4 in. air space	0.46	0.58	0.49	0.78	0.59	0.71	1.20	0.59	0.56	0.54	0.50	0.57	0.55	1.03	0.96
14 1/2 in. air space	0.35	0.50	0.39	0.70	0.50	0.62	1.19	0.52	0.48	0.44	0.40	0.48	0.46	1.02	0.95
15 1/4 in. argon space	0.38	0.52	0.41	0.72	0.52	0.64	1.12	0.54	0.50	0.47	0.43	0.50	0.48	0.95	0.87
16 1/2 in. argon space	0.30	0.46	0.35	0.66	0.45	0.58	1.14	0.48	0.44	0.40	0.36	0.44	0.42	0.97	0.89
Double Glazing, $E = 0.10$ on surface 2 or 3															
17 1/4 in. air space	0.43	0.56	0.46	0.76	0.57	0.69	1.17	0.57	0.54	0.51	0.48	0.54	0.52	1.00	0.93
18 1/2 in. air space	0.32	0.48	0.36	0.67	0.47	0.59	1.17	0.49	0.45	0.42	0.38	0.46	0.43	1.00	0.93
19 1/4 in. argon space	0.35	0.50	0.39	0.70	0.50	0.62	1.09	0.52	0.48	0.44	0.40	0.48	0.46	0.92	0.84
20 1/2 in. argon space	0.27	0.44	0.32	0.64	0.43	0.55	1.11	0.46	0.42	0.37	0.33	0.42	0.39	0.94	0.86
Triple Glazing															
21 1/4 in. air spaces	0.38	0.52	0.41	0.72	0.52	0.64	1.13	0.54	0.50	0.47	0.43	0.50	0.48	0.93	0.88
22 1/2 in. air spaces	0.31	0.47	0.35	0.67	0.46	0.59	1.10	0.49	0.45	0.41	0.37	0.44	0.42	0.91	0.85
23 1/4 in. argon spaces	0.34	0.49	0.38	0.69	0.49	0.61	1.08	0.51	0.47	0.43	0.40	0.47	0.45	0.89	0.83
24 1/2 in. argon spaces	0.29	0.45	0.34	0.65	0.44	0.57	1.07	0.47	0.43	0.39	0.35	0.43	0.40	0.88	0.82
Triple Glazing, $E = 0.40$ on surface 2, 3, 4, or 5															
25 1/4 in. air spaces	0.35	0.50	0.39	0.70	0.50	0.62	1.09	0.52	0.48	0.44	0.40	0.47	0.45	0.90	0.84
26 1/2 in. air spaces	0.27	0.44	0.32	0.64	0.43	0.55	1.06	0.46	0.42	0.37	0.33	0.41	0.39	0.87	0.81
27 1/4 in. argon spaces	0.30	0.46	0.35	0.66	0.45	0.58	1.04	0.48	0.44	0.40	0.36	0.43	0.41	0.85	0.79
28 1/2 in. argon spaces	0.24	0.42	0.30	0.61	0.40	0.53	1.03	0.44	0.40	0.35	0.31	0.39	0.36	0.84	0.78
Triple Glazing, $E = 0.20$ on surface 2, 3 ,4, or 5															
29 1/4 in. air spaces	0.33	0.48	0.37	0.68	0.48	0.60	1.07	0.50	0.46	0.43	0.39	0.46	0.44	0.88	0.82
30 1/2 in. air spaces	0.24	0.42	0.30	0.61	0.40	0.53	1.04	0.44	0.40	0.35	0.31	0.39	0.36	0.85	0.79
31 1/4 in. argon spaces	0.27	0.44	0.32	0.64	0.43	0.55	1.02	0.46	0.42	0.37	0.33	0.41	0.39	0.82	0.76
32 1/2 in. argon spaces	0.21	0.39	0.27	0.59	0.37	0.50	1.01	0.41	0.37	0.32	0.28	0.36	0.34	0.81	0.75
Triple Glazing, $E = 0.10$ on surface 2, 3, 4, or 5															
33 1/4 in. air spaces	0.32	0.48	0.36	0.67	0.47	0.59	1.05	0.49	0.45	0.42	0.38	0.45	0.43	0.86	0.80
34 1/2 in. air spaces	0.22	0.40	0.28	0.60	0.38	0.51	1.03	0.42	0.38	0.33	0.29	0.37	0.35	0.84	0.78
35 1/4 in. argon spaces	0.26	0.43	0.31	0.63	0.42	0.54	1.00	0.45	0.41	0.36	0.32	0.40	0.38	0.80	0.74
36 1/2 in. argon spaces	0.19	0.38	0.26	0.57	0.36	0.48	0.99	0.40	0.36	0.30	0.26	0.35	0.32	0.79	0.73
Triple Glazing, $E = 0.40$ on surfaces 3 and 5															
37 1/4 in. air spaces	0.33	0.48	0.37	0.68	0.48	0.60	1.06	0.50	0.46	0.43	0.39	0.46	0.44	0.87	0.81
38 1/2 in. air spaces	0.24	0.42	0.30	0.61	0.40	0.53	1.04	0.44	0.40	0.35	0.31	0.39	0.36	0.85	0.79
39 1/4 in. argon spaces	0.27	0.44	0.32	0.64	0.43	0.55	1.02	0.46	0.42	0.37	0.33	0.41	0.39	0.82	0.76
40 1/2 in. argon spaces	0.21	0.39	0.27	0.59	0.37	0.50	1.00	0.41	0.37	0.32	0.28	0.36	0.34	0.80	0.74
Triple Glazing, $E = 0.20$ on surfaces 3 and 5															
41 1/4 in. air spaces	0.29	0.45	0.34	0.65	0.44	0.57	1.03	0.47	0.43	0.39	0.35	0.43	0.40	0.84	0.78
42 1/2 in. air spaces	0.20	0.39	0.27	0.58	0.36	0.49	1.01	0.41	0.37	0.31	0.27	0.35	0.33	0.81	0.75
43 1/4 in. argon spaces	0.23	0.41	0.29	0.60	0.39	0.52	0.98	0.43	0.39	0.34	0.30	0.38	0.36	0.79	0.72
44 1/2 in. argon spaces	0.16	0.35	0.24	0.55	0.33	0.46	0.96	0.38	0.34	0.28	0.24	0.32	0.30	0.77	0.71
Triple Glazing, $E = 0.10$ on surfaces 3 and 5															
45 1/4 in. air spaces	0.27	0.44	0.32	0.64	0.43	0.55	1.02	0.46	0.42	0.37	0.33	0.41	0.39	0.82	0.76
46 1/2 in. air spaces	0.17	0.36	0.24	0.56	0.34	0.47	0.99	0.39	0.34	0.29	0.25	0.33	0.31	0.79	0.73
47 1/4 in. argon spaces	0.21	0.39	0.27	0.59	0.37	0.50	0.95	0.41	0.37	0.32	0.28	0.36	0.34	0.76	0.70
48 1/2 in. argon spaces	0.14	0.34	0.22	0.53	0.31	0.44	0.94	0.36	0.32	0.26	0.22	0.31	0.28	0.75	0.69
Quadruple Glazing, $E = 0.10$ on surfaces 3 and 5															
49 1/4 in. air spaces	0.23	0.41	0.29	0.60	0.39	0.52	0.97	0.43	0.39	0.34	0.30	0.38	0.36	0.78	0.71
50 1/2 in. air spaces	0.15	0.35	0.23	0.54	0.32	0.45	0.93	0.37	0.33	0.27	0.23	0.31	0.29	0.74	0.68
51 1/4 in. argon spaces	0.17	0.36	0.24	0.56	0.34	0.47	0.92	0.39	0.34	0.29	0.25	0.33	0.31	0.73	0.67
52 1/2 in. argon spaces	0.12	0.32	0.21	0.52	0.29	0.43	0.90	0.35	0.31	0.24	0.20	0.29	0.27	0.71	0.65
53 1/4 in. krypton spaces	0.12	0.32	0.21	0.52	0.29	0.43	0.89	0.35	0.31	0.24	0.20	0.29	0.27	0.69	0.63
54 1/2 in. krypton spaces	0.11	0.31	0.20	0.51	0.29	0.42	0.89	0.34	0.30	0.23	0.19	0.28	0.26	0.69	0.63

FIGURE 8-49   Glazing *U* values

Overall Coefficients of Heat Transmission of Various Fenestration Products (Btu/h·ft²·°F) (Concluded)

Reinforced Vinyl/Aluminum-Clad Wood								Wood Vinyl								Insulated Fiberglass/Vinyl						
Operable		Fixed		Double Door		Sloped Skylight		Operable		Fixed		Double Door		Sloped Skylight		Operable		Fixed		Double Door		
Metal	Insul.	Metal	Insul.	Metal	Insul.	Metal	Insul.	Metal	Insul.	Metal	Insul.	Metal	Insul.	Metal	Insul.	Metal	Insul.	Metal	Insul.	Metal	Insul.	ID
0.98	—	1.05	—	0.99	—	1.50	—	0.94	—	1.04	—	0.98	—	1.47	—	0.86	—	1.02	—	0.93	—	1
0.85	—	0.89	—	0.85	—	1.31	—	0.81	—	0.88	—	0.84	—	1.27	—	0.74	—	0.86	—	0.79	—	2
0.92	—	0.97	—	0.92	—	1.51	—	0.87	—	0.96	—	0.91	—	1.48	—	0.80	—	0.94	—	0.86	—	3
																						4
0.60	0.57	0.58	0.56	0.57	0.55	0.88	0.86	0.56	0.54	0.57	0.56	0.56	0.54	0.85	0.82	0.50	0.47	0.55	0.54	0.52	0.50	5
0.55	0.52	0.51	0.49	0.52	0.49	0.87	0.85	0.51	0.48	0.51	0.49	0.50	0.48	0.84	0.81	0.45	0.42	0.49	0.47	0.46	0.44	6
0.57	0.54	0.54	0.52	0.54	0.52	0.84	0.81	0.53	0.50	0.53	0.51	0.53	0.51	0.80	0.77	0.47	0.44	0.51	0.49	0.48	0.46	7
0.53	0.50	0.49	0.46	0.49	0.47	0.84	0.81	0.49	0.46	0.48	0.46	0.48	0.46	0.80	0.77	0.43	0.40	0.46	0.44	0.44	0.42	8
0.56	0.52	0.52	0.50	0.52	0.50	0.82	0.79	0.52	0.49	0.52	0.50	0.51	0.49	0.79	0.75	0.46	0.43	0.50	0.48	0.47	0.45	9
0.49	0.46	0.44	0.41	0.45	0.43	0.81	0.78	0.45	0.42	0.43	0.41	0.44	0.41	0.78	0.74	0.40	0.36	0.41	0.39	0.40	0.37	10
0.51	0.48	0.46	0.44	0.47	0.45	0.76	0.72	0.47	0.44	0.46	0.43	0.46	0.44	0.72	0.68	0.42	0.38	0.44	0.42	0.42	0.40	11
0.47	0.43	0.40	0.38	0.42	0.40	0.77	0.73	0.43	0.40	0.40	0.37	0.41	0.38	0.73	0.69	0.37	0.34	0.38	0.36	0.37	0.34	12
0.53	0.50	0.49	0.46	0.49	0.47	0.78	0.75	0.49	0.46	0.48	0.46	0.48	0.46	0.75	0.71	0.43	0.40	0.46	0.44	0.44	0.42	13
0.46	0.42	0.39	0.37	0.41	0.39	0.78	0.74	0.42	0.39	0.39	0.36	0.40	0.37	0.74	0.70	0.37	0.33	0.37	0.35	0.36	0.34	14
0.48	0.44	0.42	0.39	0.43	0.41	0.70	0.67	0.44	0.41	0.41	0.39	0.42	0.40	0.67	0.63	0.38	0.35	0.40	0.37	0.38	0.36	15
0.43	0.39	0.35	0.32	0.38	0.35	0.72	0.68	0.39	0.36	0.35	0.32	0.36	0.34	0.69	0.65	0.33	0.30	0.33	0.31	0.33	0.30	16
0.51	0.48	0.46	0.44	0.47	0.45	0.76	0.72	0.47	0.44	0.46	0.43	0.46	0.44	0.72	0.68	0.42	0.38	0.44	0.42	0.42	0.40	17
0.44	0.40	0.37	0.34	0.39	0.37	0.76	0.72	0.40	0.37	0.36	0.34	0.37	0.35	0.72	0.68	0.35	0.31	0.35	0.32	0.34	0.32	18
0.46	0.42	0.39	0.37	0.41	0.39	0.68	0.64	0.42	0.39	0.39	0.36	0.40	0.37	0.64	0.60	0.37	0.33	0.37	0.35	0.36	0.34	19
0.41	0.37	0.32	0.30	0.35	0.33	0.69	0.66	0.37	0.34	0.32	0.29	0.34	0.31	0.66	0.62	0.32	0.28	0.30	0.28	0.30	0.28	20
0.46	0.41	0.41	0.39	0.43	0.40	0.71	0.67	0.43	0.39	0.41	0.38	0.42	0.39	0.67	0.63	0.37	0.34	0.39	0.37	0.38	0.35	21
0.42	0.37	0.35	0.33	0.37	0.35	0.68	0.64	0.38	0.34	0.35	0.32	0.36	0.34	0.65	0.60	0.33	0.29	0.33	0.31	0.32	0.30	22
0.44	0.39	0.38	0.35	0.40	0.37	0.66	0.62	0.40	0.36	0.38	0.35	0.39	0.36	0.63	0.58	0.35	0.31	0.36	0.33	0.35	0.32	23
0.40	0.35	0.34	0.31	0.36	0.33	0.65	0.61	0.37	0.33	0.34	0.31	0.35	0.32	0.62	0.57	0.32	0.28	0.32	0.29	0.31	0.28	24
0.44	0.39	0.39	0.36	0.40	0.38	0.67	0.63	0.41	0.37	0.39	0.36	0.39	0.37	0.64	0.59	0.35	0.32	0.37	0.34	0.35	0.33	25
0.39	0.34	0.32	0.29	0.35	0.32	0.65	0.61	0.36	0.31	0.32	0.29	0.33	0.31	0.61	0.56	0.30	0.27	0.30	0.28	0.30	0.27	26
0.41	0.36	0.35	0.32	0.37	0.34	0.63	0.59	0.38	0.33	0.34	0.31	0.36	0.33	0.59	0.54	0.32	0.29	0.33	0.30	0.32	0.29	27
0.37	0.32	0.30	0.27	0.32	0.29	0.62	0.58	0.34	0.29	0.29	0.26	0.31	0.28	0.58	0.53	0.29	0.25	0.27	0.25	0.27	0.25	28
0.43	0.38	0.37	0.34	0.39	0.36	0.65	0.61	0.40	0.35	0.37	0.34	0.38	0.35	0.62	0.57	0.34	0.31	0.35	0.33	0.34	0.31	29
0.37	0.32	0.30	0.27	0.32	0.29	0.63	0.59	0.34	0.29	0.29	0.26	0.31	0.28	0.59	0.54	0.29	0.25	0.27	0.25	0.27	0.25	30
0.39	0.34	0.32	0.29	0.35	0.32	0.60	0.56	0.36	0.31	0.32	0.29	0.33	0.31	0.57	0.52	0.30	0.27	0.30	0.28	0.30	0.27	31
0.35	0.30	0.27	0.24	0.30	0.27	0.59	0.55	0.32	0.27	0.27	0.24	0.29	0.26	0.56	0.51	0.27	0.23	0.25	0.22	0.25	0.23	32
0.42	0.37	0.36	0.34	0.38	0.35	0.64	0.60	0.39	0.35	0.36	0.33	0.37	0.34	0.60	0.55	0.34	0.30	0.34	0.32	0.33	0.31	33
0.36	0.31	0.28	0.25	0.31	0.28	0.62	0.58	0.33	0.28	0.28	0.25	0.30	0.27	0.58	0.53	0.27	0.24	0.26	0.23	0.26	0.23	34
0.38	0.33	0.31	0.28	0.34	0.31	0.58	0.54	0.35	0.31	0.31	0.28	0.33	0.30	0.55	0.50	0.30	0.26	0.29	0.27	0.29	0.26	35
0.34	0.29	0.25	0.22	0.29	0.26	0.57	0.53	0.31	0.26	0.25	0.22	0.27	0.24	0.54	0.49	0.25	0.22	0.23	0.21	0.24	0.21	36
0.43	0.38	0.37	0.34	0.39	0.36	0.65	0.61	0.40	0.35	0.37	0.34	0.38	0.35	0.61	0.56	0.34	0.31	0.35	0.33	0.34	0.31	37
0.37	0.32	0.30	0.27	0.32	0.29	0.63	0.59	0.34	0.29	0.29	0.26	0.31	0.28	0.59	0.54	0.29	0.25	0.27	0.25	0.27	0.25	38
0.39	0.34	0.32	0.29	0.35	0.32	0.60	0.56	0.36	0.31	0.32	0.29	0.33	0.31	0.57	0.52	0.30	0.27	0.30	0.28	0.30	0.27	39
0.35	0.30	0.27	0.24	0.30	0.27	0.58	0.54	0.32	0.27	0.27	0.24	0.29	0.26	0.55	0.50	0.27	0.23	0.25	0.22	0.25	0.23	40
0.40	0.35	0.34	0.31	0.36	0.33	0.62	0.58	0.37	0.33	0.34	0.31	0.35	0.32	0.58	0.53	0.32	0.28	0.32	0.29	0.31	0.28	41
0.35	0.29	0.26	0.23	0.29	0.26	0.59	0.55	0.31	0.27	0.26	0.23	0.28	0.25	0.56	0.51	0.26	0.23	0.24	0.22	0.24	0.22	42
0.37	0.31	0.29	0.26	0.32	0.29	0.56	0.52	0.33	0.29	0.28	0.25	0.30	0.27	0.53	0.48	0.28	0.24	0.27	0.24	0.27	0.24	43
0.32	0.27	0.23	0.20	0.26	0.23	0.55	0.50	0.29	0.24	0.22	0.19	0.25	0.22	0.51	0.46	0.23	0.20	0.21	0.18	0.22	0.19	44
0.39	0.34	0.32	0.29	0.35	0.32	0.60	0.56	0.36	0.31	0.32	0.29	0.33	0.31	0.57	0.52	0.30	0.27	0.30	0.28	0.30	0.27	45
0.33	0.28	0.24	0.21	0.27	0.24	0.57	0.53	0.29	0.25	0.23	0.20	0.26	0.23	0.54	0.49	0.24	0.21	0.22	0.19	0.22	0.20	46
0.35	0.30	0.27	0.24	0.30	0.27	0.54	0.49	0.32	0.27	0.27	0.24	0.29	0.26	0.50	0.45	0.27	0.23	0.25	0.22	0.25	0.23	47
0.31	0.26	0.21	0.18	0.25	0.22	0.53	0.49	0.27	0.23	0.21	0.18	0.23	0.21	0.49	0.44	0.22	0.19	0.19	0.17	0.20	0.18	48
0.37	0.31	0.29	0.26	0.32	0.29	0.56	0.51	0.33	0.29	0.28	0.25	0.30	0.27	0.52	0.47	0.28	0.24	0.27	0.24	0.27	0.24	49
0.31	0.26	0.22	0.19	0.26	0.23	0.52	0.48	0.28	0.23	0.22	0.19	0.24	0.21	0.48	0.43	0.23	0.19	0.20	0.17	0.21	0.18	50
0.33	0.28	0.24	0.21	0.27	0.24	0.51	0.47	0.29	0.25	0.23	0.20	0.26	0.23	0.48	0.42	0.24	0.21	0.22	0.19	0.22	0.20	51
0.29	0.24	0.19	0.16	0.23	0.20	0.49	0.45	0.26	0.22	0.19	0.16	0.22	0.19	0.46	0.41	0.21	0.18	0.17	0.15	0.19	0.16	52
0.29	0.24	0.19	0.16	0.23	0.20	0.47	0.43	0.26	0.22	0.19	0.16	0.22	0.19	0.44	0.39	0.21	0.18	0.17	0.15	0.19	0.16	53
0.29	0.24	0.18	0.15	0.23	0.20	0.47	0.43	0.25	0.21	0.18	0.15	0.21	0.18	0.44	0.39	0.20	0.17	0.16	0.14	0.18	0.15	54

FIGURE 8-49   (continued)

Notes

1. All heat transmission coefficients in this table include film resistances and are based on winter conditions of 0°F outdoor air and 70°F indoor air temperature, with 15 mph outdoor air velocity and zero solar flux. With the exception of single glazing, small changes in the interior and exterior temperatures will not significantly affect overall U-factors. The coefficients are for vertical position except skylight values, which are for 20° from horizontal with heat flow up.
2. Glazing layer surfaces are numbered from the outside to the inside. Double, triple, and quadruple refer to the number of glazing panels. All data are based on 1/8 in. glass, unless otherwise noted. Thermal conductivities are: 0.53 Btu/(h · ft · °F) for glass, and 0.11 Btu/(h · ft · °F) for acrylic and polycarbonate .
3. Standard spacers are metal. Insulation means wood, fiberglass, or butyl. Edge-of-glass effects assumed to extend over the 2-1/2 in. band around perimeter of each glazing unit as seen in Figure 3.
4. Product sizes are described in Figure 3 and frame U-factors are from Table 2.

5. Interpolation procedure for estimating U_{cg} for vertically oriented gas space widths between 0.25 and 0.5 in. The most general equation is:

$$U_t = \left[\frac{1}{U_{1/4}} - \frac{n_g}{h + 48\,k_g} + \frac{n_g}{h + (12\,k_g/t)} \right]^{-1}$$

$$h = -36\,k_g + \left[\frac{24\,n_g\,k_g\,U_{1/4}\,U_{1/2}}{U_{1/4} - U_{1/2}} + 144\,k_g^{2} \right]^{0.5}$$

where

n_g = number of gaps (*e.g.*, double-pane windows have one gap)
k_g = gas conductivity Btu/(h · ft · °F)
t = gas space width of window, in.
$U_{1/4}$ = U-factor for identical glazing with 1/4 in. gas spaces
$U_{1/2}$ = U-factor for identical glazing with 1/2 in. gas spaces

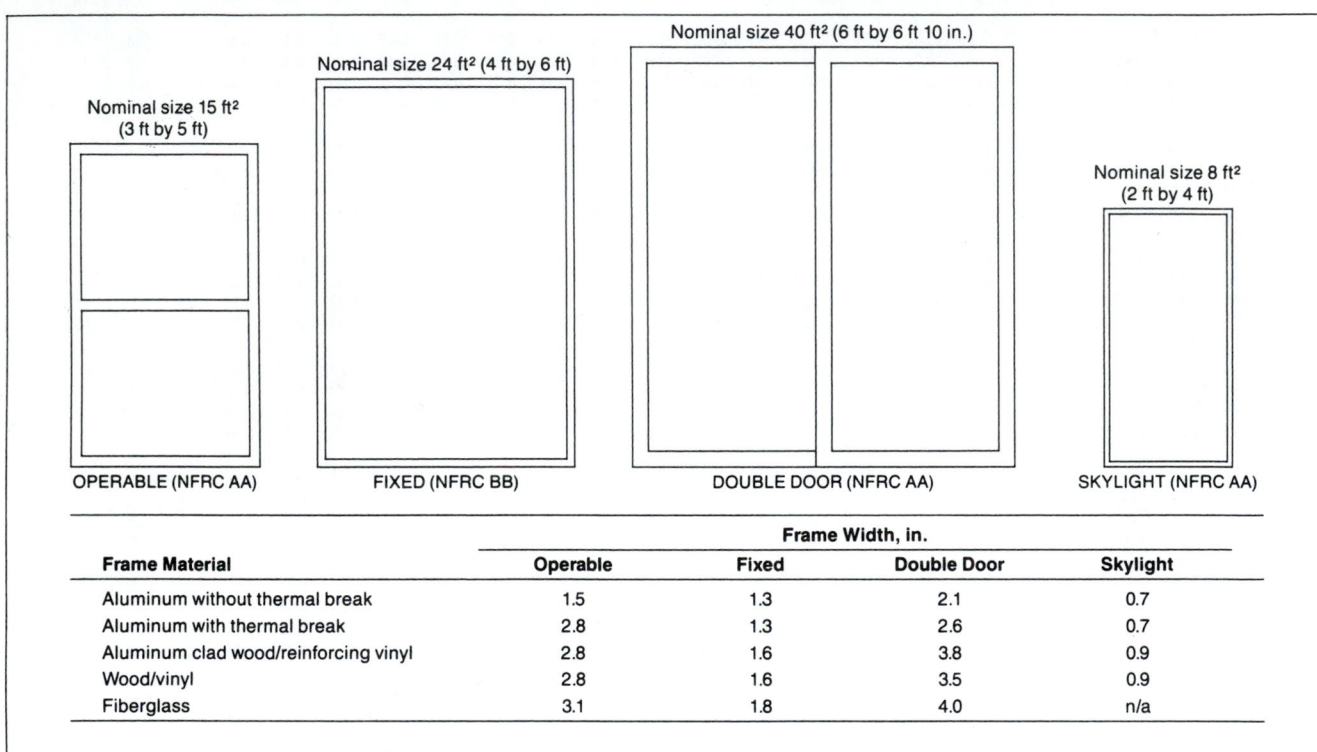

Frame Material	Frame Width, in.			
	Operable	Fixed	Double Door	Skylight
Aluminum without thermal break	1.5	1.3	2.1	0.7
Aluminum with thermal break	2.8	1.3	2.6	0.7
Aluminum clad wood/reinforcing vinyl	2.8	1.6	3.8	0.9
Wood/vinyl	2.8	1.6	3.5	0.9
Fiberglass	3.1	1.8	4.0	n/a

FIGURE 8-49   *(continued)*

SI UNITS **Overall Coefficients of Heat Transmission of Various Fenestration Products in W/(m² · K)**

Operator Type	Glass Only			Aluminum w/o Thermal Break				Aluminum with Thermal Break							
	Center Glass	Edge of Glass		Oper-able	Fixed	Double Door	Sloped Skylight	Operable		Fixed		Double Door		Sloped Skylight	
Spacer Type		Metal	Insul.	All	All	All	All	Metal	Insul.	Metal	Insul.	Metal	Insul.	Metal	Insul.
Glazing ID and Type															
Single Glazing															
1 3 mm glass	6.30	6.30	—	7.37	6.63	7.16	9.88	6.09	—	6.33	—	6.24	—	9.96	—
2 6.4 mm acrylic/polycarbonate	5.28	5.28	—	6.53	5.69	6.27	8.86	5.31	—	5.40	—	5.38	—	8.94	—
3 3 mm acrylic/polycarbonate	5.79	5.79	—	6.95	6.16	6.71	9.94	5.70	—	5.86	—	5.81	—	10.02	—
4 glass block	3.40														
Double Glazing															
5 6.4 mm airspace	3.24	3.71	3.34	4.93	3.90	4.55	6.70	3.82	3.64	3.60	3.42	3.73	3.62	5.82	5.51
6 12.7 mm airspace	2.78	3.40	2.91	4.58	3.51	4.18	6.65	3.50	3.30	3.21	3.01	3.37	3.25	5.77	5.45
7 6.4 mm argon space	2.95	3.52	3.07	4.71	3.66	4.32	6.47	3.62	3.43	3.36	3.16	3.50	3.39	5.59	5.25
8 12.7 mm argon space	2.61	3.28	2.76	4.45	3.36	4.04	6.47	3.38	3.17	3.06	2.86	3.23	3.10	5.59	5.25
Double Glazing, $E = 0.40$ on surface 2 or 3															
9 6.4 mm airspace	2.84	3.44	2.96	4.62	3.56	4.23	6.37	3.54	3.34	3.26	3.06	3.41	3.29	5.49	5.15
10 12.7 mm airspace	2.27	3.04	2.45	4.18	3.06	3.75	6.33	3.13	2.92	2.76	2.55	2.96	2.82	5.45	5.10
11 6.4 mm argon space	2.44	3.16	2.60	4.32	3.21	3.89	6.04	3.26	3.05	2.91	2.70	3.09	2.96	5.17	4.80
12 12.7 mm argon space	2.04	2.88	2.26	4.01	2.87	3.56	6.09	2.97	2.75	2.57	2.35	2.77	2.64	5.21	4.85
Double Glazing, $E = 0.20$ on surface 2 or 3															
13 6.4 mm airspace	2.61	3.28	2.76	4.45	3.36	4.04	6.19	3.38	3.17	3.06	2.86	3.23	3.10	5.31	4.95
14 12.7 mm airspace	1.99	2.83	2.21	3.96	2.82	3.52	6.14	2.93	2.71	2.52	2.30	2.73	2.59	5.26	4.90
15 6.4 mm argon space	2.16	2.96	2.35	4.10	2.97	3.66	5.76	3.05	2.83	2.67	2.45	2.87	2.73	4.88	4.50
16 12.7 mm argon space	1.70	2.62	1.97	3.74	2.57	3.28	5.86	2.72	2.50	2.27	2.05	2.50	2.36	4.98	4.60
Double Glazing, $E = 0.10$ on surface 2 or 3															
17 6.4 mm airspace	2.44	3.16	2.60	4.32	3.21	3.89	6.04	3.26	3.05	2.91	2.70	3.09	2.96	5.17	4.80
18 12.7 mm airspace	1.82	2.71	2.06	3.83	2.67	3.37	6.04	2.81	2.58	2.37	2.15	2.59	2.45	5.17	4.80
19 6.4 mm argon space	1.99	2.83	2.21	3.96	2.82	3.52	5.62	2.93	2.71	2.52	2.30	2.73	2.59	4.74	4.35
20 12.7 mm argon space	1.53	2.49	1.83	3.61	2.42	3.14	5.71	2.60	2.37	2.12	1.89	2.36	2.22	4.84	4.45
Triple Glazing															
21 6.4 mm airspace	2.16	2.96	2.35	4.10	2.97	3.66	5.81	3.05	2.83	2.67	2.45	2.83	2.71	4.81	4.52
22 12.7 mm airspace	1.76	2.67	2.02	3.79	2.62	3.33	5.67	2.77	2.54	2.32	2.10	2.51	2.39	4.67	4.37
23 6.4 mm argon space	1.93	2.79	2.16	3.92	2.77	3.47	5.57	2.89	2.67	2.47	2.25	2.65	2.53	4.58	4.27
24 12.7 mm argon space	1.65	2.58	1.92	3.70	2.52	3.23	5.53	2.68	2.46	2.22	1.99	2.42	2.30	4.53	4.22
Triple Glazing, $E = 0.40$ on surface 2,3,4, or 5															
25 6.4 mm airspace	1.99	2.83	2.21	3.96	2.82	3.52	5.62	2.93	2.71	2.52	2.30	2.70	2.58	4.62	4.32
26 12.7 mm airspace	1.53	2.49	1.83	3.61	2.42	3.14	5.48	2.60	2.37	2.12	1.89	2.33	2.20	4.48	4.17
27 6.4 mm argon space	1.70	2.62	1.97	3.74	2.57	3.28	5.38	2.72	2.50	2.27	2.05	2.47	2.34	4.39	4.08
28 12.7 mm argon space	1.36	2.36	1.69	3.48	2.27	2.99	5.34	2.48	2.25	1.97	1.74	2.19	2.07	4.34	4.03
Triple Glazing, $E = 0.20$ on surface 2,3,4, or 5															
29 6.4 mm airspace	1.87	2.75	2.11	3.87	2.72	3.42	5.53	2.85	2.62	2.42	2.20	2.61	2.48	4.53	4.22
30 12.7 mm airspace	1.36	2.36	1.69	3.48	2.27	2.99	5.38	2.48	2.25	1.97	1.74	2.19	2.07	4.39	4.08
31 6.4 mm argon space	1.53	2.49	1.83	3.61	2.42	3.14	5.24	2.60	2.37	2.12	1.89	2.33	2.20	4.24	3.93
32 12.7 mm argon space	1.19	2.23	1.56	3.34	2.12	2.85	5.19	2.35	2.12	1.82	1.59	2.05	1.93	4.19	3.88
Triple Glazing, $E = 0.10$ on surface 2,3,4, or 5															
33 6.4 mm airspace	1.82	2.71	2.06	3.83	2.67	3.37	5.43	2.81	2.58	2.37	2.15	2.56	2.44	4.43	4.13
34 12.7 mm airspace	1.25	2.28	1.60	3.39	2.17	2.90	5.34	2.39	2.16	1.87	1.64	2.10	1.97	4.34	4.03
35 6.4 mm argon space	1.48	2.45	1.78	3.56	2.37	3.09	5.14	2.56	2.33	2.07	1.84	2.28	2.16	4.15	3.83
36 12.7 mm argon space	1.08	2.14	1.47	3.25	2.02	2.75	5.10	2.27	2.04	1.72	1.49	1.96	1.84	4.10	3.78
Triple Glazing, $E = 0.40$ on surfaces 3 and 5															
37 6.4 mm airspace	1.87	2.75	2.11	3.87	2.72	3.42	5.48	2.85	2.62	2.42	2.20	2.61	2.48	4.48	4.17
38 12.7 mm airspace	1.36	2.36	1.69	3.48	2.27	2.99	5.38	2.48	2.25	1.97	1.74	2.19	2.07	4.39	4.08
39 6.4 mm argon space	1.53	2.49	1.83	3.61	2.42	3.14	5.24	2.60	2.37	2.12	1.89	2.33	2.20	4.24	3.93
40 12.7 mm argon space	1.19	2.23	1.56	3.34	2.12	2.85	5.14	2.35	2.12	1.82	1.59	2.05	1.93	4.15	3.83
Triple Glazing, $\bar{E} = 0.20$ on surfaces 3 and 5															
41 6.4 mm airspace	1.65	2.58	1.92	3.70	2.52	3.23	5.34	2.68	2.46	2.22	1.99	2.42	2.30	4.34	4.03
42 12.7 mm airspace	1.14	2.19	1.51	3.30	2.07	2.80	5.19	2.31	2.08	1.77	1.54	2.01	1.88	4.19	3.88
43 6.4 mm argon space	1.31	2.32	1.65	3.43	2.22	2.94	5.05	2.44	2.21	1.92	1.69	2.15	2.02	4.05	3.74
44 12.7 mm argon space	0.91	2.01	1.34	3.12	1.87	2.61	4.95	2.15	1.92	1.57	1.34	1.82	1.70	3.95	3.64
Triple Glazing, $E = 0.10$ on surfaces 3 and 5															
45 6.4 mm airspace	1.53	2.49	1.83	3.61	2.42	3.14	5.24	2.60	2.37	2.12	1.89	2.33	2.20	4.24	3.93
46 12.7 mm airspace	0.97	2.05	1.38	3.16	1.92	2.66	5.10	2.19	1.96	1.62	1.39	1.87	1.74	4.10	3.78
47 6.4 mm argon space	1.19	2.23	1.56	3.34	2.12	2.85	4.90	2.35	2.12	1.82	1.59	2.05	1.93	3.91	3.59
48 12.7 mm argon space	0.80	1.92	1.25	3.03	1.77	2.51	4.86	2.06	1.83	1.47	1.24	1.73	1.61	3.86	3.54
Quadruple Glazing, $E = 0.10$ on surfaces 3 and 5															
49 6.4 mm airspace	1.31	2.32	1.65	3.43	2.22	2.94	5.00	2.44	2.21	1.92	1.69	2.15	2.02	4.00	3.69
50 12.7 mm airspace	0.85	1.96	1.29	3.07	1.82	2.56	4.81	2.10	1.88	1.52	1.29	1.78	1.65	3.81	3.49
51 6.4 mm argon space	0.97	2.05	1.38	3.16	1.92	2.66	4.76	2.19	1.96	1.62	1.39	1.87	1.74	3.76	3.45
52 12.7 mm argon space	0.68	1.83	1.17	2.94	1.67	2.42	4.66	1.98	1.75	1.37	1.15	1.64	1.51	3.67	3.35
53 6.4 mm krypton spaces	0.68	1.83	1.17	2.94	1.67	2.42	4.57	1.98	1.75	1.37	1.15	1.64	1.51	3.57	3.26
54 12.7 mm krypton spaces	0.62	1.78	1.12	2.89	1.62	2.37	4.57	1.94	1.71	1.32	1.10	1.59	1.47	3.57	3.26

Note: Skylight values are for 20° slope from horizontal.

FIGURE 8-49 *(continued)*

SI UNITS   Overall Coefficients of Heat Transmission of Various Fenestration Products in W/(m² · K) (*Concluded*)

| Reinforced Vinyl/Aluminum-Clad Wood | | | | | | | | Wood Vinyl | | | | | | | | Insulated Fiberglass/Vinyl | | | | | | |
| Operable | | Fixed | | Double Door | | Sloped Skylight | | Operable | | Fixed | | Double Door | | Sloped Skylight | | Operable | | Fixed | | Double Door | | ID |
Metal	Insul.	Metal	Insul.	Metal	Insul.	Metal	Insul.	Metal	Insul.	Metal	Insul.	Metal	Insul.	Metal	Insul.	Metal	Insul.	Metal	Insul.	Metal	Insul.	
5.56	—	5.97	—	5.60	—	7.74	—	5.31	—	5.93	—	5.57	—	7.57	—	4.87	—	5.79	—	5.26	—	1
4.86	—	5.06	—	4.82	—	6.74	—	4.61	—	5.02	—	4.77	—	6.57	—	4.20	—	4.89	—	4.49	—	2
5.21	—	5.51	—	5.21	—	7.80	—	4.96	—	5.48	—	5.17	—	7.63	—	4.53	—	5.34	—	4.87	—	3
																						4
3.42	3.24	3.29	3.18	3.26	3.15	4.55	4.42	3.20	3.06	3.26	3.16	3.20	3.09	4.37	4.22	2.85	2.69	3.15	3.05	2.96	2.84	5
3.13	2.93	2.90	2.78	2.93	2.80	4.50	4.37	2.91	2.75	2.88	2.76	2.86	2.74	4.32	4.17	2.57	2.40	2.77	2.66	2.63	2.50	6
3.24	3.05	3.05	2.93	3.05	2.93	4.32	4.17	3.02	2.86	3.03	2.91	2.98	2.87	4.14	3.97	2.67	2.51	2.91	2.81	2.75	2.63	7
3.02	2.82	2.76	2.63	2.80	2.67	4.32	4.17	2.80	2.63	2.74	2.61	2.73	2.60	4.14	3.97	2.46	2.29	2.63	2.51	2.51	2.38	8
3.16	2.97	2.95	2.83	2.97	2.84	4.23	4.07	2.95	2.78	2.93	2.81	2.90	2.78	4.05	3.87	2.60	2.43	2.82	2.71	2.67	2.55	9
2.80	2.59	2.47	2.34	2.55	2.42	4.18	4.02	2.58	2.40	2.45	2.31	2.47	2.34	4.01	3.82	2.25	2.07	2.35	2.22	2.26	2.12	10
2.91	2.70	2.62	2.48	2.68	2.54	3.91	3.73	2.69	2.52	2.59	2.46	2.60	2.47	3.73	3.53	2.36	2.18	2.49	2.36	2.39	2.25	11
2.65	2.43	2.28	2.14	2.38	2.25	3.95	3.78	2.43	2.25	2.26	2.12	2.30	2.17	3.78	3.58	2.11	1.92	2.16	2.02	2.10	1.96	12
3.02	2.82	2.76	2.63	2.80	2.67	4.04	3.87	2.80	2.63	2.74	2.61	2.73	2.60	3.87	3.68	2.46	2.29	2.63	2.51	2.51	2.38	13
2.61	2.40	2.23	2.09	2.34	2.20	4.00	3.82	2.40	2.21	2.21	2.07	2.26	2.12	3.82	3.63	2.08	1.89	2.11	1.97	2.06	1.92	14
2.72	2.51	2.38	2.24	2.47	2.33	3.63	3.43	2.51	2.32	2.35	2.22	2.39	2.25	3.45	3.24	2.18	2.00	2.25	2.12	2.18	2.04	15
2.43	2.21	1.99	1.84	2.13	1.99	3.72	3.53	2.21	2.02	1.97	1.82	2.04	1.91	3.55	3.33	1.90	1.71	1.87	1.73	1.85	1.71	16
2.91	2.70	2.62	2.48	2.68	2.54	3.91	3.73	2.69	2.52	2.59	2.46	2.60	2.47	3.73	3.53	2.36	2.18	2.49	2.36	2.39	2.25	17
2.50	2.28	2.09	1.94	2.22	2.08	3.91	3.73	2.28	2.10	2.06	1.92	2.13	1.99	3.73	3.53	1.97	1.78	1.97	1.83	1.94	1.79	18
2.61	2.40	2.23	2.09	2.34	2.20	3.49	3.29	2.40	2.21	2.21	2.07	2.26	2.12	3.32	3.09	2.08	1.89	2.11	1.97	2.06	1.92	19
2.31	2.09	1.84	1.70	2.01	1.86	3.58	3.39	2.10	1.91	1.82	1.67	1.91	1.78	3.41	3.19	1.79	1.60	1.73	1.59	1.73	1.58	20
2.63	2.35	2.35	2.20	2.42	2.26	3.66	3.46	2.44	2.20	2.34	2.18	2.36	2.21	3.48	3.24	2.12	1.93	2.23	2.10	2.13	1.99	21
2.37	2.09	2.02	1.86	2.13	1.96	3.52	3.32	2.19	1.93	2.01	1.84	2.07	1.91	3.34	3.09	1.87	1.67	1.89	1.76	1.84	1.70	22
2.48	2.20	2.16	2.01	2.25	2.09	3.42	3.22	2.30	2.05	2.15	1.99	2.19	2.04	3.25	3.00	1.97	1.78	2.04	1.90	1.97	1.82	23
2.30	2.01	1.92	1.76	2.04	1.88	3.38	3.17	2.11	1.86	1.91	1.74	1.98	1.82	3.20	2.95	1.80	1.60	1.80	1.66	1.76	1.62	24
2.52	2.24	2.21	2.06	2.30	2.13	3.47	3.27	2.33	2.09	2.20	2.03	2.24	2.08	3.29	3.05	2.01	1.82	2.08	1.95	2.01	1.87	25
2.22	1.94	1.82	1.66	1.96	1.79	3.33	3.12	2.04	1.78	1.81	1.64	1.89	1.73	3.15	2.90	1.73	1.53	1.70	1.56	1.68	1.53	26
2.33	2.05	1.97	1.81	2.09	1.92	3.24	3.03	2.15	1.90	1.96	1.79	2.02	1.86	3.06	2.81	1.83	1.64	1.85	1.71	1.80	1.66	27
2.11	1.82	1.68	1.52	1.83	1.67	3.19	2.98	1.93	1.67	1.67	1.49	1.76	1.60	3.01	2.76	1.62	1.42	1.56	1.42	1.56	1.41	28
2.45	2.16	2.11	1.96	2.21	2.05	3.38	3.17	2.26	2.01	2.10	1.94	2.15	1.99	3.20	2.95	1.94	1.75	1.99	1.85	1.93	1.78	29
2.11	1.82	1.68	1.52	1.83	1.67	3.24	3.03	1.93	1.67	1.67	1.49	1.76	1.60	3.06	2.81	1.62	1.42	1.56	1.42	1.56	1.41	30
2.22	1.94	1.82	1.66	1.96	1.79	3.10	2.88	2.04	1.78	1.81	1.64	1.89	1.73	2.92	2.66	1.73	1.53	1.70	1.56	1.68	1.53	31
2.00	1.71	1.53	1.37	1.71	1.54	3.05	2.83	1.81	1.56	1.52	1.35	1.64	1.47	2.87	2.61	1.51	1.31	1.42	1.27	1.43	1.28	32
2.41	2.13	2.06	1.91	2.17	2.01	3.28	3.07	2.22	1.97	2.05	1.89	2.11	1.95	3.11	2.85	1.90	1.71	1.94	1.80	1.89	1.74	33
2.04	1.75	1.58	1.42	1.75	1.58	3.19	2.98	1.85	1.60	1.57	1.40	1.68	1.52	3.01	2.38	1.55	1.35	1.46	1.32	1.47	1.33	34
2.19	1.90	1.77	1.61	1.92	1.75	3.00	2.79	2.00	1.75	1.76	1.59	1.85	1.69	2.83	2.57	1.69	1.49	1.66	1.52	1.64	1.49	35
1.92	1.64	1.43	1.27	1.62	1.46	2.96	2.74	1.74	1.48	1.42	1.25	1.55	1.39	2.78	2.52	1.44	1.24	1.32	1.18	1.35	1.20	36
2.45	2.16	2.11	1.96	2.21	2.05	3.33	3.12	2.26	2.01	2.10	1.94	2.15	1.99	3.15	2.90	1.94	1.75	1.99	1.85	1.93	1.78	37
2.11	1.82	1.68	1.52	1.83	1.67	3.24	3.03	1.93	1.67	1.67	1.49	1.76	1.60	3.06	2.81	1.62	1.42	1.56	1.42	1.56	1.41	38
2.22	1.94	1.82	1.66	1.96	1.79	3.10	2.88	2.04	1.78	1.81	1.64	1.89	1.73	2.92	2.66	1.73	1.53	1.70	1.56	1.68	1.53	39
2.00	1.71	1.53	1.37	1.71	1.54	3.00	2.79	1.81	1.56	1.52	1.35	1.64	1.47	2.83	2.57	1.51	1.31	1.42	1.27	1.43	1.28	40
2.30	2.01	1.92	1.76	2.04	1.88	3.19	2.98	2.11	1.86	1.91	1.74	1.98	1.82	3.01	2.76	1.80	1.60	1.80	1.66	1.76	1.62	41
1.96	1.68	1.48	1.32	1.67	1.50	3.05	2.83	1.78	1.52	1.47	1.30	1.59	1.43	2.87	2.61	1.48	1.28	1.37	1.23	1.39	1.24	42
2.07	1.79	1.63	1.47	1.79	1.62	2.91	2.69	1.89	1.63	1.62	1.45	1.72	1.56	2.73	2.47	1.58	1.39	1.51	1.37	1.51	1.37	43
1.81	1.53	1.29	1.13	1.50	1.33	2.82	2.60	1.63	1.37	1.28	1.11	1.42	1.26	2.64	2.38	1.33	1.14	1.18	1.04	1.23	1.08	44
2.22	1.94	1.82	1.66	1.96	1.79	3.10	2.88	2.04	1.78	1.81	1.64	1.89	1.73	2.92	2.66	1.73	1.53	1.70	1.56	1.68	1.53	45
1.85	1.56	1.34	1.18	1.54	1.37	2.96	2.74	1.66	1.41	1.33	1.15	1.46	1.30	2.78	2.52	1.37	1.17	1.22	1.08	1.27	1.12	46
2.00	1.71	1.53	1.37	1.71	1.54	2.77	2.55	1.81	1.56	1.52	1.35	1.64	1.47	2.59	2.33	1.51	1.31	1.42	1.27	1.43	1.28	47
1.74	1.45	1.19	1.03	1.41	1.25	2.72	2.50	1.55	1.30	1.18	1.01	1.33	1.17	2.55	2.28	1.26	1.07	1.08	0.94	1.14	1.00	48
2.07	1.79	1.63	1.47	1.79	1.62	2.86	2.64	1.89	1.63	1.62	1.45	1.72	1.56	2.69	2.42	1.58	1.39	1.51	1.37	1.51	1.37	49
1.77	1.49	1.24	1.08	1.46	1.29	2.68	2.46	1.54	1.33	1.23	1.06	1.38	1.21	2.50	2.24	1.30	1.10	1.13	0.99	1.18	1.04	50
1.85	1.56	1.34	1.18	1.54	1.37	2.63	2.41	1.66	1.41	1.33	1.15	1.46	1.30	2.45	2.19	1.37	1.17	1.22	1.08	1.27	1.12	51
1.66	1.38	1.09	0.93	1.33	1.16	2.53	2.32	1.48	1.22	1.08	0.91	1.25	1.09	2.36	2.10	1.19	1.00	0.98	0.84	1.06	0.91	52
1.66	1.38	1.09	0.93	1.33	1.16	2.44	2.22	1.48	1.22	1.08	0.91	1.25	1.09	2.26	2.00	1.19	1.00	0.98	0.84	1.06	0.91	53
1.62	1.34	1.04	0.89	1.29	1.12	2.44	2.22	1.44	1.19	1.03	0.86	1.20	1.04	2.26	2.00	1.16	0.96	0.94	0.80	1.02	0.87	54

FIGURE 8-49   (*continued*)

Notes

1. All heat transmission coefficients in this table include film resistances and are based on winter conditions of −18°C outdoor air and 21°C indoor air temperature, with 24 km/h outdoor air velocity and zero solar flux. With the exception of single glazing, small changes in the interior and exterior temperatures will not significantly affect overall U-factors. The coefficients are for vertical position except skylight values, which are for 20° from horizontal with heat flow up.
2. Glazing layer surfaces are numbered from the outside to the inside. Double, triple, and quadruple refer to the number of glazing panels. All data are based on 1/8 in. glass, unless otherwise noted. Thermal conductivities are: 0.917 W(m·K) for glass, and 0.19 W/(m·K) for acrylic and polycarbonate.
3. Standard spacers are metal. Insulation means wood, fiberglass, or butyl. Edge-of-glass effects assumed to extend over the 65 mm band around perimeter of each glazing unit as seen in Figure 3.
4. Product sizes are described in Figure 3 and frame U-factors are from Table 2.

5. Interpolation procedure for estimating U_{cg} for vertically oriented gas space widths between 6 mm and 13 mm. The most general equation is:

$$U_t = \left[\frac{1}{U_6} - \frac{n_g}{h + 157.5\,k_g} + \frac{n_g}{h + (1000\,k_g/t)} \right]^{-1}$$

$$h = -118.1\,k_g + \left[\frac{78.74\,n_g\,k_g\,U_6\,U_{13}}{U_6 - U_{13}} + 1550\,k_g^2 \right]^{0.5}$$

where

n_g = number of gaps (*e.g.*, double-pane windows have one gap)
k_g = gas conductivity W/(m·K)
t = gas space width of window, mm
U_6 = U-factor for identical glazing with 6 mm gas spaces
U_{13} = U-factor for identical glazing with 13 mm gas spaces

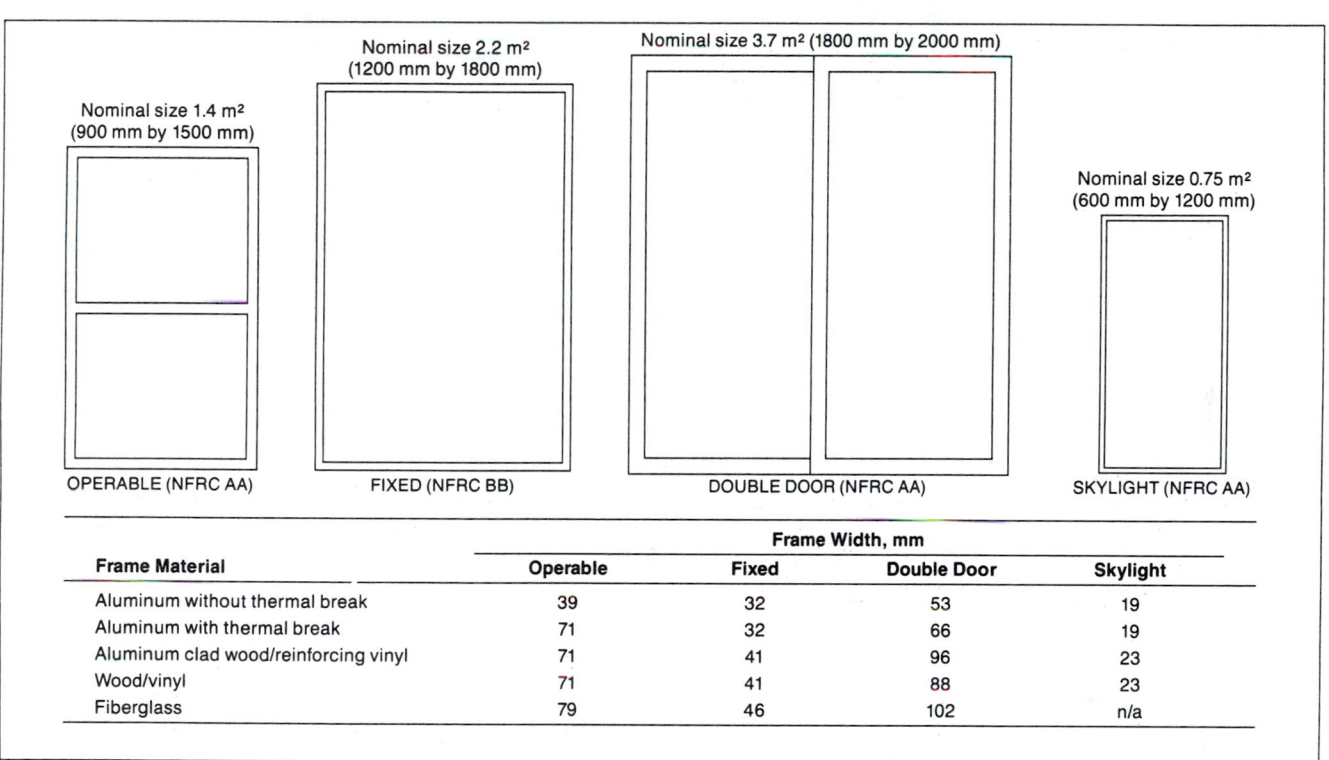

Frame Material	Frame Width, mm			
	Operable	Fixed	Double Door	Skylight
Aluminum without thermal break	39	32	53	19
Aluminum with thermal break	71	32	66	19
Aluminum clad wood/reinforcing vinyl	71	41	96	23
Wood/vinyl	71	41	88	23
Fiberglass	79	46	102	n/a

Standard Fenestration Units

FIGURE 8-49   *(continued)*

Transmission Coefficients U for Wood and Steel Doors, Btu/h·ft²·°F

Nominal Door Thickness, in.	Description	No Storm Door	Wood Storm Door[c]	Metal Storm Door[d]
Wood Doors[a,b]				
1-3/8	Panel door with 7/16-in. panels[e]	0.57	0.33	0.37
1-3/8	Hollow core flush door	0.47	0.30	0.32
1-3/8	Solid core flush door	0.39	0.26	0.28
1-3/4	Panel door with 7/16-in. panels[e]	0.54	0.32	0.36
1-3/4	Hollow core flush door	0.46	0.29	0.32
1-3/4	Panel door with 1-1/8-in. panels[e]	0.39	0.26	0.28
1-3/4	Solid core flush door	0.40	—	0.26
2-1/4	Solid core flush door	0.27	0.20	0.21
Steel Doors[b]				
1-3/4	Fiberglass or mineral wool core with steel stiffeners, no thermal break[f]	0.60	—	—
1-3/4	Paper honeycomb core without thermal break[f]	0.56	—	—
1-3/4	Solid urethane foam core without thermal break[a]	0.40	—	—
1-3/4	Solid fire rated mineral fiberboard core without thermal break[f]	0.38	—	—
1-3/4	Polystyrene core without thermal break (18 gage commercial steel)[f]	0.35	—	—
1-3/4	Polyurethane core without thermal break (18 gage commercial steel)[f]	0.29	—	—
1-3/4	Polyurethane core without thermal break (24 gage residential steel)[f]	0.29	—	—
1-3/4	Polyurethane core with thermal break and wood perimeter (24 gage residential steel)[f]	0.20	—	—
1-3/4	Solid urethane foam core with thermal break[a]	0.20	—	0.16

Note: All U-factors for exterior doors in this table are for doors with no glazing, except for the storm doors which are in addition to the main exterior door. Any glazing area in exterior doors should be included with the appropriate glass type and analyzed as a window (see Chapter 27). Interpolation and moderate extrapolation are permitted for door thicknesses other than those specified.
[a] Values are based on a nominal 32 by 80 in. door size with no glazing.

[b] Outside air conditions: 15 mph wind speed, 0 °F air temperature; inside air conditions: natural convection, 70 °F air temperature.
[c] Values for wood storm door are for approximately 50% glass area.
[d] Values for metal storm door are for any percent glass area.
[e] 55% panel area.
[f] ASTM C 236 hotbox data on a nominal 3 by 7 ft door size with no glazing.

Transmission Coefficients U for Wood and Steel Doors, W/(m²·K)

Nominal Door Thickness, mm	Description	No Storm Door	Wood Storm Door[c]	Metal Storm Door[d]
Wood Doors[a,b]				
35	Panel door with 11-mm panels[e]	3.24	1.87	2.10
35	Hollow core flush door	2.67	1.70	1.82
35	Solid core flush door	2.21	1.48	1.59
35	Panel door with 11-mm panels[e]	3.07	1.82	2.04
45	Hollow core flush door	2.61	1.65	1.82
45	Panel door with 29-mm panels[e]	2.21	1.48	1.59
45	Solid core flush door	2.27	—	1.48
57	Solid core flush door	1.53	1.14	1.19
Steel Doors[b]				
45	Fiberglass or mineral wool core with steel stiffeners, no thermal break[f]	3.41	—	—
45	Paper honeycomb core without thermal break[f]	3.18	—	—
45	Solid urethane foam core without thermal break[a]	2.27	—	—
45	Solid fire rated mineral fiberboard core without thermal break[f]	2.16	—	—
45	Polystyrene core without thermal break (18 gage commercial steel)[f]	1.99	—	—
45	Polyurethane core without thermal break (18 gage commercial steel)[f]	1.65	—	—
45	Polyurethane core without thermal break (24 gage residential steel)[f]	1.65	—	—
45	Polyurethane core with thermal break and wood perimeter (24 gage residential steel)[f]	1.14	—	—
45	Solid urethane foam core with thermal break[a]	1.14	—	0.91

Note: All U-factors for exterior doors in this table are for doors with no glazing, except for the storm doors which are in addition to the main exterior door. Any glazing area in exterior doors should be included with the appropriate glass type and analyzed as a window (see Chapter 27). Interpolation and moderate extrapolation are permitted for door thicknesses other than those specified.
[a] Values are based on a nominal 810 mm by 2030 mm door size with no glazing.

[b] Outside air conditions: 6.7 m/s wind speed, −18 °C air temperature; inside air conditions: natural convection, 21 °C air temperature.
[c] Values for wood storm door are for approximately 50% glass area.
[d] Values for metal storm door are for any percent glass area.
[e] 55% panel area.
[f] ASTM C 236 hotbox data on a nominal 910 mm by 2130 mm door size with no glazing

FIGURE 8-50   Door U values

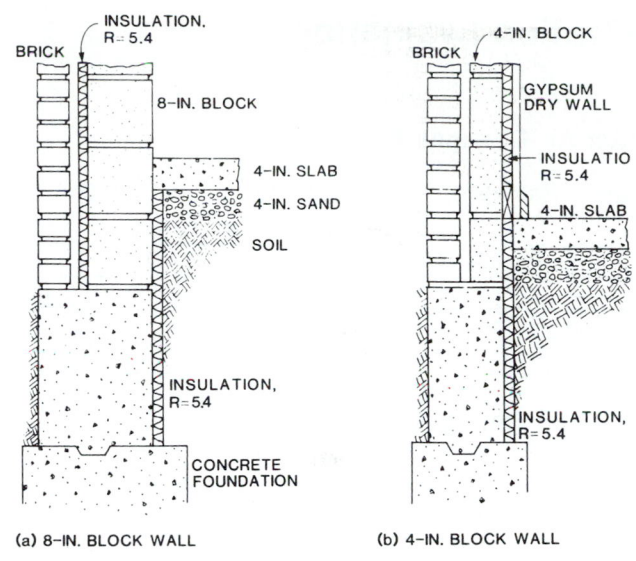

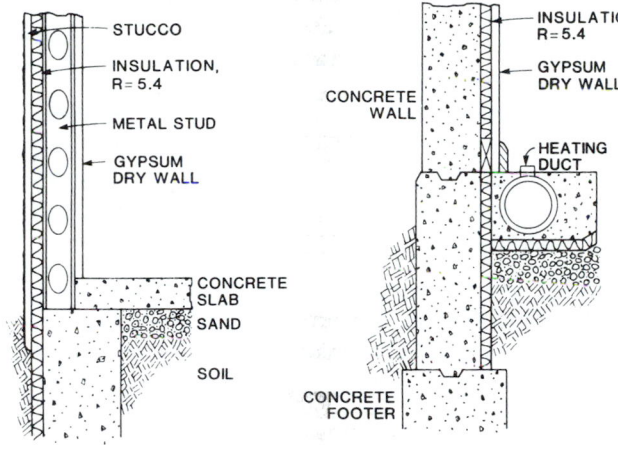

(a) 8-IN. BLOCK WALL   (b) 4-IN. BLOCK WALL

(c) METAL STUD WALL   (d) CONCRETE WALL

Slab-on-Grade Foundation Insulation

Heat Loss Coefficient F_2 of Slab Floor Construction, Btu/h·°F per ft of Perimeter

Construction	Insulation[a]	2950	5350	7433
		Degree Days (65°F Base)		
8-in. block wall, brick facing	Uninsulated	0.62	0.68	0.72
	$R = 5.4$ from edge to footer	0.48	0.50	0.56
4-in block wall, brick facing	Uninsulated	0.80	0.84	0.93
	$R = 5.4$ from edge to footer	0.47	0.49	0.54
Metal stud wall, stucco	Uninsulated	1.15	1.20	1.34
	$R = 5.4$ from edge to footer	0.51	0.53	0.58
Poured concrete wall with duct near perimeter[b]	Uninsulated	1.84	2.12	2.73
	$R = 5.4$ from edge to footer, 3 ft under floor	0.64	0.72	0.90

[a]R-value units in °F·ft²·h/Btu·in.
[b]Weighted average temperature of the heating duct was assumed at 110°F during the heating season (outdoor air temperature less than 65°F).

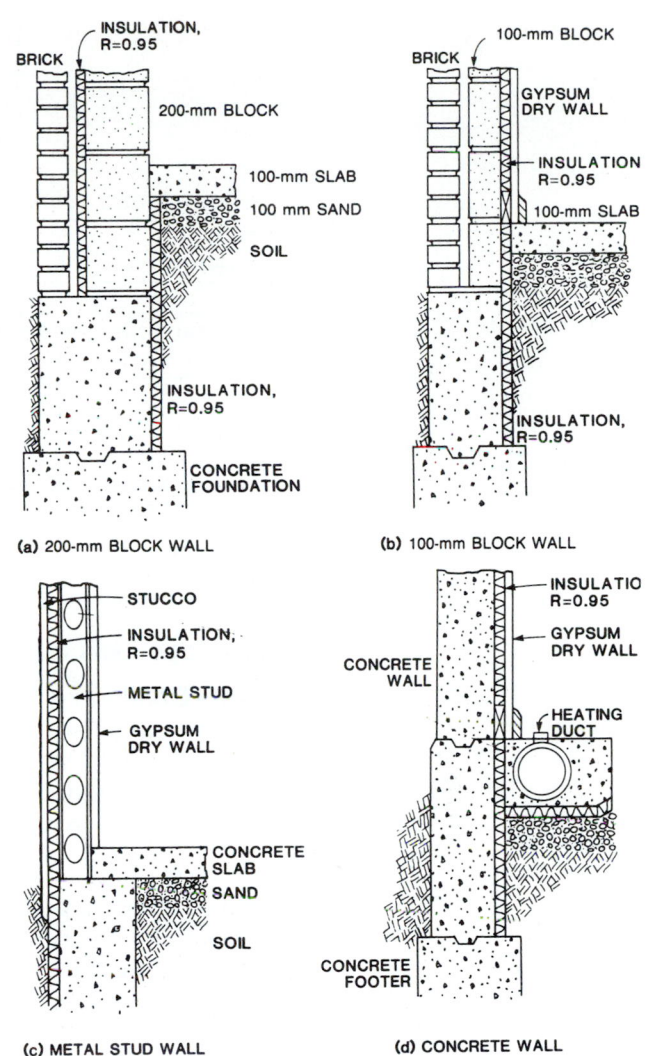

(a) 200-mm BLOCK WALL   (b) 100-mm BLOCK WALL

(c) METAL STUD WALL   (d) CONCRETE WALL

Slab-on-Grade Foundation Insulation

Heat Loss Coefficient F_2 of Slab Floor Construction, W/(m²·K) per metre of Perimeter

Construction	Insulation	4130 K·d/yr	2970 K·d/yr	1640 K·d/yr
		Kelvin Days (18°C Base)		
200 mm block wall, brick facing	Uninsulated	1.07	1.17	1.24
	$R = 0.95$ K·m²/W from edge to footer	0.83	0.86	0.97
100 mm block wall, brick facing	Uninsulated	1.38	1.45	1.61
	$R = 0.85$ from edge to footer	0.81	0.85	0.93
Metal stud wall, stucco	Uninsulated	1.99	2.07	2.32
	$R = 0.95$ from edge to footer	0.88	0.92	1.00
Poured concrete wall with duct near perimeter[a]	Uninsulated	3.18	3.67	4.72
	$R = 0.95$ from edge to footer, 910 mm under floor	1.11	1.24	1.56

[a]Weighted average temperature of the heating duct was assumed at 43°C during the heating season (outdoor air temperature less than 18°C).

FIGURE 8-51   Slab on grade U values

Infiltration Evaluation

Winter Air Changes Per Hour

Floor Area	900 or less	900-1500	1500-2100	over 2100
Best	0.4	0.4	0.3	0.3
Average	1.2	1.0	0.8	0.7
Poor	2.2	1.6	1.2	1.0
For each fire place add:			Best 0.1	Average 0.2	Poor 0.6

Summer Air Changes Per Hour

Floor Area	900 or less	900-1500	1500-2100	over 2100
Best	0.2	0.2	0.2	0.2
Average	0.5	0.5	0.4	0.4
Poor	0.8	0.7	0.6	0.5

Envelope Evaluation

Best - Continuous infiltration barrier, all cracks and penetrations sealed, tested leakage of windows and doors less then 0.25 CFM per running foot of crack, vents and exhaust fans dampered, recessed ceiling lights gasketed or taped, no combustion air required or combustion air from outdoors, no duct leakage.

Average - Plastic vapor barrier, major cracks and penetrations sealed, tested leakage of windows and doors between 0.25 and 0.50 CFM per running foot of crack, electrical fixtures which penetrate the envelope not taped or gasketed, vents and exhaust fans dampered, combustion air from indoors, intermittent ignition and flue damper, some duct leakage to unconditioned space.

Poor - No infiltration barrier or plastic vapor barrier, no attempt to seal cracks and penetrations, tested leakage of windows and doors greater than 0.50 CFM per running foot of crack, vents and exhaust fans not dampered, combustion air from indoors, standing pilot, no flue damper, considerable duct leakage to unconditioned space.

Fireplace Evaluation

Best - Combustion air from outdoors, tight glass doors and damper.
Average - Combustion air from indoors, tight glass doors or damper.
Poor - Combustion air from indoors, no glass doors or damper.

Notes

1. One, two or three story, or split level; any wind exposure.
2. Floor plan aspect ratio between 1:1 and 3:1.
3. Glass plus door areas between 10% and 30% of the wall area.
4. Allowance for one kitchen and two bathroom exhaust fans, dryer vent, recessed lighting fixtures, pipe and duct penetrations.

FIGURE 8-52   Infiltration

Location	Latitude Degrees	WINTER 97½% Design db	WINTER Heating D.D. Below 65°F	SUMMER 2½% Design db	SUMMER Coincident Design wb	SUMMER Grains Difference 55% RH	SUMMER Grains Difference 50% RH	Daily Range
ALABAMA								
Alexander City	33	22	....	93	76	37	44	21 M
Anniston AP	33	22	2810	94	76	35	42	21 M
Auburn	32	22	....	93	76	37	44	21 M
Birmingham AP	33	21	2710	94	75	30	37	21 M
Decatur	34	16	3050	93	74	25	32	22 M
Dothan AP	31	27	1400	92	76	39	46	20 M
Florence AP	34	21	3199	94	74	23	30	22 M
Gadsden	34	20	3000	94	75	30	37	22 M
Huntsville AP	34	16	3190	93	74	25	33	23 M
Mobile AP	30	29	1620	93	77	44	51	18 M
Mobile CO	30	29	1620	93	77	44	51	16 M
Montgomery AP	32	25	2250	95	76	33	40	21 M
Selma-Craig AFB	32	26	2160	95	77	38	47	21 M
Talladega	33	22	....	94	76	33	42	21 M
Tuscaloosa AP	33	23	2590	96	76	32	39	22 M
ALASKA								
Anchorage AP	61	−18	10860	68	58	0	0	15 L
Barrow (S)	71	−41	20265	53	50	0	0	12 L
Fairbanks AP (S)	64	−47	14290	78	60	0	0	24 M
Juneau AP	58	1	9080	70	58	0	0	15 L
Kodiak	57	13	8860	65	56	0	0	10 L
Nome AP	64	−27	14170	62	55	0	0	10 L
ARIZONA								
Douglas AP	31	31	2630	95	63	0	0	31 H
Flagstaff AP	35	4	7290	82	55	0	0	31 H
Fort Huachuca AP (S)	31	28	2551	92	62	0	0	27 H
Kingman AP	35	25	....	100	64	0	0	30 H
Nogales	31	32	2150	96	64	0	0	31 H
Phoenix AP (S)	33	34	1680	107	71	0	0	27 H
Prescott AP	34	9	....	94	60	0	0	30 H
Tuscon AP (S)	32	32	1700	102	66	0	0	26 H
Winslow AP	35	10	4780	95	60	0	0	32 H
Yuma AP	32	39	970	109	72	0	0	27 H
ARKANSAS								
Blytheville AFB	36	15	3760	94	77	42	49	21 M
Camden	33	23	....	96	76	32	39	21 M
El Dorado AP	33	23	2300	96	76	32	39	21 M
Fayetteville AP	36	12	3840	94	73	18	25	23 M
Fort Smith AP	35	17	3290	98	76	29	36	24 M
Hot Springs	34	23	2729	97	77	37	44	22 M
Jonesboro	35	15	....	94	77	42	49	21 M
Little Rock AP (S)	34	20	3170	96	77	39	46	22 M
Pine Bluff AP	34	22	2588	97	77	37	44	22 M
Texarkana AP	33	23	2530	96	77	39	46	21 M
CALIFORNIA								
Bakersfield AP	35	32	2150	101	69	0	0	32 H
Barstow	34	29	2203	104	68	0	0	37 H
Blythe AP	33	33	....	110	71	0	0	28 H
Burbank AP	34	39	1700	91	68	0	2	25 M
Chico	39	30	2835	101	68	0	0	36 H
Concord	38	27	3035	97	68	0	0	32 H
Covina	34	35	....	95	68	0	0	31 H
Crescent City AP	41	33	....	65	59	0	0	18 M
Downey	34	40	....	89	70	9	16	22 M
El Cajon	32	44	....	80	69	17	24	30 H
El Centro AP (S)	32	38	925	110	74	0	4	34 H
Escondido	33	41	....	85	68	4	11	30 H
Eureaka/Arcata AP	41	33	4640	65	59	0	0	11 L
Fairfield-Travis AFB	38	30	2725	95	67	0	0	34 H
Fresno AP (S)	36	30	2610	100	69	0	0	34 H
Hamilton AFB	38	32	3311	84	66	0	3	28 H
Laguna Beach	33	43	....	80	68	13	20	18 M
Livermore	37	27	3035	97	68	0	0	24 M
Lompoc, Vandenburg AFB	34	38	3451	70	61	0	0	20 M
Long Beach AP	33	43	1803	80	68	13	20	22 M
Los Angeles AP (S)	34	43	1960	80	68	13	20	15 L
Los Angeles CO (S)	34	40	1960	89	70	9	16	20 M
Merced-Castle AFB	37	31	2470	99	69	0	0	36 H
Modesto	37	30	....	98	68	0	0	36 H
Monterey	36	38	2750	71	61	0	0	20 M
Napa	38	32	....	96	68	0	0	30 H
Needles AP	34	33	....	110	71	0	0	27 H
Oakland AP	37	36	2940	80	63	0	0	19 M
Oceanside	33	43	....	80	68	13	20	13 L
Ontario	34	33	2009	99	69	0	5	36 H

AP = Airport
CO = City Office
S = Solar Data Available

NOTE: The tables for Canada are based on °F. To convert degree day Fahrenheit to degree day Celsius, multiply by 5/9 or 0.5556. For example, Edmonton Alberta has 9703 degree days (°F).

Degree days (°C)= 9703 × 5/9 = 5391

AP = Airport
CO = City Office
S = Solar Data Available

Reprinted with permission of ACCA, *Manual J* 1986.

FIGURE 8-53 Outdoor design conditions for United States and Canada (Design grains based on an inside design temperature of 75°F)

Top table

Location	Latitude Degrees	WINTER 97½% Design db	Heating D.D. Below 65°F	SUMMER 2½% Design db	Coincident Design wb	Grains Difference 55% RH	Grains Difference 50% RH	Daily Range
Grand Junction AP (S)	39	7	5660	94	59	0	0	29 H
Greeley	40	4	...	94	60	0	0	29 H
LaJunta AP	38	3	5132	98	68	0	0	31 H
Leadville	39	−14	...	81	51	0	0	30 H
Pueblo AP	38	0	5480	95	61	0	0	31 H
Sterling	40	−2	...	93	62	0	0	30 H
Trinidad AP	37	3	5330	91	61	0	0	32 H
CONNECTICUT								
Bridgeport AP	41	9	5617	84	71	22	29	18 M
Hartford, Brainard Field	41	7	6170	88	73	28	35	22 M
New Haven AP	41	7	5890	84	73	33	40	17 M
New London	41	9	...	85	72	26	33	16 M
Norwalk	41	9	...	84	71	22	29	19 M
Norwich	41	7	...	86	73	31	38	18 M
Waterbury	41	2	6672	85	71	20	27	21 M
Windsor Locks, Bradley Field (S)	42	4	6350	88	72	22	29	22 M
DELAWARE								
Dover AFB	39	15	4700	90	75	36	43	18 M
Wilmington AP	39	14	4930	89	74	41	38	20 M
DISTRICT OF COLUMBIA								
Andrews AFB	38	14	...	90	74	30	37	18 M
Washington National AP	38	17	4240	91	74	29	36	18 M
FLORIDA								
Belle Glade	26	44	...	91	76	41	48	16 M
Cape Kennedy AP	28	38	711	88	78	59	66	15 L
Daytona Beach AP	29	35	879	90	77	49	56	15 L
Fort Lauderdale	26	46	244	91	78	53	60	15 L
Fort Myers AP	26	44	442	92	78	51	58	18 M
Fort Pierce	27	42	...	90	78	55	62	15 L
Gainesville AP (S)	29	31	1081	93	77	45	51	18 M
Jacksonville AP	30	32	1230	94	77	42	49	19 M
Key West AP	24	57	110	90	78	55	62	9 L
Lakeland CO (S)	28	41	661	91	76	51	48	17 M

AP = Airport
CO = City Office
S = Solar Data Available

Reprinted with permission of ACCA, *Manual J* 1986.

Bottom table

Location	Latitude Degrees	WINTER 97½% Design db	Heating D.D. Below 65°F	SUMMER 2½% Design db	Coincident Design wb	Grains Difference 55% RH	Grains Difference 50% RH	Daily Range
Oxnard	34	36	...	80	64	0	0	19 M
Palmdale AP	34	22	2929	101	65	0	0	35 H
Palm Springs	33	35	...	110	70	0	0	35 H
Pasadena	34	35	1694	95	68	0	0	29 H
Petaluma	38	29	...	90	66	0	0	31 H
Pomona CO	34	30	2166	99	69	0	0	36 H
Redding AP	40	31	...	102	67	0	0	32 H
Redlands	34	33	...	99	69	0	0	33 H
Richmond	38	36	...	80	63	0	0	17 M
Riverside-March AFB (S)	33	32	2162	98	68	0	0	37 H
Sacramento AP	38	32	2700	98	70	0	2	36 H
Salinas AP	36	32	...	70	60	0	0	24 M
San Bernardino, Norton AFB	34	33	1978	99	69	0	24	38 H
San Diego AP	32	44	1500	80	69	17	2	12 L
San Fernando	34	39	...	91	68	0	0	38 H
San Francisco AP	37	38	3040	77	63	0	3	20 M
San Francisco CO	37	40	3040	71	62	0	3	14 L
San Jose AP	37	36	2416	81	65	0	3	26 H
San Luis Obispo	35	35	2472	88	70	10	17	26 H
Santa Ana AP	33	39	1675	85	68	4	11	28 H
Santa Barbara MAP	34	36	2470	77	66	7	14	24 M
Santa Cruz	37	38	...	71	61	0	0	28 H
Santa Maria AP (S)	34	33	2930	76	63	2	2	23 M
Santa Monica CO	34	43	...	80	68	13	20	16 M
Santa Paula	34	35	...	86	67	0	4	36 H
Santa Rosa	38	29	3065	95	67	0	0	34 H
Stockton AP	37	30	2806	97	68	0	0	37 H
Ukiah	39	29	...	95	68	0	0	40 H
Visalia	36	30	...	100	69	0	0	38 H
Vreka	41	17	...	92	64	0	0	38 H
Yuba City	39	31	...	101	67	0	0	36 H
COLORADO								
Alamosa AP	37	−6	8529	82	57	0	0	35 H
Boulder	40	0	...	91	59	0	0	27 H
Colorado Springs AP	38	2	6410	88	57	0	0	30 H
Denver AP	39	1	6150	91	59	0	0	28 H
Durango	37	−1	...	87	59	0	0	30 H
Fort Collins	40	−1	...	91	59	0	0	28 H

AP = Airport
CO = City Office
S = Solar Data Available

Reprinted with permission of ACCA, *Manual J* 1986.

FIGURE 8-53 *(continued)*

Location	WINTER Latitude Degrees	WINTER 97½% Design db	WINTER Heating D.D. Below 65°F	SUMMER 2½% Design db	SUMMER Coincident Design wb	SUMMER Grains Difference 55% RH	SUMMER Grains Difference 50% RH	SUMMER Daily Range
IDAHO								
Boise AP (S)	43	10	5830	94	64	0	0	31 H
Burley	42	2	.....	95	61	0	0	35 H
Coeur D'Alene AP	47	– 1	6660	86	61	0	0	31 H
Idaho Falls AP	43	– 6	7890	87	61	0	0	38 H
Lewiston AP	46	6	5500	93	64	0	0	32 H
Moscow	46	0	.....	87	62	0	0	32 H
Mountain Home AFB	43	12	6120	97	63	0	0	36 H
Pocatello AP	43	– 1	7030	91	60	0	0	35 H
Twin Falls AP (S)	42	2	6730	95	61	0	0	34 H
ILLINOIS								
Aurora	41	– 1	6660	91	76	41	48	20 M
Belleville, Scott AFB	38	6	4480	92	76	39	49	21 M
Bloomington	40	– 2	.....	90	74	30	37	21 M
Carbondale	37	7	.....	93	77	44	51	21 M
Champaign/Urbana	40	2	5800	92	74	27	34	21 M
Chicago, Midway AP	41	0	6160	91	73	23	29	20 M
Chicago, O'Hare AP	42	– 4	6640	89	74	31	38	20 M
Chicago CO	41	2	6640	91	74	29	36	15 L
Danville	40	1	5538	90	74	30	37	21 M
Decatur	39	2	5480	91	74	29	36	21 M
Dixon	41	– 2	.....	90	74	30	37	23 M
Elgin	42	– 2	.....	88	74	33	40	21 M
Freeport	42	– 4	.....	89	73	26	33	24 M
Galesburg	41	– 2	6005	91	75	35	42	22 M
Greenville	39	4	.....	92	75	33	40	21 M
Joliet	41	0	6180	90	74	30	37	20 M
Kankakee	41	1	6040	90	74	30	37	21 M
LaSalle/Peru	41	– 2	.....	91	75	35	42	22 M
Macomb	40	0	.....	92	76	39	46	22 M
Moline AP	41	– 4	6410	91	75	35	42	23 M
Mt. Vernon	38	5	.....	92	75	33	40	21 M
Peoria AP	40	– 4	6070	89	74	31	38	22 M
Quincy AP	40	3	5267	93	76	37	44	22 M
Rantoul, Chanute AFB	40	1	5966	91	74	29	36	21 M
Rockford	42	– 4	6840	89	73	26	33	24 M
Springfield AP	39	2	5530	92	74	27	34	21 M
Waukegan	42	– 3	.....	89	74	31	37	21 M

AP = Airport
CO = City Office
S = Solar Data Available

Reprinted with permission of ACCA, *Manual J* 1986.

Location	WINTER Latitude Degrees	WINTER 97½% Design db	WINTER Heating D.D. Below 65°F	SUMMER 2½% Design db	SUMMER Coincident Design wb	SUMMER Grains Difference 55% RH	SUMMER Grains Difference 50% RH	SUMMER Daily Range
Miami AP (S)	25	47	200	90	77	49	56	15 L
Miami Beach CO	25	48	200	89	77	51	58	10 L
Ocala	29	34	.....	93	77	44	51	18 M
Orlando AP	28	38	720	93	76	37	44	17 M
Panama City, Tyndall AFB	30	33	1390	90	77	49	56	14 L
Pensacola CO	30	29	1480	93	77	44	51	14 L
St. Augustine	29	35	1051	89	78	57	64	16 M
St. Petersburg	28	40	670	91	77	47	54	16 M
Sanford	28	38	.....	93	76	37	44	17 M
Sarasota	27	42	.....	92	77	45	52	17 M
Tallahassee AP (S)	30	30	1520	92	76	39	46	19 M
Tampa AP (S)	28	40	700	91	77	47	54	17 M
West Palm Beach AP	26	45	270	91	78	53	60	16 M
GEORGIA								
Albany, Turner AFB	31	29	1760	95	76	33	40	20 M
Americus	32	25	.....	94	76	35	42	20 M
Athens	34	22	2929	92	74	27	34	21 M
Atlanta AP (S)	33	22	2990	92	74	27	34	19 M
Augusta AP	33	23	2410	95	76	33	40	19 M
Brunswick	31	32	1531	89	78	57	64	18 M
Columbus, Lawson AFB	32	24	2380	93	76	37	44	21 M
Dalton	34	22	.....	93	76	37	44	22 M
Dublin	32	25	.....	93	76	37	44	20 M
Gainesville	34	21	.....	91	74	29	36	21 M
Griffin (S)	33	22	.....	90	75	36	43	21 M
La Grange	33	23	.....	91	75	35	42	21 M
Macon AP	32	25	2160	93	76	37	44	22 M
Marietta, Dobbins AFB	34	21	3080	92	74	27	34	21 M
Moultrie	31	30	.....	95	77	40	47	20 M
Rome AP	34	22	3326	93	76	37	44	23 M
Savannah-Travis AP	32	27	1850	93	77	44	51	20 M
Valdosta-Moody AFB	31	31	1520	94	77	43	49	20 M
Waycross	31	29	.....	94	77	43	49	20 M
HAWAII								
Hilo AP (S)	19	62	0	83	72	30	37	15 L
Honolulu AP	21	63	0	86	73	31	38	12 L
Kaneohe Bay MCAS	21	66	0	84	74	40	47	12 L
Wahiawa	21	59	10	85	72	26	33	14 L

AP = Airport
CO = City Office
S = Solar Data Available

Reprinted with permission of ACCA, *Manual J* 1986.

FIGURE 8-53 *(continued)*

Location	Latitude Degrees	WINTER		SUMMER				
		97½% Design db	Heating D.D. Below 65°F	2½% Design db	Coincident Design wb	Grains Difference 55% RH	Grains Difference 50% RH	Daily Range
Ottumwa AP	41	−4	....	91	74	29	36	22 M
Sioux City AP	42	−7	6960	92	74	27	34	24 M
Waterloo	42	−10	7370	89	75	38	45	23 M
KANSAS								
Atchison	39	2	....	93	76	37	44	23 M
Chanute AP	34	7	4566	97	74	19	26	23 M
Dodge City AP (S)	37	5	....	97	69	0	0	25 M
El Dorado	37	7	4990	98	73	11	18	24 M
Emporia	38	5	....	97	74	19	26	25 M
Garden City AP	38	4	....	96	69	0	0	28 H
Goodland AP	39	0	6140	96	65	0	0	31 H
Great Bend	38	4	....	98	73	11	18	28 H
Hutchinson AP	38	8	4680	99	72	3	10	28 H
Liberal	37	7	....	96	68	0	0	28 H
Manhattan, Fort Riley (S)	39	3	5306	95	75	28	35	24 M
Parsons	37	9	4158	97	74	19	26	23 M
Russel AP	38	4	....	98	73	11	18	29 H
Salina	38	5	4970	100	74	13	20	26 H
Topeka AP	39	4	5210	96	75	26	33	24 M
Wichita AP	37	7	4640	98	73	11	18	23 M
KENTUCKY								
Ashland	38	10	4555	91	74	29	36	22 M
Bowling Green AP	37	10	4280	92	75	33	40	21 M
Corbin AP	37	9	....	92	73	21	28	23 M
Covington AP	39	6	5260	90	72	19	26	22 M
Hopkinsville, Campbell AFB	36	10	4290	92	75	33	40	21 M
Lexington AP (S)	38	8	4760	91	73	25	30	22 M
Louisville AP	38	10	4610	93	74	25	33	23 M
Madisonville	37	10	....	93	75	31	38	22 M
Owensboro	37	10	4200	94	75	30	37	23 M
Pudacah AP	37	12	3650	95	75	28	35	20 M
LOUISIANA								
Alexandria AP	31	27	2000	94	77	43	49	20 M
Baton Rouge AP	30	29	1610	93	77	44	51	19 M
Bogalusa	30	28	....	93	77	44	51	19 M

AP = Airport
CO = City Office
S = Solar Data Available

Reprinted with permission of ACCA, *Manual J* 1986.

Location	Latitude Degrees	WINTER		SUMMER				
		97½% Design db	Heating D.D. Below 65°F	2½% Design db	Coincident Design wb	Grains Difference 55% RH	Grains Difference 50% RH	Daily Range
INDIANA								
Anderson	40	6	5580	92	75	33	40	22 M
Bedford	38	5	....	92	75	33	40	22 M
Bloomington	39	5	4860	92	75	33	40	22 M
Columbus, Bakalar AFB	39	7	5132	92	75	33	40	22 M
Crawfordsville	40	3	....	91	74	29	36	22 M
Evansville AP	38	9	4500	93	75	31	38	22 M
Fort Wayne AP	41	1	6220	89	72	20	27	24 M
Goshen AP	41	1	....	89	73	26	33	23 M
Hobart	41	2	....	88	73	26	33	21 M
Huntington	40	1	....	89	72	20	27	22 M
Indianapolis AP (S)	39	2	5630	90	74	30	37	22 M
Jeffersonville	38	10	....	93	74	25	33	23 M
Kokomo	40	0	5590	90	73	24	31	22 M
Lafayette	40	3	5820	91	73	23	30	22 M
LaPorte	41	3	....	90	74	30	37	22 M
Marion	40	0	....	90	73	24	31	23 M
Muncie	40	2	5950	90	73	24	31	22 M
Peru, Bunker Hill AFB	40	−1	....	88	73	28	35	22 M
Richmond AP	39	2	....	90	74	30	37	22 M
Shelbyville	39	3	....	91	74	29	36	22 M
South Bend AP	41	1	6460	89	73	26	33	22 M
Terre Haute AP	39	4	5360	92	74	27	34	22 M
Valparaise	41	3	....	90	74	30	37	22 M
Vincennes	38	6	....	92	74	27	34	22 M
IOWA								
Ames (S)	42	−6	....	90	74	30	37	23 M
Burlington AP	40	−3	6120	91	75	35	43	22 M
Cedar Rapids AP	41	−5	6600	88	75	38	45	23 M
Clinton	41	−3	....	90	75	36	33	23 M
Council Bluffs	41	−3	6610	91	75	35	42	22 M
Des Moines AP	41	−5	6610	91	74	29	36	23 M
Dubuque	42	−7	7380	88	73	28	35	22 M
Fort Dodge	42	−7	7070	88	74	33	40	23 M
Iowa City	41	−6	6404	89	76	44	51	22 M
Keokuk	40	0	6404	92	75	33	40	22 M
Marshalltown	42	−7	6850	90	75	36	43	23 M
Mason City AP	43	−11	7790	88	74	33	40	24 M
Newton	41	−5	....	91	74	29	36	23 M

AP = Airport
CO = City Office
S = Solar Data Available

Reprinted with permission of ACCA, *Manual J* 1986.

FIGURE 8-53 *(continued)*

MICHIGAN / MINNESOTA

Location	Latitude Degrees	WINTER Heating D.D. Below 65°F	WINTER 97½% Design db	SUMMER 2½% Design db	Coincident Design wb	Grains Difference 55% RH	Grains Difference 50% RH	Daily Range
MICHIGAN								
Adrian	41	.....	3	88	72	22	29	23 M
Alpena AP	45	8510	−6	85	70	15	22	27 H
Battle Creek AP	42	6580	5	88	72	22	29	23 M
Benton Harbor AP	42	.....	5	88	72	22	29	20 M
Detroit	42	6290	6	88	72	22	29	20 M
Escanaba	45	8481	−7	83	69	13	20	17 M
Flint AP	42	7200	1	87	72	23	30	25 M
Grand Rapids AP	42	6890	5	88	72	22	29	24 M
Holland	42	.....	6	86	71	19	26	22 M
Jackson AP	42	.....	5	88	72	22	29	23 M
Kalamazoo	42	.....	5	88	72	22	29	23 M
Lansing AP	42	6940	1	87	72	23	30	24 M
Marquette CO	46	8390	−8	81	69	16	23	18 M
Mt. Pleasant	43	.....	4	87	72	23	30	24 M
Muskegon AP	43	6700	6	84	70	16	23	21 M
Pontiac	42	.....	4	87	72	23	30	21 M
Port Huron	43	6564	4	87	72	23	30	21 H
Saginaw AP	43	7120	4	87	72	23	30	23 M
Sault Ste. Marie AP (S)	46	9050	−8	81	69	16	23	23 M
Traverse City AP	44	7700	1	86	71	19	26	22 M
Yipsilanti	42	6424	5	89	71	15	22	22 M
MINNESOTA								
Albert Lea	43	.....	−12	87	72	23	30	24 M
Alexandria AP	45	.....	−16	88	72	22	29	24 M
Bemidji AP	47	10203	−26	85	69	9	16	24 M
Brainerd	46	.....	−16	87	71	18	25	24 M
Duluth AP	46	9890	−16	82	68	10	17	22 M
Fairbault	44	.....	−12	88	72	22	27	24 M
Fergus Falls	46	.....	−17	88	72	22	27	24 M
International Falls AP	48	10600	−25	83	68	8	15	26 M
Mankato	44	8310	−12	88	72	22	29	24 M
Minneapolis / St. Paul AP	44	8250	−12	89	73	26	33	22 M
Rochester AP	44	8295	−12	87	72	23	30	24 M
St. Cloud AP (S)	45	8890	−11	88	72	22	29	24 M
Virginia	47	.....	−21	83	68	8	15	23 M

AP = Airport
CO = City Office
S = Solar Data Available

Reprinted with permission of ACCA, *Manual J* 1986.

Location	Latitude Degrees	WINTER Heating D.D. Below 65°F	WINTER 97½% Design db	SUMMER 2½% Design db	Coincident Design wb	Grains Difference 55% RH	Grains Difference 50% RH	Daily Range
Houma	29	.....	35	93	78	50	57	15 L
Lafayette AP	30	1550	30	94	78	49	56	18 M
Lake Charles AP (S)	30	1490	31	93	77	44	51	17 M
Minden	32	2310	25	96	76	32	38	20 M
Monroe AP	32	.....	25	96	76	32	38	20 M
Natchitoches	31	.....	26	95	77	40	47	20 M
New Orleans AP	30	1400	33	92	78	51	58	16 M
Shreveport AP (S)	32	2160	25	96	76	32	39	20 M
MAINE								
Augusta AP	44	7826	−3	85	70	15	23	22 M
Bangor, Dow AFB	44	8220	−6	83	68	8	15	22 M
Caribou AP (S)	46	9770	−13	81	67	7	14	21 M
Lewiston	44	7690	−2	85	70	15	22	22 M
Millinocket AP	45	8533	−9	83	68	8	15	22 M
Portland (S)	43	7570	−1	84	71	23	29	22 M
Waterville	44	.....	−4	84	69	11	18	22 M
MARYLAND								
Baltimore AP	39	4680	13	91	75	35	42	21 M
Baltimore CO	39	.....	17	89	76	43	51	17 M
Cumberland	39	5070	10	89	74	31	38	22 M
Frederick AP	39	5030	12	91	75	35	42	22 M
Hagerstown	39	5130	12	91	74	29	36	22 M
Salisbury (S)	38	4220	16	91	75	35	42	18 M
MASSACHUSETTS								
Boston AP (S)	42	5630	9	88	71	16	23	16 M
Clinton	42	.....	2	87	71	18	25	17 M
Fall River	41	5774	9	84	71	22	29	18 M
Framingham	42	.....	6	86	71	19	26	17 M
Gloucester	42	.....	5	86	71	19	26	15 L
Greenfield	42	6195	−2	85	71	20	27	23 M
Lawrence	42	.....	0	87	72	23	30	22 M
Lowell	42	6060	1	88	72	22	29	21 M
New Bedford	41	5400	9	82	71	26	33	19 M
Pittsfield AP	42	7580	−3	84	70	16	23	23 M
Springfield, Westover AFB	42	5840	0	87	71	18	25	19 M
Tauton	41	.....	9	86	72	25	32	18 M
Worcester AP	42	6970	4	84	70	16	23	18 M

AP = Airport
CO = City Office
S = Solar Data Available

Reprinted with permission of ACCA, *Manual J* 1986.

FIGURE 8-53 *(continued)*

Top table

Location	Latitude Degrees	WINTER 97½% Design db	WINTER Heating D.D. Below 65°F	SUMMER 2½% Design db	SUMMER Coincident Design wb	SUMMER Grains Difference 55% RH	SUMMER Grains Difference 50% RH	Daily Range
MONTANA								
Billings AP	45	-10	7150	91	64	0	0	31 H
Bozeman	45	-14	....	87	60	0	0	32 H
Butte AP	46	-17	9730	83	56	0	0	35 H
Cut Bank AP	48	-20	9033	85	61	0	0	35 H
Glasgow AP (S)	48	-18	9000	89	63	0	0	29 H
Glendive	47	-13	....	92	64	0	0	29 H
Great Falls AP (S)	47	-15	7670	88	60	0	0	28 H
Havre	48	-11	8880	90	64	0	0	33 H
Helena AP	46	-16	8190	88	60	0	0	32 H
Kalispell AP	48	-7	8150	87	61	0	0	34 H
Lewiston AP	47	-16	8586	87	61	0	0	30 H
Livingston AP	45	-14	....	87	60	0	0	32 H
Miles City AP	46	15	7810	95	66	0	0	30 H
Missoula AP	46	-6	8000	88	61	0	0	36 H
NEBRASKA								
Beatrice	40	-2	....	95	74	21	28	24 M
Chadron AP	42	-3	7100	94	65	0	0	30 H
Columbus	41	-2	....	95	73	16	23	M
Fremont	41	-2	....	95	74	21	28	22 M
Grand Island AP	41	-3	6440	94	71	6	13	28 H
Hastings	40	-3	6070	94	71	6	13	27 H
Kearney	40	-4	....	93	70	2	9	28 H
Lincoln CO (S)	40	-2	6050	95	74	21	28	24 M
McCook	40	-2	....	95	69	0	0	28 H
Norfolk	42	-4	7010	93	74	25	32	30 H
North Platte AP (S)	41	-4	6680	94	69	0	2	28 H
Omaha AP	41	-3	6290	91	75	35	42	22 M
Scottsbluff AP	41	-3	6670	92	65	0	0	31 H
Sidney AP	41	-3	7030	92	65	0	0	31 H
NEVADA								
Carson City	39	9	5753	91	59	0	0	42 H
Elko AP	40	-2	7430	92	59	0	0	42 H
Ely AP (S)	39	-4	7710	87	56	0	0	39 H
Las Vegas AP (S)	36	28	2610	106	65	0	0	30 H
Lovelock AP	40	12	....	96	63	0	0	42 H
Reno AP (S)	39	10	6150	92	60	0	0	45 H
Reno CO	39	11	....	93	60	0	0	45 H

AP = Airport
CO = City Office
S = Solar Data Available

Reprinted with permission of ACCA, *Manual J* 1986.

Bottom table

Location	Latitude Degrees	WINTER 97½% Design db	Heating D.D. Below 65°F	SUMMER 2½% Design db	Coincident Design wb	Grains Difference 55% RH	Grains Difference 50% RH	Daily Range
Wilmar	45	-11	....	88	72	22	29	24 M
Winona	44	-10	....	88	73	28	35	24 M
MISSISSIPPI								
Biloxi, Keesler AFB	30	31	1500	92	79	58	65	16 M
Clarksdale	34	19	....	94	77	42	49	21 M
Columbus AFB	33	20	2890	93	77	44	51	22 M
Greenville AFB	33	20	2580	93	77	44	51	21 M
Greenwood	33	20	....	93	77	44	51	21 M
Hattiesburg	31	27	1840	94	77	42	49	21 M
Jackson AP	32	25	2260	95	76	33	40	21 M
Laurel	31	27	....	94	77	42	49	21 M
McComb AP	31	26	....	94	76	35	42	18 M
Meridian AP	32	23	2340	95	76	33	40	22 M
Natchez	31	27	1800	94	78	49	56	21 M
Tupelo	34	19	....	94	77	42	49	22 M
Vicksburg CO	32	26	2040	95	78	47	54	21 M
MISSOURI								
Cape Girardeau	37	13	....	95	75	28	35	21 M
Columbia AP (S)	39	4	5070	94	74	23	30	22 M
Farmington AP	37	8	....	93	75	31	38	22 M
Hannibal	39	3	5512	93	76	37	44	22 M
Jefferson City	38	7	4620	95	74	21	28	23 M
Joplin AP	37	10	4090	97	73	13	20	24 M
Kansas City AP	39	6	4750	96	74	20	27	20 M
Kirksville AP	40	0	....	93	74	25	32	24 M
Mexico	39	4	....	94	74	23	30	22 M
Moberly	39	3	....	94	74	23	30	23 M
Poplar Bluff	36	16	3910	95	76	32	40	22 M
Rolla	38	9	....	91	75	35	42	22 M
St. Joseph AP	39	2	5440	93	76	37	44	23 M
St. Louis AP	38	6	4900	94	75	30	37	21 M
St. Louis CO	38	8	....	94	75	30	37	18 M
Sedalia, Whiteman AFB	38	4	5012	92	76	39	46	22 M
Sikeston	36	15	....	95	76	33	40	21 M
Springfield AP	37	9	4900	93	74	25	32	23 M

AP = Airport
CO = City Office
S = Solar Data Available

Reprinted with permission of ACCA, **Manual J** 1986.

FIGURE 8-53 *(continued)*

FIGURE 8-53 (continued) — Table (lower-left portion)

Location	Latitude Degrees	WINTER — 97½% Design db	Heating D.D. Below 65°F	SUMMER — 2½% Design db	Coincident Design wb	Grains Difference 55% RH	Grains Difference 50% RH	Daily Range
Tonopah AP	38	10	5900	92	59	0	0	40 H
Winnemucca AP	40	3	6760	94	60	0	0	42 H
NEW HAMPSHIRE								
Berlin	44	−9	8270	84	69	11	18	22 M
Claremont	43	−4	7850	86	70	14	21	24 M
Concord AP	43	−3	7360	87	70	12	19	26 H
Keene	43	−7	7460	87	70	12	19	24 M
Laconia	43	−5	7560	86	70	14	21	25 M
Manchester, Grenier AFB	43	−3	7100	88	71	16	23	24 M
Portsmouth, Pease AFB	43	2	6710	85	71	20	27	22 M
NEW JERSEY								
Atlantic City CO	39	13	4810	89	74	31	38	18 M
Long Branch	40	13	....	90	73	24	31	18 M
Newark AP	40	14	4900	91	73	23	30	20 M
New Brunswick	40	10	5400	89	73	26	33	19 M
Paterson	40	10	5360	91	73	23	30	21 M
Phillipsburg	40	6	....	89	72	20	27	21 M
Trenton CO	40	14	4980	88	74	33	40	19 M
Vineland	39	11	....	89	74	31	38	19 M
NEW MEXICO								
Alamagordo Holloman AFB	32	19	3240	96	64	0	0	30 H
Albuquerque AP (S)	35	16	4250	94	61	0	0	27 H
Artesia	32	19	....	100	67	0	0	30 H
Carlsbad AP	32	19	2835	100	67	0	0	28 H
Clovis AP	34	13	4200	93	65	0	0	28 H
Farmington AP	36	6	5713	89	62	0	0	30 H
Gallup	35	5	....	88	58	0	0	32 H
Grants	35	4	....		58	0	0	32 H
Hobbs AP	32	18	....	99	66	0	0	29 H
Las Cruces	32	20	3194	96	64	0	0	30 H
Los Alamos	35	9	....	87	60	0	0	32 H
Raton AP	36	1	6228	89	60	0	0	34 H
Roswell, Walker AFB	33	18	3680	98	66	0	0	33 H

AP = Airport
CO = City Office
S = Solar Data Available

Reprinted with permission of ACCA, Manual J 1986.

FIGURE 8-53 (continued) — Table (upper-right portion)

Location	Latitude Degrees	WINTER — 97½% Design db	Heating D.D. Below 65°F	SUMMER — 2½% Design db	Coincident Design wb	Grains Difference 55% RH	Grains Difference 50% RH	Daily Range
Santa Fe CO	35	10	6120	88	61	0	0	28 H
Silver City AP	32	10	3705	94	60	0	0	30 H
Socorro AP	34	17	....	95	62	0	0	30 H
Tucumcari AP	35	13	4047	97	66	0	0	28 H
NEW YORK								
Albany AP (S)	42	−1	6900	88	72	22	29	23 M
Albany CO	42	1	....	88	72	22	29	20 M
Auburn	43	2	....	87	71	18	25	22 M
Batavia	43	5	....	87	71	18	25	22 M
Binghamton AP	42	1	7340	83	69	13	20	20 M
Buffalo AP	43	6	6960	85	70	15	22	21 M
Cortland	42	0	....	85	71	20	27	23 M
Dunkirk	42	9	6851	85	72	26	33	18 M
Elmira AP	42	1	....	86	71	19	26	24 M
Geneva (S)	42	2	....	87	71	18	20	22 M
Glen Falls	42	−5	7270	85	71	20	27	23 M
Gloversville	43	−2	....	86	71	19	26	23 M
Hornell	42	0	....	85	70	15	22	24 M
Ithaca (S)	42	0	7052	85	71	20	27	24 M
Jamestown	42	3	6849	86	70	14	21	20 M
Kingston	42	2	....	88	72	22	29	22 M
Lockport	43	7	6724	86	72	25	32	21 M
Massena AP	45	−8	....	83	69	13	20	20 M
Newburg-Stewart AFB	41	4	6336	88	71	22	29	21 M
NYC-Central Park (S)	40	15	4880	89	73	26	33	17 M
NYC-Kennedy AP	40	15	5219	87	72	23	30	16 M
NYC-La Guardia AP	40	15	4811	89	73	26	33	16 M
Niagara Falls AP	43	7	6688	86	72	25	32	20 M
Olean	42	2	....	84	71	22	29	23 M
Oneonta	42	−4	....	83	69	13	20	24 M
Oswego CO	43	7	6792	83	71	24	31	20 M
Plattsburg AFB	44	−8	8044	83	69	13	20	22 M
Poughkeepsie	41	6	5820	89	74	31	38	21 M
Rochester AP	43	5	6760	88	71	16	23	22 M
Rome-Griffiss AFB	43	−5	7331	85	70	15	22	22 M
Schenectady (S)	42	1	6650	87	72	23	30	22 M
Suffolk County AFB	40	10	5951	83	71	24	31	16 M
Syracuse AP	43	2	6720	87	71	18	25	20 M

AP = Airport
CO = City Office
S = Solar Data Available

Reprinted with permission of ACCA, Manual J 1986.

SUMMER / WINTER (continued section — Ohio/Oklahoma)

Location	Latitude Degrees	WINTER 97½% Design db	WINTER Heating D.D. Below 65°F	SUMMER 2½% Design db	SUMMER Coincident Design wb	SUMMER Grains Difference 55% RH	SUMMER Grains Difference 50% RH	Daily Range
Cambridge	40	7	.....	90	74	30	37	23 M
Chillicothe	39	6	.....	92	74	27	34	22 M
Cincinnati CO	39	6	4830	90	72	19	26	21 M
Cleveland AP (S)	41	5	6200	88	72	22	29	22 M
Columbus AP (S)	40	5	5670	90	73	24	31	24 M
Dayton AP	39	4	5620	89	72	20	27	20 M
Defiance	41	4	.....	91	73	23	30	24 M
Findlay AP	41	3	.....	90	73	24	31	24 M
Fremont	41	1	.....	88	73	28	35	24 M
Hamilton	39	5	.....	90	72	19	26	22 M
Lancaster	39	5	.....	91	73	23	30	23 M
Lima	40	4	5870	91	73	23	30	24 M
Mansfield AP	40	5	6403	87	72	23	30	22 M
Marion	40	5	.....	91	73	23	30	23 M
Middletown	39	5	.....	90	72	19	26	22 M
Newark	40	5	5655	92	73	21	28	23 M
Norwalk	41	1	.....	88	73	28	35	22 M
Portsmouth	38	10	4410	92	74	27	34	22 M
Sandusky CO	41	6	5796	91	72	17	24	21 M
Springfield	40	3	4284	89	73	26	33	21 M
Stubenville	40	5	.....	86	71	19	26	22 M
Toledo AP	41	1	6430	88	73	28	35	25 M
Warren	41	6	.....	87	71	18	25	23 M
Wooster	40	6	.....	86	71	19	26	22 M
Youngstown AP	41	4	6370	86	71	19	26	23 M
Zanesville AP	40	7	.....	90	74	30	37	23 M
OKLAHOMA								
Ada	34	14	3190	97	74	19	26	23 M
Altus AFB	34	16	.....	100	73	7	14	25 M
Ardmore	34	17	3060	98	74	17	24	24 M
Bartlesville	36	10	.....	98	74	17	24	23 M
Chickasha	35	14	.....	98	74	17	24	24 M
Enid-Vance AFB	36	13	3971	100	74	13	20	24 M
Lawton AP	34	16	3250	99	74	15	22	24 M
McAlester	34	19	3255	96	74	20	27	23 M
Muskogee AP	35	15	.....	98	75	23	30	23 M
Norman	35	13	3247	96	74	20	27	24 M

AP = Airport
CO = City Office
S = Solar Data Available

Reprinted with permission of ACCA, *Manual J* 1986.

SUMMER / WINTER (continued section — New York/North Carolina/North Dakota/Ohio)

Location	Latitude Degrees	WINTER 97½% Design db	WINTER Heating D.D. Below 65°F	SUMMER 2½% Design db	SUMMER Coincident Design wb	SUMMER Grains Difference 55% RH	SUMMER Grains Difference 50% RH	Daily Range
Utica	43	-6	7200	85	71	20	27	22 M
Watertown	44	-6	7300	83	71	24	31	20 M
NORTH CAROLINA								
Ashville AP	35	14	4130	87	72	23	30	21 M
Charlotte AP	35	22	3200	93	74	25	32	20 M
Durham	36	20	.....	92	75	33	40	20 M
Elizabeth City AP	36	19	3207	91	77	47	54	18 M
Fayetteville, Pope AFB	35	20	3080	92	76	39	46	20 M
Goldsboro, Seymour-Johnson AFB	35	21	3124	91	76	41	48	18 M
Greensboro AP (S)	36	18	3810	91	73	23	30	21 M
Greenville	35	21	.....	91	76	41	48	19 M
Henderson	36	15	.....	92	76	39	46	20 M
Hickory	35	18	.....	90	72	19	26	21 M
Jacksonville	34	24	.....	90	78	55	62	18 M
Lumberton	34	21	.....	92	76	39	46	20 M
New Bern AP	35	24	.....	90	78	55	62	18 M
Raleigh/ Durham AP (S)	35	20	3440	92	75	33	40	20 M
Rocky Mount	36	21	.....	91	76	41	48	19 M
Wilmington AP	34	26	2380	91	78	53	60	18 M
Winston-Salem AP	36	20	3650	91	73	23	30	20 M
NORTH DAKOTA								
Bismark AP (S)	46	-19	8960	91	68	0	2	27 H
Devil's Lake	48	-21	9901	88	68	0	7	25 M
Dickinson AP	46	-17	8942	90	66	0	0	25 M
Fargo AP	46	-18	9250	89	71	15	22	25 M
Grands Forks AP	48	-22	9930	87	70	12	19	26 H
Jamestown AP	47	-18	.....	90	69	2	9	25 M
Minot AP	48	-20	9610	89	67	0	0	25 M
Williston	48	-21	9180	88	67	0	1	25 M
OHIO								
Akron-Canton AP	41	6	6140	86	71	19	26	21 M
Ashtabula	41	9	.....	85	72	26	33	18 M
Athens	39	6	.....	92	74	27	34	22 M
Bowling Green	41	2	.....	89	73	26	33	23 M

AP = Airport
CO = City Office
S = Solar Data Available

Reprinted with permission of ACCA, *Manual J* 1986.

FIGURE 8-53 (*continued*)

WINTER / SUMMER (top table)

Location	Latitude Degrees	97½% Design db	Heating D.D. Below 65°F	2½% Design db	Coincident Design wb	Grains Difference 55% RH	Grains Difference 50% RH	Daily Range
State College (S)	40	7	6160	87	71	18	25	23 M
Sunbury	40	7	.....	89	72	20	27	22 M
Uniontown	39	9	.....	88	73	28	35	22 M
Warren	41	4	.....	86	71	19	26	24 M
West Chester	40	13	.....	89	74	31	38	20 M
Williamsport AP	41	7	5950	89	72	20	27	23 M
York	40	12	.....	91	74	29	36	22 M
RHODE ISLAND								
Newport (S)	41	9	5800	85	72	26	33	16 M
Providence AP	41	9	5950	86	72	25	32	19 M
SOUTH CAROLINA								
Anderson	34	23	.....	92	74	27	34	21 M
Charleston AFB (S)	32	27	2070	91	78	53	60	18 M
Charleston CO	32	28	2146	92	78	51	58	13 L
Columbia AP	34	24	2520	95	75	28	35	22 M
Florence AP	34	25	2480	92	77	45	52	21 M
Georgetown	33	26	2228	90	78	55	62	18 M
Greenville AP	34	22	3070	91	74	29	36	21 M
Greenwood	34	22	2890	93	74	25	32	21 M
Orangeburg	33	24	.....	95	75	28	35	20 M
Rock Hill	35	23	.....	94	74	23	30	20 M
Spartanburg AP	35	22	.....	91	74	29	36	20 M
Sumter-Shaw AFB	34	25	2453	92	76	39	46	21 M
SOUTH DAKOTA								
Aberdeen AP	45	-15	8620	91	72	17	24	27 H
Brookings	44	-13	.....	92	72	15	22	25 M
Huron AP	44	-14	8220	93	72	13	20	28 H
Mitchel	43	-10	.....	93	71	7	14	28 H
Pierre AP	44	-10	7550	95	71	4	11	29 H
Rapid City AP (S)	44	- 7	7370	92	65	0	0	28 H
Sioux Falls AP	43	-11	7840	91	72	17	24	24 M
Watertown AP	45	-15	8390	91	72	17	24	26 H
Yankton	43	- 7	.....	91	72	17	24	25 M
TENNESSEE								
Athens	33	18	.....	92	73	21	28	22 M
Bristol-Tri City AP	36	14	4143	89	72	20	27	22 M

AP = Airport
CO = City Office
S = Solar Data Available

Reprinted with permission of ACCA, *Manual J* 1986.

WINTER / SUMMER (bottom table)

Location	Latitude Degrees	97½% Design db	Heating D.D. Below 65°F	2½% Design db	Coincident Design wb	Grains Difference 55% RH	Grains Difference 50% RH	Daily Range
Oklahoma City AP	35	13	3700	97	74	19	26	23 M
Ponca City	36	9	3850	97	74	19	26	24 M
Seminole	35	15	.....	96	74	20	27	23 M
Stillwater (S)	36	13	.....	96	74	20	27	24 M
Tulsa AP	36	13	3730	98	75	23	30	22 M
Woodward	36	10	.....	97	73	13	20	26 H
OREGON								
Albany	44	22	.....	89	66	0	0	31 H
Astoria AP (S)	46	29	5190	71	62	0	4	16 M
Baker AP	44	6	.....	89	61	0	0	30 H
Bend	44	4	.....	87	60	0	0	33 H
Corvallis (S)	44	22	4854	89	66	0	0	31 H
Eugene AP	44	22	4740	89	66	0	0	31 H
Grants Pass	42	24	4375	96	68	0	0	33 H
Klamath Falls AP	42	9	6810	87	60	0	0	36 H
Medford AP (S)	42	23	4880	94	67	0	0	35 H
Pendleton AP	45	5	4700	93	64	0	0	29 H
Portland AP	45	23	4635	85	67	0	7	23 M
Portland CO	45	24	.....	86	67	0	4	21 M
Roseburg AP	43	23	4491	90	66	0	0	30 H
Salem AP	45	23	4760	88	66	0	0	31 H
The Dalles	45	19	.....	89	68	3	10	28 H
PENNSYLVANIA								
Allentown AP	40	9	5810	88	72	22	29	22 M
Altoona CO	40	5	6192	87	71	18	25	23 M
Butler	40	6	.....	87	72	23	30	22 M
Chambersburg	40	8	5170	90	74	30	37	23 M
Erie AP	42	9	6540	85	72	26	33	18 M
Harrisburg AP	40	11	5280	91	74	29	36	21 M
Johnstown	40	2	7804	83	70	19	26	23 M
Lancaster	40	8	5560	90	74	30	37	22 M
Meadville	41	4	.....	85	70	15	32	21 M
New Castle	41	7	5800	88	72	22	30	23 M
Philadelphia AP	39	14	5180	90	74	30	37	21 M
Pittsburgh AP	40	5	5950	86	71	19	26	22 M
Pittsburgh CO	40	7	.....	88	71	16	22	19 M
Reading CO	40	13	4960	89	72	20	22	19 M
Scranton/ Wilkes-Barre	41	5	6160	87	71	18	25	19 M

AP = Airport
CO = City Office
S = Solar Data Available

Reprinted with permission of ACCA, *Manual J* 1986.

FIGURE 8-53 (*continued*)

Location	WINTER			SUMMER				
	Latitude Degrees	97½% Design db	Heating D.D. Below 65°F	2½% Design db	Coincident Design wb	Grains Difference 55% RH	Grains Difference 50% RH	Daily Range
Chattanooga AP	35	18	3380	93	74	25	32	22 M
Clarksville	36	12	....	93	74	25	32	21 M
Columbia	35	15	....	94	74	23	30	21 M
Dyersburg	36	15	....	94	77	42	49	21 M
Greenville	35	16	....	90	72	19	26	22 M
Jackson AP	35	16	3350	95	75	28	35	21 M
Knoxville AP	35	19	3510	92	73	21	28	21 M
Memphis AP	35	18	3210	95	76	33	48	21 M
Murfreesboro	35	14	....	94	74	23	30	22 M
Nashville AP (S)	36	14	3610	94	74	23	30	21 M
Tullahoma	35	13	3577	93	73	19	26	22 M
TEXAS								
Abilene AP	32	20	2620	99	71	0	5	22 M
Alice AP	27	34	....	98	77	35	42	20 M
Amarillo AP	35	11	4140	95	67	33	40	26 H
Austin AP	30	28	1720	98	74	23	30	22 M
Bay City	29	33	....	94	77	42	49	16 M
Beaumont	30	31	1370	93	78	50	57	19 M
Beeville	28	33	1189	97	77	37	44	18 M
Big Springs AP (S)	32	20	600	97	69	0	0	26 H
Brownsville AP (S)	25	39	600	93	77	44	51	18 M
Brownwood	31	22	2437	99	73	9	16	22 M
Bryan AP	30	29	1640	96	76	32	39	20 M
Corpus Christi AP	27	35	930	94	78	49	56	19 M
Corsicana	32	25	....	98	75	23	30	21 M
Dallas AP	32	22	2320	100	75	20	27	20 M
Del Rio, Laughlin AFB	29	31	1520	98	73	11	18	24 M
Denton	33	22	....	99	74	15	22	22 M
Eagle Pass	28	32	1423	99	73	9	16	24 M
El Paso AP (S)	31	24	2680	98	64	0	0	27 H
Fort Worth AP (S)	32	22	2390	99	74	15	22	22 M
Galveston AP	29	36	1270	89	79	63	70	10 L
Greenville	33	22	....	99	74	15	22	21 M
Harlingen	26	39	693	94	77	42	49	19 M
Houston AP	29	32	1410	94	77	42	49	18 M
Houston CO	29	33	....	95	77	40	47	18 M
Huntsville	30	27	....	98	75	23	30	20 M
Killeen-Gray AFB	31	25	....	97	73	13	20	22 M
Lamesa	32	17	....	96	69	0	0	26 H

AP = Airport
CO = City Office
S = Solar Data Available

Reprinted with permission of ACCA, *Manual J* 1986.

Location	WINTER			SUMMER				
	Latitude Degrees	97½% Design db	Heating D.D. Below 65°F	2½% Design db	Coincident Design wb	Grains Difference 55% RH	Grains Difference 50% RH	Daily Range
Laredo AFB	27	36	800	101	73	6	13	23 M
Longview	32	24	....	97	76	30	37	20 M
Lubbock AP	33	15	3570	96	69	0	0	26 H
Lufkin AP	31	29	1940	97	76	30	37	20 M
McAllen	26	39	....	95	77	40	47	21 M
Midland AP (S)	32	21	2600	98	69	0	0	26 H
Mineral Wells AP	32	22	....	99	74	15	22	22 M
Palestine CO	31	27	....	98	76	29	36	20 M
Pampa	35	12	....	96	67	0	0	26 H
Pecos	31	21	....	98	69	0	0	27 H
Plainview	34	13	....	96	68	0	0	26 H
Port Arthur AP	30	31	1447	93	78	50	57	19 M
San Angelo, Goodfellow AFB	31	22	2220	99	71	0	5	24 M
San Antonio AP (S)	29	30	1560	97	73	13	20	19 M
Sherman-Perrin AFB	33	20	2837	98	75	23	30	22 M
Snyder	32	18	....	98	70	0	1	26 M
Temple	31	27	....	99	74	15	22	22 M
Tyler AP	32	24	....	97	76	30	37	21 M
Vernon	34	17	....	100	73	7	14	24 M
Victoria AP	28	32	1160	96	77	39	46	18 M
Waco AP	31	26	2040	99	75	21	28	22 M
Wichita Falls AP	34	18	2900	101	73	6	13	24 M
UTAH								
Cedar City AP	37	5	5680	91	60	0	0	32 H
Logan	41	2	6750	91	61	0	0	33 H
Moab	38	11	....	98	60	0	0	30 H
Ogden AP	41	5	6012	91	61	0	0	33 H
Price	39	5	....	91	60	0	0	33 H
Provo	40	6	5720	96	62	0	0	32 H
Richfield	38	5	....	91	60	0	0	34 H
St. George CO	37	21	....	101	65	0	0	33 H
Salt Lake City AP (S)	40	8	5990	95	62	0	0	32 H
Vernal AP	40	0	....	89	60	0	0	32 H
VERMONT								
Barre	44	-11	....	81	69	16	23	23 M
Burlington AP (S)	44	-7	8030	85	70	15	22	23 M
Rutland	43	-8	7440	84	70	16	23	23 M

AP = Airport
CO = City Office
S = Solar Data Available

Reprinted with permission of ACCA, *Manual J* 1986.

FIGURE 8-53 (*continued*)

Location	WINTER Latitude Degrees	WINTER 97½% Design db	WINTER Heating D.D. Below 65°F	SUMMER 2½% Design db	SUMMER Coincident Design wb	SUMMER Grains Difference 55% RH	SUMMER Grains Difference 50% RH	Daily Range
VIRGINIA								
Charlottesville	38	18	4220	91	74	29	36	23 M
Danville AP	36	16	3510	92	73	21	28	21 M
Fredericksburg	38	14	....	93	75	31	38	21 M
Harrisonburg	38	16	4166	91	72	17	24	23 M
Lynchburg AP	37	16	3440	90	74	30	37	21 M
Norfolk AP	36	22	....	91	76	41	48	18 M
Petersburg	37	17	....	92	76	39	46	20 M
Richmond AP	37	17	3910	92	76	39	46	21 M
Roanoke AP	37	16	4150	91	72	17	24	23 M
Staunton	38	16	4307	91	72	17	24	23 M
Winchester	39	10	4780	90	74	30	37	21 M
WASHINGTON								
Aberdeen	47	28	5316	77	62	0	0	16 M
Bellingham AP	48	15	5420	77	65	2	9	19 M
Bremerton	47	25	5432	78	64	0	3	20 M
Ellensburg AP	47	6	....	91	64	0	0	34 H
Everett-Paine AFB	47	25	5940	76	64	0	7	20 M
Kennewick	46	11	....	96	67	0	0	30 H
Longview	46	24	5064	85	67	0	7	30 H
Moses Lake, Larson AFB	47	7	....	94	65	0	0	32 H
Olympia AP	47	22	5236	83	65	0	0	32 H
Port Angeles	48	27	....	69	61	0	3	18 M
Seattle-Boeing Fld	47	26	....	81	66	1	8	24 M
Seattle CO (S)	47	27	4424	82	66	0	7	19 M
Seattle-Tacoma AP (S)	47	26	5190	80	64	0	0	22 M
Spokane AP (S)	47	2	6770	90	63	0	0	28 M
Tacoma-McChord AFB	47	24	5287	82	65	0	2	22 M
Walla Walla AP	46	7	4800	94	66	0	0	27 H
Wenatchee	47	11	....	96	66	0	0	32 H
Yakima AP	46	5	5950	93	65	0	0	36 H
WEST VIRGINIA								
Beckley	37	4	5615	81	69	16	23	22 M
Bluefield AP	37	4	5000	81	69	16	23	22 M
Charleston AP	38	11	4510	90	73	24	31	20 M

Location	WINTER Latitude Degrees	WINTER 97½% Design db	WINTER Heating D.D. Below 65°F	SUMMER 2½% Design db	SUMMER Coincident Design wb	SUMMER Grains Difference 55% RH	SUMMER Grains Difference 50% RH	Daily Range
Clarksville	39	10	4590	90	73	24	31	21 M
Elkins AP	38	6	5680	84	70	16	23	22 M
Huntington CO	38	10	4340	91	74	29	36	22 M
Martinsburg AP	39	10	5231	90	74	30	37	22 H
Morgantown AP	39	8	5100	87	73	29	36	22 H
Parkersburg CO	39	11	4780	90	74	30	37	21 M
Wheeling	40	5	5220	86	71	19	26	21 M
WISCONSIN								
Appleton	44	-9	....	86	72	25	32	23 M
Ashland	46	-16	....	82	68	10	17	23 M
Beloit	42	-3	....	90	75	36	43	24 M
Eau Claire AP	44	-11	7970	89	73	26	33	23 M
Fond du Lac	43	-8	....	86	72	25	32	23 M
Green Bay AP	44	-9	8100	85	72	26	33	23 M
LaCrosse AP	43	-9	7530	88	73	28	35	22 M
Madison AP (S)	43	-7	7720	88	73	28	35	22 M
Manitowoc	44	-7	....	86	72	25	32	21 M
Marinette	45	-11	....	84	71	32	29	20 M
Milwaukee AP	43	-4	7470	87	73	29	36	21 M
Racine	42	-2	....	88	73	28	35	20 M
Sheboygan	43	-6	....	86	73	31	38	20 M
Stevens Point	43	-11	....	89	73	26	33	22 M
Waukesha	43	-5	....	87	73	29	36	22 M
Wausau AP	44	-12	8490	88	72	22	29	23 M
WYOMING								
Casper AP	42	-5	7510	90	57	0	0	31 H
Cheyenne AP	41	-1	7370	86	58	0	0	30 H
Cody AP	44	-13	....	86	60	0	0	32 H
Evanston	41	-3	....	84	55	0	0	32 H
Lander AP (S)	42	-11	7870	88	61	0	0	32 H
Laramie AP (S)	41	-6	7560	81	56	0	0	28 H
Newcastle	43	-12	....	87	63	0	0	30 H
Rawlins	41	-4	....	83	57	0	0	40 H
Rock Springs AP	41	-3	8430	84	55	0	0	32 H
Sheridan AP	44	-8	7740	91	62	0	0	32 H
Torrington	42	-8	....	91	62	0	0	30 H

AP = Airport
CO = City Office
S = Solar Data Available

Reprinted with permission of ACCA, *Manual J* 1986.

Reprinted with permission of ACCA, *Manual J* 1986.

FIGURE 8-53 (*continued*)

| Location | Latitude Degrees | WINTER | | SUMMER | | | | |
		97½% Design db	Heating D.D. Below 65°F	2½% Design db	Coincident Design wb	Grains Difference 55% RH	Grains Difference 50% RH	Daily Range
ALBERTA								
Calgary AP	51	-23	9703	81	61	0	0	25 M
Edmonton AP	53	-25	10268	82	65	0	2	23 M
Grande Prairie AP	55	-33	11129	80	63	0	0	23 M
Jasper	52	-26	10112	80	62	0	0	28 H
Lethbridge AP (S)	49	-22	8644	87	63	0	0	28 H
McMurray AP	56	-38	12462	82	65	0	2	26 H
Medicine Hat AP	50	-24	8852	90	65	0	0	28 H
Red Deer AP	52	-26	10302	81	64	0	0	25 M
BRITISH COLUMBIA								
Dawson Creek	55	-33	10800	79	63	0	0	26 H
Fort Nelson AP (S)	58	-40	10874	81	63	0	0	23 M
Kamloops CO	50	-15	6799	91	65	0	0	29 H
Nanaimo (S)	49	20	5554	80	65	0	4	21 M
New Westminister	49	18	....	81	67	5	12	19 M
Penticton AP	49	4	6522	89	67	0	0	31 H
Prince George AP (S)	53	-28	9755	80	62	0	0	26 H
Prince Rupert CO	54	2	2029	63	57	0	0	12 L
Trail	49	0	6711	89	65	0	0	33 H
Vancouver AP (S)	49	19	5515	77	66	8	15	17 M
Victoria CO	48	23	5579	73	62	0	0	16 M
MANITOBA								
Brandon	49	-27	10828	86	70	14	21	25 M
Churchill AP (S)	58	-39	16728	77	64	0	3	18 M
Dauphin AP	51	-28	....	84	70	16	23	23 M
Flin Flon	54	-37	12414	81	66	2	9	19 M
Portage la Prairie AP	49	-24	10800	86	72	25	32	22 M
The Pas AP (S)	53	-33	12281	82	67	4	11	20 M
Winnipeg AP (S)	49	-27	10679	86	71	19	26	22 M
NEW BRUNSWICK								
Campbellton CO	48	-14	....	82	67	4	11	21 M
Chatham AP	47	-10	....	85	68	5	12	22 M
Edmundston CO	47	-16	9796	83	68	8	15	21 M
Fredericton AP (S)	45	-11	8671	85	69	9	11	23 M
Moncton AP (S)	46	-8	8711	82	69	14	21	23 M
Saint John AP	45	-8	8453	77	65	3	10	19 M
NEWFOUNDLAND								
Corner Brook	48	0	8978	73	63	0	5	17 M
Gander AP	48	-1	9254	79	65	0	7	19 M
Goose Bay AP (S)	53	-24	11887	81	64	0	0	19 M
St. John's AP (S)	47	7	8991	75	65	6	13	18 M
Stephenville AP	48	4	8717	74	64	3	10	14 L
NORTHWEST TERR.								
Fort Smith AP (S)	60	-45	....	81	64	0	0	24 M
Frobisher AP (S)	63	-41	17876	63	51	0	0	14 L
Inuvik (S)	68	-53	....	77	60	0	0	21 M
Resolute AP (S)	74	-47	22673	54	46	0	0	10 L
Yellowknife AP	62	-46	15634	77	61	0	0	16 M
NOVA SCOTIA								
Amherst	45	-6	8400	81	68	11	18	21 M
Halifax AP (S)	44	5	7361	76	65	3	10	16 M
Kentville (S)	45	1	7792	83	68	8	15	22 M
New Glasgow	45	-5	....	79	68	14	21	20 M
Sydney AP	46	3	8049	80	68	12	19	19 M
Truro CO	45	-5	8226	80	69	18	25	22 M
Yarmouth AP	43	9	7340	72	64	6	13	15 L
ONTARIO								
Belleville	44	-7	7709	84	72	28	35	20 M
Chatham	42	3	6503	87	73	29	36	19 M
Cornwall	45	-9	8200	87	72	23	30	21 M
Hamilton	43	1	6821	86	72	25	32	21 M
Kaupuskasing AP (S)	49	-28	11560	83	69	13	20	23 M
Kenora AP	49	-28	10796	82	69	14	21	19 M
Kingston	44	-7	7724	84	72	28	35	20 M
Kitchener	43	-2	7566	85	72	26	33	23 M
London AP	43	0	7349	85	73	32	39	21 M
North Bay AP	46	-18	9677	81	67	6	13	20 M
Oshawa	43	-3	7600	86	72	25	32	20 M

AP - Airport
CO - City Office
S - Solar Data Available

FIGURE 8-53   (continued)

Location	Latitude Degrees	WINTER		SUMMER				
		97½% Design db	Heating D.D. Below 65°F	2½% Design db	Coincident Design wb	Grains Difference 55% RH	Grains Difference 50% RH	Daily Range
Ottawa AP (S)	45	-13	8693	87	71	18	25	21 M
Owen Sound	44	-2	....	82	70	20	27	21 M
Peterborough	44	-9	8309	85	71	20	27	21 M
St. Catharines	43	3	6537	85	72	26	33	20 M
Sarnia	42	3	7061	86	72	25	32	19 M
Sault Ste. Marie AP	46	-13	9500	82	69	14	21	22 M
Sudbury AP	46	-19	9600	83	67	3	10	22 M
Thunder Bay AP	48	-24	10405	83	68	8	15	24 M
Timmins AP	48	-29	11400	84	68	6	13	25 M
Toronto AP (S)	43	-1	6827	87	72	23	30	20 M
Windsor AP	42	4	6579	88	73	28	35	20 M
PRINCE EDWARD ISLAND								
Charlottetown AP (S)	46	-4	8486	78	68	16	23	16 M
Summerside AP	46	-4	8440	79	68	14	21	16 M
QUEBEC								
Bagotville AP	48	-23	....	83	68	8	15	21 M
Chicoutimi	48	-22	10104	83	68	8	15	20 M
Drummondville	45	-14	8700	85	71	20	27	21 M
Granby	45	-14	8400	85	71	20	27	21 M
Hull	45	-14	8700	87	71	18	25	21 M
Megantic AP	45	-16	....	83	70	19	26	20 M
Montreal AP (S)	45	-10	8213	85	72	26	33	17 M
Quebec AP	46	-14	8937	84	70	16	23	20 M
Rimouski	48	-12	9906	79	66	4	11	18 M
St. Jean	45	-11	8500	86	72	25	32	20 M
St. Jeirome	45	-13	9285	86	71	19	26	23 M
Sept. Iles AP (S)	50	-21	....	73	61	0	0	17 M
Shawinigan	46	-14	9380	84	70	16	23	21 M
Sherbrooke CO	45	-21	8490	84	71	22	29	20 M
Thetford Mines	46	-14	9815	84	70	16	23	21 M
Trois Rivieres	46	-13	9306	85	70	15	22	23 M
Val d'Or AP	48	-27	11169	83	68	8	15	22 M
Valleyfield	45	-10	8300	86	72	25	32	20 M
SASKATCHEWAN								
Estevan AP	49	-25	9950	89	68	0	5	26 H
Moose Jaw AP	50	-25	9894	89	67	0	0	27 H
North Battleford AP	52	-30	11082	85	66	0	0	23 M
Prince Albert AP	53	-35	11630	84	66	0	2	25 M
Regina AP	50	-29	10806	88	68	0	7	26 H
Saskatoon AP (S)	52	-31	10856	86	66	0	0	26 H
Swift Current AP (S)	50	-25	9849	90	66	0	0	25 M
Yorkton AP	51	-30	11362	84	68	6	13	23 M
YUKON TERRITORY								
Whitehorse AP (S)	60	-43	12475	77	58	0	0	22 M

AP - Airport
CO - City Office
S - Solar Data Available

FIGURE 8-53 (continued)

QUESTIONS

8-1. Describe the factors used in calculating heat loss.

8-2. What are the relationships of *R*, *C*, *k*, and *U* values?

8-3. How is the outside design temperature determined?

8-4. What is meant by *infiltration heat loss,* where does it occur and how is it calculated?

8-5. Why must the designer check the architect's specifications and drawings?

8-6. What is meant by the term *degree days,* and how will it affect the amount of fuel used per year?

8-7. List some of the ways in which the heat loss of a building can be controlled.

8-8. Why is limiting of window areas important in keeping the heat loss low?

Design Exercises

8-9. The *U* value of an uninsulated wall is 0.29 and of the same wall insulated, 0.08. What is the resistance value of the insulation?

8-10. Using resistance values, what is the *U* value of the wall construction shown in Fig. E8-10?

8-11. What would be the *U* value for the wall used in Exercise 8-10 if 3-in. (75 mm) batt insulation is installed in the studs?

8-12. Using resistance values, what is the *U* value of the wall construction shown in Fig. E8-12?

8-13. What would be the *U* value for the wall used in Exercise 8-12 if 1-in. (25 mm) urethane and ⅜-in. gypsum board were glued to the inside of the wall?

8-14. Using resistance values, what is the *U* value of the wall construction shown in Fig. E8-14?

8-15. Calculate the heat loss for the residence in Appendix D. Assume the residence will be built in Richmond, Virginia.

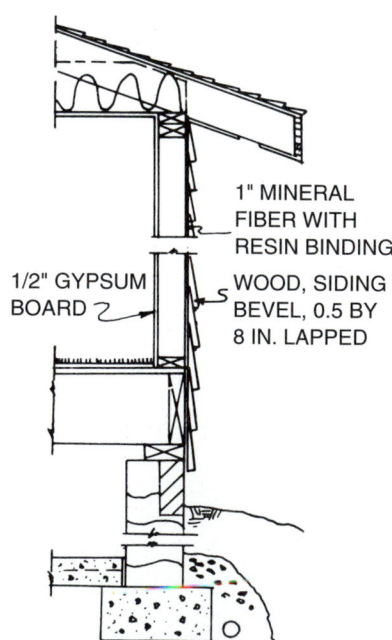

1" MINERAL FIBER WITH RESIN BINDING

1/2" GYPSUM BOARD

WOOD, SIDING BEVEL, 0.5 BY 8 IN. LAPPED

FIGURE E8-10

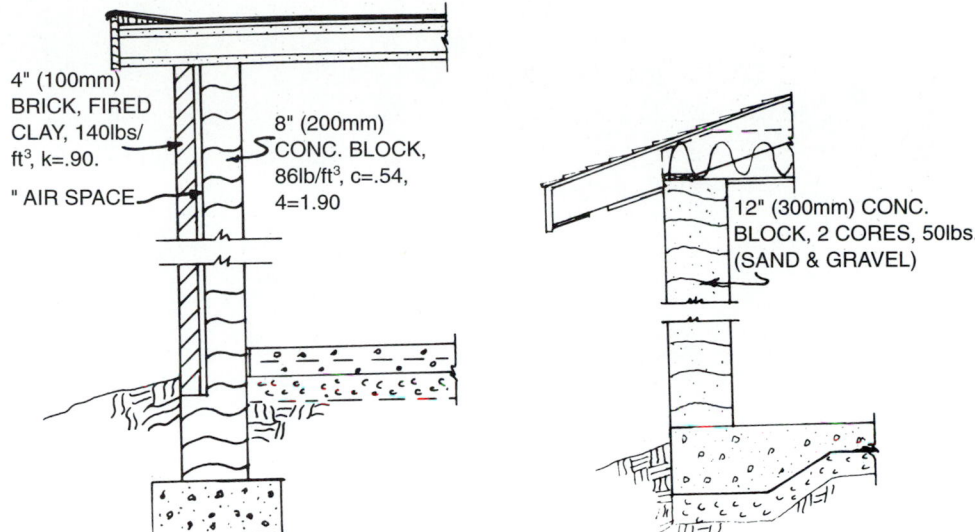

FIGURE E8-12

FIGURE E8-14

8-16. Compare fuel costs, based on the cost of fuels in *your* locale, for the heat loss calculated in Exercise 8-15.

8-17. Calculate the heat loss for a top-floor apartment and an apartment which is below the top floor, using the apartment building in Appendix A. Assume the apartment building will be built in Seattle, Washington.

CHAPTER 9

Hot Water Heating Systems and Design

9-1   TYPES OF SYSTEMS

For heat, the water is heated in a boiler to the preset temperature. Then a circulating pump is automatically cut on, and the hot water is circulated through the system of pipes, passing through any of a variety of convector types (Sec. 9-4). As it circulates, it gives off the heat, primarily through the convectors; then the pipes return the water to the boiler to be reheated and circulated again. Most of the convectors used can be regulated slightly by opening or closing adjustable dampers. A compression tank is included in the system to adjust for the varying pressure in the system, since water expands as it is heated.

Circulating hot water systems are also referred to as *hydronic* heating systems, and the four different hot water piping systems commonly used are series loop, one-pipe and two-pipe systems, and radiant panels.

Series Loop Systems

Most commonly used for residences and small buildings, the convectors in the series loop system are fed by a single pipe which goes through the convector and makes a loop around the building, or one portion of the building (Fig. 9-1 and 9-2). The pipe acts as supply and return with the water getting cooler as it progresses through the system. To the designer, this means that larger convectors are required to obtain the same amount of heat at the end of the loop as compared to convectors at the beginning, because the water is cooler. Also, the longer the run of piping and the more convectors it serves, the cooler the water will become. For more even heat, the building may be broken into zones (Fig. 9-3), each with its own series loop system activated by its individual thermostat. It is economical to install, but any control of individual convectors is minimal. Only heating is suppled with this system.

One-Pipe Systems

As shown in Figs. 9-4 and 9-5, the one-pipe system has a single pipe going around the building, or a zone of the building. A portion of the hot water is diverted through a special tee at each convector so that a portion of the hot water is diverted through the convector where it gives off heat to the room, thereby cooling the water. Then the water is returned to the supply pipe. Upon entering the supply pipe, it will slightly reduce the temperature of the water in that pipe. The primary advantage of this system over the series loop is that each individual convector may be controlled by a valve to turn it on, off, or in between. The convectors may be placed above the pipe (upfeed) or below it (downfeed). The upfeed is more effective since the water tends to be diverted more easily in that direction. This system is more expensive than the series loop since it requires additional piping, fittings, and valves. It may also be zoned for larger buildings where the temperature drop over a long pipe run would be excessive.

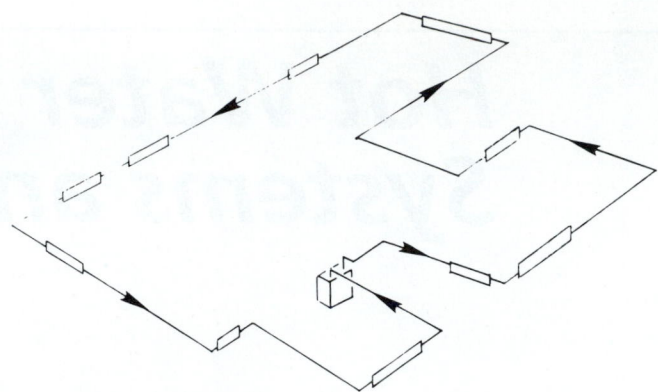

FIGURE 9-1 Series loop isometric

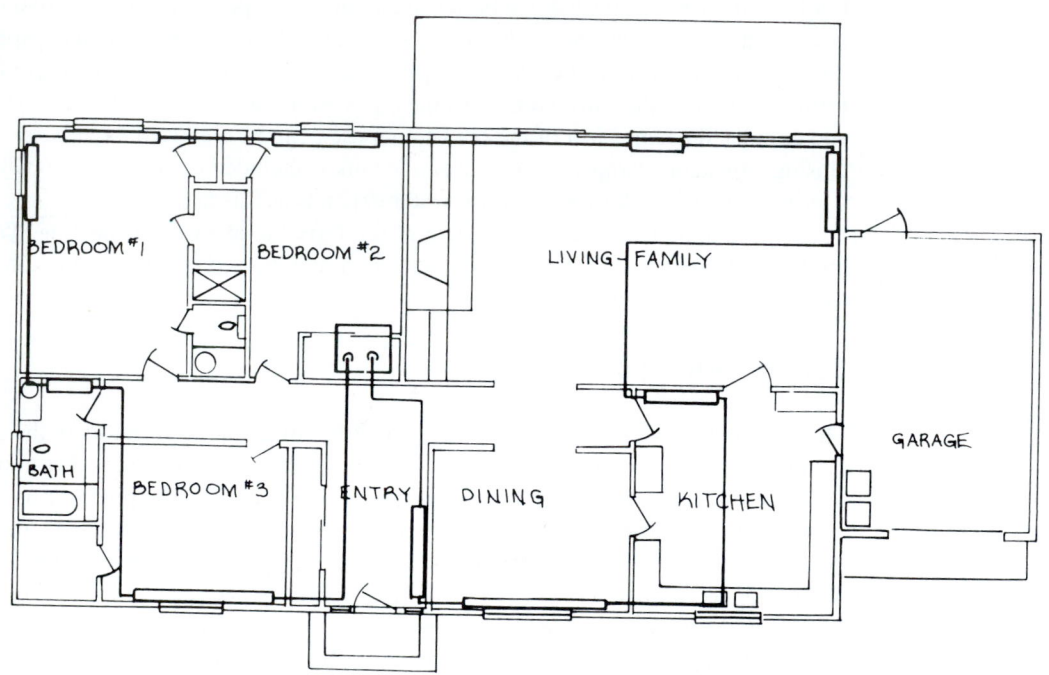

FIGURE 9-2 Series loop plan

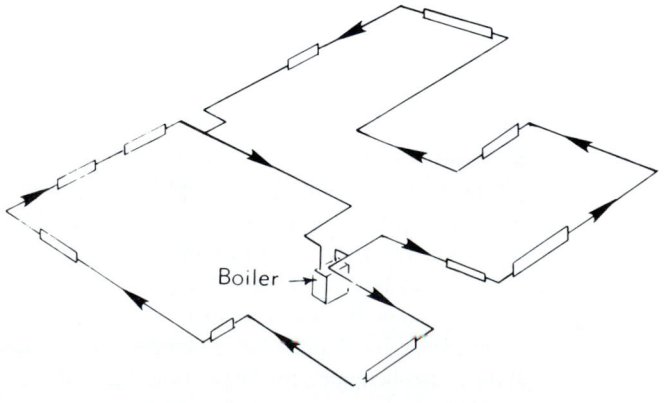

FIGURE 9-3 Series loop (zoned) isometric

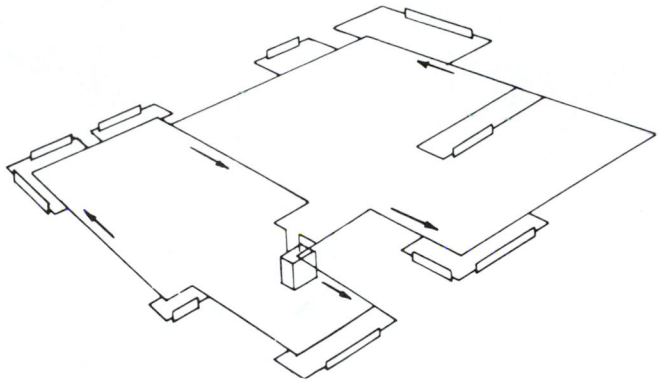

FIGURE 9-4   One-pipe (zoned) isometric

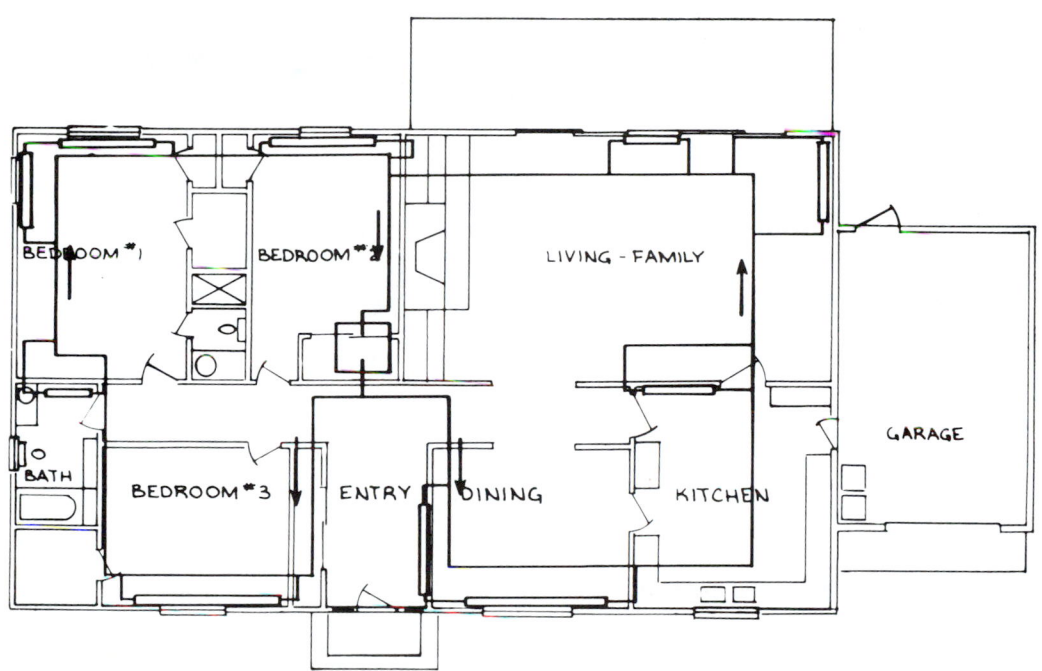

FIGURE 9-5   One-pipe plan

Two-Pipe Systems

For large installations, the supply of hot water needs to be kept separate from that water which has been cooled by passing through a convector. To accomplish this, one pipe is used to supply the hot water, which then empties into another pipe used for return. This type of system provides the water at as high a temperature as possible. The return may be classified as *reverse return* (Fig. 9-6) or *direct return* (Fig. 9-7). The reverse return results in a more even flow of water because the supply and return are of equal length, resulting in equal friction losses. The direct return requires slightly less piping. Individual convector control is available by the installation of valves. This is the best system available and also the most expensive. It would be unnecessary to put this system in a residence.

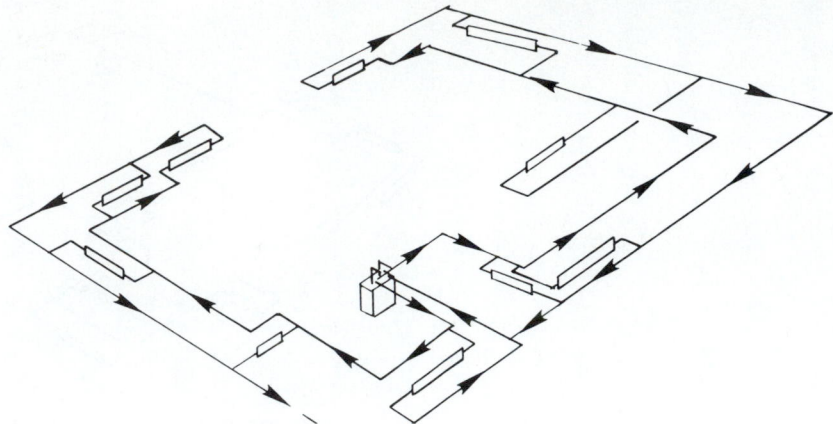

FIGURE 9-6   Two-pipe (zoned) reverse return

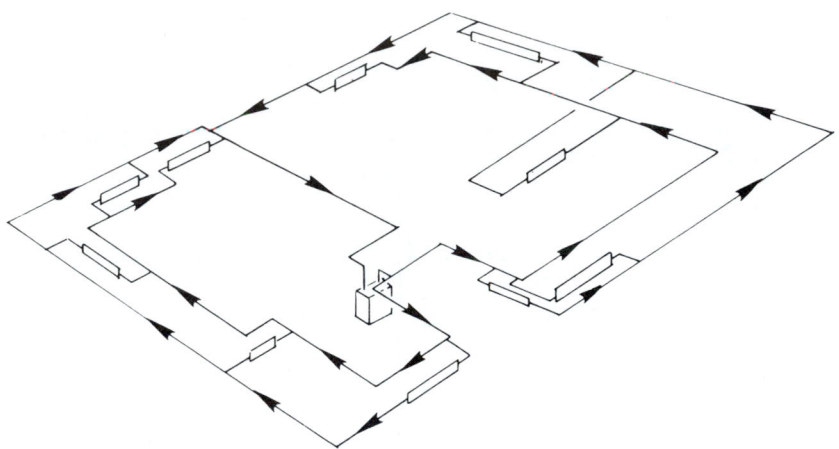

FIGURE 9-7   Two-pipe (zoned) direct return

Radiant Panels

The hot water may be circulated through pipes located, usually, in the floor or ceiling of a building. The pipes are laid in a coil or grid arrangement (Fig. 9-8); for a floor system they are embedded in concrete, whereas for a ceiling they are attached to the framing and plastered or drywalled over. They may also be used on walls, but this is not common. The basic disadvantage with this type of system is that while it provides uniform heat over an entire room, heat is not uniformly lost; most of it is lost at exterior walls, usually at the doors and windows. This situation tends to cause drafts and a feeling of being cold near the open-

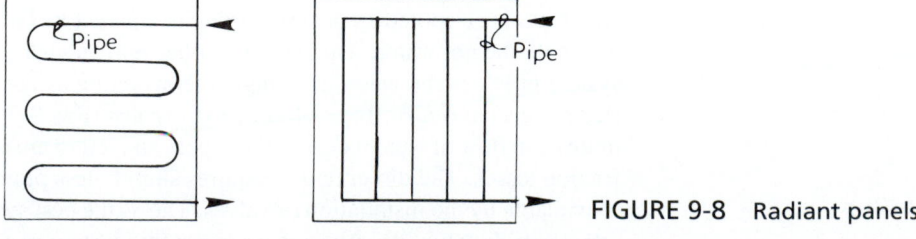

FIGURE 9-8   Radiant panels

ings. The floor panels have a tendency to make the occupant's feet hot and uncomfortable, while the heat coming down from the ceiling tends to stay high (since heat rises), and a person's legs under a table will probably feel cool. (This is also true of electric ceiling and wall radiant panels.)

9-2 PIPING AND FITTINGS

The piping used for hot water heating is usually copper, but steel pipe is sometimes used. Copper is preferred because it is lightweight and easy to work with—characteristics described in Sec. 2-4. The types of copper piping used for hot water heating systems are Type L pipe and tubing and Type M pipe.

The most commonly used fittings for the system are the same as those used in plumbing systems.

9-3 BOILER AND SYSTEM CONTROLS

Boiler

The boiler furnace (Fig. 9-9) heats the water for circulation through the system. It may be rectangular or square (occasionally even round) and made of steel or cast iron. A boiler is rated by the amount of heat it can produce in an hour. The maximum amount of heat the system can put out is limited first by the size of the boiler selected. Boiler efficiency increases when the boiler runs for long periods of time, so the unit selected should not be oversized or it will run intermittently and thus be less efficient. Hot water boilers may use oil, gas, propane gas, coal, or electricity as fuel. In residences oil and gas boilers are most commonly used.

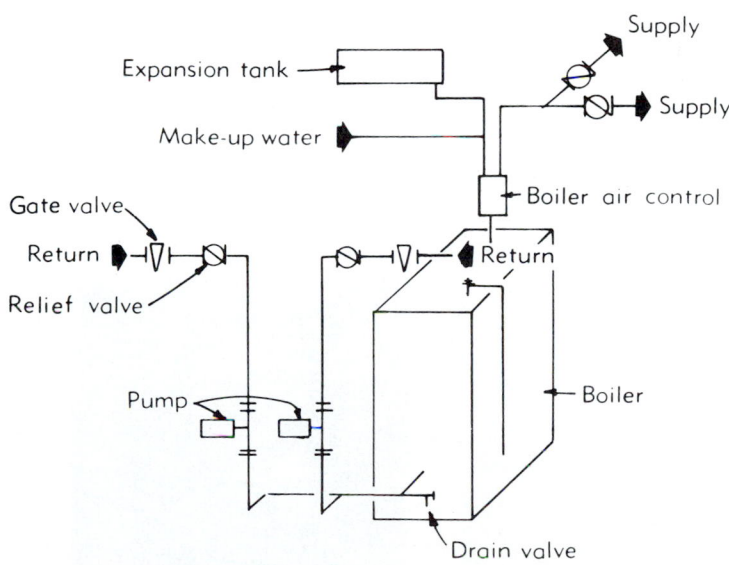

FIGURE 9-9 Typical boiler installation

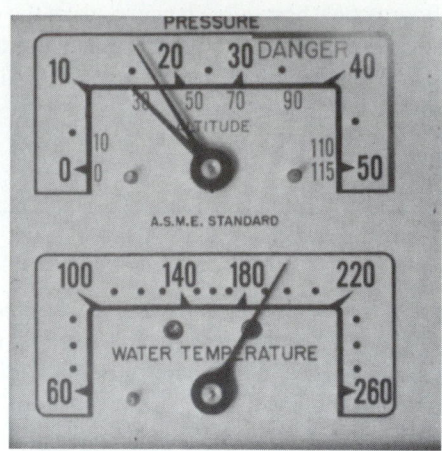

FIGURE 9-10   Boiler thermostat

For a hot water heating system, the temperature that the water is heated to in the boiler is of prime importance since it has a direct relationship to the amount of heat which the radiation or convector units will give off. Typically, the thermostat should be set at about 180°F (82°C) but temperatures as high as 220°F (105°C) are possible. The thermostat which controls the water temperature (Fig. 9-10) is located somewhere on the boiler (usually in plain sight, sometimes behind a small sheet-metal cover plate), and should be checked, after installation, by the designer.

Thermostat

The thermostat, which controls the air temperature in the space (Fig. 9-11), is placed in the building. With hot water heat, there will be one thermostat for each zone. The temperature desired is set on the thermostat, and when the temperature in the room falls below the desired temperature, the thermostat turns on the circulating pump. When the desired temperature is reached, the thermostat will turn the pump off. The thermostat has a temperature differential within which it will call for heat and then cut it off. Typically, this temperature differential is about 2°F (1°C), which means that the thermostat will call for heat at 68°F (20°C) and shut it off at 70°F (21°C). This differential may be adjusted on a small calibration setting inside the thermostat housing (Fig. 9-12). The less the differential, the more comfortable the space will feel. But it also means that the boiler will be turned on

FIGURE 9-11   Thermostat

FIGURE 9-12   Thermostat calibration

and off more frequently (to maintain a close tolerance of temperature), which will lower the efficiency of the boiler operation.

Thermostats used for hot water systems are usually either the single-setting thermostat (Fig. 9-11) or a timer thermostat such as the day-night thermostat in Fig. 9-13.

The single-setting thermostat is the one most commonly used. The desired temperature is set and the thermostat will operate the cycles of the heating system. The temperature may be changed at any time by simply adjusting the temperature setting on the thermostat.

A day-night thermostat permits the setting of one temperature for the daytime and another (usually lower) for nighttime. With this thermostat, a clock is set for a given time (say, 8 A.M.), and at that time the temperature will go to the daytime setting (perhaps 72°F or

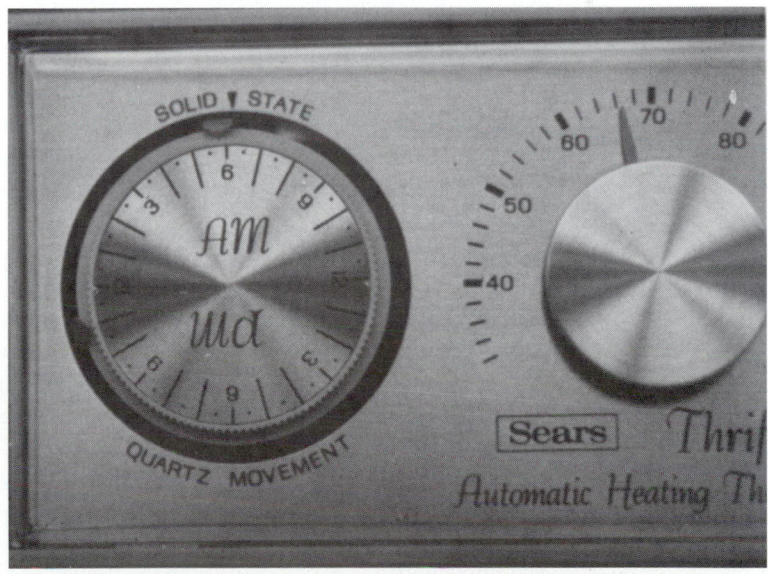

FIGURE 9-13   Day-night thermostat

22°C); then at the other time set (say, 5 P.M.) the temperature will go to the nighttime setting (perhaps 60°F or 16°C). The times and temperatures are set on the thermostat and may be changed as desired.

The day-night thermostat is used in residences, stores, offices, apartments, institutions, and commercial and industrial projects. In a residence, the temperature differential may be 3°F or 4°F (2°C or 3°C), set to begin to cool just before bedtime and to heat up as the occupants arise. In apartments where the heat is furnished and paid for by the apartment owner and not the renter, this type of thermostat is commonly used to control heat costs for the owner. In this case, the thermostat used would have two parts, one located in the apartments to sense the temperature and the second located in a place the renters cannot enter which controls the temperature desired and the times of operation. Since many stores and offices are open only limited hours, perhaps 8 A.M. to 5 P.M., it would be foolish to maintain the temperature needed during hours of operation through the hours in which the space is not in use. Rather than depend on someone to turn down the temperature before they leave (which the individual may forget to do) and to turn up the temperature in the morning (no one will begin work until it warms up), the day-night thermostat takes care of these functions automatically. In locations where the possibility of people tampering with the thermostats is a problem, the unit may have a cover which locks it or may be designed so that the temperature settings can be changed only with a special keying device.

Thermostats are usually placed about 5 ft (1.5 m) up on a wall. The location should be carefully checked to be certain that the thermostat will provide a true representation of the temperature in the spaces it serves. Guidelines used in thermostat location include the following:

1. Always mount it on an inside wall (on an outside wall, the cold air outside will affect the readings).

2. Keep it away from the cold air drafts (such as near a door or window), away from any possible warm air drafts (near a radiation unit, fireplace, or stove), and away from any direct sunlight.

Expansion Tank

The expansion tank (also called a *compression tank*) allows for the expansion of the water in the system as it is heated. It is located above the boiler.

Automatic Filler Valve

When the pressure in the system drops below 12 psi (82.7 kPa), this valve opens, allowing more water into the system; then the check valve automatically closes as the pressure increases. This maintains a minimum water level in the system.

Circulating Pump

The hot water is circulated from the boiler through the pipes and back to the boiler by a pump. The pump provides fast distribution of hot water through the system, thus delivering heat as quickly as possible. The circulating pump size is based on the delivery of water required, in gpm (L/s), and the amount of friction head allowed. Circulating pumps are not used on gravity systems.

Flow Control Valve

The flow control valve closes to stop the flow of hot water when the pump stops, so that the hot water will not flow through the system by gravity. Any gravity flow of the water would cause the temperature in the room to continue to rise.

Boiler Relief Valve

When the pressure in the system exceeds 30 psi (207 kPa), a spring-loaded valve opens, allowing some of the water to bleed out of the system and allowing the pressure to drop. The valve should be located where the discharge will not cause any damage.

Boiler Rating

Boiler manufacturers list both gross and net Btuh (watts) ratings of the boiler. The gross Btuh (watts) is the heat input to the boiler by the fuel, and the net Btuh (watts) is the usable heat output which is available for use in the hot water system.

9-4   HOT WATER HEATING DEVICES

Some type of heating device must be used to distribute the heat efficiently from the hot water to the room being heated. These devices are classified as *radiant* and *convector,* according to the way in which they transfer the heat.

Radiant heating units have the heat transfer surface exposed so that the heat is transferred by radiation to the surrounding objects and by natural convection to the surrounding air.

Convector heating units have the heat transfer surface enclosed in a cabinet or other enclosure (Fig. 9-14 and 9-15). The transfer of heat occurs primarily through convection as air flows through the enclosure, and past the heat transfer surface, by gravity. For institutional and commercial projects, the convectors used may have several rows of large finned-tube radiators in the cabinet.

Unit heaters use a fan to distribute the air through the space. The unit usually consists of a heating coil and a fan enclosed in a cabinet. The heat is supplied by hot water. Often this type of unit is suspended from the ceiling in areas such as a warehouse, a storage area, or any large room situation, especially one with high ceilings. Such a unit is sized according to heat output, and the layout of these units to provide adequate coverage of the space with warm air must be carefully checked. This type of unit is also effective when the room is deep with a relatively small outside wall area for a convector or radiation unit. A typical situation would be that of a motel room where the unit heater is placed under the glass area

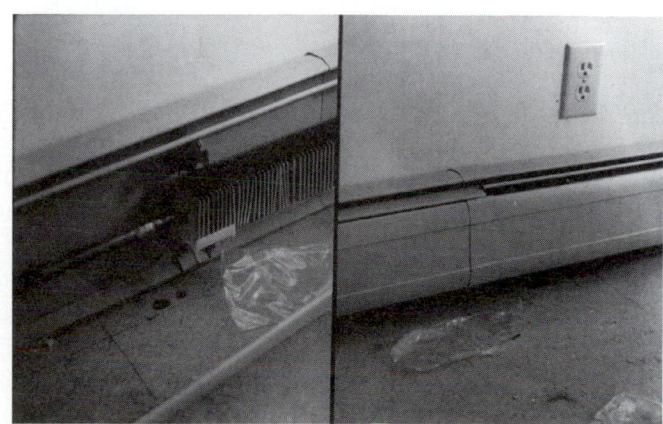

FIGURE 9-14   Convector heating device

FIGURE 9-15   Unit heater

in the exterior wall and the air is fan-blown through the room. Unit heaters may be recessed, surface-mounted, or suspended from the ceiling or wall, or recessed or surface-mounted on the floor.

9-5   HEATING DEVICE RATINGS

All radiant, convector, and unit heaters are rated in terms of the amount of heat they will give off in an hour; the capacity ratings are given in Btuh (watts) for small devices and in MBH (thousands of Btu) or kilowatts (thousands of watts) for larger devices. The heating capacities vary, depending on the type of pipe, the size of the finned tube, the number of fins per foot of radiation, and the temperature of the water. Each manufacturer's specifications (or technical data report) should be checked for heating capacity of a particular unit.

Baseboard radiation units may be cast-iron radiation units or finned-tubed convectors. A typical manufacturer's specification rating is shown in Fig. 9-16. The various types of elements available from this manufacturer are shown in the left column and the heat ratings, for the various water temperatures, on the right. Notice that the heating capacity of the unit increases as the water temperature increases. Comparing the copper-aluminum elements, the capacity for the 2¾-in. × 5-in. × 0.020 × 40/ft, 1¼-in. tube element (the sec-

	Hot Water Output* 1 gal., (3.8 L) flow rate (for 5 gal. (19 L) flow rate use factor 1.067)									
Copper-Aluminum Elements	220°F	105°C	210°F	99°C	200°F	93°C	190°F	88°C	180°F	82°C
2¾″ ×3¾″ × .011 × 50/ft. 1″ tube	1240	1190	1120	1076	1020	980	930	890	840	800
2¾″ × 5″ × .020 × 40/ft. 1¼″ tube	1250	1200	1120	1076	1030	990	940	900	850	815
2¾″ × 5″ × .020 × 50/ft. 1¼″ tube	1340	1285	1200	1150	1100	1050	1000	960	910	870
Steel Elements										
2¾″ × 5 × 24 ga. × 40/ft. 1″ tube (IPS)	1020	980	920	880	840	800	770	735	690	660
2¾″ × 5 × 24 ga. × 40/ft. 1¼″ tube (IPS)	1040	1000	940	900	860	825	780	745	710	680

*Based on 65° (18°C) entering air.
°F values are per foot
°C values are per meter

FIGURE 9-16   Finned tube ratings

ond entry on left) at 180°F is 850 Btuh, while at 200°F the capacity is 1,030 Btuh. The capacity is increased by 180 Btuh simply by increasing the temperature of water that will flow through the element. This 180-Btuh increase represents a 21% gain in the heating capacity at no increase in the cost of boiler or heating device, only an increase in the temperature of the water. The effects of the fins and tube on the heating capacity of an element can be seen by comparing the first and third listings under the copper-aluminum elements. The smaller top element (2¾-in × 3¾-in. × 0.011 × 50/ft 1-in. tube) has a capacity of 840 Btuh at 180°F, while the third listing (2¾-in. × 5-in. × 0.020 × 50/ft 1¼-in. tube) has a capacity of 910 Btuh. This is an increase of 70 Btuh or about 8.33%. The copper-aluminum elements may be compared with the steel elements by checking the second copper-aluminum listing against the second steel element listing. The second copper-aluminum listing (2¾-in. × 5-in. × 0.020 × 40/ft 1¼-in tube) has a capacity of 850 Btuh at 180°F, and the similar steel element listing has a capacity of 710 Btuh at 180°F. The copper-aluminum element capacity is 140 Btuh or 19.7% higher than that of the steel element.

As initial selections of heating devices are made, the various ratings must be checked. It may be necessary to use different sizes of heating devices in different rooms (usually depending on the amount of wall space available for mounting the devices), but once a basic decision is made as to type of material (steel or copper-aluminum), this will usually be used throughout.

To make an economical selection, the designer should also consider the relative costs of the devices per foot and then compare these costs to the heating capacities of the units. Prices vary considerably, but unless the steel elements, installed, cost about 20% less than the copper-aluminum elements, they will not be as economical to install. It is part of the designer's responsibility to provide the best system at the lowest cost.

9-6   HOT WATER HEATING SYSTEM DESIGN

The first thing which must be done is to decide whether the hot water heating system will be a series loop, one-pipe, two-pipe, or radiant panel system. The following step-by-step approach will work for any of the systems (except radiant panels) which are discussed in Sec. 9-1. Completed heat loss tabulations calculated for the residences in Chapter 8 are in Fig. 8-35.

Step-by-Step Approach

1. Determine the heat loss of each room and list them (Fig. 9-17).

   Heat losses for this residence are determined in Chapter 8.

2. Next, determine whether a series loop or a one-pipe or a two-pipe system will be used.

   For this design a series loop system has been selected.

3. Using a floor plan, locate the approximate position of the heating devices in each room.

   The heating devices have been located on the plan in Fig. 9-18.

4. Locate the boiler on the plan (Fig. 9-19).

5. Determine how many zones will be used in the design.

   In this design, two zones will be used, as shown in Fig. 9-20.

Trunk	32,480	BTUH	Zone 2	
Zone 1			Bedroom 3	3483
Entry		4061	Bath	1092
Dining		3056	Bedroom 1	4326
Kitchen		4160	Bedroom 2	2772
Living		9327	Int. Bath	203
		20,604		11,876

SI UNITS				
Trunk	9,546		Zone 2	
Zone 1			Bedroom 3	1013
Entry		1287	Bath	322
Dining		898	Bedroom 1	1267
Kitchen		1205	Bedroom 2	807
Living		2690	Int. Bath	57
		6080		3466

FIGURE 9-17   Zone heat loss totals

6. Determine the actual length of the longest zone (circuit).

From Fig. 9-21, the actual length of the longest zone is taken as 145 lineal ft (44.3 m).

7. Assume an average pipe size for the system. *This pipe size is a preliminary selection which will be rechecked later.*

Assume ¾-in. (20 mm) copper tubing for this design.

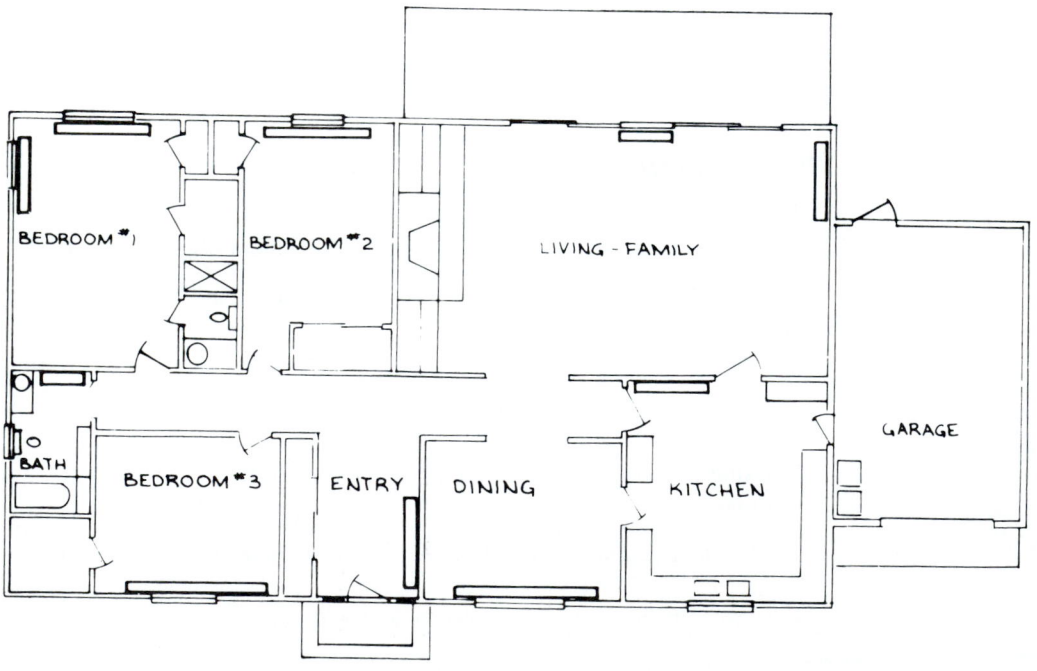

FIGURE 9-18   Series loop system

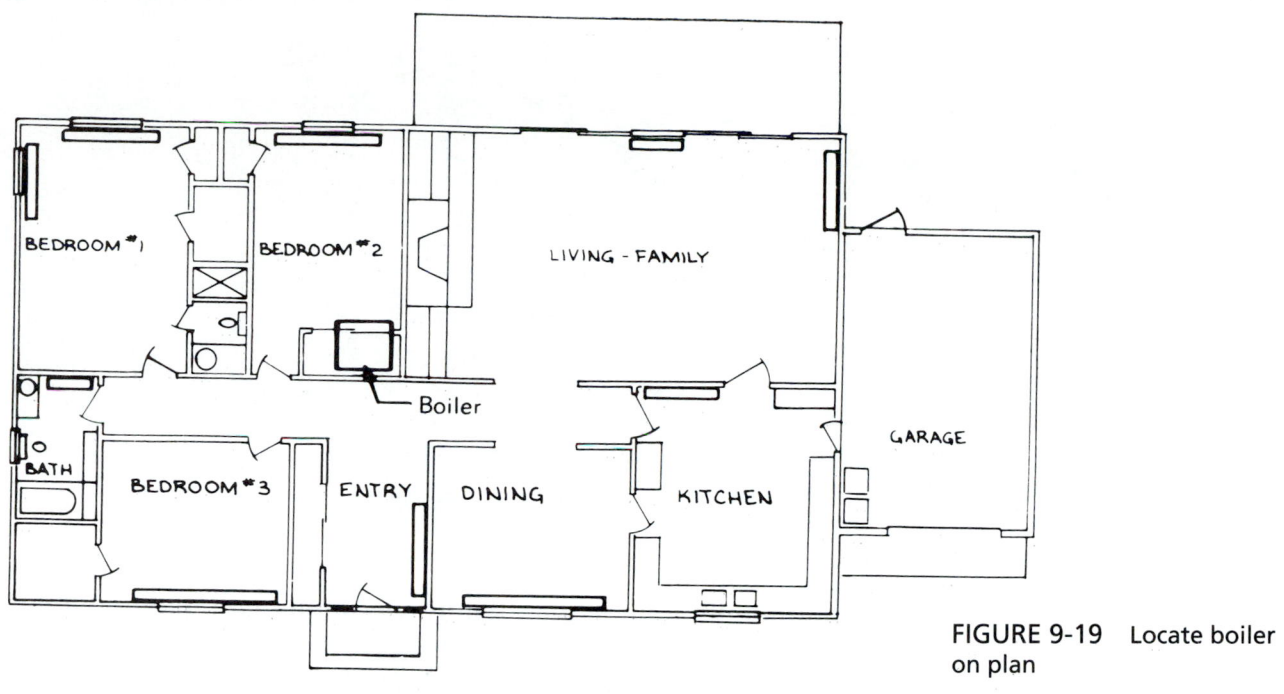

BEDROOM #1

BEDROOM #2

LIVING - FAMILY

GARAGE

Boiler

BATH

BEDROOM #3

ENTRY

DINING

KITCHEN

FIGURE 9-19 Locate boiler on plan

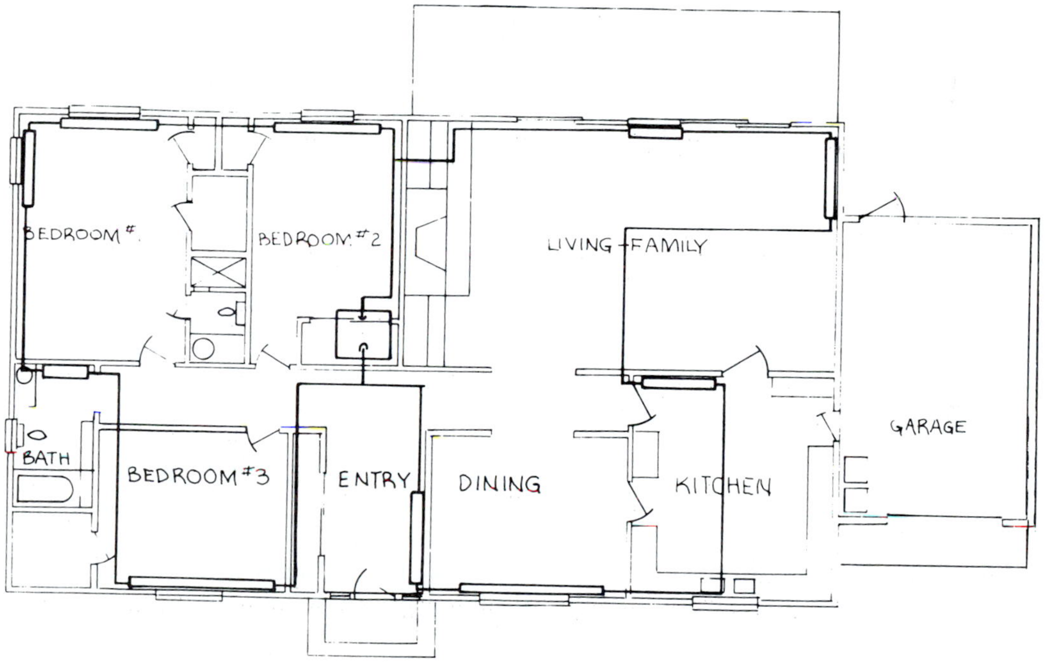

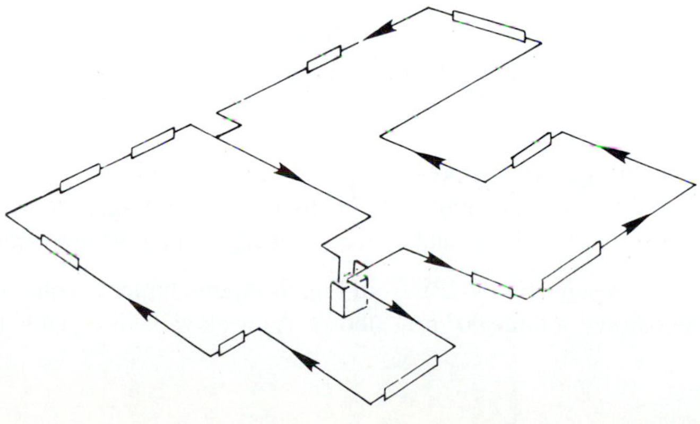

BEDROOM #1

BEDROOM #2

LIVING - FAMILY

GARAGE

BATH

BEDROOM #3

ENTRY

DINING

KITCHEN

FIGURE 9-20 Determine zones

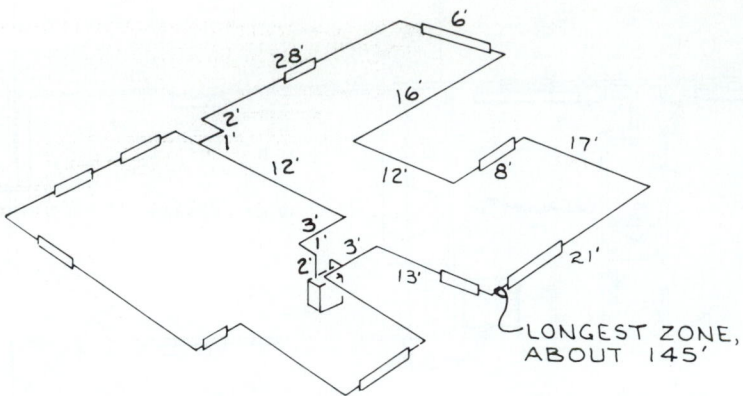

FIGURE 9-21   Determine actual length of zones

8. Determine the velocity of the water in the system from Fig. 9-32 for steel pipe and from Fig. 9-33 for copper tubing, based on the heat to be conveyed per hour in the zone or on of hot water flow (gpm or L/s). Also, note the friction.

$$°\text{F flow rate} = \frac{\text{design heat loss}}{20°\text{F} \times 60 \text{ min} \times 8 \text{ lb/gal}}$$

SI

$$°\text{C flow rate} = \frac{\text{design heat loss}}{11°\text{C} \times 60 \text{ min} \times 60 \text{ seconds} \times 13.75 \text{ kg/L}}$$

Note: The Btuh requirement may be converted to gpm by dividing Btuh by 9600 (based on a 20°F temperature drop). Zone 1 requires 20,604 Btuh or 2.1 gpm.

SI

The watts requirement may be converted to L/s by dividing watts by 38,000 (based on an 11°C temperature drop). Zone 1 requires 6,080 W or 0.16 L/s.

Based on this design (Fig. 9-32), with the largest zone serving 20,604 Btuh, (Fig. 9-44) a ¾-in. Type L pipe would have a velocity of 1.5 ft per second (fps). Also from the chart it is determined that:

$$\text{Friction} = 1.4 \text{ ft per 100 ft}$$
$$= 168 \text{ millinches, per ft}$$

SI

Based on this design, the largest zone serving 6080 W, a 20 mm Type L pipe would have a velocity of 0.5 m/s. Also from the chart it is determined that:

$$\text{Friction} = 185 \text{ Pa/m (pascals per meter)}$$

9. List the fittings which the hot water will pass through in the complete circuit [beginning with the heating unit and going through the longest circuit (zone) and back to the heating unit] (Fig. 9-22).

10. Determine the equivalent elbows for each fitting from Fig. 9-34. Each fitting is converted into the equivalent length of pipe by first converting the fitting into an equivalent number of 90° elbows and then converting the elbows into equivalent length.

In this design (Fig. 9-23), from Fig. 9-34, one boiler is equal to four 90° copper elbows or three 90° iron elbows. A check valve is equal to twenty 90° cop-

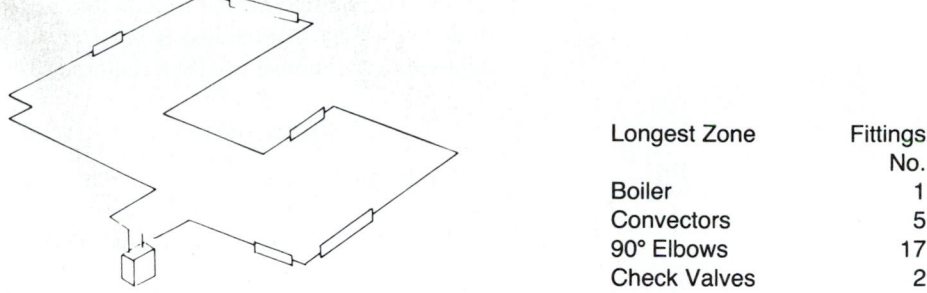

Longest Zone	Fittings No.
Boiler	1
Convectors	5
90° Elbows	17
Check Valves	2

FIGURE 9-22   Fittings

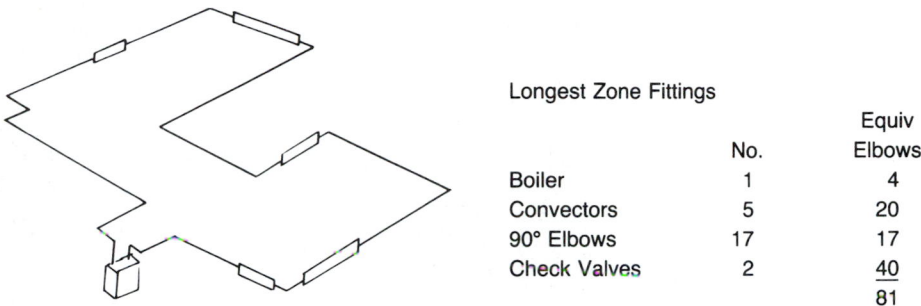

Longest Zone Fittings

	No.	Equiv Elbows
Boiler	1	4
Convectors	5	20
90° Elbows	17	17
Check Valves	2	40
		81

FIGURE 9-23   Equivalent elbows

per elbows or fifteen 90° iron elbows. The equivalent length will be found in step 11.

11. Determine the total equivalent length of the fittings from Fig. 9-34. This table is based on the pipe size and the velocity. The pipe was tentatively selected in step 7 and the velocity noted in step 8.

   In this design, a ¾-in. pipe and 1.5 fps have been selected. From Fig. 9-35, there is an equivalent length of 1.8 ft of pipe for every elbow. The equivalent length of the fittings is 46 ft (81 elbows × 1.8).

   SI In this design, a 20 mm pipe and 0.5 m/s have been selected. From Fig. 9-43, there is an equivalent length of 0.6 m of pipe for every elbow. The equivalent length of the fittings is 48.6 m (81 elbows × 0.6).

12. Add the length of the longest circuit (zone), step 6, to the total equivalent length of fittings for the circuit, step 11.

   In this design, the longest circuit is 145 lineal ft (step 6), and the equivalent length of fittings is 146 lineal ft (step 11), for a total equivalent length of 291 lineal ft.

   SI In this design, the longest circuit is 44.2 m (step 6), and the equivalent length of fittings is 48.6 m (step 11), for a total equivalent length of 92.8 m.

13. Determine the pressure head, using the friction head selected, step 8, and the total equivalent length of pipe, step 12, and Fig. 9-35.

   In this design, the friction head is about 1.4 ft per 100 ft (step 8) and the total equivalent length is 291 lineal ft (step 12). The pressure head required is determined by multiplying the friction loss per 100 ft times the equivalent length of the pipe. Note that the loss is *per 100 ft,* and the equivalent length must be adjusted to the number of 100 ft required (291 lf = 2.91).

$$1.4 \text{ ft per 100 ft} \times 2.91 = 4.1 \text{ ft pressure head}$$

In this design the friction head is about 195 Pa/m (step 8) and the total equivalent length is 92.8 m (step 12). The pressure head required is determined by multiplying the friction loss per meter times the equivalent length of the pipe.

$$185 \text{ Pa/m} \times 92.8 = 17{,}186 \text{ Pa} = 17.2 \text{ kPa}$$

$$17.2 \text{ kPa} \div 9.8 \text{ kPa/m} = 1.76 \text{ m pressure head}$$

14. Select the rest of the pipe sizes required for the system, using the friction head selected and the heat it must supply, from Fig. 9-33.

Using 1.4 ft per 100 ft:
Zone 2: 11,876 Btuh (1.1 gpm) = ¾-in. pipe required
Main: 32,480 Btuh (2.9 gpm) = 1-in. main (Fig. 9-33)

Return: After the two zones join, same as main = 1-in. pipe

Using 185 Pa/m

Zone 2: 3466 W (0.09 l/s) = 20 mm pipe required

Main: 9,546 W (0.25 L/s) = 25 mm main

return: after the two zones join, same as main = 25 mm

15. Select the pump size required, using the gpm necessary with the required pressure head, from Fig. 9-36.

In this design, the gpm is noted in step 8 as 2.1 gpm and the pressure head in step 13 as 4.1 ft. The pump chart is shown in Fig. 9-36. Find the gpm along the bottom and move vertically; now find the pressure head along the left and move horizontally until they meet. On this chart, an A pump would provide a pressure head of about 6.0 ft. (Do not use this chart for actual design; secure the manufacturer's performance ratings.)

In this design the L/s is noted in step 8 as 0.16 L/s, and the pressure head in step 13 as 1.72 m. The pump chart is shown in Fig. 9-36. Find the L/s along the top and move vertically; now find the pressure head along the right and move horizontally until they meet. On this chart an A pump would provide a pressure head of about 1.8 m. (Do not use this chart for actual design; secure the manufacturer's performance ratings.)

16. Find the temperature drop for the total circuit based on the total friction head in millinches (friction times total equivalent length) times the design temperature drop for the system divided by the pressure head (pressure head in millinches equals pressure head in feet times 12,000).

$$\frac{\text{Friction head (millinches)} \times \text{Total equivalent length (ft)} \times \text{Design temperature}}{\text{Pressure head (ft)} \times 12{,}000}$$

In this design, the friction head is 168 millinches, the total equivalent length is 291 ft, the design temperature drop is 20°F, and the pressure head of the pump is 6.0 ft.

	BTUH Req.	L.F. of Ext. Wall		W Req.	L.M. of Ext. Wall
			SI UNITS		
Entry	4061	—	Entry	1287	—
Dining	3056	12	Dining	898	3.6
Kitchen	4160	—	Kitchen	1205	—
Living Room	9327	10	Living Room	2690	3.0

FIGURE 9-24   Exterior wall available

$$\frac{168 \times 291 \text{ ft} \times 20°F}{6.0 \text{ ft} \times 12{,}000} = 13.58 \text{ (use 14.0)}°F \text{ temperature drop in circuit}$$

SI

$$\frac{\text{Friction head (Pa/m)} \times \text{Total equivalent length (m)} \times \text{Design temperature}}{\text{Pump pressure head (m)} \times \text{Pa/m}}$$

In this design the friction head is 195 Pa/m, the total equivalent length is 48.6 m, the design temperature drop is 11°C, and the pressure head is 1.8 m.

$$\frac{185 \times 92.8 \times 11°C}{1.8 \times 9800} = 10.7 \text{ (use 11°C)}$$

17. The next phase of the design is to determine how many lineal feet (meters) of exterior wall are available for radiation units of some type (Fig. 9-24). In a room with little or no exterior wall available, it may be necessary to use interior wall space.

18. Calculate the temperature drop for each radiation unit served through the circuit, and record it in the tabulated form (Fig. 9-25) under "System Temperature Drop." The formula used to calculate the temperature drop for each radiation unit(s) is:

$$\frac{\text{Radiation unit(s) heating capacity}}{\text{Circuit (zone) capacity}} \times \text{Temperature drop in total system}$$

19. Calculate the temperature in the circuit main, in a series loop, or after the water returns from the radiation unit(s) served. Tabulate this temperature in the main in the form under "Enter." *This is the entering temperature for each of the radiation units* (Fig. 9-26).

   Note: If designing a series loop system, go directly to step 21a.

SI UNITS

ENTRY	$\frac{4061}{20{,}604} \times 14 = 2.8°F$		ENTRY	$\frac{1287}{6080} \times 11 = 2.3°C$	
DINING	$\frac{3056}{20{,}604} \times 14 = 2.1°F$		DINING	$\frac{898}{6080} \times 11 = 1.6°C$	
KITCHEN	$\frac{4160}{20{,}604} \times 14 = 2.8°F$		KITCHEN	$\frac{1205}{6080} \times 11 = 2.2°C$	
LIVING ROOM	$\frac{9327}{20{,}604} \times 14 = 6.3°F$		LIVING ROOM	$\frac{2690}{6080} \times 11 = 4.9°C$	

FIGURE 9-25   Calculate temperature drop

	ENTER TEMP °F	TEMP LOSS IN CONVECTOR °F
ENTRY	200.0	2.8
DINING	197.2	2.1
KITCHEN	195.1	2.8
LIVING ROOM	192.3	6.3

SI UNITS

	ENTER TEMP °C	TEMP LOSS IN CONVECTOR °C
ENTRY	93°C	2.3
DINING	90	1.6
KITCHEN	89.1	2.2
LIVING ROOM	86.9	4.9

FIGURE 9-26   Calculate temperature in main

20. *Note: This step is used only in the one-pipe system. For all other systems go to step 21.*

Calculate the temperature drop of the water which is diverted through the radiation unit(s) and through the fitting used. The amount of water diverted depends on the size of the fitting used and location of the main above or below the radiation unit; Fig. 9-31 is used for this percentage of diversion.

$$\text{Radiation unit(s) temp. drop} = \frac{\text{Temp. drop in circuit (zone)}}{\text{Percent diversion}} \times 100$$

Note: The 100 is a constant and changes the percentage to a decimal equivalent.

Tabulate the temperature drop for the radiation unit(s) in the form (Fig. 9-27).

21. Calculate the average temperature.

a. *Series loop only.* Calculate the average temperature in the radiation unit(s) by subtracting one-half of the temperature drop from the entering tempera-

	ENTER	LEAVE	AVE.
ENTRY	200.0	197.2	198.6
DINING	197.2	195.1	196.2
KITCHEN	195.1	192.5	193.8
LIVING ROOM (3.2°F loss)	192.5	189.3	190.9
LIVING ROOM (3.1°F loss)	189.3	186.2	187.7

SI UNITS

	ENTER	LEAVE	AVE.
ENTRY	93.0	90.7	91.8
DINING	90.7	89.1	89.9
KITCHEN	89.1	86.9	88.0
LIVING ROOM (2.5°C)	86.9	84.4	85.6
LIVING ROOM (2.4°C)	84.4	82	83.2

FIGURE 9-27   Calculate average temperature drop

ture of the circuit main. Tabulate the average temperature in the form under "Average Radiation Temperature."

b. *All except series loop.* Calculate the average temperature in the radiation unit(s) by subtracting one-half of the temperature drop which occurs in the radiation units (from step 16) from the entering temperature of the circuit main. Tabulate the average temperature in the form.

22. Next, the heat emission rate of radiation required is determined by dividing the heat loss for each room by the available length of exterior wall.

23. Select the type of radiation unit(s) to be used. 2¾″ × 3¾″ × 0.11 × 50/Ft. 1″ tube (Fig. 9-28)

24. Determine the length or amount of radiation unit(s) required to evenly heat each room by dividing the heat loss of the room by the heat emission rate of the radiation unit(s) selected. Tabulate the lengths on the form (Fig. 9-29).

25. Size the compression tank by determining the entire volume of water in the system and then finding the amount of expansion based on a temperature rise of 150°F (from 50°F entering to 200°F heated) times (0.004lt − 0.0466) the net coefficient of ex-

	Hot Water Output* 1 gal., (3.8 L) flow rate (for 5 gal., (19 L) flow rate use factor 1.067)									
Copper-Aluminum Elements	220°F	105°C	210°F	99°C	200°F	93°C	190°F	88°C	180°F	82°C
2¾″ × 3¾″ × .011 × 50/ft. 1″ tube	1240	1190	1120	1076	1020	980	930	890	840	800
2¾″ × 5″ × .020 × 40/ft. 1¼″ tube	1250	1200	1120	1076	1030	990	940	900	850	815
2¾″ × 5″ × .020 × 50/ft. 1¼″ tube	1340	1285	1200	1150	1100	1050	1000	960	910	870

FIGURE 9-28   Select radiation unit(s)

	BTUH REQ.	L.F. OF EXT.WALL	AVE,. TEMP	AVE.HEAT EMISSION	LENGTH REQ.
ENTRY	4061	—	198.6	1007	4.03′
DINING	3056	12	196.2	986	3.1′
KITCHEN	4160	—	193.8	964	4.32′
LIVING ROOM	4664	10	190.9	938	4.97′
LIVING ROOM	4663	—	187.7	909	5.13′

SI UNITS

	WATTS REQ.	L.F. OF EXT.WALL	AVE,. TEMP	AVE.HEAT EMISSION	LENGTH REQ.
ENTRY	1287	—	91.8	958	1.4 m
DINING	898	12	89.9	924	1.0 m
KITCHEN	1205	—	88.0	890	1.4 m
LIVING ROOM	1345	10	84.4	836	1.6 m
LIVING ROOM	1345	—	83.2	813	1.7 m

Note:The lengths vary slightly due to rounding off.

FIGURE 9-29   Determine lengths of radiation unit(s)

Volume of Water
in Standard Pipe and Tube

Nominal Pipe Size In.	Standard Steel Pipe			Type L Copper Tube	
	Schedule No.	Inside Dia In.	Gallons per Lin Ft	Inside Dia In.	Gallons per Lin Ft
3/8	—	—	—	0.430	0.0075
1/2	40	0.622	0.0157	0.545	0.0121
5/8	—	—	—	0.666	0.0181
3/4	40	0.824	0.0277	0.785	0.0251
1	40	1.049	0.0449	1.025	0.0429
1 1/4	40	1.380	0.0779	1.265	0.0653
1 1/2	40	1.610	0.106	1.505	0.0924
2	40	2.067	0.174	1.985	0.161
2 1/2	40	2.469	0.249	2.465	0.248
3	40	3.068	0.384	2.945	0.354
3 1/2	40	3.548	0.514	3.425	0.479
4	40	4.026	0.661	3.905	0.622
5	40	5.047	1.04	4.875	0.970
6	40	6.065	1.50	5.845	1.39
8	30	8.071	2.66	7.725	2.43
10	30	10.136	4.19	9.625	3.78
12	30	12.090	5.96	11.565	5.46

Reprinted with permission from ASHRAE, *Applications Handbook,* 1991.

FIGURE 9-30   Volume of water in standard pipe or tube

pansion of the water. The compression tank must be sized to accommodate this water expansion. The volume of water in the system is determined from the table in Fig. 9-30.

9-7   BOILERS

In residences, the boiler (furnace) which provides the heat (either hot water or forced air) is commonly located in the basement or crawl space or in a first-floor utility room, but it may also be in the attic. Boilers are most commonly fired by natural gas or oil, and occasionally by electricity, coal, or bottled gas.

The boiler selected must be large enough to supply all of the heat loss in the spaces as calculated. The boiler must also have enough extra capacity to compensate for heat loss which will occur in the pipes which circulate the hot water through the system (called pipe loss or pipe tax). In addition, an extra reserve capacity is required so that the boiler can provide heat quickly to warm up a "cooled off" space. This "pickup" allowance also provides some extra heating capacity when it is necessary to increase the temperature of a space several degrees, not merely to maintain a temperature. For example, a store which closes at 5 P.M. may turn its thermostat back to 60°F before closing. Then in the morning as the thermostat is raised to 68° or 70°F, it is necessary to have some extra capacity in the boiler to provide this extra heat as quickly as possible. Many buildings have automatic, timed thermostats which are used to regulate the temperatures throughout the day.

Typically, the designer allows about 33⅓ % for the heat loss in the pipes and an additional 15% to 20% as a pick-up allowance, in addition to the calculated heat loss of the building. For example, in the residence being designed:

Calculated heat loss	32,480 Btuh
Piping loss (33⅓ %)	+10,825
	43,305
Pick-up allowance (20%)	+8,660
Total	51,965 Btuh

SI

Calculated heat loss	9,546 w
Piping loss (33⅓ %)	+3,182
	12,728
Pick-up allowance (20%)	2,545
Total	15,273 w

Spaces which are heated intermittently, such as a church or an auditorium, will have to have a much larger pick-up allowance, perhaps as much as 50%. This is because the temperature will often be allowed to drop quite low (perhaps 50°F) when the space is not in use, and then when the heat is raised to 70°F, it must heat up quickly.

While most designers simply allow a certain percentage for piping loss, it can be calculated—but only once the piping layout is finalized and the installation decided upon and specified. From the piping layout, it would be necessary to determine the length of pipe and the various temperatures of the air through which the pipe travels. Where the pipe travels through an unheated basement or attic, the heat loss would be much greater than the heat loss of the same pipe traveling through a heated space.

Once the piping loss and the pick-up allowance have been taken into account, the boiler size may be selected. The boiler selected should be as close to the total Btuh as possible to provide the most efficient operation. Keep in mind that the outside design temperature used will not actually occur very often, that the boiler will be operating at less than full capacity during most of the time, and that the pickup allowance also gives a little extra capacity. Also, oversizing the boiler will probably be of little value during any unusually long cold spells with temperatures well below the outside design temperature unless the radiation or convector units selected are able to put out additional heat.

In large projects (institutional and commercial), quite often two boilers are used. Since boilers are more efficient the longer they operate, the first boiler would be designed to satisfy about 60% of the heat required, and the second would supply the balance. In this way, the first unit would come on and supply the heat required for about 75% of the days that heat is needed. Only during colder weather would the second unit turn on. The more efficient operation of the smaller first unit (when compared with the operation of a single boiler) pays for the added cost of a second unit. Also, this provides a backup unit in case one of the boilers requires repairs.

9-8   INSTALLED AND EXISTING SYSTEMS

No matter how carefully a system is designed and laid out, there is the possibility that at least a portion of the building is not being heated satisfactorily. This, of course, is also true of many existing systems. Quite often the problems in these systems can be corrected easily and relatively inexpensively after a careful analysis of the system.

First, carefully take notes as to exactly what the problem is, what time of day it is most noticeable, what the outside temperature is during the time that sufficient heat is not being properly distributed, and whether the boiler is running continuously or intermittently.

Sample complaint: Temperature in house 60° to 62°F during night, it raised to desired temperature during the day.

Heating equipment: hot water, working satisfactorily

Outside temperature—10°F

Wind: light

Boiler water temperature: 180°F

For example, in this complaint, note first that the general complaint is that the temperature of the entire house dropped to about 60° to 62°F during the night. During the day, it warmed up to the 70°F reading desired. The outside temperature that night reached a low of about −10°F, well below the design temperature of 0°F used for Albany, N.Y. (the location of the residence).

The first observation is that the outside temperature during the night was about 10°F below the outside design temperature and the inside temperature registers about 8°F below the thermostat setting. So, the problem is how to get more heat out of the baseboard convector radiators installed in the house. The first possibility is to check the temperature to which the boiler is heating the hot water. Each boiler has a thermostat which sets the highest temperature to which it can heat the water, and then the boiler will shut off while the circulating pumps continue to push hot water through the system.

It is most important to know whether or not the boiler was operating continuously or intermittently. Quite often the occupant confuses the fact that the pipes were hot and the hot water was being circulated with the actual boiler operation. It may be necessary to obtain more information about the boiler operation before the system is corrected. If the boiler is operating continuously, it indicates that no matter how hard the boiler works, it cannot heat the water to the thermostat setting. When the boiler operates intermittently, it means that the boiler has heated the water to the thermostat setting and is waiting for cooler water to circulate to it before starting again. Intermittent operation is most commonly found and will be discussed first.

Typically, the boiler thermostat is set at 170° or 180°F during installation. Keeping in mind that the amount of heat delivered by the convectors will increase with hotter water (Sec. 9-5, Fig. 9-16), the solution may be to increase the thermostat setting of the boiler. The hot water, in a closed heating system such as this, can be heated as high as 220°F. Generally, for every increase of 10°F in water temperature, the heat output of the radiation or convector units is increased about 10%. Intermittent operation is the most common problem and, fortunately, this is the most common solution.

If the temperature *is* set high and the boiler is operating intermittently, the next solution is to provide additional heat in each room by some means. In most rooms, the length of fin-tubed radiators may be increased, thus increasing the amount of heat which is put into the room. Many rooms have convector covers all along a wall, but check inside the cover; a great deal of that length may be plain tube and not finned tube. In such a case, a plumber can add some fin-tubed element. It may be necessary to add extra lengths of finned tube on an interior wall or to put a convector or unit heater up in a wall where baseboard space is not available. Any revision or addition of this type must be carefully planned and may be costly. Unfortunately, I don't known of anyone who has marketed a snap-on finned tube which the occupants or owners could install themselves in such situations.

9-9   SYSTEM INSTALLATION

The piping required for the hot water heating system will run in the walls, floors, and ceiling spaces. The heating devices may require recessed wall spaces or extra structural support if hung from the ceiling. The heating unit is often located in a basement or mechanical room.

Radiant heating panels in the floor slab require that the pipes be installed before the slab is placed. When located in the ceiling, they are attached to the framing. The pipe needs to be checked for watertightness before it is covered.

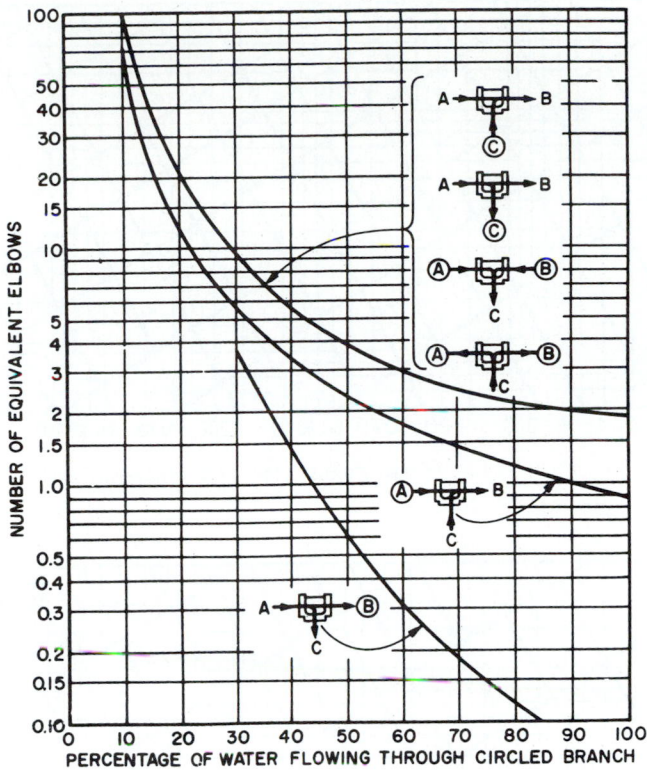

Notes: 1. The chart is based on straight tees, that is, branches A, B, and C are the same size.

2. Head loss in desired circuit is obtained by selecting proper curve according to illustrations, determining the flow at the circled branch, and multiplying the head loss for the same size elbow at the flow rate in the circled branch by the equivalent elbows indicated.

3. When the size of an outlet is reduced the equivalent elbows shown in the chart do not apply. The maximum loss for any circuit for any flow will not exceed 2 elbow equivalents at the maximum flow (gpm) occurring in any branch of the tee.

4. The top curve of the chart is the average of 4 curves, one for each of the tee circuits illustrated.

Elbow Equivalents of Tees at
Various Flow Conditions[1,4]

Reprinted with permission from ASHRAE, *Fundamentals Handbook,* 1993.

FIGURE 9-31   Elbow equivalents of tees

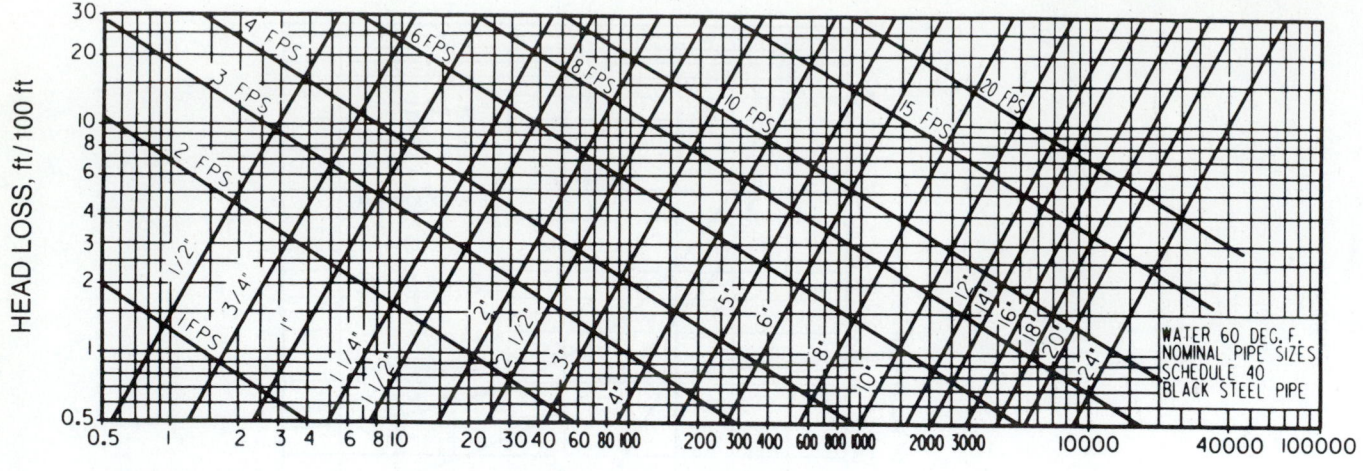

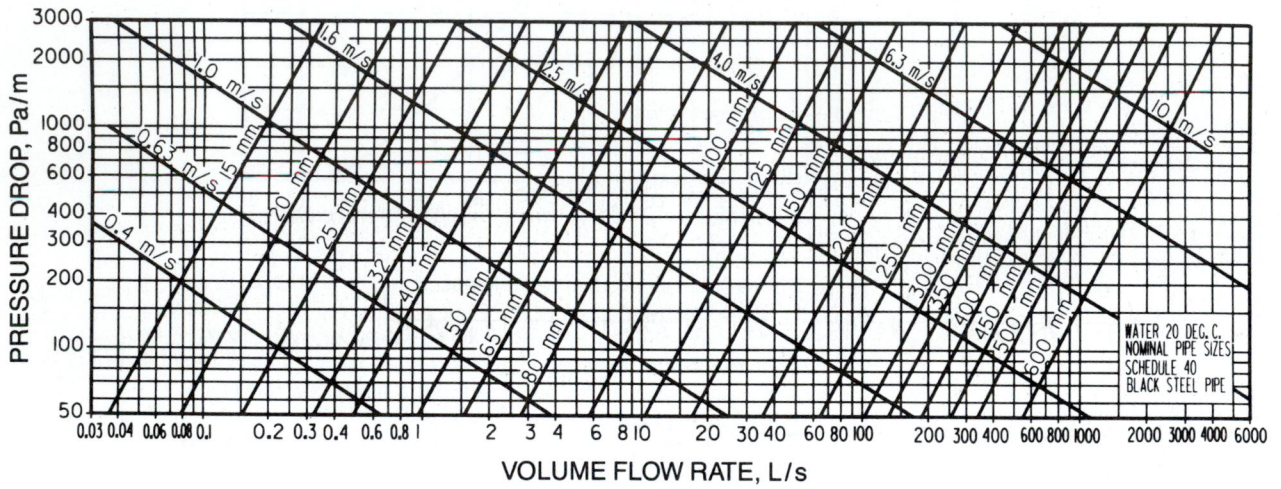

Reprinted with permission form ASHRAE, *Fundamentals Handbook,* 1993.

FIGURE 9-32   Friction loss for water in commercial steel pipe

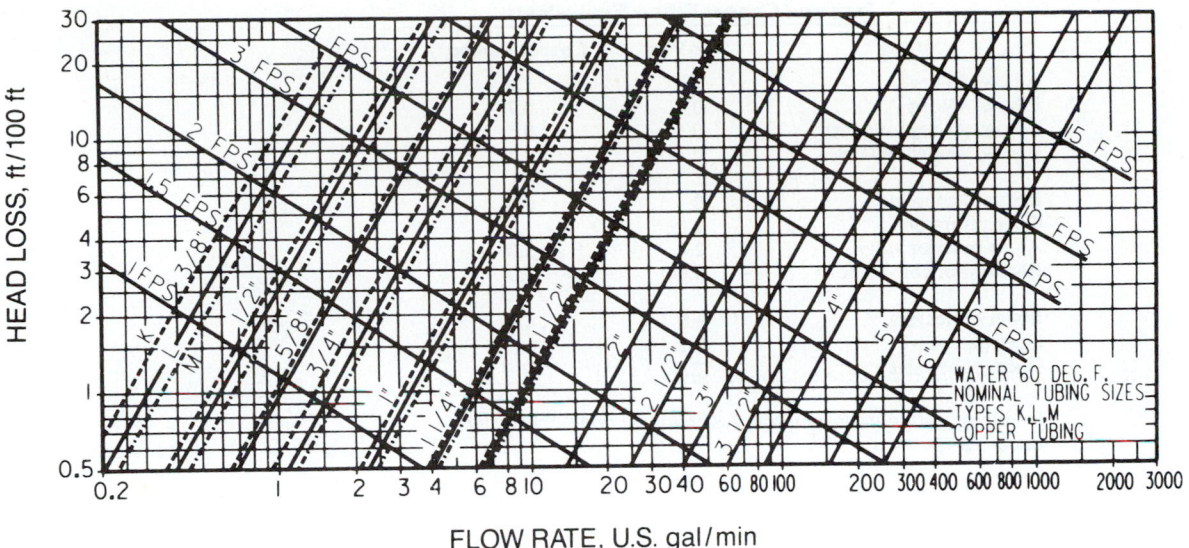

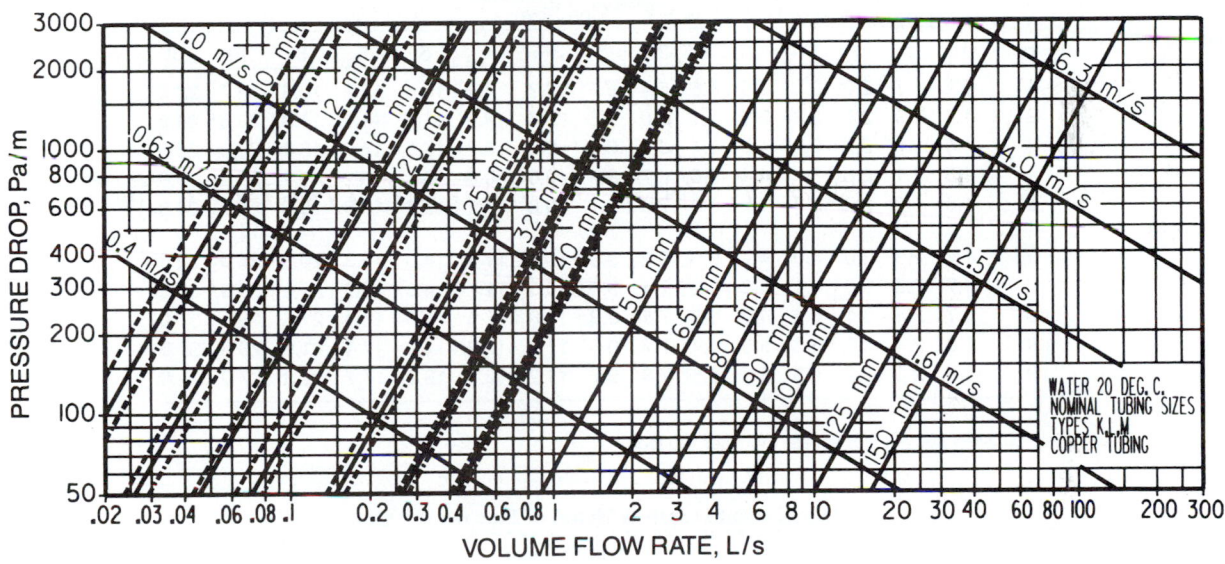

FIGURE 9-33   Friction loss for water in copper tubing

Iron and Copper Elbow Equivalents

Fitting	Iron Pipe	Copper Tubing
Elbow, 90-deg	1.0	1.0
Elbow, 45-deg	0.7	0.7
Elbow, 90-deg long turn. . .	0.5	0.5
Elbow, welded, 90-deg	0.5	0.5
Reduced coupling	0.4	0.4
Open return bend.	1.0	1.0
Angle radiator valve.	2.0	3.0
Radiator or convector	3.0	4.0
Boiler or heater	3.0	4.0
Open gate valve	0.5	0.7
Open globe valve	12.0	17.0

Reprinted with permission from ASHRAE, *Fundamentals Handbook,* 1993.

FIGURE 9-34   Elbow equivalents

Reprinted with permission from ASHRAE, *Fundamentals Handbook,* 1993.

Equivalent Length of Pipe for 90-Deg Elbows

Vel. Fps.	½	¾	1	1¼	1½	2	2½	3	3½	4	5	6	8	10	12
						Pipe Size									
1	1.2	1.7	2.2	3.0	3.5	4.5	5.4	6.7	7.7	8.6	10.5	12.2	15.4	18.7	22.2
2	1.4	1.9	2.5	3.3	3.9	5.1	6.0	7.5	8.6	9.5	11.7	13.7	17.3	20.8	24.8
3	1.5	2.0	2.7	3.6	4.2	5.4	6.4	8.0	9.2	10.2	12.5	14.6	18.4	22.3	26.5
4	1.5	2.1	2.8	3.7	4.4	5.6	6.7	8.3	9.6	10.6	13.1	15.2	19.2	23.2	27.6
5	1.6	2.2	2.9	3.9	4.5	5.9	7.0	8.7	10.0	11.1	13.6	15.8	19.8	24.2	28.8
6	1.7	2.3	3.0	4.0	4.7	6.0	7.2	8.9	10.3	11.4	14.0	16.3	20.5	24.9	29.6
7	1.7	2.3	3.0	4.1	4.8	6.2	7.4	9.1	10.5	11.7	14.3	16.7	21.0	25.5	30.3
8	1.7	2.4	3.1	4.2	4.9	6.3	7.5	9.3	10.8	11.9	14.6	17.1	21.5	26.1	31.0
9	1.8	2.4	3.2	4.3	5.0	6.4	7.7	9.5	11.0	12.2	14.9	17.4	21.9	26.6	31.6
10	1.8	2.5	3.2	4.3	5.1	6.5	7.8	9.7	11.2	12.4	15.2	17.7	22.2	27.0	32.0

Note: 120 millinches per foot equals 1 foot per 100 feet.

Equivalent Length in Metres of Pipe for 90° Elbows

Velocity, m/s	15	20	25	32	40	50	65	90	100	125	150	200	250	300
							Pipe Size, mm							
0.33	0.4	0.5	0.7	0.9	1.1	1.4	1.6	2.0	2.6	3.2	3.7	4.7	5.7	6.8
0.67	0.4	0.6	0.8	1.0	1.2	1.5	1.8	2.3	2.9	3.6	4.2	5.3	6.3	7.6
1.00	0.5	0.6	0.8	1.1	1.3	1.6	1.9	2.5	3.1	3.8	4.5	5.6	6.8	8.0
1.33	0.5	0.6	0.8	1.1	1.3	1.7	2.0	2.5	3.2	4.0	4.6	5.8	7.1	8.4
1.67	0.5	0.7	0.9	1.2	1.4	1.8	2.1	2.6	3.4	4.1	4.8	6.0	7.4	8.8
2.00	0.5	0.7	0.9	1.2	1.4	1.8	2.2	2.7	3.5	4.3	5.0	6.2	7.6	9.0
2.35	0.5	0.7	0.9	1.2	1.5	1.9	2.2	2.8	3.6	4.4	5.1	6.4	7.8	9.2
2.67	0.5	0.7	0.9	1.3	1.5	1.9	2.3	2.8	3.6	4.5	5.2	6.5	8.0	9.4
3.00	0.5	0.7	0.9	1.3	1.5	1.9	2.3	2.9	3.7	4.5	5.3	6.7	8.1	9.6
3.33	0.5	0.8	0.9	1.3	1.5	1.9	2.4	3.0	3.8	4.6	5.4	6.8	8.2	9.8

FIGURE 9-35   Equivalent length of pipe

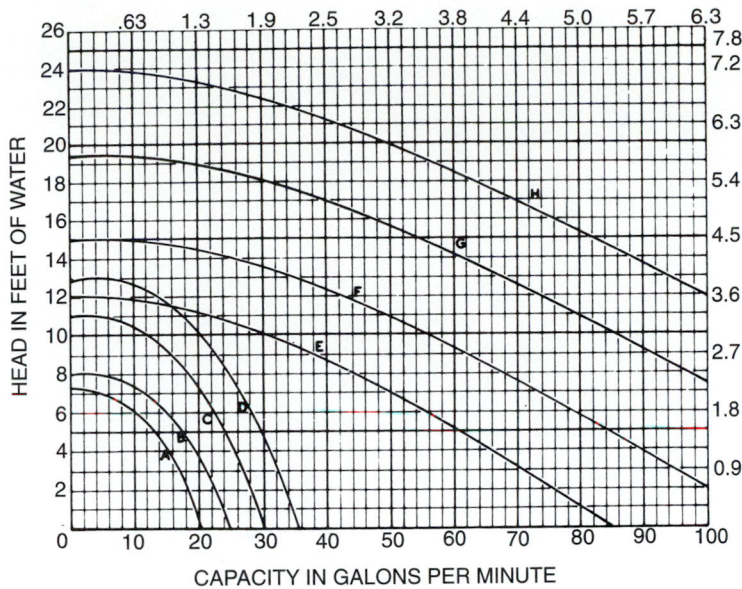

HEAD IN FEET OF WATER

CAPACITY IN GALONS PER MINUTE

Note: Metrics added by author.

FIGURE 9-36   Pump performance

QUESTIONS

9-1. What are the differences between series loop and one-pipe hot water systems, and what are the advantages of each?

9-2. Discuss the two-pipe system, how it works and its advantages and disadvantages.

9-3. Why are multiple heating circuits (zones) often used?

9-4. How does a radiant hot water heating panel work?

9-5. What types of heating devices are available to distribute heat to a space?

9-6. What is the one variable which affects the amount of heat given out by a finned-tube convector?

9-7. What type of system is most commonly used in a residence?

Design Exercises

9-8. Design a series loop system, two zones, for the residence in Appendix D, based on the heat loss calculations from Exercise 8-15.

9-9. Design a series loop system, one zone, for the apartments in Appendix A, based on the heat loss calculations from Exercise 8-17.

CHAPTER 10

Forced Air System Design (Heat)

10-1 FORCED AIR SYSTEMS (HEATING)

A forced air heating system has a motor-driven fan that is used to circulate filtered, heated air from a central heating unit through supply ducts to each of the rooms. As the hot air is delivered through the ducts and into the room through the supply outlet, cooler air from the space (room) is being returned through return grilles, into ducts, and back through the central heating unit to be heated and sent back to the space. The ducts may be circular or rectangular, and their size depends on the amount of heated air which must flow through them to maintain the desired temperature of the room. A variety of duct systems, or basic designs, may be used; several of the most common are shown in Fig. 10-1.

Forced air heating systems are economical and generally easy to install. Filters are put in the system to reduce the amount of dust in the air. The furnace (heating unit) may be located wherever convenient in the building, including the basement, crawl space, attic, utility room, or garage, and in larger buildings it is sometimes located on the roof or in the structural systems.

Humidifiers (discussed in Sec. 6-5) are recommended for use in most heating systems. They provide extra comfort at a minimal cost, making them a good investment. The humidifier will fit right into the duct system as it is being installed.

10-2 DUCTS AND FITTINGS

The ductwork is used to take the forced warm air from the furnace to the supply outlet and the cooler air from the return register back to the furnace.

The most commonly used materials for ducts are galvanized steel and aluminum. Both materials are relatively lightweight and easily shaped to whatever size duct is required, either round or rectangular. Minimum metal thicknesses required of these materials vary according to the size of the duct required and are shown in Fig. 10-2. These ducts may have to be insulated as discussed in Sec. 10-3. Flexible connections (Fig. 10-3) allow ducts to move and expand and contract without transmitting noises to the space.

Another very popular ductwork material is glass fiber molded duct board. These ducts are available in a large variety of sizes, both round and rectangular. Its principal advantage is that for installations which need insulating (Sec. 10-3), it is less expensive and installed in one operation, as opposed to metal ducts which are installed in one operation and insulated in a second operation. The round fiber ducts are compatible with standard round metal ducts and may be used as part of a system which also uses metal ducts. The round fiber duct makes use of metal fittings to connect, reduce, and make elbows as shown in Fig. 10-4. Another advantage of this type of duct is its excellent acoustic properties to ensure a quiet sys-

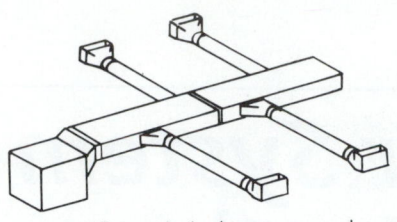
Extended plenum supply

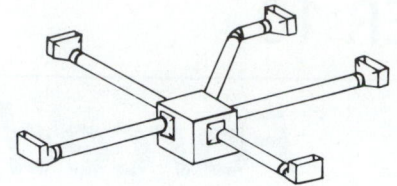
Individual supply system

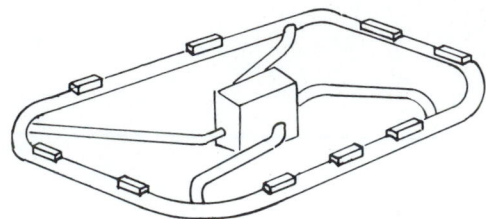
Perimeter-loop system

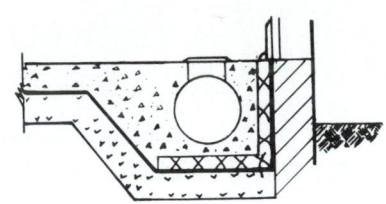

FIGURE 10-1   Typical heating duct systems

Round Ducts Diameter, In.	Minimum Thickness		Minimum Weight of Tinplate
	Galv. Iron, U.S. Gage	Aluminum, B&S Gage	
Less than 14	30	26	
14 or more	28	24	IX (135 lb)
Rectangular Ducts Width, In.	Minimum Thickness		Minimum Weight of Tinplate
	Galv. Iron, U.S. Gage	Aluminum, B&S Gage	
Ducts Enclosed in Partitions			
14 or less	30	26	
Over 14	28	24	IX (135 lb)
Ducts Not Enclosed in Partitions			
Less than 14	28	24	—
14 or more	26	23	—

Note: The table is in accordance with Standard 90B of the National Board of Fire Underwriters.® Industry practice is to use heavier gage metals where maximum duct widths exceed 24 in (see also NBFU No 90A). Reprinted with permission from ASHRAE, Systems Handbook, 1976

FIGURE 10-2   Metal duct thicknesses

FIGURE 10-3   Fiber duct connections

FIGURE 10-4   Fiber duct connection

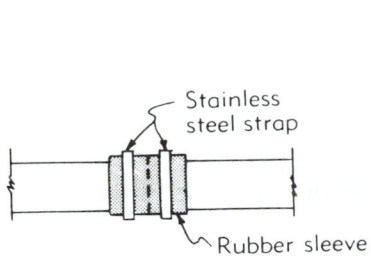

Stainless
steel strap

Rubber sleeve

FIGURE 10-5   Duct connection

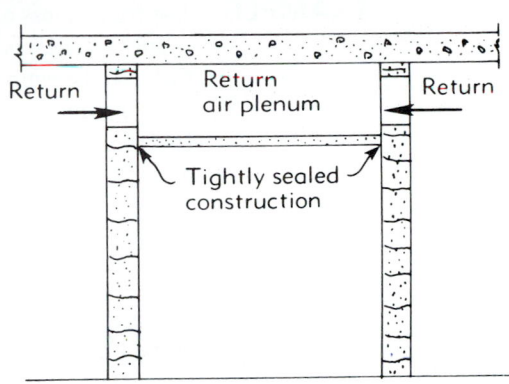

Return — Return air plenum — Return

Tightly sealed construction

FIGURE 10-6   Return air plenum

tem. Fiber duct is also sometimes used for the section of ductwork connecting the main trunk supply and the return lines to the furnace.

Where the ducts will be in and under a concrete slab, cement, plastic, tile, reinforced fiberglass, and coated steel round ducts are most commonly used. Ducts placed below a slab must be made from a material which is not subject to moisture transmission or corrosion by concrete, will not float as the concrete is poured, and is noncombustible. The duct is available with inside diameters from 4 to 36 in. (100 to 900 mm). The ducts are joined with an impermeable rubber sleeve and two stainless steel straps (Fig. 10-5). Tees, wyes, elbows, reducers, and end caps are also available.

Oftentimes the ducts are made an integral part of the construction of the building, especially the returns which handle cooler air. In large buildings (hospitals, nursing homes, office buildings) the ceiling area over a corridor is sometimes used as an air return (also called an *air plenum*). A typical section through a corridor with an air plenum above it is shown in Fig. 10-6. The ceiling used should be tight-fitting so that the air will not "leak" out of the space. Quite often gypsum board is used. The spaces between joists may also be used for air returns, with the bottom usually formed from sheet metal nailed to the bottom of the joists (Fig. 10-7). If the joist runs through cold spaces, it may be desirable to insulate

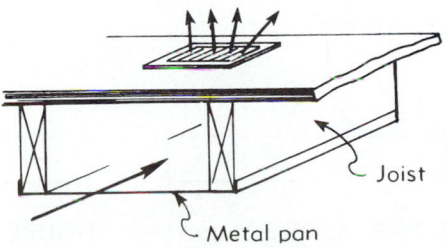

Metal pan   Joist

FIGURE 10-7   Joists as ducts

the underside of the return. Also, the designer must check to be certain that the joists run in the direction which the return air must run.

A wide variety of fittings may be used to make all of the reductions, branch take-offs, turns, and bends required in many duct systems. Ideally, the best system layout has the fewest and simplest duct fittings since fittings restrict the flow of the forced air and increase the friction in the system. The friction of the various fittings used must be included in the duct size calculations. In order to calculate the amount of pressure lost in a fitting, it is estimated as the number of feet of straight run that would equal the friction loss in that fitting. It is called *equivalent length.*

EXAMPLE   The friction loss of boot fitting A (which is the piece used to connect the branch duct to the supply register) is listed in Fig. 10-34 as A-30. This means that the friction loss in the best fitting is the equivalent length of 30 ft (9.1 m) of straight duct.

During the duct design, it will be necessary to make preliminary selections of fittings so that the duct design will be as accurate as possible. Fittings and their equivalent lengths are given in Fig. 10-34.

Supply ducts should be equipped with an adjustable locking-type damper (Fig. 10-8) so that the air volume can be controlled. The damper should be located in an accessible spot in the branch duct as far from the supply outlet as possible. This allows a measure of control over the flow of the air and also allows a branch to be shut off when desired.

Splitter dampers (Fig. 10-9) are used to direct part of the air into the branch where it is taken off the trunk. They do not give precise volume control.

Squeeze dampers (Fig. 10-10) are placed in a duct to provide a means to control the air volume in the duct.

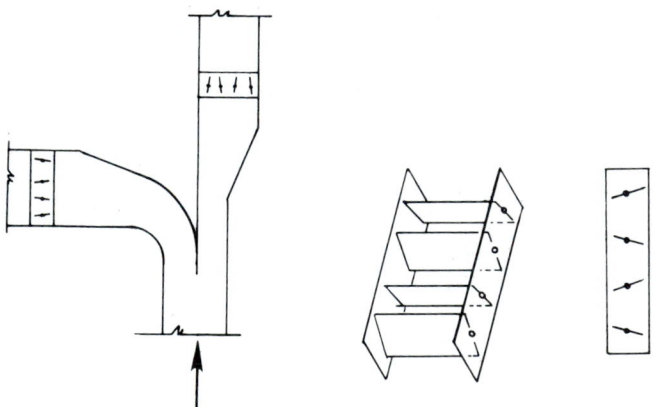

FIGURE 10-8   Adjustable dampers

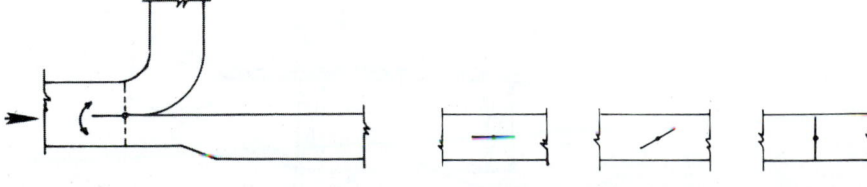

FIGURE 10-9   Splitter damper   FIGURE 10-10   Squeeze dampers

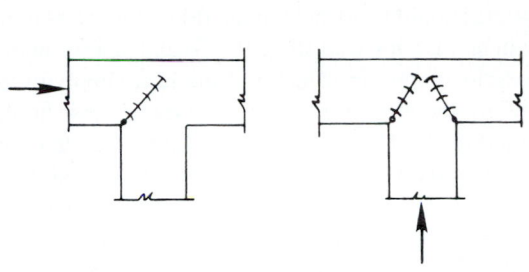

FIGURE 10-11   Turning vanes

FIGURE 10-12   Boiler-duct connection

Turning vanes may be used to direct the flow of air smoothly around a corner or a bend, as shown in Fig. 10-11.

The entire system should be checked to be certain that proper attention has been given to the elimination of as much noise as possible from the system. The following suggestions will help keep noise to a minimum (see also Chapter 19).

1. The furnace and metal ducts should be connected with a flexible fire-resistant fabric. In this manner, any noises or vibrations will not be transmitted directly through the system (Fig. 10-12).

2. All electrical conduits and pipes should have flexible connections to the furnace.

3. Do not locate the return air immediately adjacent to the furnace.

4. Do not install a fan directly below the return air grille.

10-3   DUCT INSULATION

Ducts which are located in heated spaces do not need to be insulated. But ducts which run through enclosed, unheated spaces or in spaces which are exposed to outdoor temperatures should be insulated. While glass fiber ducts are made of an insulating material (the glass fiber), sheet metal ducts must be wrapped in an insulation. Thicknesses of insulation recommended are:

Supply Ducts

1. When located in enclosed, unheated spaces—1 in. (25 mm) of insulation.

2. When located in a space in which it is exposed to outdoor temperatures—2 in. (50 mm) of insulation.

Return Ducts

When not located in a heated space, use 1 in. (25 mm) of insulation.

The most commonly used insulation on residential and small commercial buildings is fiberglass with a facing of reinforced aluminum foil vapor-barrier which goes to the outside. In large buildings, the ducts may be sprayed with an insulating/fire-resistant coating.

10-4   SUPPLY AND RETURN LOCATIONS

This information given on the location of supply and return ducts is for installations where only heating will be supplied or in areas where heat is required much of the time, while any cooling requirements are small. (For example, in an upstate New York residence, heat will probably be required regularly for about six to seven months, October 15-May 15, and air conditioning intermittently for two months.)

The supply registers should be located in the floor, 4 in. (100 mm) out from the baseboard, or be very low in an exterior wall (Fig. 10-13) and near or under windows, in exterior walls and near exterior doors. In effect, put the heat-supplying registers as close as possible to the spots where the most heat is lost. These registers should have both vertical and horizontal dampers (Fig. 10-14) so that the air will be directed downward at the floor by the horizontal dampers and diffused to the sides by the vertical dampers.

Returns are often located on interior walls, in hallways, and in exposed corners. A low baseboard location is required for a return on the floor, and a centrally located return, perhaps one for a small residence and two for a large one, provides satisfactory results. If there is only one central return, it is important to put it in a location where it will be able to draw return air from as much of the building as possible. This is why they are frequently located in hallways. More expensive, individual room exhausts may be used if the designer feels that the layout of rooms may cause an uneven return of air or that the air flow from the rooms to the return (under doorways or through adjoining rooms) may cause a problem. For example, if there is a central return in the hallway and the door to a bedroom is closed all night, it will be very difficult for air to flow from the room to the return, unless the bottom of the door is trimmed (undercut) up about an inch or so. Unless air is drawn from the room as the supply keeps bringing air into the room, a slight pressure is built up in the room, reducing the amount of warm air coming through the supply outlets, and the room will tend to be cool. The same situation will occur in any other room closed off from the return. If the designer is aware that such a problem may exist, the solution would be to put a separate return in the room.

The returns are covered with grilles which are put over the opening primarily to "cover it up" so there won't be a big hole in the wall. There are no movable dampers. The type of grille used will determine the required grille size (in conjunction with the cfm required for the system). This is because air can only flow through the openings in the grille (called the *face openings*) and not through the material. Therefore, the percentage of face opening for the grille used must be determined from the manufacturer's specifications.

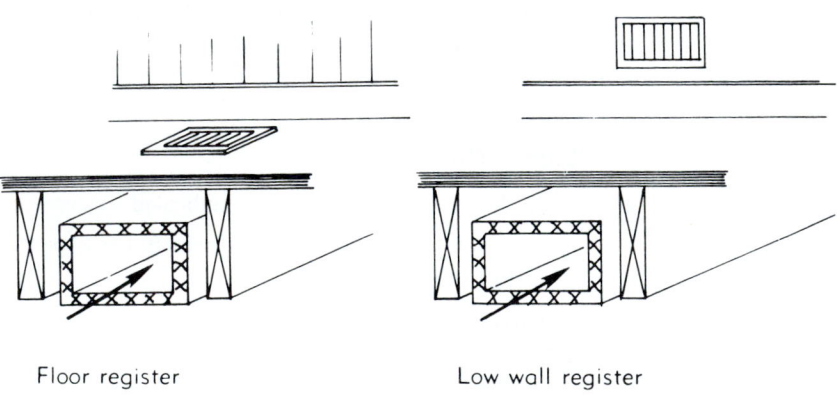

Floor register   Low wall register

FIGURE 10-13   Register location

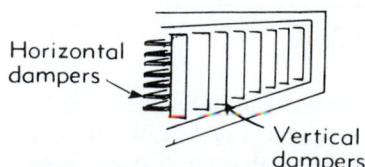

Horizontal dampers

Vertical dampers   FIGURE 10-14   Register dampers

The register must be proportionally larger than the return duct, based on the percentage of face opening of the grille required. For example, if a grille with 50% face opening is used on a 20-in. × 14-in. duct carrying 1,400 cfm, it will be necessary to have a grille twice as large as the duct branch (or trunk).

10-5 FURNACE LOCATION

Furnaces for forced air systems are available in various designs so that they may be located in the basement, crawl space, attic, or first floor of the buildiing. In addition, each is usually classified in terms of the direction in which the air is delivered. The three basic types are *upflow, counterflow,* and *horizontal,* and there are several variations depending on the actual installation (Fig. 10-15). A lowboy-style (upflow) furnace (Fig. 10-15) is installed in the basement of a building, and the air is delivered to the building through ducts from the top of the boiler.

The highboy (upflow) furnace (Fig. 10-15) is used primarily in single-level homes when the ducts are placed in the attic. This may be because of the slab on grade construction or because there is limited crawl space. The furnace may be located in a closet or recessed area in the wall. The air enters the furnace through a low side entry or through the bottom, and it leaves through the top.

The counterflow (downflow) furnace (Fig. 10-15) is used primarily where it is preferable to have the furnace in the building, on the first floor, and not in a crawl space. The air enters the furnace at the top, and it leaves through the bottom. It is usually used with crawl spaces and slabs on grade. While it may be used in a building with a basement, such buildings usually have the furnace located in the basement.

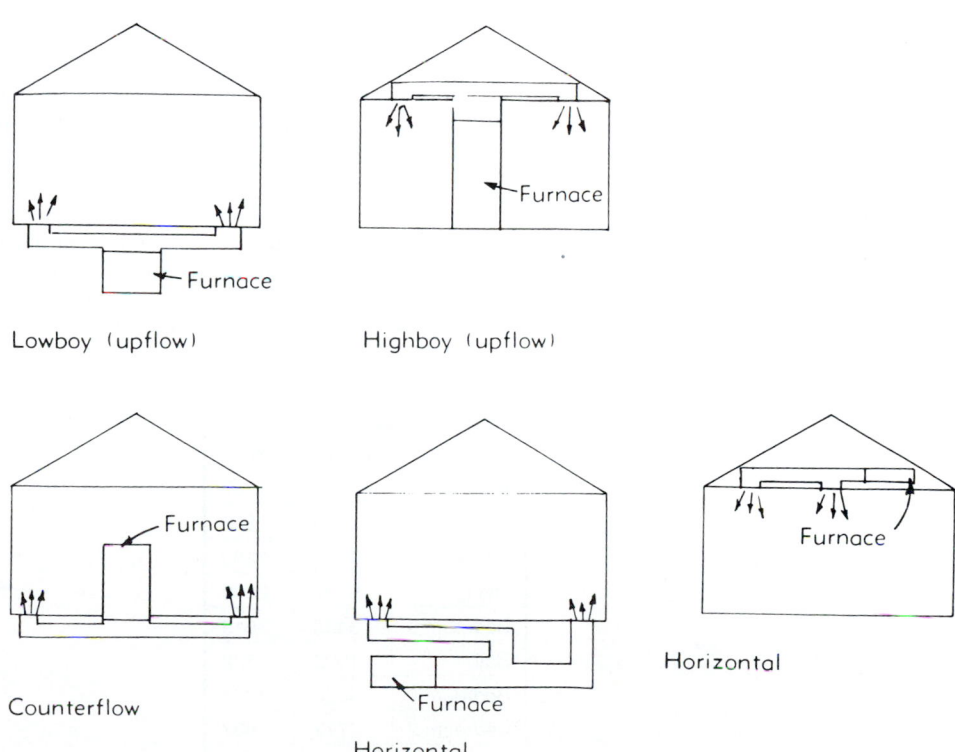

FIGURE 10-15 Boiler locations

The horizontal furnace (Fig. 10-15) is used primarily in homes with crawl spaces or concrete slab floors (i.e., no basement). The furnace may be set off the ground in the crawl space, placed in the attic, or suspended from posts in an attic or utility room. The air enters the furnace at one end, and it leaves through the opposite end.

Combinations of systems are often used in larger homes and in many industrial, commercial, or institutional buildings. In a two-story home, it may be desirable to heat the first floor with an upflow furnace in the basement and the second floor with a horizontal furnace in the attic.

10-6 DUCT DESIGN

Step-by-Step Approach

1. Determine the heat loss of each individual room and list them.

   From Chapter 8 the heat loss for each room is tabulated in Fig. 10-16.

2. Next, determine the location and number of supply outlets (in this case, located on outside walls under windows; read Sec. 10-4 for a complete discussion) and return air intakes. Sketch the proposed locations on the floor plan as shown in Fig. 10-17. As a general rule, in residential design, no one supply outlet should supply more than 8,000 Btuh (2,340 W).

   Fig. 10-17 has a proposed layout for the ductwork. The ducts are shown as single lines and the outlets as rectangles, and the Btuh for each outlet is listed (the watts for each room are in parentheses).

   Fig. 10-18 has a tabulation of each supply outlet and the Btuh.

3. Note on the sketch the types of fittings which will be used and the actual length of each duct. The fittings may be selected from Fig. 10-34, which also gives the equivalent length for each type of fitting. The plan in Fig. 10-19a shows the fittings and their equivalent lengths.

   Now, on the sketch, note the actual length of each run. Then add the equivalent and actual lengths for each run to obtain the total equivalent length for each run from the bonnet to the supply outlet (Fig. 10-19b).

   The tabulated total equivalent length of each run is tabulated in Fig. 10-19.

	BTUH	Watts
Entry	4061	1287
Dining	3056	898
Kitchen	4160	1205
Living	9327	2690
Bedroom 3	3483	1013
Bath	1092	322
Bedroom 1	4326	1267
Bedroom 2	2772	807
Int. Bath	203	57
	32,480	9546

FIGURE 10-16   Room heat loss

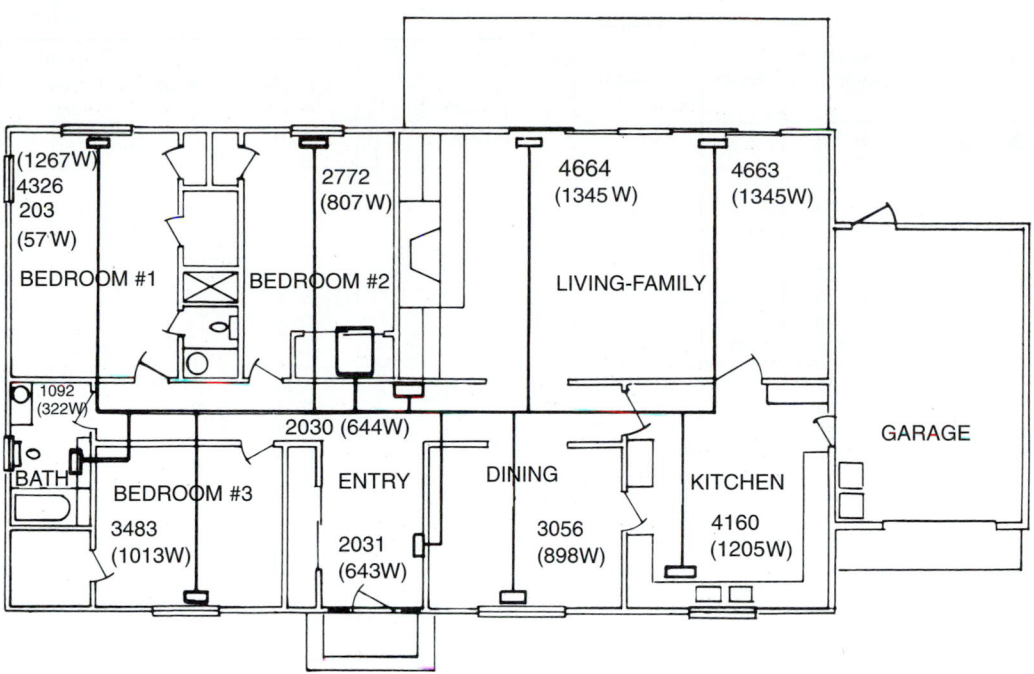

FIGURE 10-17   Duct schematic

Supply Outlet	Heat Loss (BTUH)	Watts
Bedroom #1	4529	1324
Bedroom #2	2772	807
Living	4664	1345
Living	4663	1345
Kitchen	4160	1205
Dining	3056	898
Entry	2031	643
Hall	2030	644
Bedroom #3	3483	1013
Bath	1092	322
Total	32,480	9546

Note: Bedroom 1 includes interior bath.

FIGURE 10-18   Heat loss for each outlet

4. A tentative blower size is selected next. For the most efficient operation and even heating it is recommended that the total cfm (L/s) should not be less than 400 cfm (190 L/s) per 12,000 Btuh (3510 Watts).

$$\frac{32,480}{12,000} \times 400 = 1083 \text{ cfm (use 1200)}$$

SI

$$\frac{9546}{3510} \times 190 \text{ L/s} = 518 \text{ L/s (use 550 L/s)}$$

Outlet	Equivalent length	Actual length	Total equivalent length
Bedroom (1)	F 35 B 15 G 30 — 80	36	116
Bedroom (2)	F 35 + 25 B 15 G 30 — 105	26	131
Living 1	F 35 B 15 + 25 G 30 — 105	33	138
2	F 35 B 15 G 30 — 80	55	135
Kitchen	F 35 B 15 G 30 — 80	46	126
Dining	F 35 + 25 B 15 G 30 — 105	27	132
Entry	F 35 + 25 B 15 E 10 G 30 — 115	16	131
Hall	F 35 B 15 G 30 — 80	10	90
Bedroom (3)	F 35 + 25 B 15 G 30 — 105	30	135
Bath	F 35 B 15 E 10 G 30 — 90	25	115

SI UNITS

Outlet	Equivalent length	Actual length	Total equivalent length
Bedroom (1)	F 10.7 B 4.6 G 9.1 — 24.4	11	35.4
Bedroom (2)	F 10.7 + 7.6 B 4.6 G 9.1 — 32	7.9	39.9
Living 1	F 10.7 B 4.6 + 7.6 G 9.1 — 32	10	42
2	F 10.7 B 4.6 G 9.1 — 24.4	16.8	41.2
Kitchen	F 10.7 B 4.6 G 9.1 — 24.4	14	38.4
Dining	F 10.7 + 7.6 B 4.6 G 9.1 — 32	8.2	40.2
Entry	F 10.7 + 7.6 B 4.6 E 3.1 G 9.1 — 35.1	5	40.1
Hall	F 10.7 B 4.6 G 9.1 — 24.4	3.1	27.5
Bedroom (3)	F 10.7 + 7.6 B 4.6 G 9.1 — 32	3.1	35.1
Bath	F 10.7 B 4.6 E 3.1 G 9.1 — 27.5	7.6	35.1

FIGURE 10-19(a)   Equivalent lengths

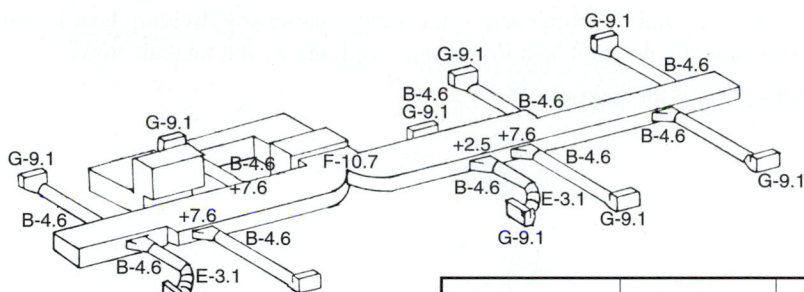

Supply Outlet	Heat Loss (BTUH)	Actual Length (ft)	Equivalent Length (ft)	Total Equiv. Length (ft)
Bedroom #1	4529	36	80	116
Bedroom #2	2772	26	105	131
Living	4664	33	105	138
Living	4663	55	80	135
Kitchen	4160	46	80	126
Dining	3056	27	105	132
Entry	2031	16	115	131
Hall	2030	10	80	90
Bedroom #3	3483	30	105	135
Bath	1092	25	90	115
Totals				

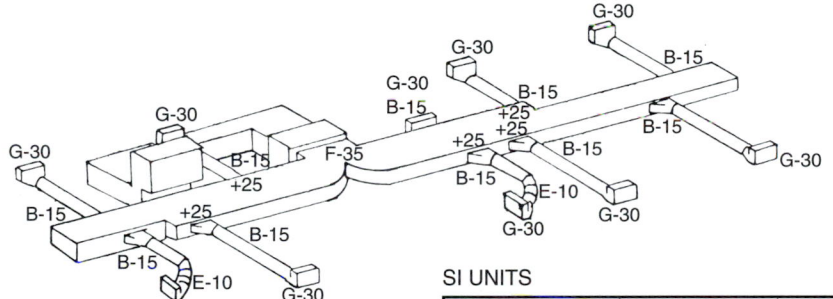

SI UNITS

Supply Outlet	Heat Loss (Watts)	Actual Length (m)	Equivalent Length	Total Equiv. Length
Bedroom #1	1,324	11	24.4	35.4
Bedroom #2	807	7.9	32	39.9
Living	1,345	10	32	42
Living	1,345	16.8	24.4	41.2
Kitchen	1,194	14	24.4	38.4
Dining	898	8.2	32	40.2
Entry	643	5.0	35.1	40.1
Hall	644	3.1	24.4	27.5
Bedroom #3	1,013	3.1	32	35.1
Bath	322	7.6	7.5	35.1
Totals	9,535			

FIGURE 10-19(b) Total equivalent lengths

5. Calculate the cfm required from each of the supply outlets by dividing heat loss of each supply outlet by the total heat loss and multiplying by the total air flow.

In this design, for bedroom 1:

$$\frac{4,529}{32,480} \times 1200 = 167$$

SI

$$\frac{1324}{9546} \times 550 = 76 \text{ L/s}$$

Following this example, calculate the required cfm for each room and tabulate as shown (Fig. 10-20).

6. Bonnet pressure.

a. The bonnet pressure is selected next. The approximate bonnet pressure required for a trunk (main duct) to carry the maximum volume of air (cfm) is found in Fig. 10-37.

Supply Outlet	Heat Loss (BTUH)	Actual Length (ft)	Equivalent Length (ft)	Total Equiv. Length (ft)	Heating CFM
Bedroom #1	4529	36	80	116	167
Bedroom #2	2772	26	105	131	103
Living	4664	33	105	138	173
Living	4663	55	80	135	173
Kitchen	4160	46	80	126	154
Dining	3056	27	105	132	113
Entry	2031	16	115	131	75
Hall	2030	10	80	90	75
Bedroom #3	3483	30	105	135	129
Bath	1092	25	90	115	40

SI UNITS

Supply Outlet	Heat Loss (Watts)	Actual Length (m)	Equivalent Length (m)	Total Equiv. Length (m)	Heating (L/s)
Bedroom #1	1324	11	24.4	35.4	77
Bedroom #2	807	7.9	32	39.9	47
Living	1345	10	32	42	78
Living	1345	16.8	24.4	41.2	78
Kitchen	1205	14	24.4	38.4	69
Dining	898	8.2	32	40.2	52
Entry	643	5	35.1	40.1	37
Hall	644	3.1	24.4	27.5	37
Bedroom #3	1013	3.1	32	35.1	59
Bath	322	7.6	27.5	35.1	19

FIGURE 10-20   Heating cfm required

In this design, the total cfm required is 1200 (the total of all the supply outlets). From Fig. 10-37, from a quiet operation in a residence (Type A), the suggested bonnet pressure is a minimum of 0.10 in. (24.8 Pa) of water.

    b. If a furnace has been selected, the rated capacity of the unit must be checked, and its pressure used. In this design, assume a furnace with 0.10 in. (24.8 Pa) of water was selected.

7. The available bonnet pressure is divided proportionally by length between the supply and return runs. In perimeter heating (Fig. 10-1), the supply runs are longer than the returns, and a 0.10 in. (24.8 Pa)-pressure might be divided with 70% on the supply and 30% on the return. For other systems, the sketch plan layout shows the approximate proportion of supply to return runs.

    The sketch plan layout in Fig. 10-17 shows that there are longer supply than return runs, so a division of 0.07 in. (17.36 Pa) for supply and 0.03 in. (7.44 Pa) for return is selected and tabulated (Fig. 10-21).

Supply Outlet	Heat Loss (BTUH)	Actual Length (Ft)	Equivalent Length (Ft)	Total Equiv. Length (Ft)	Heating CFM	Supply Air Pressure
Bedroom #1	4529	36	80	116	167	0.07
Bedroom #2	2772	26	105	131	103	0.07
Living	4664	33	105	138	173	0.07
Living	4663	55	80	135	173	0.07
Kitchen	4160	46	80	126	154	0.07
Dining	3056	27	105	132	113	0.07
Entry	2031	16	115	131	75	0.07
Hall	2030	10	80	90	75	0.07
Bedroom #3	3483	30	105	135	129	0.07
Bath	1092	25	90	115	40	0.07

SI UNITS

Supply Outlet	Heat Loss (Watts)	Actual Length (m)	Equivalent Length (m)	Total Equiv. Length (m)	Heating (L/s)	Supply Air Pressure
Bedroom #1	1,324	11	24.4	35.4	77	17.36
Bedroom #2	807	7.9	32	39.9	47	17.36
Living	1,345	10	32	42	78	17.36
Living	1,345	16.8	24.4	41.2	78	17.36
Kitchen	1,205	14	24.4	38.4	69	17.36
Dining	898	8.2	32	40.2	52	17.36
Entry	643	5	35.1	40.1	37	17.36
Hall	644	3.1	24.4	27.5	37	17.36
Bedroom #3	1,013	3.1	32	35.1	59	17.36
Bath	322	7.6	27.5	35.1	19	17.36

FIGURE 10-21 Supply air pressure

Note: If later during the design the proportion selected does not provide satisfactory results, it can be reapportioned between supply and return.

8. Next, the supply outlet size and its pressure loss are selected from the manufacturer's engineering data. A typical example of the manufacturer's data is shown in Fig. 13-14. Typically, the pressure loss will range from 0.01 to 0.02 in., and many designers simply allow 0.02 in. (4.8 Pa).

   In this design, the pressure loss has been tabulated in Fig. 10-22.

9. Next, the actual pressure which is available for duct loss is obtained by subtracting the supply outlet loss (step 8) from the total pressure available for supply runs (Fig. 10-22).

Supply Outlet	Heat Loss (BTUH)	Actual Length (ft)	Equivalent Length (ft)	Total Equiv. Length (ft)	Heating CFM	Supply Air Pressure	Supply Outlet Pressure Loss
Bedroom #1	4529	36	80	116	167	0.07	0.02
Bedroom #2	2772	26	105	131	103	0.07	0.02
Living	4664	33	105	138	173	0.07	0.02
Living	4663	55	80	135	173	0.07	0.02
Kitchen	4160	46	80	126	154	0.07	0.02
Dining	3056	27	105	132	113	0.07	0.02
Entry	2031	16	115	131	75	0.07	0.02
Hall	2030	10	80	90	75	0.07	0.02
Bedroom #3	3483	30	105	135	129	0.07	0.02
Bath	1092	25	90	115	40	0.07	0.02

SI UNITS

Supply Outlet	Heat Loss (Watts)	Actual Length (m)	Equivalent Length (m)	Total Equiv. Length (m)	Heating (L/s)	Supply Air Pressure	Supply Outlet Pressure Loss
Bedroom #1	1,324	11	24.4	35.4	77	17.36	4.8
Bedroom #2	807	7.9	32	39.9	47	17.36	4.8
Living	1,345	10	32	42	78	17.36	4.8
Living	1,345	16.8	24.4	41.2	78	17.36	4.8
Kitchen	1,205	14	24.4	38.4	69	17.36	4.8
Dining	898	8.2	32	40.2	52	17.36	4.8
Entry	643	5	35.1	40.1	37	17.36	4.8
Hall	644	3.1	24.4	27.5	37	17.36	4.8
Bedroom #3	1,013	3.1	32	35.1	59	17.36	4.8
Bath	322	7.6	27.5	35.1	19	17.36	4.8

FIGURE 10-22   Supply outlet pressure

In this design, the pressure for supply runs is 0.07 in. Beginning with the bedroom 1 supply run, from step 8 the supply outlet loss is 0.02 in., so for this run, the pressure available for actual duct loss is 0.07 in. − 0.02 in. = 0.05 in. Calculate the actual pressure available for duct loss for each of the supply runs, and tabulate the information as shown (Fig. 10-23).

SI

In this design, the pressure for supply runs is 16.8 Pa. Beginning with bedroom 1 supply run, from step 8 the supply outlet loss is 4.8 Pa, so for this run, the pressure available for actual duct loss is 16.8 − 4.8 = 12 Pa. Calculate the actual pressure available for each of the supply runs, and tabulate the information as shown in Fig. 10-23.

10. In order to make a workable table of duct sizes, the pressure drop for duct loss must be calculated as the pressure drop per *100 ft* (per meter) of the duct. Since the

Supply Outlet	Heat Loss (BTUH)	Actual Length (ft)	Equivalent Length (ft)	Total Equiv. Length (ft)	Heating CFM	Supply Air Pressure	Supply Outlet Pressure Loss	Pressure Avail. for Duct Loss
Bedroom #1	4529	36	80	116	167	0.07	0.02	0.05
Bedroom #2	2772	26	105	131	103	0.07	0.02	0.05
Living	4664	33	105	138	173	0.07	0.02	0.05
Living	4663	55	80	135	173	0.07	0.02	0.05
Kitchen	4160	46	80	126	154	0.07	0.02	0.05
Dining	3056	27	105	132	113	0.07	0.02	0.05
Entry	2031	16	115	131	75	0.07	0.02	0.05
Hall	2030	10	80	90	75	0.07	0.02	0.05
Bedroom #3	3483	30	105	135	129	0.07	0.02	0.05
Bath	1092	25	90	115	40	0.07	0.02	0.05

SI UNITS

Supply Outlet	Heat Loss (Watts)	Actual Length (m)	Equivalent Length (m)	Total Equiv. Length (m)	Heating (L/s)	Supply Air Pressure	Supply Outlet Pressure Loss	Pressure Avail. for Duct Loss
Bedroom #1	1,324	11	24.4	35.4	77	17.36	4.8	12.56
Bedroom #2	807	7.9	32	39.9	47	17.36	4.8	12.56
Living	1,345	10	32	42	78	17.36	4.8	12.56
Living	1,345	16.8	24.4	41.2	78	17.36	4.8	12.56
Kitchen	1,205	14	24.4	38.4	69	17.36	4.8	12.56
Dining	898	8.2	32	40.2	52	17.36	4.8	12.56
Entry	643	5	35.1	40.1	37	17.36	4.8	12.56
Hall	644	3.1	24.4	27.5	37	17.36	4.8	12.56
Bedroom #3	1,013	3.1	32	35.1	59	17.36	4.8	12.56
Bath	322	7.6	27.5	35.1	19	17.36	4.8	12.56

FIGURE 10-23   Available duct pressure

various supply runs are all different lengths, it is necessary to find the allowable *pressure drop per 100 ft* (per meter) based on the *allowable duct loss* and the *total equivalent length* of the supply run. This allowable pressure drop is calculated by using the equation:

$$\text{Allowable pressure drop per 100 ft} = \frac{\text{Allowable duct loss} \times 100}{\text{Total equivalent length}}$$

In this design, the bedroom 1 supply run has an allowable duct loss of .05 in. (step 9) and an equivalent length of 116 ft (step 3). Using the formula to determine the allowable pressure drop per 100 ft.

$$\text{Allowable pressure drop per 100 ft} = \frac{0.05 \text{ in.} \times 100 \text{ ft}}{116 \text{ ft}} = 0.43 \text{ in.}$$

SI

The allowable pressure drop is calculated by using the equation:

$$\text{Allowable pressure drop} = \frac{\text{Supply static pressure available}}{\text{Total equivalent length of branch supply run}}$$

In this design, the bedroom 1 supply run has an allowable duct loss of 12.0 Pa and an equivalent length of 35.4 m (step 3). Using the formula to determine the allowable pressure drop:

$$\text{Allowable pressure drop} = \frac{12}{35.4} = 0.34$$

Calculate the allowable pressure drop for each supply run, and tabulate the information as shown (Fig. 10-24).

11. The branch duct sizes may be determined by using the table in Fig. 10-35 or by using an air duct calculator (Fig. 10-25), if one is available. The table in Fig. 10-35 gives the size of the round duct based on the cfm required and the allowable pressure drop per 100 ft (step 12). The size of the round duct may be converted to an equivalent rectangular duct size which will handle the required cfm within the allowable pressure drop.

   In this design, the bedroom 1 supply run has an allowable duct loss of 0.043 in. per 100 ft (step 12) and a required cfm of 170 (step 5). Using the table in Fig. 10-35 a 9.0 in. round duct would be selected. Fig. 10-26 shows the use of the table. Begin by finding the pressure drop per 100 ft on the left; then the required cfm along the bottom; then move horizontally to the right and vertically from the bottom to read the round duct size required.

   From Fig. 10-35, determine the round duct sizes for each of the supply runs, and tabulate the information as shown in Fig. 10-26.

   The air duct calculator is used to obtain required round duct sizes by aligning the required air volume and the allowable pressure drop (friction) per 100 ft of duct and reading the round duct diameter directly from the calculator.

   SI

   In this design, the bedroom 1 supply run has an allowable duct loss of 0.35 (step 12) and a required L/s of 77 (step 7). Using the table in Fig. 10-35, a 250 mm round duct would be selected. Begin by finding the pressure drop on the left and the L/s along the bottom; then find where they intersect and read the duct size to the right of the intersection.

Supply Outlet	Heat Loss (BTUH)	Actual Length (ft)	Equivalent Length (ft)	Total Equiv. Length (ft)	Heating CFM	Supply Air Pressure	Supply Outlet Pressure Loss	Pressure Avail. for Duct Loss	Pressure Avail. Per 100 Ft
Bedroom #1	4529	36	80	116	167	0.07	0.02	0.05	0.043
Bedroom #2	2772	26	105	131	103	0.07	0.02	0.05	0.038
Living	4664	33	105	138	173	0.07	0.02	0.05	0.036
Living	4663	55	80	135	173	0.07	0.02	0.05	0.037
Kitchen	4160	46	80	126	154	0.07	0.02	0.05	0.040
Dining	3056	27	105	132	113	0.07	0.02	0.05	0.038
Entry	2031	16	115	131	75	0.07	0.02	0.05	0.038
Hall	2030	10	80	90	75	0.07	0.02	0.05	0.056
Bedroom #3	3483	30	105	135	129	0.07	0.02	0.05	0.037
Bath	1092	25	90	115	40	0.07	0.02	0.05	0.043

SI UNITS

Supply Outlet	Heat Loss (Watts)	Actual Length (m)	Equivalent Length (m)	Total Equiv. Length (m)	Heating (L/s)	Supply Air Pressure	Supply Outlet Pressure Loss	Pressure Avail. for Duct Loss	Pressure Avail. Per Branch
Bedroom #1	1,324	11	24.4	35.4	77	17.36	4.8	12.56	0.35
Bedroom #2	807	7.9	32	39.9	47	17.36	4.8	12.56	0.31
Living	1,345	10	32	42	78	17.36	4.8	12.56	0.30
Living	1,345	16.8	24.4	41.2	78	17.36	4.8	12.56	0.30
Kitchen	1,205	14	24.4	38.4	69	17.36	4.8	12.56	0.33
Dining	898	8.2	32	40.2	52	17.36	4.8	12.56	0.31
Entry	643	5.0	35.1	40.1	37	17.36	4.8	12.56	0.31
Hall	644	3.1	24.4	27.5	37	17.36	4.8	12.56	0.46
Bedroom #3	1,013	3.1	32	35.1	59	17.36	4.8	12.56	0.36
Bath	322	7.6	27.5	35.1	19	17.36	4.8	12.56	0.36

FIGURE 10-24   Pressure available

12. The round duct sizes selected in step 11 may be changed to equivalent rectangular duct sizes, carrying the same cfm (L/s) required while maintaining the allowable pressure drops. Using the table in Fig. 10-36, the 9-0-in. round duct for bedroom 1 can be changed to a rectangular duct of 8 in. × 8 in., 11 in. × 7 in., or 14 in. × 6 in. and others.

The round duct sizes are found along the side of the table. The rectangular sizes are to the right.

The actual size selected depends to a large extent on the space available for installation of the duct.

From Fig. 10-36 determine a rectangular duct size for each supply run which could be used instead of the round duct previously selected. Tabulate the rectangular duct sizes as shown (Fig. 10-27).

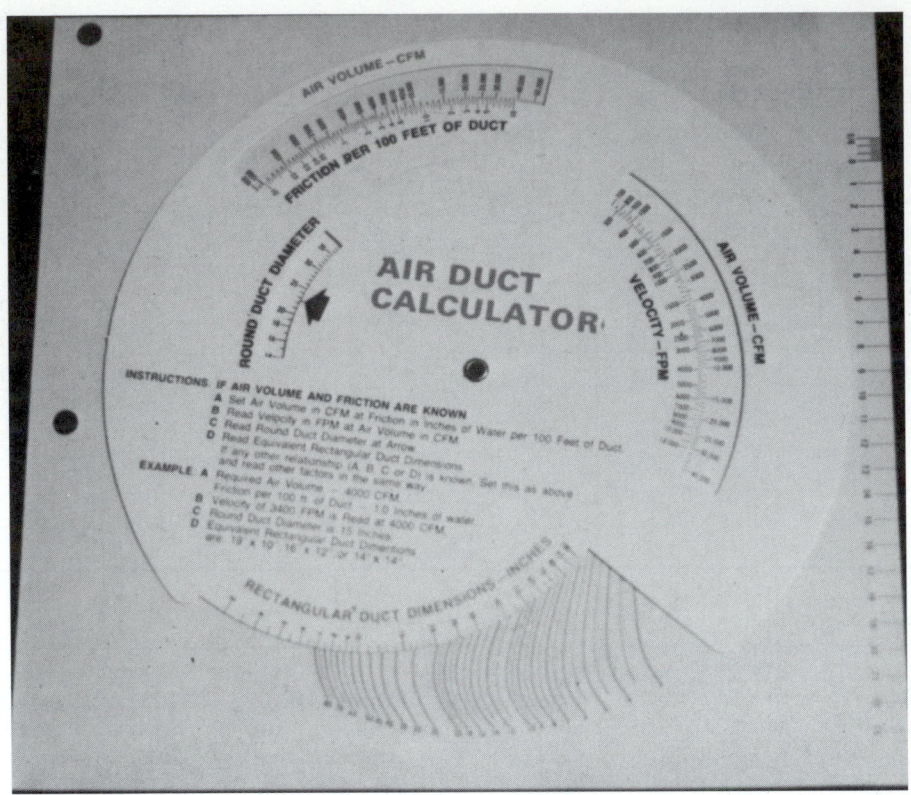

FIGURE 10-25   Air duct calculator

13. Add the air volumes of all of the branch supply runs from each trunk duct. When there is more than one trunk duct (such as in this design), keep the totals for each trunk duct separate.

> In this design, there are two trunk ducts (one toward the living room and kitchen, the other toward the bedrooms). Tabulate and total the air volume for each trunk duct as shown (Fig. 10-28).

14. For best air flow distribution through the ducts, it is important that the friction losses per 100 ft be approximately equal in both trunk ducts.

> Using the table in Fig. 10-35, select the trunk sizes required. When there is more than one trunk duct, size each separately using the required air volume (cfm) and the allowable duct friction (step 10).

> The bedroom branch trunk must handle 439 cfm with an allowable pressure drop of 0.37 in. per 100 ft (0.05 times 100 divided by 135 ft.). Using Fig. 10-35, the bedroom trunk duct will be a 14 inch round duct.

> The living space trunk duct has 763 cfm and an allowable 0.040 in. per 100 ft pressure drop. Using Fig. 10-35, the living space trunk duct will be a 16-in. round duct.

> The bedroom trunk must handle 202 L/s with an allowable pressure drop of 0.31 Pa. Using Fig. 10-35, the bedroom trunk will be a 315 mm round duct.

> The living space trunk duct has 351 L/s with an allowable pressure of 0.30 Pa. Using Fig. 10-35, the living space trunk will be a 400 mm round duct.

Supply Outlet	Heat Loss (BTUH)	Actual Length (ft)	Equivalent Length (ft)	Total Equiv. Length (Ft)	Heating CFM	Supply Air Pressure	Supply Outlet Pressure Loss	Pressure Avail. for Duct Loss	Pressure Avail. Per 100 Ft	Round Duct Size (In)
Bedroom #1	4529	36	80	116	167	0.07	0.02	0.05	0.043	9.0
Bedroom #2	2772	26	105	131	103	0.07	0.02	0.05	0.038	7.0
Living	4664	33	105	138	173	0.07	0.02	0.05	0.036	9.0
Living	4663	55	80	135	173	0.07	0.02	0.05	0.037	9.0
Kitchen	4160	46	80	126	154	0.07	0.02	0.05	0.040	8.0
Dining	3056	27	105	132	113	0.07	0.02	0.05	0.038	8.0
Entry	2031	16	115	131	75	0.07	0.02	0.05	0.038	7.0
Hall	2030	10	80	90	75	0.07	0.02	0.05	0.056	6.0
Bedroom #3	3483	30	105	135	129	0.07	0.02	0.05	0.037	8.0
Bath	1029	25	90	115	40	0.07	0.02	0.05	0.043	6.0

SI UNITS

Supply Outlet	Heat Loss (Watts)	Actual Length (m)	Equivalent Length (m)	Total Equiv. Length (m)	Heating (L/s)	Supply Air Pressure	Supply Outlet Pressure Loss	Pressure Avail. for Duct Loss	Pressure Avail. Per Branch	Round Duct Size (mm)
Bedroom #1	1,324	11	24.4	35.4	77	17.36	4.8	12.56	0.35	250
Bedroom #2	807	7.9	32	39.9	47	17.36	4.8	12.56	0.31	200
Living	1,345	10	32	42	78	17.36	4.8	12.56	0.30	250
Living	1,345	16.8	24.4	41.2	78	17.36	4.8	12.56	0.30	250
Kitchen	1,205	14	24.4	38.4	69	17.36	4.8	12.56	0.33	200
Dining	898	8.2	32	40.2	52	17.36	4.8	12.56	0.31	200
Entry	643	5	35.1	40.1	37	17.36	4.8	12.56	0.31	160
Hall	644	3.1	24.4	27.5	37	17.36	4.8	12.56	0.46	160
Bedroom #3	1,013	3.1	32	35.1	59	17.36	4.8	12.56	0.36	200
Bath	322	7.6	27.5	35.1	19	17.36	4.8	12.56	0.36	160

FIGURE 10-26   Tabulated round duct sizes

Since the trunk ducts have been sized separately, based on air flow and allowable pressure drop, they are balanced.

The round duct sizes may be converted to rectangular duct sizes using Fig. 10-36. Once again, the space available for installation will have a large influence on the shape of the rectangular duct selected. Tabulate the trunk duct sizes as shown in Fig. 10-29.

15. Review the supply trunk duct to determine if it is desirable to reduce the size of the duct as each duct leaves the trunk. Generally, such a reduction is suggested so that the velocity of air through the duct will not drop too low. Using the remaining air vol-

Supply Outlet	Heat loss (BTUH)	Actual Length (Ft)	Equivalent Length (Ft)	Total Equiv. Length (Ft)	Heating CFM	Supply Air Pressure	Supply Outlet Pressure Loss	Pressure Avail. for Duct Loss	Pressure Avail. Per 100 Ft	Round Duct Size (In.)	Rectangular Duct Size (In.)
Bedroom #1	4,529	36	80	116	167	0.07	0.02	0.05	0.060	9.0	6 × 12
Bedroom #2	2,772	26	105	131	103	0.07	0.02	0.05	0.054	7.0	6 × 8
Living	4,664	33	105	138	173	0.07	0.02	0.05	0.051	9.0	6 × 12
Living	4,663	55	80	135	173	0.07	0.02	0.05	0.052	9.0	6 × 12
Kitchen	4,160	46	80	126	154	0.07	0.02	0.05	0.056	8.0	6 × 9
Dining	3,056	27	105	132	113	0.07	0.02	0.05	0.053	8.0	6 × 9
Entry	2,031	16	115	131	75	0.07	0.02	0.05	0.054	7.0	6 × 7
Hall	2,030	10	80	90	75	0.07	0.02	0.05	0.078	6.0	6 × 5
Bedroom #3	3,483	30	105	135	129	0.07	0.02	0.05	0.052	8.0	5 × 9
Bath	1,092	25	90	115	40	0.07	0.02	0.05	0.061	0.6	6 × 5

SI UNITS

Supply Outlet	Heat loss (Watts)	Actual Length (m)	Equivalent Length (m)	Total Equiv. Length (m)	Heating (L/s)	Supply Air Pressure	Supply Outlet Pressure Loss	Pressure Avail. for Duct Loss	Pressure Avail. Per Branch	Round Duct Size (mm)	Rectangular Duct Size (mm)
Bedroom #1	1,324	11	24.4	35.4	77	17.36	4.8	12.56	0.35	250	150 × 400
Bedroom #2	807	7.9	32	39.9	47	17.36	4.8	12.56	0.31	200	150 × 225
Living	1,345	10	32	42	78	17.36	4.8	12.56	0.30	250	150 × 400
Living	1,345	16.8	24.4	41.2	78	17.36	4.8	12.56	0.30	250	150 × 400
Kitchen	1,205	14	24.4	38.4	69	17.36	4.8	12.56	0.33	200	150 × 225
Dining	898	8.2	32	40.2	52	17.36	4.8	12.56	0.31	200	150 × 225
Entry	643	5.0	35.1	40.1	37	17.36	4.8	12.56	0.31	160	100 × 200
Hall	644	3.1	24.4	27.5	37	17.36	4.8	12.56	0.46	160	100 × 200
Bedroom #3	1,013	3.1	32	35.1	59	17.36	4.8	12.56	0.36	200	150 × 225
Bath	322	7.6	27.5	35.1	19	17.36	4.8	12.56	0.36	160	100 × 200

FIGURE 10-27   Tabulated rectangular duct sizes

ume in the trunk duct, after each branch, and the allowable pressure drop of duct (step 10), select the reduced trunk sizes from Fig. 10-35.

Using tables in Fig. 10-35 and 10-36, determine the reduced trunk duct sizes for both the bedroom and the living space ducts, and tabulate the sizes as shown in Fig. 10-31.

16. Next, the design turns to the return ducts. The first step is to select an allowable return air pressure drop. This was previously decided (step 10), but it is reviewed at this point as the designer reviews the supply trunk and branch duct sizes to see if he or

SI UNITS

Duct		CFM
Main		
	A-B	1,200
Bedroom		
Branch	B-C	439
	C-D	336
	D-E	207
	E-F	167
Living		
Branch	B-G	763
	G-H	688
	H-I	613
	I-J	500
	J-K	327
	K-L	173

Duct		L/s
Main		
	A-B	550
Bedroom		
Branch	B-C	202
	C-D	155
	D-E	96
	E-F	77
Living		
Branch	B-G	351
	G-H	314
	H-I	277
	I-J	225
	J-K	147
	K-L	78

FIGURE 10-28 Trunk duct volumes

she might want to increase the duct sizes (reducing the allowable air pressure drop) or perhaps to reduce the duct sizes (increasing the allowable air pressure drop for the supply runs). Remember that a change in allowable air pressure drop for the supply will affect the allowable air pressure drop for the return runs.

A review of the supply duct sizes is made, and it is decided that they will not be revised. Therefore, the original decision, in step 7, to use 0.03 in. as the allowable pressure drop for the return, is unchanged, and the return design will be based on it.

17. The return trunk duct size is determined by the air volume it must handle and the allowable pressure drop. This allowable pressure drop must be converted to allowable pressure drop per 100 ft or Pa/m, just as was done for the supply runs. Using the air volume and the allowable pressure drop, the round return duct size may be selected from Fig. 10-35.

The rectangular duct of equivalent size may be found in Fig. 10-36. The return trunk duct will have to handle an air volume equal to what the supply trunk duct handled; from step 5 this was 1,200 cfm. The allowable pressure drop selected for the return was 0.03 in. (step 7). First, the allowable pressure drop must be converted into the allowable pressure drop *per* 100 ft: (0.03 in. × 100 ft) ÷ 82 ft = 0.036 in. per 100 ft allowable pressure drop.

Using this information (1,200 cfm, 0.036 in. per 100 ft) select the round return duct size from the table in Fig. 10-35 and its equivalent rectangular size from Fig. 10-36. Tabulate the information as shown (Fig. 10-36).

SI

The return duct will have to handle an air volume equal to what the supply trunk handled; from step 5 this was 550 L/s. The allowable pressure drop selected was 7.44 Pa (step 7). First the allowable drop must be converted into the allowable pressure drop available for this length of run:

$$(7.44 \text{ Pa} \div 25.1\text{m}) = 0.30 \text{ Pa}.$$

Using this information (550 L/s, 0.30 Pa), select the round duct size from Fig. 10-35 and its equivalent rectangular duct size from Fig. 10-36. Tabulate the information as shown in Fig. 10-30.

Duct		CFM	Supply Static Pressure*	Total Equiv. Length (Ft)	Supply Pressure Avail. Per 100 Ft	Round Duct Size (In.)	Rectangular Duct Size (In.)
Main							
	A-B	1,200	0.05	138	0.049	18.0	12 × 24
Bedroom Branch	B-C	439	0.05	135	0.037	14.0	8 × 22
	C-D	336				12.0	8 × 14
	D-E	207				10.0	7 × 11
	E-F	167				8.0	7 × 9
Living Branch	B-G	763	0.05	138	0.048	16.0	8 × 28
	G-H	688				16.0	8 × 28
	H-I	613				14.0	8 × 22
	I-J	500				14.0	8 × 22
	J-K	327				10.0	8 × 12
	K-L	173				9.0	8 × 8

*Allows 0.02 for supply outlet

SI UNITS

Duct		L/s	Supply Static Pressure* (Pa)	Total Total Equiv. Length (m)	Supply Pressure Avail. (Pa)	Round Duct Size (mm)	Rectangular Duct Size (mm)
Main							
	A-B	550	12.56	42.1	0.30	500	200 × 1300
Bedroom Branch	B-C	202	12.56	40	0.31	315	200 × 450
	C-D	155				315	200 × 450
	D-E	96				250	200 × 275
	E-F	77				250	200 × 275
Living Branch	B-G	351	12.56	42.1	0.30	400	200 × 750
	G-H	314				400	200 × 750
	H-I	277				400	200 × 750
	I-J	225				400	200 × 750
	J-K	147				315	200 × 450
	K-L	78				250	200 × 275

*Allows 4.8 Pa for supply outlet.
Note: Rounding off causes some variations in L/s.

FIGURE 10-29   Tabulated trunk sizes

If the return trunk sizes are too large, they may be reduced in size by:

a. Reapportioning the pressure drop available so there is less drop allowed for the supply runs and more pressure allowed for the return runs. This may require a revision of supply branch and trunk duct sizes.

b. Checking to see if the return air grille can be located closer to the furnace, which would reduce the actual length of duct. This increases the allowable pressure drop, resulting in a smaller duct size.

18. The size of the blower on the furnace is determined from the total air volume (the total cfm to be delivered) and the total static pressure requirements.

Duct	Cfm	Actual Length	Equivalent Length	Total Equivalent Length	Supply Pressure Avail. per 100 ft	Round Duct Size (in.)	Rectangular Duct Size (in.)
Return	1,200	12	70	82	0.036	18.0	12 × 24

SI UNITS

Duct	L/s	Actual Length	Equivalent Length	Total Equivalent Length	Supply Pressure Avail. (Pa)	Round Duct Size	Rectangular Duct Size
Return	550	3.7	21.4	25.1	0.30	500	200 × 1300

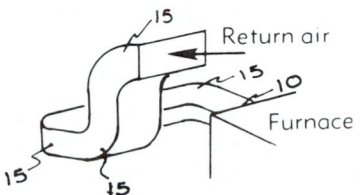

FIGURE 10-30   Return duct sizes

FIGURE 10-31   Ducts run in trusses

The total static pressure (bonnet pressure) of a furnace-blower combination unit is the total of the actual pressure drops in the supply and return ducts. Most residential and small commercial buildings have furnace-blower combination units.

When the blowers are selected separately from the furnace, the total static pressure is the sum of the actual pressure losses in the supply and return ducts, the filter loss, the casing loss, and losses through any devices which are put on the system, such as air washers and purifiers.

In this design, the pressure drop, on which the designs of ducts were based, has been selected to provide a standard blower-furnace combination. The actual drop was not calculated since each duct selected is large enough to handle the air volume within the allowable pressure drop.

10-7   SYSTEM INSTALLATION

Forced air systems need to have a place to run the ductwork. In wood frame construction, the ductwork may be run in a crawl space, basement, or in an attic space. In truss construction, the ducts may be run in the open spaces if the spaces are large enough (Fig.

FIGURE 10-32   Heating unit outside

FIGURE 10-33   Heating unit in attic

10-31). With poured concrete slabs, the duct may either be placed in the slab or the attic or ceiling area.

The heating unit may be placed on the ground outside the building (Fig. 10-32), in a crawl space, basement, or attic (Fig. 10-33). Small units have also been placed in "closets," while large units may be part of a large mechanical equipment room. In commercial construction, the units may be found on the roof of the project.

Large ductwork that must run vertically in the building will need to have a space to run. Returns located high in the wall will easily be 2 feet by 2 feet, and if a system on the first floor is serving a second floor, a large supply duct will need to go up to feed the outlets. Often these are located in a furred-out area of a closet.

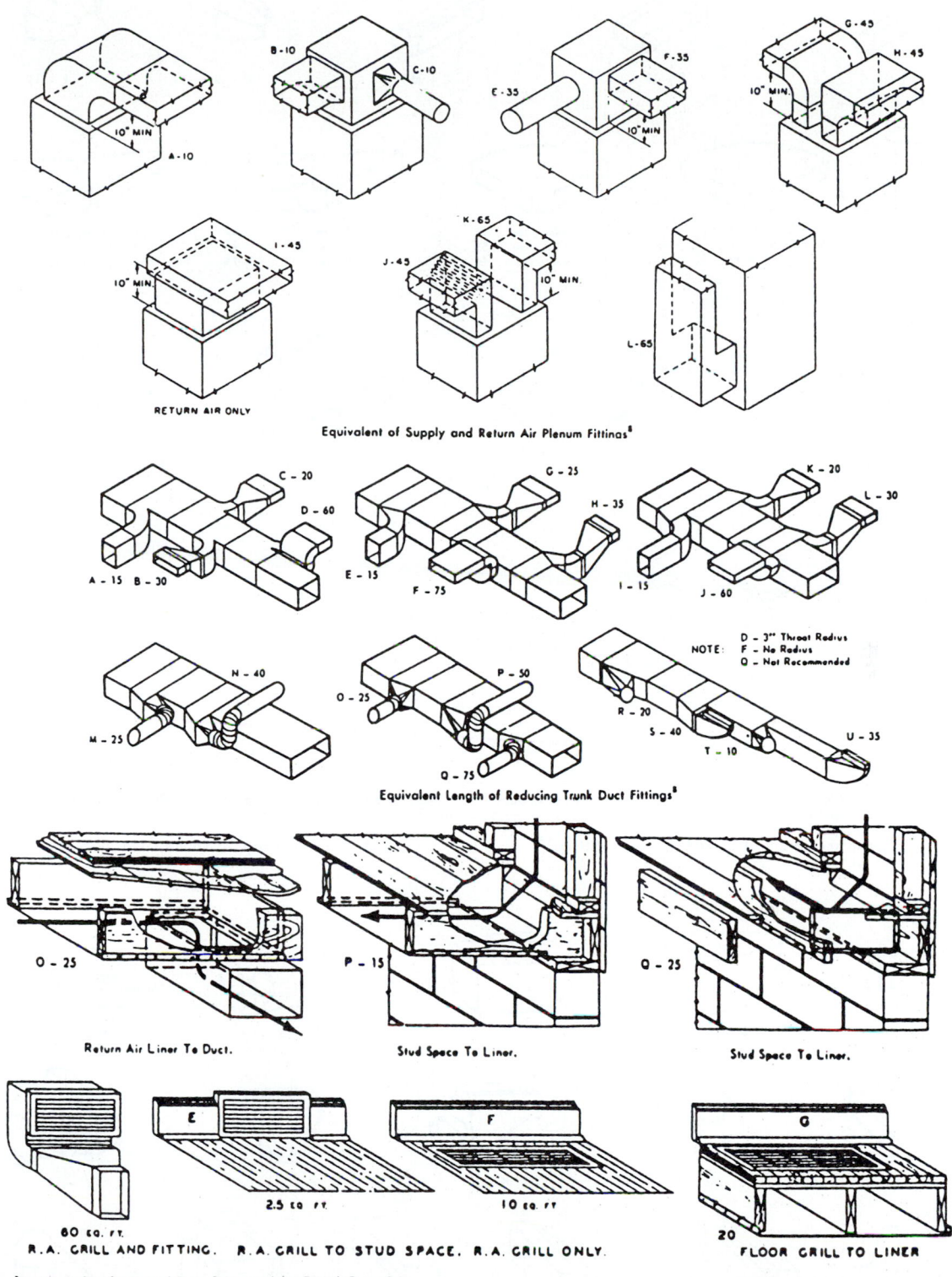

Equivalent of Supply and Return Air Plenum Fittings[a]

Equivalent Length of Reducing Trunk Duct Fittings[a]

Equivalent Length of Angles and Elbows for Individual and Branch Ducts[a,b]

[a] Inside radius for A and B = 3 in., and for F and G = 5 in.

FIGURE 10-34 Equivalent fittings

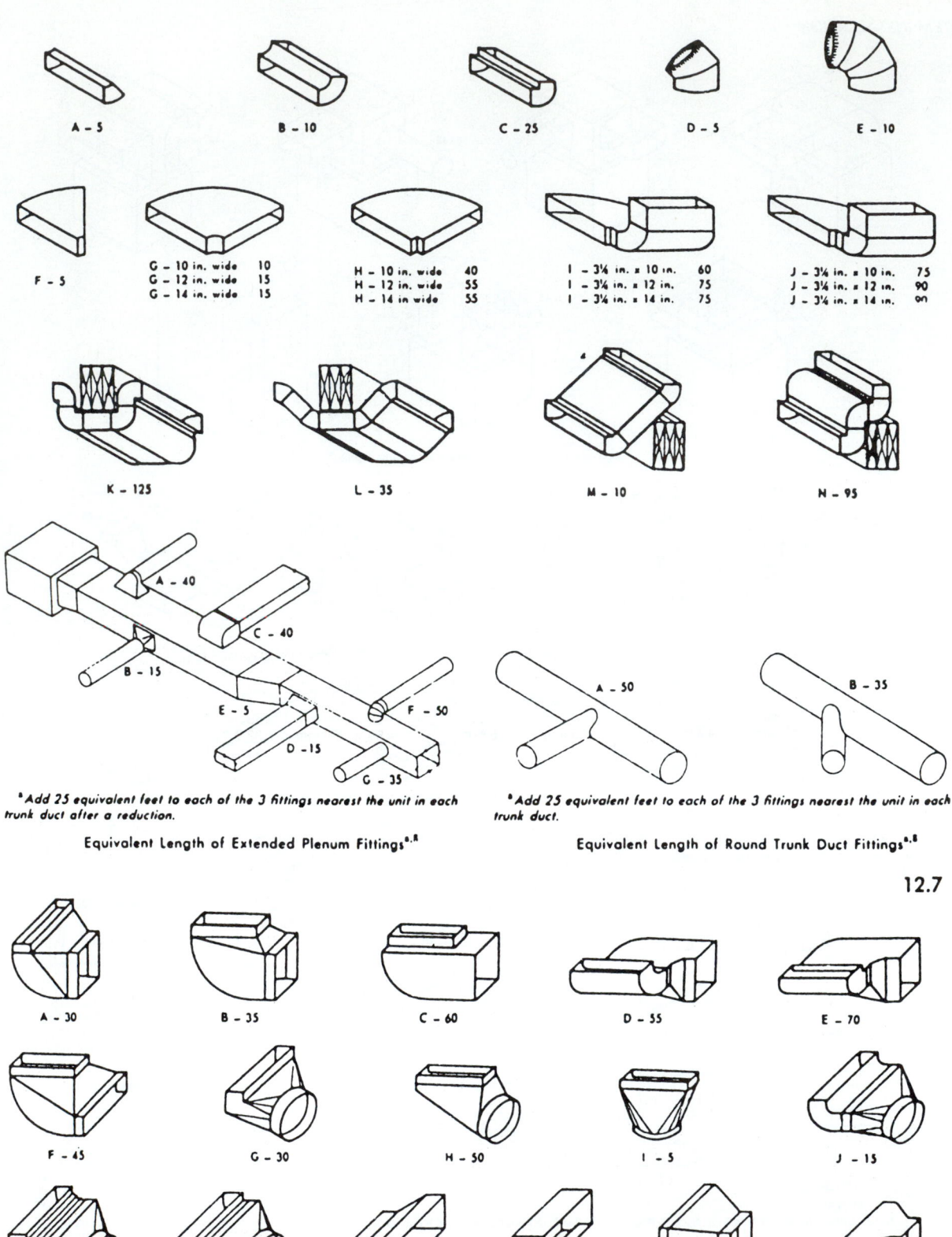

A – 5   B – 10   C – 25   D – 5   E – 10

F – 5

G – 10 in. wide   10
G – 12 in. wide   15
G – 14 in. wide   15

H – 10 in. wide   40
H – 12 in. wide   55
H – 14 in wide   55

I – 3¼ in. x 10 in.   60
I – 3¼ in. x 12 in.   75
I – 3¼ in. x 14 in.   75

J – 3¼ in. x 10 in.   75
J – 3¼ in. x 12 in.   90
J – 3¼ in. x 14 in.   90

K – 125   L – 35   M – 10   N – 95

A – 40
C – 40
B – 15
E – 5   F – 50
D – 15
G – 35

A – 50   B – 35

[a]Add 25 equivalent feet to each of the 3 fittings nearest the unit in each trunk duct after a reduction.

Equivalent Length of Extended Plenum Fittings[a,B]

[B]Add 25 equivalent feet to each of the 3 fittings nearest the unit in each trunk duct.

Equivalent Length of Round Trunk Duct Fittings[a,B]

12.7

A – 30   B – 35   C – 60   D – 55   E – 70

F – 45   G – 30   H – 50   I – 5   J – 15

K – 30   L – 30   M – 5   N – 15   O – 15   P – 5

[a]These values may also be used for floor diffuser boxes.

Equivalent Length of Boot Fittings[a,B]

FIGURE 10-34   (continued)

SI UNITS

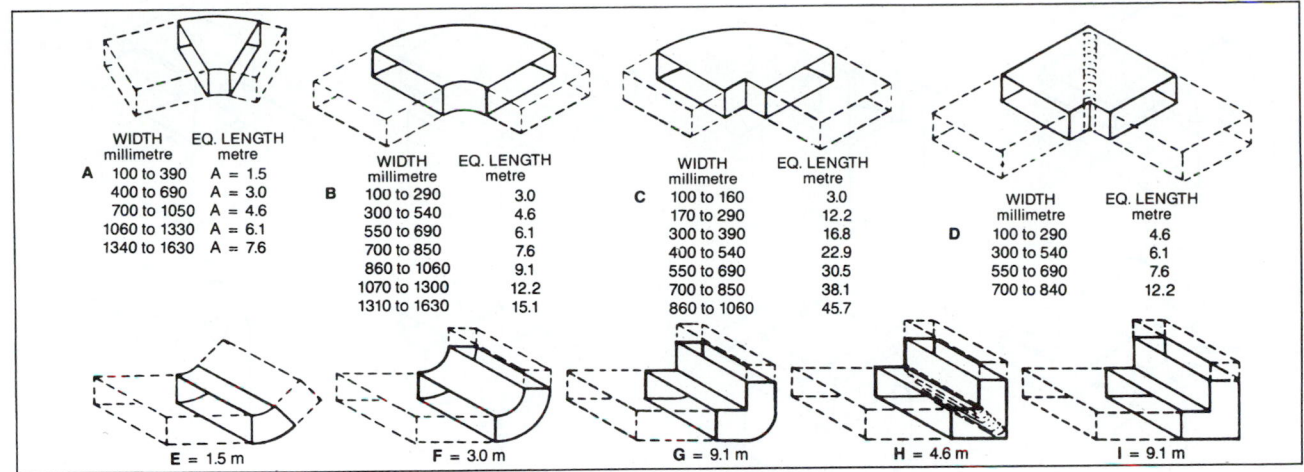

Equivalent Length of Angles and Elbows for Trunk Ducts (ACCA 1984)

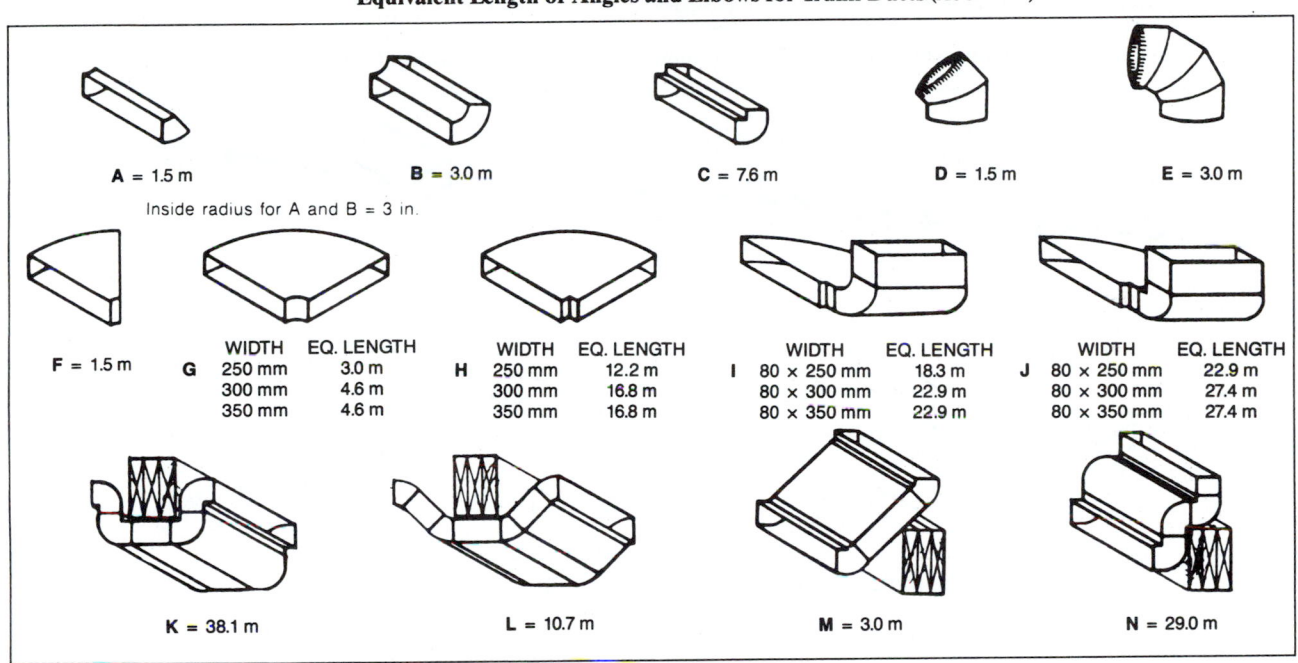

Equivalent Length of Angles and Elbows for Branch Ducts (ACCA 1984)

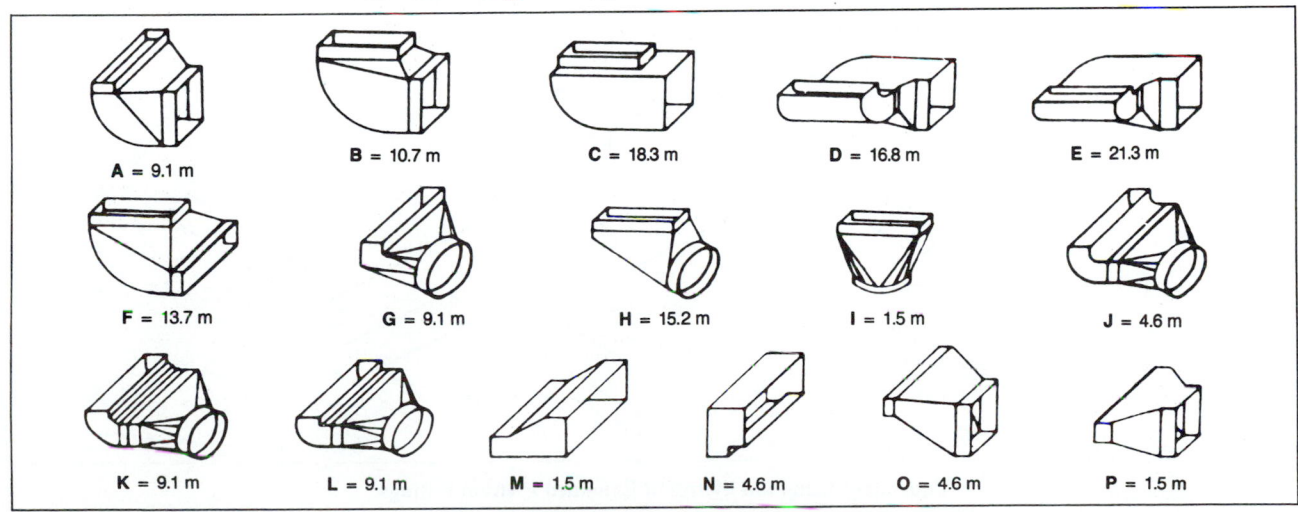

Equivalent Length of Boot Fittings

FIGURE 10-34 (*continued*)

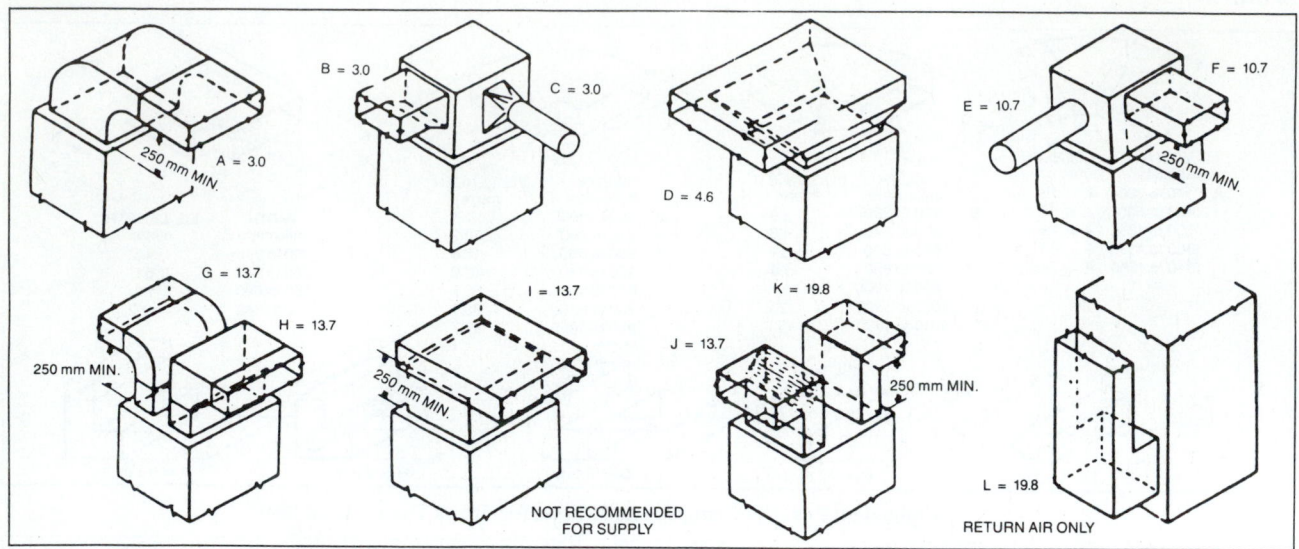

Equivalent Length in Metres of Supply and Return Air Plenum Fittings

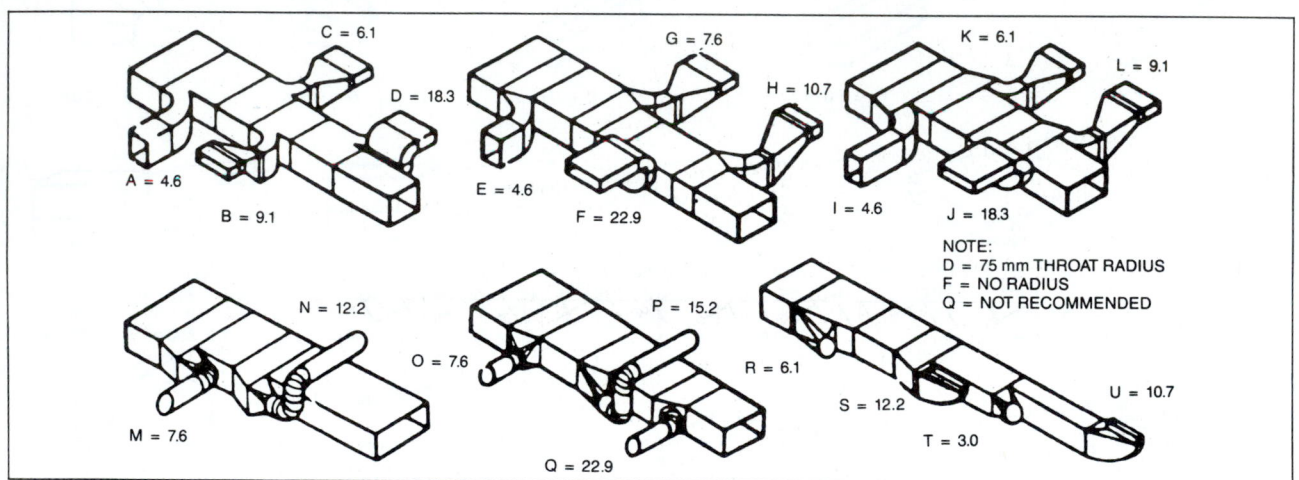

Equivalent Length in Metres of Reducing Trunk Duct Fittings

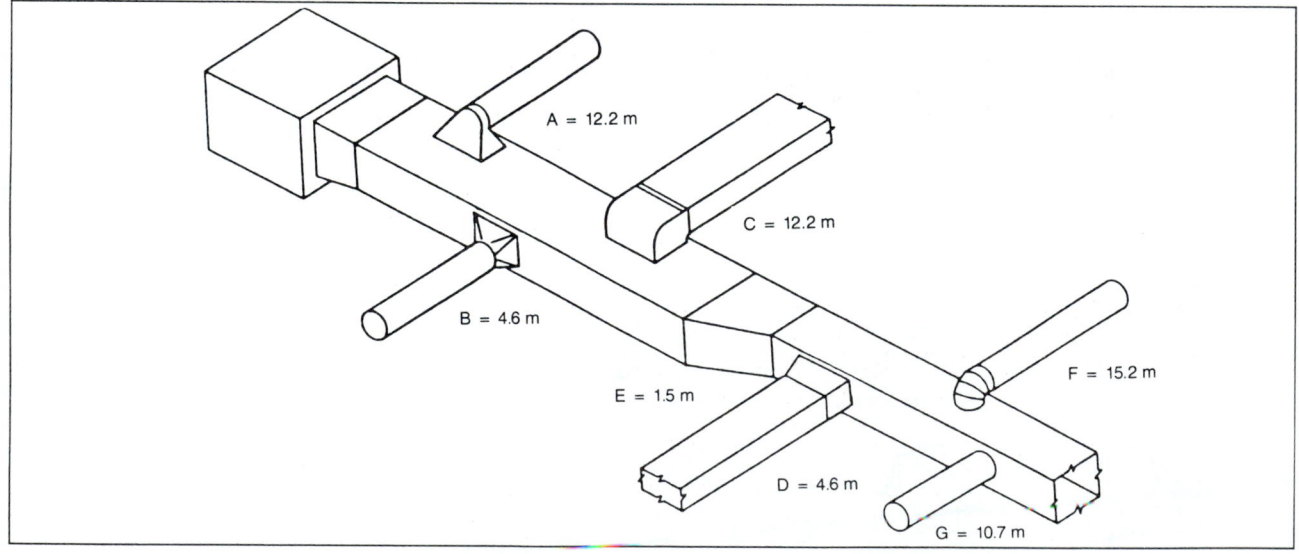

Equivalent Length in Metres of Extended Plenum Fittings

FIGURE 10-34 *(continued)*

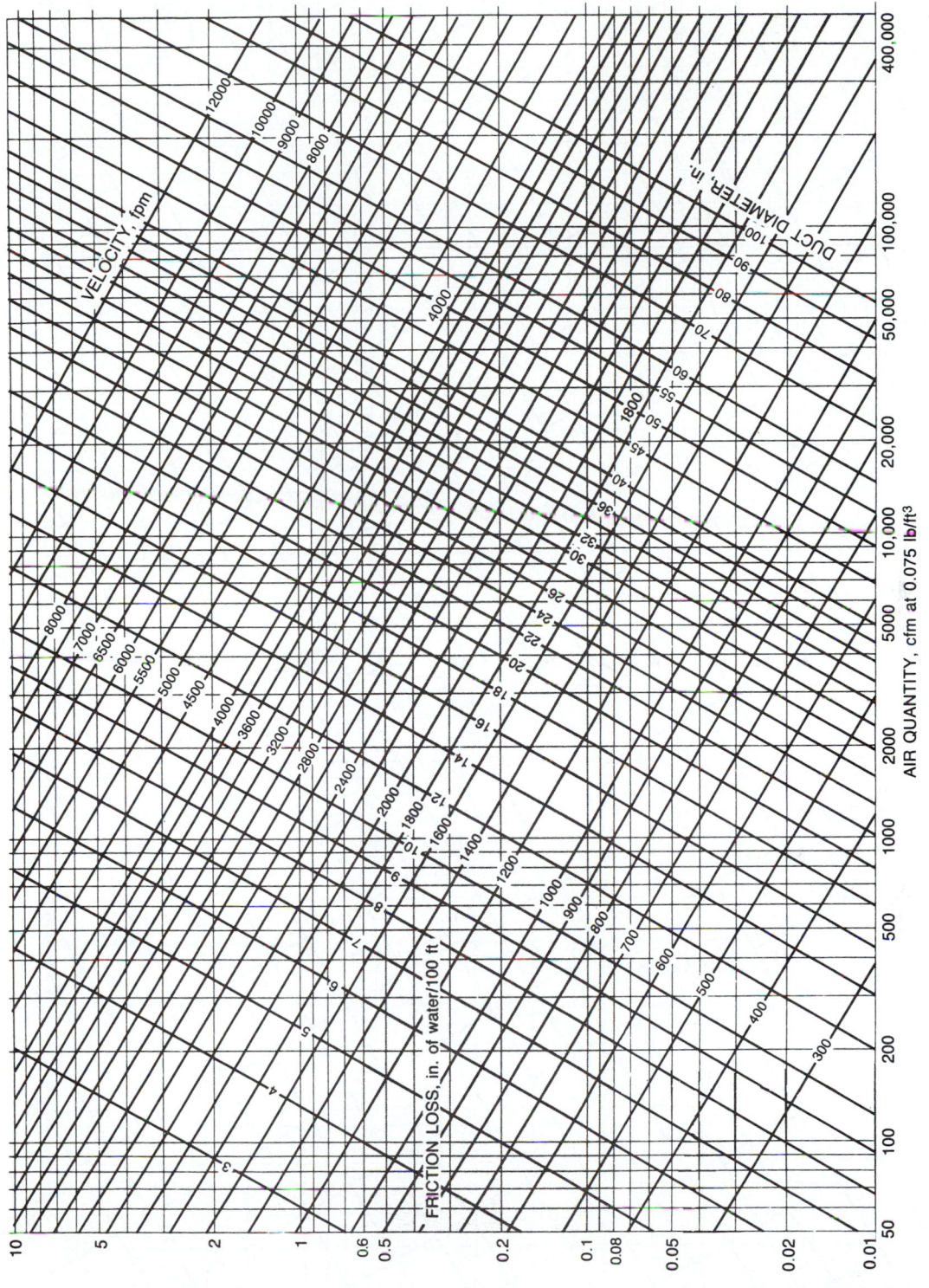

Reprinted with permission from ASHRAE, *Fundamentals Handbook*, 1993.

FIGURE 10-35   Round duct sizes

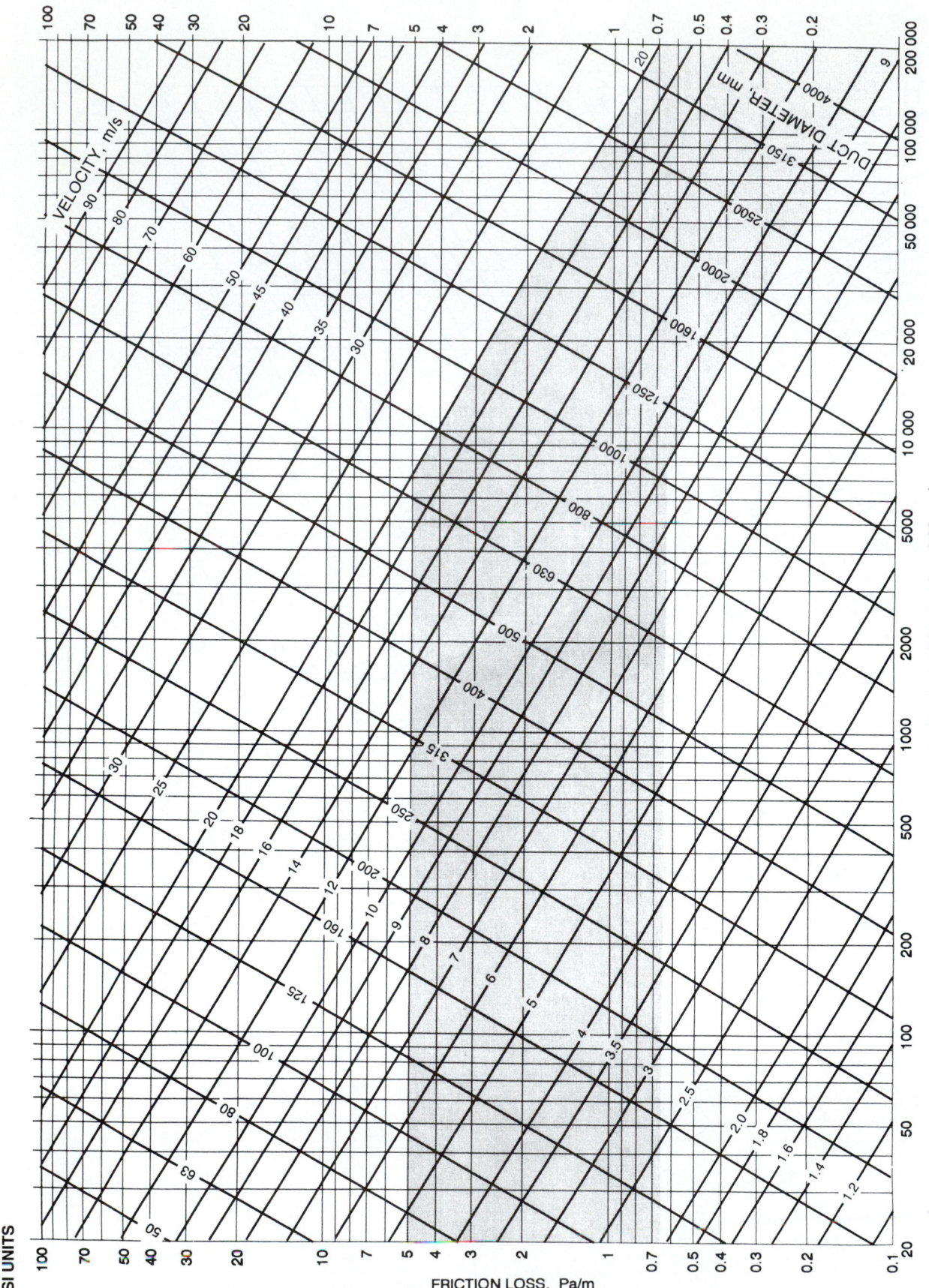

SI UNITS

VELOCITY, m/s

DUCT DIAMETER, mm

AIR QUANTITY, L/s at 1.20 kg/m³ (ε = 0.09 mm)

FRICTION LOSS, Pa/m

FIGURE 10-35   SI Units (continued)

Equivalent Rectangular Duct Dimension

Duct Diameter, in.	Rectangular Size, in.	Aspect Ratio													
		1.00	1.25	1.50	1.75	2.00	2.25	2.50	2.75	3.00	3.50	4.00	5.00	6.00	7.00	8.00
6	Width	—	6													
	Height	—	5													
7	Width	6	8													
	Height	6	6													
8	Width	7	9	9	11											
	Height	7	7	6	6											
9	Width	8	9	11	11	12	14									
	Height	8	7	7	6	6	6									
10	Width	9	10	12	12	14	14	15	17							
	Height	9	8	8	7	7	6	6	6							
11	Width	10	11	12	14	14	16	18	17	18	21					
	Height	10	9	8	8	7	7	7	6	6	6					
12	Width	11	13	14	14	16	16	18	19	21	21	24				
	Height	11	10	9	8	8	7	7	7	7	6	6				
13	Width	12	14	15	16	18	18	20	19	21	25	24	30			
	Height	12	11	10	9	9	8	8	7	7	7	6	6			
14	Width	13	14	17	18	18	20	20	22	24	25	28	30	36		
	Height	13	11	11	10	9	9	8	8	8	7	7	6	6		
15	Width	14	15	17	18	20	20	23	25	24	28	28	35	36	42	
	Height	14	12	11	10	10	9	9	9	8	8	7	7	6	6	
16	Width	15	16	18	19	20	23	23	25	27	28	32	35	42	42	48
	Height	15	13	12	11	10	10	9	9	9	8	8	7	7	6	6
17	Width	16	18	20	21	22	25	25	28	27	32	32	35	42	49	48
	Height	16	14	13	12	11	11	10	10	9	9	8	7	7	7	6
18	Width	16	19	21	23	24	25	28	28	30	32	36	40	42	49	56
	Height	16	15	14	13	12	11	11	10	10	9	9	8	7	7	7
19	Width	17	20	21	23	24	27	28	30	30	35	36	40	48	49	56
	Height	17	16	14	13	12	12	11	11	10	10	9	8	8	7	7
20	Width	18	20	23	25	26	27	30	30	33	35	40	45	48	56	56
	Height	18	16	15	14	13	12	12	11	11	10	10	9	8	8	7
21	Width	19	21	24	26	28	29	30	33	33	39	40	45	54	56	64
	Height	19	17	16	15	14	13	12	12	11	11	10	9	9	8	8
22	Width	20	23	26	26	28	32	33	36	36	39	44	50	54	56	64
	Height	20	18	17	15	14	14	13	13	12	11	11	10	9	8	8
23	Width	21	24	26	28	30	32	35	36	39	42	44	50	54	63	64
	Height	21	19	17	16	15	14	14	13	13	12	11	10	9	9	8
24	Width	22	25	27	30	32	34	35	39	39	42	48	55	60	63	72
	Height	22	20	18	17	16	15	14	14	13	12	12	11	10	9	9
25	Width	23	25	29	30	32	36	38	39	42	46	48	55	60	70	72
	Height	23	20	19	17	16	16	15	14	14	13	12	11	10	10	9
26	Width	24	26	30	32	34	36	38	41	42	46	52	55	66	70	72
	Height	24	21	20	18	17	16	15	15	14	13	13	11	11	10	9
27	Width	25	28	30	33	36	38	40	41	45	49	52	60	66	70	80
	Height	25	22	20	19	18	17	16	15	15	14	13	12	11	10	10
28	Width	26	29	32	35	36	38	43	44	45	49	56	60	66	77	80
	Height	26	23	21	20	18	17	17	16	15	14	14	12	11	11	10
29	Width	27	30	33	35	38	41	43	44	48	53	56	65	72	77	88
	Height	27	24	22	20	19	18	17	16	16	15	14	13	12	11	11
30	Width	27	31	35	37	40	43	45	47	48	53	60	65	72	77	88
	Height	27	25	23	21	20	19	18	17	16	15	15	13	12	11	11
31	Width	28	31	35	39	40	43	45	50	51	56	60	70	78	84	88
	Height	28	25	23	22	20	19	18	18	17	16	15	14	13	12	11
32	Width	29	33	36	39	42	45	48	50	54	56	60	70	78	84	96
	Height	29	26	24	22	21	20	19	18	18	16	15	14	13	12	12
33	Width	30	34	38	40	44	47	50	52	54	60	64	75	78	91	96
	Height	30	27	25	23	22	21	20	19	18	17	16	15	13	13	12
34	Width	31	35	39	42	44	47	50	52	57	60	64	75	84	91	96
	Height	31	28	26	24	22	21	20	19	19	17	16	15	14	13	12
35	Width	32	36	39	42	46	50	53	55	57	63	68	75	84	91	104
	Height	32	29	26	24	23	22	21	20	19	18	17	15	14	13	13
36	Width	33	36	41	44	48	50	53	55	60	63	68	80	90	98	104
	Height	33	29	27	25	24	22	21	20	20	18	17	16	15	14	13
38	Width	35	39	44	47	50	54	58	61	63	67	72	85	96	105	112
	Height	35	31	29	27	25	24	23	22	21	19	18	17	16	15	14

*Shaded area not recommended.

FIGURE 10-36   Rectangular ducts

Equivalent Rectangular Duct Dimension (*Continued*).

Duct Diameter, in.	Rectangular Size, in.	Aspect Ratio														
		1.00	1.25	1.50	1.75	2.00	2.25	2.50	2.75	3.00	3.50	4.00	5.00	6.00	7.00	8.00
40	Width	37	41	45	49	52	56	60	63	66	70	76	90	96	105	120
	Height	37	33	30	28	26	25	24	23	22	20	19	18	16	15	15
42	Width	38	43	48	51	56	59	63	66	69	74	80	90	102	112	120
	Height	38	34	32	29	28	26	25	24	23	21	20	18	17	16	15
44	Width	40	45	50	54	58	61	65	69	72	81	84	95	108	119	128
	Height	40	36	33	31	29	27	26	25	24	23	21	19	18	17	16
46	Width	42	48	53	56	60	65	68	72	75	84	88	100	114	126	136
	Height	42	38	35	32	30	29	27	26	25	24	22	20	19	18	17
48	Width	44	49	54	60	62	68	70	74	78	88	92	105	120	126	136
	Height	44	39	36	34	31	30	28	27	26	25	23	21	20	18	17
50	Width	46	51	57	61	66	70	75	77	81	91	96	110	120	133	144
	Height	46	41	38	35	33	31	30	28	27	26	24	22	20	19	18
52	Width	48	54	59	63	68	72	78	83	84	95	100	115	126	140	152
	Height	48	43	39	36	34	32	31	30	28	27	25	23	21	20	19
54	Width	49	55	62	67	70	77	80	85	90	98	104	120	132	147	160
	Height	49	44	41	38	35	34	32	31	30	28	26	24	22	21	20
56	Width	51	58	63	68	74	79	83	88	93	102	108	125	138	147	160
	Height	51	46	42	39	37	35	33	32	31	29	27	25	23	21	20
58	Width	53	60	66	70	76	81	85	91	96	105	112	130	144	154	168
	Height	53	48	44	40	38	36	34	33	32	30	28	26	24	22	21
60	Width	55	61	68	74	78	83	90	94	99	109	116	130	144	161	
	Height	55	49	45	42	39	37	36	34	33	31	29	26	24	23	
62	Width	57	64	71	75	82	88	93	96	102	112	120	135	150	168	
	Height	57	51	47	43	41	39	37	35	34	32	30	27	25	24	
64	Width	59	65	72	79	84	90	95	99	105	116	124	140	156		
	Height	59	52	48	45	42	40	38	36	35	33	31	28	26		
66	Width	60	68	75	81	86	92	98	105	108	119	128	145	162		
	Height	60	54	50	46	43	41	39	38	36	34	32	29	27		
68	Width	62	70	77	82	90	95	100	107	111	123	132	150	168		
	Height	62	56	51	47	45	42	40	39	37	35	33	30	28		
70	Width	64	71	80	86	92	99	105	110	114	126	136	155			
	Height	64	57	53	49	46	44	42	40	38	36	34	31			
72	Width	66	74	81	88	94	101	108	113	117	130	140	160			
	Height	66	59	54	50	47	45	43	41	39	37	35	32			
74	Width	68	76	84	91	98	104	110	116	123	133	144	165			
	Height	68	61	56	52	49	46	44	42	41	38	36	33			
76	Width	70	78	86	93	100	106	113	118	126	137	148	165			
	Height	70	62	57	53	50	47	45	43	42	39	37	33			
78	Width	71	80	89	95	102	110	115	121	129	140	152				
	Height	71	64	59	54	51	49	46	44	43	40	38				
80	Width	73	83	90	98	104	113	118	124	132	144	156				
	Height	73	66	60	56	52	50	47	45	44	41	39				
82	Width	75	84	93	100	108	115	123	129	135	147	160				
	Height	75	67	62	57	54	51	49	47	45	42	40				
84	Width	77	86	95	103	110	117	125	132	138	151	164				
	Height	77	69	63	59	55	52	50	48	46	43	41				
86	Width	79	88	98	105	112	119	128	135	141	154	168				
	Height	79	70	65	60	56	53	51	49	47	44	42				
88	Width	80	90	99	107	116	124	130	138	144	158					
	Height	80	72	66	61	58	55	52	50	48	45					
90	Width	82	93	102	110	118	126	133	140	147	161					
	Height	82	74	68	63	59	56	53	51	49	46					
92	Width	84	94	104	112	120	128	138	143	150	165					
	Height	84	75	69	64	60	57	55	52	50	47					
94	Width	86	96	107	116	124	131	140	146	153	168					
	Height	86	77	71	66	62	58	56	53	51	48					
96	Width	88	99	108	117	126	135	143	151	159						
	Height	88	79	72	67	63	60	57	55	53						
98	Width	90	100	111	119	128	137	145	154	162						
	Height	90	80	74	68	64	61	58	56	54						
100	Width	91	103	113	123	132	140	148	157	165						
	Height	91	82	75	70	66	62	59	57	55						
102	Width	93	105	116	124	134	142	153	160	168						
	Height	93	84	77	71	67	63	61	58	56						
104	Width	95	106	117	128	136	146	155	162							
	Height	95	85	78	73	68	65	62	59							

'Shaded area not recommended.

FIGURE 10-36   (*continued*)

Equivalent Rectangular Duct Dimension (*Concluded*)

Duct Diameter, in.	Rectangular Size, in.	Aspect Ratio														
		1.00	1.25	1.50	1.75	2.00	2.25	2.50	2.75	3.00	3.50	4.00	5.00	6.00	7.00	8.00
106	Width	97	109	120	130	140	149	158	165							
	Height	97	87	80	74	70	66	63	60							
108	Width	99	110	122	131	142	151	160	168							
	Height	99	88	81	75	71	67	64	61							
110	Width	101	113	125	135	144	153	163								
	Height	101	90	83	77	72	68	65								
112	Width	102	115	126	137	146	158	165								
	Height	102	92	84	78	73	70	66								
114	Width	104	116	129	140	150	160									
	Height	104	93	86	80	75	71									
116	Width	106	119	131	142	152	162									
	Height	106	95	87	81	76	72									
118	Width	108	121	134	144	154	164									
	Height	108	97	89	82	77	73									
120	Width	110	123	135	147	158										
	Height	110	98	90	84	79										

*Shaded area not recommended.

FIGURE 10-36   (*continued*)

Circular Equivalents of Rectangular Duct for Equal Friction and Capacity[a]

Lgth Adj.[b]	Length of One Side of Rectangular Duct (a), mm																			
	100	125	150	175	200	225	250	275	300	350	400	450	500	550	600	650	700	750	800	900
100	109																			
125	122	137																		
150	133	150	164																	
175	143	161	177	191																
200	152	172	189	204	219															
225	161	181	200	216	232	246														
250	169	190	210	228	244	259	273													
275	176	199	220	238	256	272	287	301												
300	183	207	229	248	266	283	299	314	328											
350	195	222	245	267	286	305	322	339	354	383										
400	207	235	260	283	305	325	343	361	378	409	437									
450	217	247	274	299	321	343	363	382	400	433	464	492								
500	227	258	287	313	337	360	381	401	420	455	488	518	547							
550	236	269	299	326	352	375	398	419	439	477	511	543	573	601						
600	245	279	310	339	365	390	414	436	457	496	533	567	598	628	656					
650	253	289	321	351	378	404	429	452	474	515	553	589	622	653	683	711				
700	261	298	331	362	391	418	443	467	490	533	573	610	644	677	708	737	765			
750	268	306	341	373	402	430	457	482	506	550	592	630	666	700	732	763	792	820		
800	275	314	350	383	414	442	470	496	520	567	609	649	687	722	755	787	818	847	875	
900	289	330	367	402	435	465	494	522	548	597	643	686	726	763	799	833	866	897	927	984
1000	301	344	384	420	454	486	517	546	574	626	674	719	762	802	840	876	911	944	976	1037
1100	313	358	399	437	473	506	538	569	598	652	703	751	795	838	878	916	953	988	1022	1086
1200	324	370	413	453	490	525	558	590	620	677	731	780	827	872	914	954	993	1030	1066	1133
1300	334	382	426	468	506	543	577	610	642	701	757	808	857	904	948	990	1031	1069	1107	1177
1400	344	394	439	482	522	559	595	629	662	724	781	835	886	934	980	1024	1066	1107	1146	1220
1500	353	404	452	495	536	575	612	648	681	745	805	860	913	963	1011	1057	1100	1143	1183	1260
1600	362	415	463	508	551	591	629	665	700	766	827	885	939	991	1041	1088	1133	1177	1219	1298
1700	371	425	475	521	564	605	644	682	718	785	849	908	964	1018	1069	1118	1164	1209	1253	1335
1800	379	434	485	533	577	619	660	698	735	804	869	930	988	1043	1096	1146	1195	1241	1286	1371
1900	387	444	496	544	590	663	674	713	751	823	889	952	1012	1068	1122	1174	1224	1271	1318	1405
2000	395	453	506	555	602	646	688	728	767	840	908	973	1034	1092	1147	1200	1252	1301	1348	1438
2100	402	461	516	566	614	659	702	743	782	857	927	993	1055	1115	1172	1226	1279	1329	1378	1470
2200	410	470	525	577	625	671	715	757	797	874	945	1013	1076	1137	1195	1251	1305	1356	1406	1501
2300	417	478	534	587	636	683	728	771	812	890	963	1031	1097	1159	1218	1275	1330	1383	1434	1532
2400	424	486	543	597	647	695	740	784	826	905	980	1050	1116	1180	1241	1299	1355	1409	1461	1561
2500	430	494	552	606	658	706	753	797	840	920	996	1068	1136	1200	1262	1322	1379	1434	1488	1589
2600	437	501	560	616	668	717	764	810	853	935	1012	1085	1154	1220	1283	1344	1402	1459	1513	1617
2700	443	509	569	625	678	728	776	822	866	950	1028	1102	1173	1240	1304	1366	1425	1483	1538	1644
2800	450	516	577	634	688	738	787	834	879	964	1043	1119	1190	1259	1324	1387	1447	1506	1562	1670
2900	456	523	585	643	697	749	798	845	891	977	1058	1135	1208	1277	1344	1408	1469	1529	1586	1696

Lgth Adj.[b]	Length of One Side of Rectangular Duct (a), mm																			
	1000	1100	1200	1300	1400	1500	1600	1700	1800	1900	2000	2100	2200	2300	2400	2500	2600	2700	2800	2900
1000	1093																			
1100	1146	1202																		
1200	1196	1256	1312																	
1300	1244	1306	1365	1421																
1400	1289	1354	1416	1475	1530															
1500	1332	1400	1464	1526	1584	1640														
1600	1373	1444	1511	1574	1635	1693	1749													
1700	1413	1486	1555	1621	1684	1745	1803	1858												
1800	1451	1527	1598	1667	1732	1794	1854	1912	1968											
1900	1488	1566	1640	1710	1778	1842	1904	1964	2021	2077										
2000	1523	1604	1680	1753	1822	1889	1952	2014	2073	2131	2186									
2100	1558	1640	1719	1793	1865	1933	1999	2063	2124	2183	2240	2296								
2200	1591	1676	1756	1833	1906	1977	2044	2110	2173	2233	2292	2350	2405							
2300	1623	1710	1793	1871	1947	2019	2088	2155	2220	2283	2343	2402	2459	2514						
2400	1655	1744	1828	1909	1986	2060	2131	2200	2266	2330	2393	2453	2511	2568	2624					
2500	1685	1776	1862	1945	2024	2100	2173	2243	2311	2377	2441	2502	2562	2621	2678	2733				
2600	1715	1808	1896	1980	2061	2139	2213	2285	2355	2422	2487	2551	2612	2672	2730	2787	2842			
2700	1744	1839	1929	2015	2097	2177	2253	2327	2398	2466	2533	2598	2661	2722	2782	2840	2896	2952		
2800	1772	1869	1961	2048	2133	2214	2292	2367	2439	2510	2578	2644	2708	2771	2832	2891	2949	3006	3061	
2900	1800	1898	1992	2081	2167	2250	2329	2406	2480	2552	2621	2689	2755	2819	2881	2941	3001	3058	3115	3170

[a] Table based on $D_e = 1.30(ab)^{0.625}/(a+b)^{0.25}$.
[b] Length of adjacent side of rectangular duct (b), mm.

FIGURE 10-36 (continued)

CFM	TYPE A	TYPE B	TYPE C
1000	0.10	0.15	0.20
1500	.12	.17	.22
2000	.13	.18	.24
2500	.14	.20	.27
3000	.16	.22	.29
3500	.17	.24	.31
4000	.18	.26	.33
4500	0.20	0.28	0.36
5000	.21	.29	.38
5500	.22	.31	.40
6000	.24	.33	.42
6500	.25	.34	.45
7000	.27	.36	.47
7500	.28	.38	.49
8000	0.29	0.39	0.51
8500	.30	.41	.53
9000	.32	.43	.56
9500	.34	.45	.58
10,000	.34	.47	.60
10,500	.37	.48	.60
11,000	.39	.50	.60
11,500	0.40	0.50	0.60
12,000	.40	.50	.60
12,500	.40	.50	.60
13,000	.40	.50	.60
14,000	.40	.50	.60

CFM = Maximum cfm in any one duct.
Type A = Systems for quiet operation in residences, churches, concert halls, broadcasting studios, funeral homes, etc.
Type B = Systems for schools, theaters, public buildings, etc.
Type C = Systems for industrial buildings.

FIGURE 10-37   Suggested bonnet and return pressures

QUESTIONS

10-1. What are the advantages of a forced air system as compared to hot water?

10-2. What types of materials are most commonly used for forced air ducts?

10-3. What can be done to be certain that the system will be as quiet as possible?

10-4. What is the purpose of putting dampers and vanes in the ductwork?

10-5. When is insulation around the ductwork recommended?

10-6. When heating is the predominant function of the forced air system, what are the recommended locations for supply and return?

10-7. What types of furnaces are available for use in forced air systems, and where might each type be located?

10-8. How may humidity be introduced into the forced air system?

Design Exercises

10-9. Design a forced hot air system for the residence in Appendix D based on the heat loss calculations from Exercise 8-15.

CHAPTER 11

Electric Heat Design

11-1 TYPES OF SYSTEMS

The advantages of an electric heating system include low installation cost, individual room control, quiet operation, and cleanliness, and when cables or panels are used, there are no exposed heating units. The primary disadvantage of electric heat is its high cost of operation in almost all areas. A specialist is required to give honest figures for comparing electric heat costs with those of other types of fuels. Electrically operated boilers (furnaces) are not discussed in this chapter since they are not different systems but rather a different forced air unit with electricity as a fuel.

Most electric ratings are given in watts and Btuh. There are 3,413 Btuh for every 1,000 watts (1 kilowatt = 1,000 watts), and often the ratings will be given or noted as MBH (thousand Btuh) and kW (kilowatts), such as 6.8 MBH and 2 kW.

Overall Systems

Baseboard
Electric baseboard units have a heating element enclosed in a metal case. This system offers individual room control, but furniture arrangement and draperies must not interfere with the operation of the units by blocking the natural flow of air. This system is economical for a builder to install and is especially popular in low-cost housing and housing built for speculation (to try to sell), although the cost of operation is generally higher than most other fuels.

Resistance Cable
With this system, electric heating cable is stapled to the drywall in a grid pattern and covered with plaster or gypsum board. Individual thermostats control the heat in each room. However, since the cable is usually installed on the ceiling, the disadvantages of heat rising and cold feet must be considered.

Drawings are not usually done for this type of heating system; instead, the amount of heat required is noted for each space. The cable system allows complete freedom in furniture and drapery placement.

Panels
The prefabricated ceiling and wall panels used in this system have the heating wire sandwiched in rubber and constructed from a variety of materials. They are only 1 in. (25 mm) thick and may be plastered, painted, or wallpapered over. The panels come in standard sizes (usually 2 ft × 4 ft or 0.6 m × 1.2 m) and some panels can be cut to fit. They are often used in hung suspension ceilings. A panel system has the same basic advantages and disadvantages as a resistance cable system, including flexibility of furniture and drapery arrangement.

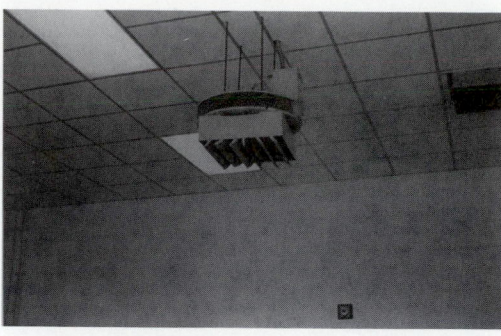

FIGURE 11-1   Ceiling-mounted unit

Unit Heaters

There are a variety of electric unit heaters available which may be used to supplement other heat sources in a space or to completely heat a room. In residences, such a unit heater is quite often installed in the bathroom, often to supplement other heat sources, when a higher temperature is necessary to feel comfortable when washing and after bathing.

Unit heaters are also commonly installed in spaces where the heat is only used periodically, such as a basement, workshop warehouse (Fig. 11-1) or garage work area.

So that the warm air will be quickly spread throughout the area to be heated, unit heaters are equipped with a fan. It is preferable if the fan switch is the type which will not start until after the unit comes on and the air has warmed to a preset temperature; in this way, cold air is not pushed around the room.

Units are generally available for use in walls (recessed into the wall), and ceiling recessed units are also available.

System Combinations

Quite often, in order to provide the best results, the design may incorporate more than one type of system. For example, a building may be predominantly heated with baseboard units, but in spaces where there is little free wall space, such as a kitchen, it may be desirable to use a unit heater in the ceiling or the wall under the cabinets. Another possibility for the kitchen might be to use resistance cables or panels in the ceiling. Bathrooms may be heated with unit heaters to provide added comfort.

Codes and Installation

While a hot water heating system is installed by the heating contractor, an electrical heating system is installed by an electrical contractor. ASHRAE publishes guides which discuss the use of the units for proper heating conditions and the installation, but the *National Electric Code* (NEC) also governs the installation of the resistance cable. The designer must have both the NEC code and ASHRAE available for reference.

11-2   BASEBOARD SYSTEM DESIGN

A baseboard electric system transfers heat to the space primarily by convection. It consists of baseboard units which may be mounted on the wall or recessed into the wall. A wall mounted unit is shown in Fig. 11-2. In selecting a baseboard unit, the designer must consider the direction in which the air will be discharged from the baseboard unit (Fig. 11-3).

FIGURE 11-2   Wall-mounted unit

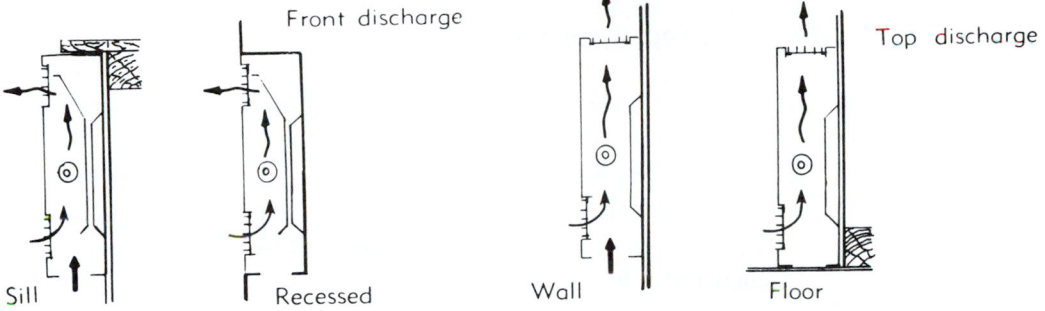

FIGURE 11-3   Baseboard air discharge

Length	Watts	Btuh
36″ (3′-0″) (0.9 m)	500	1,707
52″ (4′-4″) (1.3 m)	750	2,560
68″ (5′-8″) (1.7 m)	1,000	3,413
100″ (8′-4″) (2.5 m)	1,500	5,120

FIGURE 11-4   Baseboard ratings

Baseboard units are rated in watts and Btuh. These ratings, which may vary for different manufacturers, are given *for the length specified* and not per lineal foot. For example, in Fig. 11-4, the very first listing specifies a 36-in. (3-ft or 0.9 m) length and a rating of 500 watts and 1,707 Btuh for the 3-ft (0.9 m) length. The manufacturer often has several models, sizes, and ratings available, and the ratings will probably vary with other manufacturers. In Figs. 11-5 and 11-6, two separate series are listed; both have the same outside dimensions and the *B* series has a rating about 100% higher for each length than the *A* series. In addition, many manufacturers make larger units, with higher ratings, which are normally used in commercial, industrial, and institutional buildings. A 3-ft (0.9 m) length of this style may range from 750 to 2,250 watts.

Length	Watts	Btuh
36″ (3′-0″) (0.9 m)	750	2,560
48″ (4′-0″) (1.2 m)	1,000	3,413
60″ (5′-0″) (1.8 m)	1,250	4,269
72″ (6′-0″) (1.8 m)	1,500	5,120
96″ (8′-0″) (2.4 m)	2,000	6,830

A Series

FIGURE 11-5   High output baseboard ratings

Length	Watts	Btuh
36″ (3′-0″) (0.9 m)	1,500	5,120
48″ (4′-0″) (1.2 m)	2,000	6,830
60″ (5′-0″) (1.5 m)	2,500	8,538
72″ (6′-0″) (1.8 m)	3,000	10,245
96″ (8′-0″) (2.4 m)	4,000	13,660

B Series

FIGURE 11-6   High output baseboard ratings

In designing an electric baseboard system, it may be necessary to use the larger units in some areas in order to provide the Btuh required.

Step-by-Step Approach

1. Using the calculations from Fig. 8-36, do a heat loss calculation on each individual space, and tabulate them as shown (Fig. 11-7).

2. List the lineal feet of exterior wall that are available for baseboard convector units for each separate space, and tabulate them as shown (Fig. 11-8).

3. Using the manufacturer's ratings (Figs. 11-4 and 11-5 are typical), determine the length and corresponding rating that will be used to provide heat to each space.

> Some walls, such as those in the living room of this design, will require two or more lengths of baseboard. All that is important is that the Btuh of heat loss calculated for each room is put back into that room.
>
> In some rooms, such as a kitchen, where there is little exterior wall space, it may be necessary to put a unit on an interior wall, or a larger unit which has a higher capacity may be used.

4. Tabulate the baseboard convector units selected for each room; list their lengths, wattage ratings, and Btuh ratings.

> Based on the heat loss calculations, the units have been selected and tabulated as shown (Fig. 11-8).
>
> *Note:* Since exterior wall space for radiant baseboard units is likely to be limited in some areas, it may be desirable to use other units or devices to supplement or take the place of the radiant baseboards. The kitchen and bathrooms quite often have unit heaters recessed in the wall. Kitchen heat may also be supplemented by a "kickspace" heater which is placed in the kickspace under the kitchen cabinets.

11-3 RESISTANCE CABLE SYSTEM DESIGN

The amount of heat given off by the cables will vary with the amount of heating cable used in the installation.

The electric cable used for ceiling installations comes in rolls; it is stapled to the ceiling and then covered with gypsum or plasterboard in accordance with the manufacturer's specifications and the code requirements. Basically, the cable must not be installed within 6 in. (150 mm) of any wall, within 8 in. (200 mm) of the edge of any junction box or outlet, or within 2 in. (50 mm) of any recessed lighting fixtures.

Cable assemblies are usually rated at 2.75 watts per lineal foot, (0.84 m) with generally available ratings from 400- to 5,000-watt lengths in 200-watt increments, but the man-

Room	Heat loss (Btuh)	Exterior wall (l.f.)	Watts (Btuh ÷ 3.413)	Radiation unit selected
BEDROOM 1	4,326			
BEDROOM 2	2,772			
LIVING	4,664			
KITCHEN	4,663			
DINING	3,056			
ENTRY & HALL	4,061			
BEDROOM 3	3,483			
BATH	1,092			
INTERIOR BATH	203			

SI UNITS

Room	Heat loss (Watts)	Exterior wall (m)		Radiation unit selected
BEDROOM 1	1,324			
BEDROOM 2	807			
LIVING	2,690			
KITCHEN	1,205			
DINING	898			
ENTRY & HALL	1,287			
BEDROOM 3	956			
BATH	322			
INTERIOR BATH	57			

Note: Watts vary from English design solution due to rounding off throughout the design.   FIGURE 11-7   Room heat loss

Room	Heat loss (Btuh)	Exterior wall (l.f.)	Watts (Btuh ÷ 3.413)	Radiation unit selected
BEDROOM 1	4,326	30.0	1,268	2-750W, 4'-4"
BEDROOM 2	2,772	11.0	882	1-1,000W, 5'-8"
LIVING	9,327	23.3	2,968	3-1000W, 5'-8"
KITCHEN	4,160	—	1,324	2-750W, 4'-4"
DINING	3,056	14.0	973	2-500W, 3'
ENTRY & HALL	4,061	—	1,292	2-750W, 4'-4"
BEDROOM 3	3,483	14.0	1,094	1-1000W,5'-8"
BATH	1,092	—	348	1-500W, 3'
INTERIOR BATH	203	—	65	USE UNIT HEATER

SI UNITS

Room	Heat loss (Watts)	Exterior wall (m)	Radiation unit selected
BEDROOM 1	1,324	9.1	2-750W
BEDROOM 2	807	3.3	1-1,000W
LIVING	2,690	7.1	2-1000W, 1-750W
KITCHEN	1,205	—	1-500W, 1-750W
DINING	898	4.2	2-500W
ENTRY & HALL	1,287	—	2-750W
BEDROOM 3	956	4.2	2-500W
BATH	322	—	1-500W
INTERIOR BATH	57	—	USE UNIT HEATER

FIGURE 11-8   Select and tabulate baseboard units

Note: Watts vary from English design solution due to rounding off throughout the design.

Btuh	Watts	Length	
		Ft.	m
1365	400	145	44.2
2047	600	218	66.5
2730	800	292	89.1
3413	1000	362	110.4
4095	1200	436	133.0
5461	1600	582	177.5
6143	1800	654	199.5
6826	2000	728	222.0
7509	2200	800	244.0
8533	2500	910	277.6
10,239	3000	1090	332.5
11,287	3600	1310	399.6
15,700	4600	1672	510.0

FIGURE 11-9   Typical cable ratings

ufacturer's specifications should be checked to determine what is available. A typical list of available lengths, watts, and Btuh from one manufacturer is shown in Fig. 11-9. The cables have insulated coverings which are resistant to medium temperatures, water absorption, and the effects of aging and chemical reactions (concrete, plaster, etc.); a polyvinyl chloride covering with a nylon jacket is most commonly used. Each separate cable has an individual thermostat, providing flexible control throughout the building.

A typical cable installation in a plastered ceiling is shown in Fig. 11-10, and installation details are shown in Fig. 11-11. The space between the rows of heating cable is generally limited to a minimum of 1.5 in. (37 mm), and some manufacturers recommend a 2-in. (50 mm) minimum spacing when drywall construction is used. Another limitation on the spacing of the cable is a 2.5-in. (62 mm) clearance required between cables under each joist (Fig. 11-12), and a review of the layout in Fig. 11-10 shows that the cable is installed parallel (in the same direction) as the joists.

To be certain the required amount of Btuh is obtained, it is sometimes desirable to specify the maximum spacing of heat cable allowed in a room. This maximum spacing may be determined by using the formula:

FIGURE 11-10   Electric heating panel for wet plaster ceilings

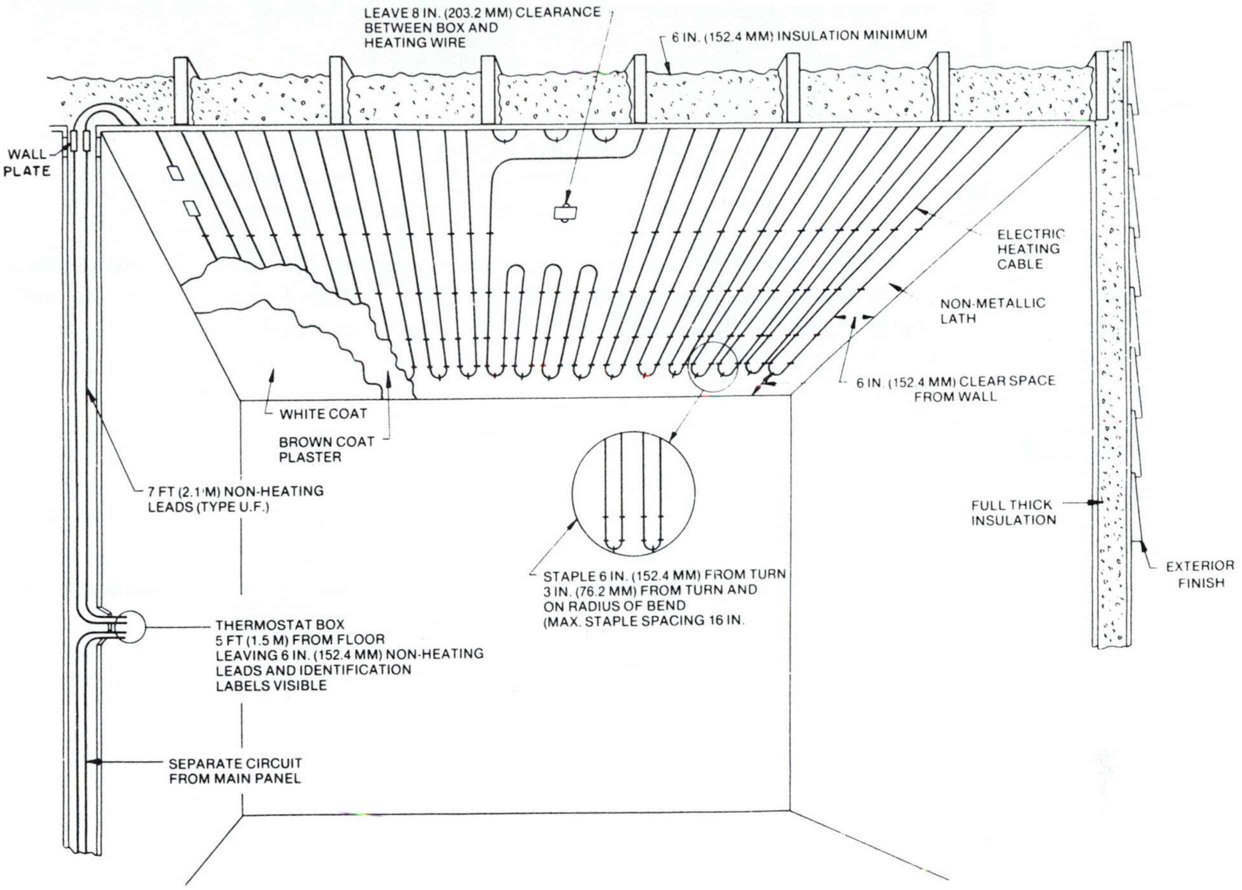

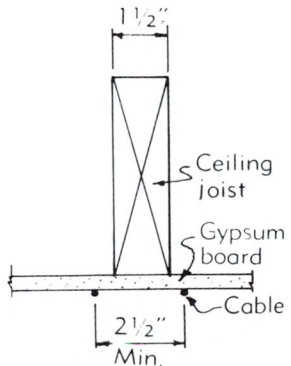

Reprinted with permission from ASHRAE, *HVAC Handbook*, 1993

FIGURE 11-11   Electric heating cable installation

FIGURE 11-12   Cable detail

$$s = 12 \, (An/C)$$

where

s = cable spacing (in.)
An = net available area for heat cables (sq ft)
C = length of a cable required to deliver required Btuh (ft)
12 = constant (used to change ft to in.)

SI The maximum spacing may be determined by using the formula:

$$s = 1000 \, (An/C)$$

s = cable spacing (mm)
An = net available area (sq m)
c = length of cable required to deliver required watts (m)
1000 = constant (used to change m to mm)

The net available area for heating cables is equal to the total ceiling area minus any area in which cable cannot be placed (due to borders, ceiling obstructions, cabinets, and any similar items). While a small lighting fixture may be neglected, if the ceiling has several, their area should be deducted.

EXAMPLE   *Given:* Assume that bedroom 1 illustrated in Fig. 11-13 is about 18 ft 3 in. by 12 ft 3 in. and requires 4,326 Btuh. From the available cable lengths given in Fig. 11-9, a 1,600-watt, 5,461-Btuh, 582-ft cable is selected. Assume one ceiling fixture which is neglected in the calculation and a 6-in. border required between the cable and the intersection of wall.
   Problem: Determine the net area available for heating cables and the maximum cable spacing.

An = 18 ft-3 in $\times$ 12 ft-3 in. $-$ (6 in. around the perimeter of room) (Fig. 11-13)
17.25 ft $\times$ 11.25 ft = 194 sq ft
s = 12 (194 sq ft/582 ft) = 4 in. maximum spacing

Bedroom 1 (Fig. 11-13) requires a maximum cable spacing of 4 in.

SI *Given:* Assume that bedroom 1 illustrated in Fig. 11-13 is about 5.57 m by 3.74 m and requires 1324 watts. From the available cable lengths in Fig. 11-19, a 1600 watt, 178 m cable is selected. Assume one ceiling fixture which is neglected in the calculation and a .15 m border required between the cable and the intersection of the wall.

An = 5.57 m $\times$ 3.74 m $-$ (.15 m around the perimeter of the room)
5.27 m $\times$ 3.44 m = 18.13 sq m
s = 1000 (18.13 sq m/177.5 m) = 102 mm maximum spacing

Bedroom 1 requires a maximum cable spacing of 102 mm.

This is the procedure used to determine cable ceiling heat requirements for a building.

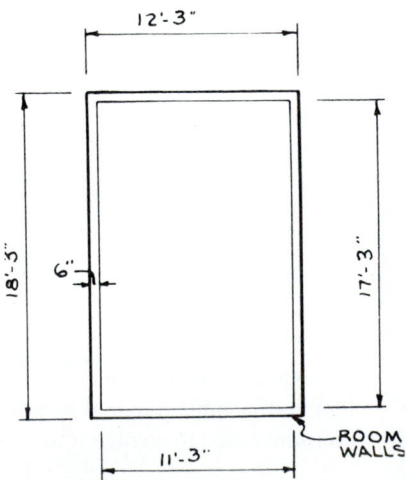

FIGURE 11-13   Cable area calculations

Step-by-Step Approach

1. Determine the heat loss of each individual room and tabulate them.

   The heat loss has already been calculated (Sec. 8-12) for the residence being designed and they are tabulated by room as shown in Fig. 11-14.

2. Select the cable required to provide the heat required for each room. Do not use a cable which will provide less heat than calculated.

   From Fig. 11-9, select the required cable for each room and tabulate each cable rating (Btuh, watts, length) as shown in Fig. 11-15.

3. Calculate the net ceiling area for each room.

   Calculate the net ceiling area for each room based on a 6-in. border. In the kitchen, deduct for the wall cabinets and the fluorescent light. Tabulate the net ceiling area for each room as shown in Fig. 11-16.

Room	Heat loss (Btuh)	Cable selected (Btuh)	Cable length (ft.)	Ceiling space available (ft.)	Ceiling area (s.f.)	Maximum spacing (in.)
BEDROOM 1	4,326					
BEDROOM 2	2,772					
LIVING	9,327					
KITCHEN	4,160					
DINING	3,056					
ENTRY & HALL	4,061					
BEDROOM 3	3,483					
BATHROOM	1,092					
INTERIOR BATH	203					

SI UNITS

Room	Heat loss	Cable selected (Watts)	Cable length (m)	Ceiling space available (sq m)	Ceiling area (sq m)	Maximum spacing (mm)
BEDROOM 1	1,324					
BEDROOM 2	807					
LIVING	2,690					
KITCHEN	1,205					
DINING	898					
ENTRY & HALL	1,287					
BEDROOM 3	956					
BATHROOM	322					
INTERIOR BATH	57					

FIGURE 11-14   Room heat loss

Room	Heat loss (Btuh)	Cable selected (Btuh)	Cable length (ft.)	Ceiling space available (ft.)	Ceiling area (s.f.)	Maximum spacing (in.)
BEDROOM 1	4,326	5,461	582			
BEDROOM 2	2,772	2,730	292			
LIVING	9,327	10,239	1,090			
KITCHEN	4,160	4,095	436			
DINING	3,056	3,413	362			
ENTRY & HALL	4,061	4,095	436			
BEDROOM 3	3,483	4,095	436			
BATHROOM	1,092	1,365	145			
INTERIOR BATH	203	USE ELECTRIC UNIT HEATER				

SI UNITS

Room	Heat loss	Cable selected (Watts)	Cable length (m)	Ceiling space available (sq m)	Ceiling area (sq m)	Maximum spacing (mm)
BEDROOM 1	1,324	1,600	177.5			
BEDROOM 2	807	1,000	110.4			
LIVING	2,690	3,000	332.5			
KITCHEN	1,205	1,200	133.0			
DINING	898	1,000	110.4			
ENTRY & HALL	1,287	1,600	177.5			
BEDROOM 3	956	1,000	110.4			
BATHROOM	322	400	44.2			
INTERIOR BATH	57	USE UNIT HEATER				

FIGURE 11-15   Select and tabulate cable lengths

Room	Heat loss (Btuh)	Cable selected (Btuh)	Cable length (ft.)	Ceiling space available (ft.)	Ceiling area (s.f.)	Maximum spacing (in.)
BEDROOM 1	4,326	5,461	582	17.25 × 11.25	194	
BEDROOM 2	2,772	2,730	292	14.3 × 11	157	
LIVING	9,327	10,239	1,090	17 × 27.6	470	
KITCHEN	4,160	4,095	436	15 × 12	160	
DINING	3,056	3,413	362	13 × 11.5	150	
ENTRY & HALL	4,061	4,095	436	7 × 11.5 + 3 × 30	170	
BEDROOM 3	3,483	4,095	436	13 × 11.5	150	
BATHROOM	1,092	1,365	145	5 × 9	45	
INTERIOR BATH	203	USE ELECTRIC UNIT HEATER				

SI UNITS

Room	Heat loss	Cable selected (Watts)	Cable length (m)	Ceiling space available (sq m)	Ceiling area (sq m)	Maximum spacing (mm)
BEDROOM 1	1,324	1,600	177.5	5.27 × 3.44	18.13	
BEDROOM 2	807	1,000	110.4	4.4 × 3.4	14.96	
LIVING	2,690	3,000	332.5	5.2 × 8.4	43.68	
KITCHEN	1,205	1,200	133.0	4.6 × 3.7	17.02	
DINING	898	1,000	110.4	4.0 × 3.5	14.0	
ENTRY & HALL	1,287	1,600	177.5	2.1 × 3.5 + 0.9 × 9.1	15.54	
BEDROOM 3	956	1,000	110.4	4.0 × 3.5	14.0	
BATHROOM	322	400	44.2	1.5 × 2.7	4.05	
INTERIOR BATH	57	USE UNIT HEATER				

FIGURE 11-16   Calculate net ceiling areas

Room	Heat loss (Btuh)	Cable selected (Btuh)	Cable length (ft.)	Ceiling space available (ft.)	Ceiling area (s.f.)	Maximum spacing (in.)
BEDROOM 1	4,326	5,461	582	17.25 × 11.25	194	4.0
BEDROOM 2	2,772	2,730	292	14.3 × 11	157	6.4
LIVING	9,327	10,239	1,090	17 × 27.6	470	5.1
KITCHEN	4,160	4,095	436	15 × 12	160	4.4
DINING	3,056	3,413	362	13 × 11.5	150	4.9
ENTRY & HALL	4,061	4,095	436	7.57 × 11.5 + 3 × 30	170	4.6
BEDROOM 3	3,483	4,095	436	13 × 11.5	150	4.1
BATHROOM	1,092	1,365	145	5 × 9	45	3.7
INTERIOR BATH	203	USE ELECTRIC UNIT HEATER				

SI UNITS

Room	Heat loss	Cable selected (Watts)	Cable length (m)	Ceiling space available (sq m)	Ceiling area (sq m)	Maximum spacing (mm)
BEDROOM 1	1,324	1,600	177.5	5.27 × 3.44	18.13	102
BEDROOM 2	807	1,000	110.4	4.4 × 3.4	14.96	135
LIVING	2,690	3,000	332.5	5.2 × 8.4	43.68	131
KITCHEN	1,205	1,200	133.0	4.6 × 3.7	17.02	128
DINING	898	1,000	110.4	4.0 × 3.5	14.0	126
ENTRY & HALL	1,287	1,600	177.5	2.1 × 3.5 + 0.9 × 9.1	15.54	87
BEDROOM 3	956	1,000	110.4	4.0 × 3.5	14.0	126
BATHROOM	322	400	44.2	1.5 × 2.7	4.05	91
INTERIOR BATH	57	USE UNIT HEATER				

FIGURE 11-17   Calculate maximum cable spacing

4. Calculate the maximum cable spacing which can be used in each space.

Using the formula $s = 12 (An/C)$, calculate the cable spacing for each room and tabulate the information as shown in Fig. 11-17.

This information is often given on the drawings in tabulated form, similar to that shown in Fig. 11-17. The location of each thermostat must be shown on the drawing and may be put on the general construction (architectural) drawings; but often the information is put on the electrical drawings since the electrician will locate and install the thermostat and tie the cable circuit into the power panel which serves it.

11-4   RADIANT PANEL SYSTEM DESIGN

The heating rates for radiant ceiling panels are generally given in watts per panel, Btuh per panel, or both. As shown in Fig. 11-18, the ratings for a 2-ft × 4-ft panel may vary from about 500 to 750 watts per panel, depending on the panel selected and the manufacturer. Since 1 kW = 3,413 Btuh, 500 watts = 1,707 Btuh and 750 watts = 2,560 Btuh. It is im-

Watts	Btuh	Size
500	1707	2'-0"x4'-0"
750	2560	2'-0"x4'-0"
560	1911	2'-0"x4'-0"
700	3019	2'-0"x5'-0"
500	1707	2'-0"x3'-0"
750	2560	2'-0"x3'-0"
1000	3413	2'-0"x3'-0"

FIGURE 11-18   Radiant ceiling panel ratings

portant that the designer get accurate ratings by checking the engineering specifications for the type of panel which will actually be used on the project.

Step-by-Step Approach

To design ceiling panels for the residence for which the heat loss was calculated in Sec. 8-13:

1. The first step in the design is to tabulate the Btuh for each space, as shown in Fig. 11-19.
2. List the square feet of ceiling area available for use in each space (Fig. 11-19).

   Any space which does not have sufficient ceiling area for the panels must have supplemental heat provided by adding wall panels or electric baseboard or wall unit heaters. Another solution would be to use another type of system, such as baseboard or unit heaters, and not put any ceiling panels in the space.

Room	Heat loss (Btuh)	Net ceiling area (s.f.)	Watts (Btuh ÷ 3413)	No. of panels
BEDROOM 1	4,326	194		
BEDROOM 2	2,772	157		
LIVING	9,327	470		
KITCHEN	4,160	160		
DINING	3,056	150		
ENTRY & HALL	4,061	170		
BEDROOM 3	3,483	150		
BATH	1,092	45		
INTERIOR BATH	203	14		

SI UNITS

Room	Heat loss (Watts)	Net ceiling area (sq m)	No. of panels
BEDROOM 1	1,324	59	
BEDROOM 2	807	48	
LIVING	2,690	124	
KITCHEN	1,194	49	
DINING	898	46	
ENTRY & HALL	1,287	52	
BEDROOM 3	956	46	
BATH	322	14	UNIT HEATER
INTERIOR BATH	57	4	UNIT HEATER

FIGURE 11-19   Tabulate Btuh

Room	Heat loss (Btuh)	Net ceiling area (s.f.)	Watts (Btuh ÷ 3413)	No. of panels
BEDROOM 1	4,326	194	1,268	
BEDROOM 2	2,772	157	882	
LIVING	9,327	470	2,968	
KITCHEN	4,160	160	1,324	
DINING	3,056	150	973	
ENTRY & HALL	4,061	170	1,292	
BEDROOM 3	3,483	150	1,094	
BATH	1,092	45	348	
INTERIOR BATH	203	14	65	

FIGURE 11-20   Tabulate watts

It may be decided to put unit heaters in certain spaces where it is desirable to have a fan circulate the air through the space. Bathrooms are a typical example of where a unit heater might be used for this reason.

3. Select the heating capacity for the space (Fig. 11-20).
4. Determine the number of ceiling panels required for each space (Fig. 11-21).

Room	Heat loss (Btuh)	Net ceiling area (s.f.)	Watts (Btuh ÷ 3413)	No. of panels
BEDROOM 1	4,326	194	1,268	1-500W, 1-750W
BEDROOM 2	2,772	157	882	1-1000W
LIVING	9,327	470	2,968	1-750W, 2-1000W
KITCHEN	4,160	160	1,324	1-500W, 1-750W
DINING	3,056	150	973	2-500W
ENTRY & HALL	4,061	170	1,292	1-500W, 1-750W
BEDROOM 3	3,483	150	1,094	2-500W
BATH	1,092	45	348	UNIT HEATER
INTERIOR BATH	203	14	65	UNIT HEATER

SI UNITS

Room	Heat loss (Watts)	Net ceiling area (sq m)	No. of panels
BEDROOM 1	1,324	59	2-750W
BEDROOM 2	807	48	1-1000W
LIVING	2,690	124	1-750W, 2-1000W
KITCHEN	1,194	49	1-500W, 1-750W
DINING	898	46	2-500W
ENTRY & HALL	1,287	52	2-750W
BEDROOM 3	956	46	2-500W
BATH	322	14	UNIT HEATER
INTERIOR BATH	57	4	UNIT HEATER

FIGURE 11-21   Tabulate number of panels

5. Determine the location of the thermostat, keeping in mind the location guides given in Sec. 9-3.

The biggest disadvantage of ceiling panels is that there is a tendency for the heat to stay high in the space (keep in mind that heat rises), requiring higher temperature settings to achieve comfort since feet will feel cold. Also, this type of heating will not heat "hidden" air spaces, such as under a table or desk; these will feel quite cool.

But from the builder's standpoint, it is an acceptable system because of its low initial cost, individual room control, and "invisibility" (there are no portions of the system exposed except the thermostat).

Ceiling panels may be installed so that the heat is effectively put into the space at the place where most of it is lost—at the windows. In general, best results are obtained with this type of unit when the rooms are generally small (such as in rental apartments) and when the glass area is limited.

11-5   SYSTEM INSTALLATION

Baseboard units that are recessed into the wall will have to be framed into the wall. Otherwise, they simply need the conductors that bring them the power and a thermostat control.

Unit heaters may be quite large, and may also be recessed into a wall. If hung from the ceiling, the structure will need to be designed to carry the extra weight. Wiring to bring power to them and a thermostat must be considered.

Resistance cable may be installed in concrete slabs and on walls but is most commonly placed in ceilings (Fig. 11-11). During construction, care must be taken not to damage the cable.

Prefabricated wall and ceiling panels do not require special system installation except for the wiring and thermostat.

QUESTIONS

11-1. Discuss the advantages and disadvantages of using electric heating systems.

11-2. What is the primary reason that electric heating systems are used?

11-3. What are the most common types of electric heat installations used?

11-4. Why is it important to have well-insulated construction when electric heat will be used?

Design Exercises

11-5. Design an electric baseboard heating system for the residence in Appendix D, based on the heat loss calculations in Exercise 8-15.

11-6. Design a resistance cable heating system for the residence in Appendix D, based on the heat loss calculations in Exercise 8-15.

11-7. Design an electric baseboard heating system for a typical apartment (not the top floor) in Appendix A, based on the heat loss calculations in Exercise 8-17.

11-8. Design a resistance cable heating system for the residence in Appendix D, based on the heat loss calculations in Exercise 8-17.

CHAPTER 12

Heat Gain

12-1 GENERAL

Heat gain deals with the amount of heat which a space will accumulate during warm weather. To adequately cool the air inside the space, it is necessary to estimate the amount of heat which will build up in the space. Many of the heat gain factors are basically the same as those considered in heat loss (Chapter 8):

1. Transmission (through walls, ceilings, and floors).
2. Infiltration (through windows and doors).

While the terms *transmission* and *infiltration* are the same terms used for heat loss, and the general principles the same, there are some important differences in the time of day considered and the method of calculation. And for *total* heat gain, several other factors must also be considered:

1. Solar radiation.
2. Heat produced in the space.
3. Latent heat.

Note: Some tables in this chapter were not available in SI units. To convert the values refer to the conversion factors given in Appendix E.

12-2 SENSIBLE AND LATENT HEAT

The heat gain of the building is a combination of *sensible* and *latent* heat, which together give the total heat gain, or total cooling load, of the building.

Sensible heat is heat which a substance absorbs, and while its temperature goes up, the substance *does not change its state*. For example, if some water (at, say, 50°F or 10°C) is placed in the sunshine, the water will absorb the heat from the sun and warm up, but it will still be water.

Latent heat is heat which is absorbed or given off by a substance *when changing its state*. As a substance changes from a solid to a liquid or from a liquid to a gas, it *absorbs* heat, and as a substance changes from a gas to a liquid or from a liquid to a solid, it *gives off* heat.

12-3 TRANSMISSION

While the greatest heat loss will probably occur at night, the greatest heat gain will occur during a sunny day. Heat gain calculations make use a heat transfer multiplier (HTM) (Fig. 12-6) which takes into account:

1. The temperature differential (between the inside of a building and the outside).

2. The type of construction (considering insulation value, mass of the type of construction, and thermal time lag) and the ability of the construction to hold heat.

Some typical heat transfer multipliers are listed in Fig. 12-6. To determine the heat gain by transmission, simply multiply the area of construction by the HTM for the type of construction.

When the HTM for a particular material is not listed in the table, it can be found by multiplying the U value by the equivalent temperature differential (ETD) (Fig. 12-11). Summer design conditions are listed in Fig. 8-53.

12-4 INFILTRATION

Heat gain *infiltration* is calculated in the same manner as heat loss infiltration—by multiplying the lineal feet of crackage by the infiltration factor. However, air leakage into structures is much smaller in the summer than in the winter. This is because winter air leakage is caused by a combination of wind direction and velocity which, when combined with the inside-outside temperature difference, creates a "chimney" effect of heat loss. This type of loss does not occur in the summer, and most localities have lower wind velocities in the summer.

Since there is much less air leakage in the summer, when calculating heat gain, ASHRAE suggests use of the air change or ventilation approach. The air change method is based on the assumption that the average air leakage will occur at a rate of one-half air change per hour. In this case, the heat gain is calculated in terms of Btuh per square foot (or watts per sq m) of gross exposed wall area, using the Btuh per square foot values given in Fig. 12-7.

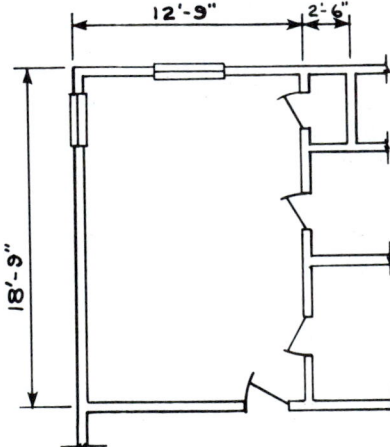

FIGURE 12-1   Room plan

EXAMPLE   Using bedroom 1 for the residence in Appendix C (Fig. 12-1), it would have an outside exposed area of 34 ft × 8 ft or 272 sq ft. Based on the Btuh per square foot values given in Fig. 12-7, the value for a 95°F outside design temperature is 1.5 Btuh per sq ft. Based on this, the heat gain for the room is 272 sq ft × 1.5 Btuh per sq ft = 408 Btuh due to infiltration.

When mechanical ventilation is used, the amount of heat gain is based on the volume of air change in cfm. From Fig. 8-32 assume average construction, over 2100 sq. ft. and 0.4 air change per hour, a room 12 ft-3 in. × 18 ft-3 in. × 8 ft high with 1,789 cu ft would ventilate 1,789 cu ft ÷ 60 min = 29.8 cfm. The heat gain is based on the Btuh of sensible heat per cfm of ventilating air shown in Fig. 12-7 and is selected from the table according to the outside design temperature used. With a 95°F outside design temperature, the heat gain would be 22 Btuh per cfm. So the heat gain of this room would be 29.8 cfm × 0.4 air changes × 22 Btuh per cfm = 262 Btuh.

SI

Using bedroom 1 for the residence in Appendix C (Fig. 12-1), it would have an outside exposed area of 10.37 m × 2.4 m, or 24.9 sq m. Based on the heat gain values given in Fig. 12-7, the value for a 35°C outside temperature is 4.7 W/sq m. Based on this, the heat gain for the room is 24.9 sq m × 4.7 W/m = 118 W due to infiltration.

When mechanical ventilation is used, the amount of heat gain is based on the volume of air change in L/s. Use the air change factor in Fig. 8-52 (0.4 changes per hour). The volume of bedroom 1 is 3.7 m × 5.6 m × 2.4 m (49.7 sq m).

The air change room volume L/s is determined by multiplying the number of air changes per hour times the room volume times 1000/3600 to convert it to L/s. In this design it is 0.4 × 49.7 × 1000/3600 = 5.5 L/s. The heat gain of this room would be 5.5 L/s × 13.7 W per L/s = 75 W.

12-5 OCCUPANCY LOADS

The amount of heat given off by people (Fig. 12-8) and appliances must also be considered in any cooling load considerations. Since the number of occupants and appliances and the amount of appliance use which will occur will vary and cannot be predicted, the cooling load they will create must be estimated.

For residences, the occupant heat release is 225 Btuh (67 W) per occupant. To estimate the number of occupants as accurately as possible, it is necessary to consider how much the owner entertains large groups of people and whether such entertaining occurs during the heat of the day or in the cooler evening. Many times this information is not available, and the designer will simply have to assume that there will be two occupants for each bedroom in the house.

The heat gain from occupants is then distributed equally among the living spaces of the house since during the time when the maximum heat gain will occur, the occupants will be using the living spaces and not the sleeping areas.

Appliance values are generally limited to the kitchen areas of the residence. The suggested heat gain load is 1,200 Btuh (350 W) for the appliances. The intermittent use of appliances, along with the use of kitchen ventilating fans, makes it difficult to calculate exactly what the heat gain value should be. Appliances which are major sources of latent and sensible heat gain loads must be vented to the exterior. It would be very difficult to provide sufficient cooling to overcome the heat gain which an unvented clothes dryer would produce.

12-6 SOLAR RADIATION

Solar radiation consists primarily of heat gain through the windows. The window heat gain for both *absorbed solar energy* and *transmitted heat gain* are calculated together by use of a heat gain multiplier. This multiplier (Fig. 12-9) takes into consideration glass orientation (the direction the window faces), type of glass, type of shading, and outside design temperature, and the result is expressed in Btuh per square foot.

Glass which is protected by permanent shading, such as a wide roof overhang, is generally calculated as glass facing north. The overhang probably only protects part of the glass area, and in making the heat gain calculations, it is necessary to separate the total area of glass which will be shaded from that portion which won't be shaded. The amount of shaded area varies with the direction the window faces, the length of the overhang, and the geographic latitude (in degrees, Fig. 8-53) of the building. The table in Fig. 12-10 lists *shade line factors*, which give the distance below the bottom of the fascia which the shadow will cover for every foot (meter) of overhang. The shade line factors shown are the average for the 5 hr of maximum solar intensity which would occur. Windows which face northeast and northwest cannot be effectively protected with roof overhangs, and they are considered to be in full sunshine with no shade.

EXAMPLE   *Given:* Using the wall section in Fig. 12-2 (from Appendix C), *note* that the top of the window is located at the same elevation (level) as the bottom of the fascia. Assume this is a 4-ft-wide (1.2 m), 4-ft-high (1.2 m), regular double-glass window, facing SE, with draperies, and that the building is located at about latitude 35°.

Problem: Determine the area in shade and the area in sunlight, and then the amount of heat gain through the window. Assume a 95°F design temperature with medium temperature range and regular double glass.

> Shade line factor (from Fig. 12-10)   = 1.4
> Roof overhang (from the wall section in Fig. 12-2) = 2 ft
> Shade distance below overhang = 2 ft × 1.4   = 2.8 ft

Since the top of the window lines up with the bottom of the fascia, the top 2.8 ft of the window are considered in shade and calculated as facing N.

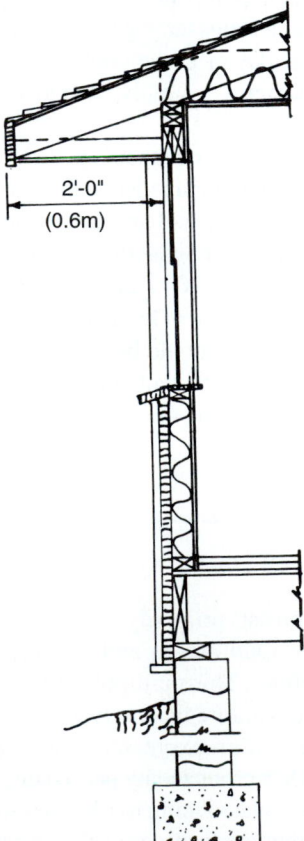

2'-0"
(0.6m)

FIGURE 12-2   Wall section

Area of window in shade:

$$2.8 \text{ ft} \times 4.0 \text{ ft} = 11.2 \text{ sq ft}$$

Area of window in sun:

$$(4.0 \text{ ft} - 2.8 \text{ ft}) \times 4.0 \text{ ft} = 1.2 \text{ ft} \times 4.0 \text{ ft} = 4.8 \text{ sq ft}$$

Calculate the heat gain based on Fig. 12-9. Using an outdoor design temperature of 95°F, the heat gain through shaded regular double glass is:

$$11.2 \text{ sq ft (in shade)} \times 17 \text{ Btuh/sq ft (N, draperies)} = 191 \text{ Btuh}$$

The heat gain through unshaded glass is:

$$4.8 \text{ sq ft (in sun)} \times 40 \text{ Btuh/sq ft (SE, draperies)} = 192 \text{ Btuh}$$

Therefore, the total heat gain through the window is:

$$191 \text{ Btuh} + 192 \text{ Btuh} = 383 \text{ Btuh}$$

 Problem: Determine the area in shade and the area in sunlight, and then the amount of heat gain through the window. Assume a 35°C design temperature with medium temperature range.

Shade line factor (from Fig. 12-10) = 1.4
Roof overhang (from the wall section in Fig. 12-2) = 0.6 m
Shade distance below overhang = 0.6m × 1.4 = 0.8m

Since the top of the window lines up with the bottom of the fascia, the top 0.8m of the window are considered in shade and calculated as facing N.

Area of window in shade:

$$0.8\text{m} \times 1.2\text{m} = 0.96 \text{ sq m (use 1.0)}$$

Area of window in sun:

$$(1.2\text{m} - 0.8) \times 1.2 = 0.4\text{m} \times 1.2 = 0.48 \text{ (0.5 sq m)}$$

Calculate the heat gain based on Fig. 12-9. Using an outdoor design temperature of 35°C, the heat gain through regular double glass is:

$$1.0 \text{ sq m (in shade)} \times 60 \text{ W/sq m (N, draperies)} = 60 \text{ watts}$$

The heat gain through unshaded glass is:

$$0.5 \text{ sq m (in sun)} \times 123 \text{ (SE, draperies)} = 62 \text{ W}$$

Therefore, the total heat gain through the window is:

$$60 + 62 = 122 \text{ w}$$

EXAMPLE   *Given:* Using the wall section in Fig. 12-3, *note* that the top of the window is 8 in. below the bottom of the fascia (overhang). Assume that this is a 3 ft-6 in.-wide, 4 ft-6 in.-high single-glass window, facing E, with draperies, and that the building is located at about latitude 35° where the outside design temperature is 95°F.

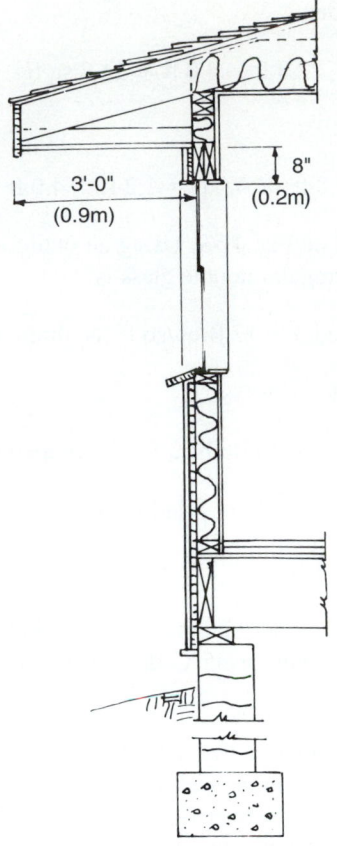

FIGURE 12-3   Wall section   FIGURE 12-4   Shade line

Problem: Determine the total heat gain through the window.

Shade line factor (from Fig. 12-10)   = 0.8
Roof overhang (from the wall section in Fig. 12-4) = 3 ft
Shade distance below overhang = 3 ft × 0.8   = 2.4 ft

Since the window begins 8 in. (0.67 ft) below the overhang, this distance must be subtracted from the shade distance to determine how much of the window is covered by the shade (Fig. 12-4).

Area of window in shade:

$$(2.4 \text{ ft} - 0.67 \text{ ft}) \times 3.5 \text{ ft} = 1.73 \text{ ft} \times 3.5 \text{ ft} = 6.1 \text{ sq ft}$$

Area of window in sun:

$$(4.5 \text{ ft} - 1.73 \text{ ft}) \times 3.5 \text{ ft} = 2.77 \text{ ft} \times 3.5 \text{ ft} = 9.7 \text{ sq ft}$$

Calculate the heat gain based on Fig. 12-9. Using an outside design temperature of 95°F, the heat gain through unshaded glass is:

$$6.1 \text{ sq ft (in shade)} \times 23 \text{ Btuh/sq ft (N, draperies)} = 140.3 \text{ (141) Btuh}$$

The heat gain through shaded glass is:

$$9.7 \text{ sq ft (in sun)} \times 56 \text{ Btuh/sq ft (E, draperies)} = 543.2 \text{ (544) Btuh}$$

Therefore, the total heat gain through the window is:

$$141 \text{ Btuh} + 544 \text{ Btuh} = 685 \text{ Btuh}$$

SI *Given:* Using the wall section in Fig. 12-3, *note* that the top of the window is 8 in. below the bottom of the fascia (overhang). Assume that this is a 1.1m-wide, 1.4 m-high *single glass* window, facing E, with draperies, and that the building is located at about latitude 35° where the outside design temperature is 35°C.
Problem: Determine the total heat gain through the window.

Shade line factor (from Fig. 12-10)	= 0.8
Roof overhang (from the wall section in Fig. 12-14)	= 0.9 m
Shade distance below overhang = 3 ft × 0.8	= 0.7 m

Since the window begins 0.2 m below the overhang, this distance must be subtracted from the shade distance to determine how much of the window is covered by the shade (Fig. 12-4).
　　Area of window in shade:

$$(0.7\text{m} - 0.2\text{m}) \times 1.1\text{m} = 0.5\text{m} \times 1.1\text{m} = 0.6 \text{ sq m}$$

Area of window in sun:

$$(1.4\text{m} - 0.5) \times 1.1\text{m} = 0.9\text{m} \times 1.1\text{m} = 1.0 \text{ sq m}$$

Calculate the heat gain based on Fig. 12-9. Using an outside design temperature of 95°F, the heat gain through unshaded single glass is:

$$0.6 \text{ sq m (in shade)} \times 73 \text{ W/sq m (N, draperies)} = 44 \text{ W}$$

The heat gain through shaded single glass is:

$$1.0 \text{ (in sun)} \times 158 \text{ W/sq m (E, draperies)} = 158 \text{ W}$$

Therefore, the total heat gain through the window is:

$$44 \text{ W} + 158 \text{ W} = 202 \text{ W}$$

12-7　LATENT HEAT GAIN

Heat gain in buildings also occurs due to latent heat. In residential cooling, the latent heat is generally estimated to be about 30 percent of the total sensible heat calculated. While this percentage is an approximation, there are numerous moisture sources in a residence. Since it would be very difficult to precisely evaluate the latent heat which will occur, ASHRAE has determined that the 30 percent results in a design system which will be comfortable.

12-8　TOTAL COOLING LOAD

The total heat gain of the building is the sum of the sensible and latent heat gains. This total cooling load in a residence is calculated as 1.3 times the total sensible heat load.

The sensible heat gain is the total of:

1. The heat transmitted through the floors, ceilings, and walls.
2. The occupants' body heat.
3. The appliance heat, including lights.
4. The heat gain through glass, by both absorption and solar radiation.
5. Infiltration.

12-9 HEAT GAIN CALCULATIONS

Heat gain calculations are similar to heat loss calculations (Chapter 8). The heat gain for the residence in Appendix C (the same one for which the heat loss was calculated) is calculated in Fig. 12-5. One major difference that must be considered is that during the summer roof temperatures outside could be as high as 140–150°F (60-66°C) when the building has a dark roof, making its ΔT value very high compared to that of the walls. This must be taken into account when selecting the HTM value.

	SI UNITS
HTM (From Fig. 12-6)	*HTM* (From Fig. 12-6)
Medium 95°F	Medium 35°C
Walls 1.9	Walls 6.0
Cold walls (medium 90°F) 1.4	Cold walls (medium 32°C) 4.4
Ceilings 1.8 (dark)	Ceilings (dark) 5.7
Floors, no heat gain	Floors, no heat gain
Infiltration 1.5 Btuh/sq. ft gross exposed wall	Infiltration 4.7 W/sq m gross exposed wall
Windows	*Windows*
Determine shade line	Determine shade line
From Appendix C—location, 35° lat.	From Appendix C—location, 35° lat.
From Appendix C—orientation	From Appendix C—orientation
Front of house faces northeast, all in sun	Front of house faces northeast, all in sun
Rear of house faces southwest	Rear of house faces southwest
Roof overhang 2'0″	Roof overhang
Shade line factor (Fig. 12-10)	Shade line factor (Fig. 12-10)
Southwest 1.4	Southwest 1.4
Draperies—all except kitchen	Draperies—all except kitchen
Window HTM (Fig. 12-9) Regular double glass	Window HTM (Fig. 12-9)
Northeast 32 (with draperies)	Northeast 101 (with draperies)
51 (without draperies)	186 (without draperies)
Southwest 17 (shade & draperies)	Southwest 60 (shade & draperies)
40 (sun & draperies)	123 (sun & draperies)
Southeast 40 (sun & draperies, no overhang)	Southeast 123 (sun & draperies, no overhang)
Occupants	*Occupants*
2 @ 225 Btuh in kitchen	2 @ 67 W in kitchen
4 @ 225 Btuh in living-family	4 @ 67 W in living-family
Appliances 1200 W in kitchen	Appliances 350 W in kitchen

FIGURE 12-5   Heat gain

Room	Portion considered	Dimensions	Height	Gross area or l.f.	Openings	Opening area	Net area or l.f.	HTM U value	ETD	Heat gain (Btuh)	Room heat gain (Btuh)
Bedroom 1 inc. 2 Closets and Bath	Exterior wall	12.7 + 18.7 + 2.5	8	272	4.0 × 4.0	28	244	1.9		464	
	—	—		—	3.0 × 4.0		—	—		—	
	Ceiling	12.7 × 18.7 +		323	—	—	323	1.8		582	
	Sm. Closet	2.5 × 4.0 +		—	—	—	—	—		—	
	Clos & Bath	5.0 × 14.7		—	—	—	—	—		—	
	Glass, S.W. Sun.	4.0 × 1.2	—	4.8	—	—	4.8	40		192	
	Shade	4.0 × 2.8	—	11.2	—	—	1.2	17		191	
	S.E. Sun	3.0 × 4.0	—	12	—	—	12	32		384	
	—		—	—	—	—	—	—		—	
	Infiltration	—		272	—	—	272	1.5		408	2221
Bedroom 2 inc.1 Closet	Ext. Wall	11.5 + 2.5	8	112	3.0 × 4.0	12	100	1.9		190	
	Ceiling	11.5 × 18.81		226	—	—	226	1.8		407	
		2.5 × 4.0	—	—	—	—	—	—		—	
	Infil	11.5 + 2.5	8	112	—	—	112	1.5		168	
	Glass, S.W. Sun	3.0 × 1.2	—	3.6	—	—	3.6	40		115	
	Shade	3.0 × 2.8	—	8.4	—	—	8.4	17		143	1023
Living (4 People)	Ext. Wall	32.25 + 7.0	8	314	2-6.0 × 6.7	80	234	1.9		445	
	Cold Wall	18.8 − 7.0	8	95	—	—	95	1.4		133	
	Ceiling	18.75 × 32.25	—	605	—	—	605	1.8		1089	
	Infil.	—	—	314	—	—	314	1.5		471	
	Glass, S.W. Sun	12 × 3.9	—	46.8	—	—	46.8	40		1872	
	Shade	12 × 2.8	—	33.6	—	—	33.6	17		571	
	4 People	—	—	—	—	—	—	—		900	5481
Kitchen (2 People)	Ext. Wall	15.8 + 7.0	8	183	4.0 × 3.0	12	171	1.9		291	
	Cold Wall	10.3	8	82	3.0 × 6.8	21	61	1.4		86	
	Ceiling	15.8 × 17.3	—	274	—	—	274	1.8		493	
	Infil	—	—	183	—	—	183	1.5		275	
	Glass, N.E. Sun	—		12			12	51		612	
	—						—				
	Door	—		21			21	9.4		197	
	Appliances	—								1200	
	2 Person	—								450	3604
Dining	Ext.Wall	14.5	8	116	6.0 × 5.0	30	86	1.9		164	
	Ceiling	14.5 × 13.3	—	193			193	1.8		375	
	Infil	—	—	116			116	1.5		174	
	Glass, N.E. Sun	—	—	30			30	32		960	1673
Entry (and Hall)	Ext.Wall	7.5 + 2.5	8	80	D1 & Sidelts 6.0 × 6.8	40	40	1.9		76	
	—	—	—	—		—	—	—		—	
	Ceiling	4.0 × 39.01		289	—	—	289	1.8		520	
	—	13.3 × (7.5 + 2.5)		—	—	—	—	—		—	
	Infil			80	—	—	80	1.5		120	
	Glass, N.E.			20	—	—	20	32		640	
	Door			20	—	—	20	12		240	1596
Bedroom 3	Ext. Wall	14.5 + 6.8 +	8	225	4.0 × 4.0	16	209	1.9		397	
	—	6.8	—	—	—	—	—	—		—	
	Ceiling	14.5 × 13.3		239	—	—	239	1.8		430	
	—	6.8 × 6.8		—	—	—	—	—		—	
	Infil			225			225	1.5		338	
	Glass, N.E.			16			16	32		512	1677
Bath	Ext. Wall	10.5	8	84	2.0 × 2.0	4	80	1.9		152	
	Ceiling	10.5 × 6.8	—	72			72	1.8		130	
	Infil			84			84	1.5		126	
	Glass, S.W.			4			4	32		128	536

CALCULATED SENSIBLE HEAT GAIN 17,811 BTUH
TOTAL HEAT GAIN = SENSIBLE HEAT GAIN × 1.3 (LATENT HEAT)
17,811 × 1.3 = 23,155 BTUH

BEDROOM 1   = 2221 × 1.3 = 2887
BEDROOM 2   = 1023 × 1.3 = 1330
LIVING   = 5481 × 1.3 = 7125
KITCHEN   = 3604 × 1.3 = 4685
DINING   = 1673 × 1.3 = 2175
ENTRY (AND HALL) = 1596 × 1.3 = 2075
BEDROOM 3   = 1677 × 1.3 = 2180
BATH   =   536 × 1.3 =   697

23,150 (differs from above due to rounding off)

FIGURE 12-5   Heat gain for building

Room	Portion considered	Dimensions	Height	Gross area or l.f.	Openings	Opening area	Net area or c.f.	HTM U value	ETD	Heat gain (W)	Room heat gain (W)
Bedroom 1	Exterior wall	3.9 + 5.7 +	2.4	25	1.2 × 1.2 +	2.6	22.4	6.0		134	
inc. 2	—	0.8	—	—	0.92 × 1.2	—	—	—		—	
Closets	Ceiling	3.9 × 5.7 +	—	25.35	—	—	25.35	5.7		145	
and Bath	Sm. closet	0.8 × 1.2 +	—	—	—	—	—	—		—	
	Clos & Bath	1.2 × 1.8	—	—	—	—	—	—		—	
	Glass, S.W. Sun.	1.2 × 0.37	—	0.45	—	—	0.45	123		55	
	Shade	1.2 × 0.85	—	1.04	—	—	1.04	60		63	
	S.E. Sun	0.9 × 1.2	—	1.12	—	—	1.12	123		138	
	—	—	—	—	—	—	—	—		—	
	Infiltration	—	—	25	—	—	25	4.7		118	653
Bedroom 2	Ext.wall	3.5 + 0.8	2.4	10.3	0.92 × 1.2	1.1	9.2	6.0		55	
inc.1 Closet	Ceiling	3.5 × 5.7 +	—	20.9	—	—	20.9	5.7		119	
		0.8 × 1.2	—	—	—	—	—	—		—	
	Infil			10.3	—	—	10.3	4.7		48	
	Glass, S.W. Sun	0.9 × 0.37	—	0.33	—	—	0.33	123		41	
	Shade	0.9 × 0.87	—	0.78	—	—	0.78	60		47	310
Living	Ext.wall	9.8 + 2.1	2.4	28.6	2-1.8 × 2	7.2	21.4	6.0		128	
(4 People)	Cold wall	3.9	2.4	9.4	—	—	9.4	4.4		41	
	Ceiling	9.8 × 5.7	—	55.9	—	—	55.9	5.7		319	
	Infil.	—	—	28.6	—	—	28.6	4.7		134	
	Glass, S.W. Sun	3.7 × 1.2	—	4.5	—	—	4.5	123		554	
	Shade	3.7 × 0.9	—	3.3	—	—	3.3	60		198	
	4 people	—	—	—	—	—	—	—		134	1508
Kitchen	Ext.wall	4.8 + 2.1	2.4	16.6	1.22 × 0.92	1.1	15.5	6.0		93	
(2 people)	Cold wall	3.1	2.4	7.4	0.92 × 2.1	1.9	5.5	4.4		24	
	Ceiling	4.8 × 5.3	—	25.4	—	—	25.4	5.7		145	
	Infil	—	—	16.6	—	—	16.6	4.7		78	
	Glass, N.E. Sun	—		1.1			1.1	186		205	
	—	—					—	—		—	
	Door	—					1.9	30		57	
	Appliances	—						—		350	
	2 person	—						—		134	1086
Dining	Ext. wall	4.4	2.4	10.6	1.83 × 1.53	2.8	7.8	6.0		47	
	Ceiling	4.4 × 4.1	—	18			18	5.7		103	
	Infil	—	—	10.6			10.6	4.7		50	
	Glass, N.E. Sun	—	—	2.8			2.8	101		283	483
Entry	Ext. wall	2.3 + 0.76	2.4	7.3	D1 & Sidelts	3.6	3.7	6.0		22	
(and hall)	—	—		—		—	—	—		—	
	Ceiling	2.3 × 0.76 +	—	30.7	—	—	30.7	5.7		175	
	—	4.1 × 0.92 + 11.9	—	—	—	—	—	—		—	
	Infil			7.3	—	—	7.3	4.7		34	
	Glass, N.E.			1.8	—	—	1.8	101		182	
	Door			1.8	—	—	1.8	38		68	481
Bedroom 3	Ext. wall	4.4 + 2.1 + 2.1	2.4	20.6	1.2 × 1.2	1.4	19.2	6.0		115	
	—			—		—	—	—		—	
	Ceiling	4.4 × 4.1 +	—	22.4		—	22.4	5.7		128	
	—	2.1 × 2.1	—	—		—	—	—		—	
	Infil			20.6			20.6	4.7		97	
	Glass, N.E.	0.85 × 2	—	1.7			1.7	101		172	512
Bath	Ext. wall	3.2	2.4	7.7			7.7	6.0		46	
	Ceiling	3.2 × 2.1	—	6.7		—	6.7	5.7		38	
	Infil			7.7	—	—	7.7	4.7		36	
	Glass, S.W.	0.61 × 0.61		0.4			0.4	123		49	169

CALCULATED SENSIBLE HEAT GAIN
TOTAL HEAT GAIN = SENSIBLE HEAT GAIN × 1.3 (LATENT HEAT)
5202 × 1.3 = 6763 Watts

BEDROOM 1	=	653 × 1.3 =	849
BEDROOM 2	=	310 × 1.3 =	403
LIVING	=	1508 × 1.3 =	1961
KITCHEN	=	1086 × 1.3 =	1412
DINING	=	483 × 1.3 =	628
ENTRY (AND HALL)	=	481 × 1.3 =	625
BEDROOM 3	=	512 × 1.3 =	666
BATH	=	169 × 1.3 =	220

6764 (differs from above due to rounding off)

FIGURE 12-5   (continued)

HEAT GAIN FACTORS FOR ROOFS, CEILINGS, FLOORS AND INFILTRATION
Btuh per Square Foot

	90		95			100		105		110
Outdoor Design Dry-Bulb Temperature	Low	Med.	Low	Med.	High	Med.	High	Med.	High	High
WALLS AND DOORS										
1. Frame and Veneer-on-Frame										
(a) Wood sheathing or 1/2" insulating sheathing	6.0	4.8	7.2	6.0	4.8	7.5	6.0	8.7	7.5	8.7
(b) 25/32" insulating sheathing or one reflective air space	4.3	3.5	5.1	4.5	3.5	5.4	4.5	6.3	5.4	6.3
(c) Same as (a) or (b) plus either 1" or 2" insulation or 2 reflective air spaces	2.9	2.4	3.6	3.1	2.4	3.7	3.1	4.4	3.7	4.4
(d) Same as (a) or (b) plus either more than 2" insulation, or 3 reflective air spaces	1.8	1.5	2.2	1.9	1.5	2.3	1.9	2.7	2.3	2.7
2. Masonry Walls, 8" Block or Brick										
(a) Plastered or plain	7.2	5.4	9.7	7.9	5.4	10.4	7.9	12.5	10.4	12.5
(b) Furred, no insulation	4.6	3.4	6.0	4.9	3.4	6.3	4.9	7.9	6.3	7.9
(c) Furred, with less than 1" insulation, or one reflective air space	3.1	2.3	4.3	3.3	2.3	4.3	3.3	5.4	4.3	5.4
(d) Furred, with 1" to 2" insulation, or two reflective air spaces	2.1	1.6	2.8	2.3	1.6	3.0	2.3	3.7	3.0	3.7
(e) Furred, with more than 2" insulation, or three reflective air spaces	1.4	1.0	1.8	1.5	1.0	1.9	1.5	2.4	1.9	2.4
3. Partitions										
(a) Frame, finished one side only, no insulation	8.5	6.0	11.4	9.1	6.0	12.0	9.1	15.0	12.0	15.0
(b) Frame, finished both sides, no insulation	4.8	3.4	6.6	5.1	3.4	6.9	5.1	8.5	6.9	8.5
(c) Frame, finished both sides, more than 1" insulation or two reflective air spaces	2.0	1.4	2.7	2.1	1.4	2.8	2.1	3.5	2.8	3.5
(d) Masonry, plastered one side, no insulation	2.6	1.2	4.4	3.0	1.2	4.7	3.0	6.6	4.7	6.6
4. Wood Doors (Consider glass areas of doors as windows)	11.4	9.4	14.0	12.0	9.4	14.0	12.0	17.0	14.0	17.0
CEILING OR ROOFS										
5. Ceilings under naturally vented attic, or vented flat roof										
(a) Uninsulated — Dark	10.0	9.1	11.0	10.0	9.1	11.4	10.0	12.5	11.4	12.5
Light	8.2	7.2	9.1	8.2	7.2	9.4	8.2	10.4	9.4	10.4
(b) Less than 2" insulation or reflective air space — Dark	4.3	3.9	4.8	4.4	3.9	4.9	4.4	5.4	4.9	5.4
Light	3.5	3.1	4.1	3.6	3.1	4.1	3.6	4.6	4.1	4.6
(c) 2" to 4" insulation, or two reflective air spaces — Dark	2.6	2.3	2.9	2.6	2.3	2.9	2.6	3.2	2.9	3.2
Light	2.1	1.9	2.4	2.2	1.9	2.5	2.2	2.8	2.5	2.8
(d) More than 4" insulation or three or more reflective air spaces — Dark	1.8	1.6	1.9	1.8	1.6	2.0	1.8	2.2	2.0	2.2
Light	1.4	1.2	1.6	1.4	1.2	1.6	1.4	1.8	1.6	1.8
6. Built-up Roof, No Ceiling										
(a) Uninsulated — Dark	17.0	16.0	19.0	18.0	16.0	20.0	18.0	22.0	20.0	22.0
Light	14.0	12.5	16.0	14.0	12.5	16.0	14.0	18.0	16.0	18.0
(b) 2" Roof insulation — Dark	8.5	7.9	9.7	8.7	7.9	9.7	8.7	11.0	9.7	11.0
Light	6.9	6.3	7.9	7.2	6.3	8.2	7.2	9.1	8.2	9.1
(c) 3" Roof insulation — Dark	6.0	5.4	6.6	6.3	5.4	6.9	6.3	7.5	6.9	7.5
Light	4.9	4.3	5.6	5.1	4.3	5.6	5.1	6.3	5.6	6.3
7. Ceilings under rooms which are not cooled	2.7	1.9	3.6	2.9	1.9	3.8	2.9	4.8	3.8	4.8
FLOORS										
8. Over rooms which are not cooled	3.4	2.4	4.6	3.6	2.4	4.8	3.6	6.0	4.8	6.0
9. Over basement, enclosed crawl space, or concrete slab on ground	0	0	0	0	0	0	0	0	0	0
10. Over open crawl space	4.8	3.4	6.6	5.1	3.4	6.9	5.1	8.5	6.9	8.5
INFILTRATION										
11. Btuh per sq. ft. of gross exposed wall area	1.1	1.1	1.5	1.5	1.5	1.9	1.9	2.2	2.2	2.6
MECHANICAL VENTILATION										
12. Btuh per cfm	16.0	16.0	22.0	22.0	22.0	27.0	27.0	32.0	32.0	38.0

Courtesy of the Hydronics Institute.

Note: The values for this table are listed in Btuh per square foot. To convert them to Watts per square meter, multiply the value given by 3.152. For example, an HTM of 6 Btuh per sq ft equals 18.9 watts per sq m (6 × 3.152).

FIGURE 12-6   Heat transfer multipliers

No. 1 through 9 - Windows and Glass Doors - Refer to Table No. 3 for Summer HTM Values.

No. 10 - Wood Doors	Summer Temperature Difference and Daily Temperature Range												
	10		15			20			25		30	35	U
	L	M	L	M	H	L	M	H	M	H	H	H	
HTM (Btuh per sq. ft.)													
A. Hollow Core	9.9	7.6	12.7	10.4	7.6	15.5	13.2	10.4	16.0	13.2	16.0	18.8	.560
B. Hollow Core & Wood Storm	5.8	4.5	7.5	6.1	4.5	9.1	7.8	6.1	9.4	7.8	9.4	11.1	.330
C. Hollow Core & Metal Storm	6.3	4.9	8.1	6.7	4.9	9.9	8.5	6.7	10.3	8.5	10.3	12.1	.360
D. Solid Core	8.1	6.3	10.4	8.6	6.3	12.7	10.9	8.6	13.2	10.9	13.2	15.5	.460
E. Solid Core & Wood Storm	5.1	3.9	6.6	5.4	3.9	8.0	6.8	5.4	8.3	6.8	8.3	9.7	.290
F. Solid Core & Metal Storm	5.6	4.4	7.2	6.0	4.4	8.8	7.6	6.0	9.2	7.6	9.2	10.8	.320
G. Panel	11.8	9.1	15.1	12.5	9.1	18.5	15.8	12.5	19.2	15.8	19.2	22.5	.670
H. Panel & Wood Storm	6.3	4.9	8.1	6.7	4.9	9.9	8.5	6.7	10.3	8.5	10.3	12.1	.360
I. Panel & Metal Storm	7.2	5.6	9.3	7.6	5.6	11.3	9.7	7.6	11.7	9.7	11.7	13.8	.410

No. 11 - Metal Doors	10		15			20			25		30	35	U
	L	M	L	M	H	L	M	H	M	H	H	H	
HTM (Btuh per sq. ft.)													
A. Fiberglass Core	10.4	8.0	13.3	11.0	8.0	16.3	13.9	11.0	16.9	13.9	16.9	19.8	.590
B. Fiberglass Core & Storm	6.5	5.0	8.3	6.8	5.0	10.1	8.7	6.8	10.5	8.7	10.5	12.3	.367
C. Polystyrene Core	8.3	6.4	10.6	8.7	6.4	13.0	11.1	8.7	13.4	11.1	13.4	15.8	.470
D. Polystyrene Core & Storm	5.6	4.3	7.2	5.9	4.3	8.7	7.5	5.9	9.1	7.5	9.1	10.7	.317
E. Urethane Core	3.3	2.6	4.3	3.5	2.6	5.2	4.5	3.5	5.4	4.5	5.4	6.4	.190
F. Urethane Core & Storm	3.0	2.3	3.8	3.2	2.3	4.7	4.0	3.2	4.9	4.0	4.9	5.7	.170

No. 12 Wood Frame Exterior Walls With Sheathing and Siding or Brick Veneer or Other Exterior Finish.	10		15			20			25		30	35	U
	L	M	L	M	H	L	M	H	M	H	H	H	
HTM (Btuh per sq. ft.)													
A. None ½" Gypsum Board (R-0.5)	4.8	3.7	6.1	5.0	3.7	7.5	6.4	5.0	7.8	6.4	7.8	9.1	.271
B. None ½" Asphalt Board (R-1.3)	3.8	3.0	4.9	4.0	3.0	6.0	5.1	4.0	6.2	5.1	6.2	7.3	.217
C. R-11 ½" Gypsum Board (R-0.5)	1.6	1.2	2.0	1.7	1.2	2.5	2.1	1.7	2.6	2.1	2.6	3.0	.090
D. R-11 ½" Asphalt Board (R-1.3) R-11 ½" Bead Brd. (R-1.8) R-13 ½" Gypsum Brd. (R-0.5)	1.4	1.1	1.8	1.5	1.1	2.2	1.9	1.5	2.3	1.9	2.3	2.7	.080
E. R-11 ½" Extr Poly Brd. (R-2.5) R-11 ¾" Bead Brd. (R-2.7) R-13 ½" Asphalt Brd. (R-1.3) R-13 ½" Bead Brd. (R-1.8)	1.3	1.0	1.7	1.4	1.0	2.1	1.8	1.4	2.1	1.8	2.1	2.5	.075
F. R-11 1" Bead Brd. (R-3.6) R-11 ¾" Extr Poly Brd. (R-3.8) R-13 ½" Extr Poly Brd (R-2.5) R-13 ¾" Bead Brd. (R-2.7)	1.2	1.0	1.6	1.3	1.0	1.9	1.7	1.3	2.0	1.7	2.0	2.4	.070
G. R-13 ¾" Extr Poly Brd. (R-3.8) R-13 1" Bead Brd (R-3.6)	1.1	.9	1.5	1.2	.9	1.8	1.5	1.2	1.9	1.5	1.9	2.2	.065
H. R-11 1" Extr Brd. (R-5.0) R-13 1" Extr Poly Brd. (R-5.0) R-19 ½" Gypsum Brd. (R-0.5)	1.1	.8	1.4	1.1	.8	1.7	1.4	1.1	1.7	1.4	1.7	2.0	.060
I. R-19 ½" Asphalt Brd. (R-1.3) R-19 ½" Bead Brd. (R-1.8)	1.0	.7	1.2	1.0	.7	1.5	1.3	1.0	1.6	1.3	1.6	1.8	.055
J. R-11 R-8 Sheathing R-13 R-8 Sheathing R-19 ½" or ¾" Extr Poly R-19 ¾" or 1" Bead Brd.	.9	.7	1.1	.9	.7	1.4	1.2	.9	1.4	1.2	1.4	1.7	.050
K. R-19 1" Extr Poly Brd (R-5.0)	.8	.6	1.0	.8	.6	1.2	1.1	.8	1.3	1.1	1.3	1.5	.045
L. R-19 R-8 Sheathing	.7	.5	.9	.7	.5	1.1	.9	.7	1.1	.9	1.1	1.3	.040
M. R-27 Wall	.7	.5	.8	.7	.5	1.0	.9	.7	1.1	.9	1.1	1.2	.037
N. R-30 Wall	.6	.4	.7	.6	.4	.9	.8	.6	.9	.8	.9	1.1	.033
O. R-33 Wall	.5	.4	.7	.6	.4	.8	.7	.6	.9	.7	.9	1.0	.030

Reprinted with permission of ACCA, *Manual J,* 1986.

Note: The values for this table are listed in Btuh per square foot. To convert them to Watts per square meter, multiply the value given by 3.152. For example, an HTM of 6 Btuh per sq ft equals 18.9 watts per sq m (6 × 3.152).

FIGURE 12-6   (*continued*)

No. 13 - Partitions Between Conditioned and Unconditioned Space - Wood Frame Partitions	Summer Temperature Difference and Daily Temperature Range												
	10		15			20			25		30	35	U
	L	M	L	M	H	L	M	H	M	H	H	H	
	HTM (Btuh per sq. ft.)												
A. None ½" Gypsum Board (R-0.5)	2.4	1.4	3.8	2.7	1.4	5.1	4.1	2.7	5.4	4.1	5.4	6.8	.271
B. None ½" Asphalt Board (R-1.3)	2.0	1.1	3.0	2.2	1.1	4.1	3.3	2.2	4.3	.3.3	4.3	5.4	.217
C. R-11 ½" Gypsum Board (R-0.5)	.8	.4	1.3	.9	.4	1.7	1.3	.9	1.8	1.3	1.8	2.2	.090
D. R-11 ½" Asphalt Board (R-1.3) R-11 ½" Bead Brd. (R-1.8) R-13 ½" Gypsum Brd. (R-0.5)	.7	.4	1.1	.8	.4	1.5	1.2	.8	1.6	1.2	1.6	2.0	.080
E. R-11 ½" Extr Poly Brd. (R-2.5) R-11 ¾" Bead Brd. (R-2.7) R-13 ½" Asphalt Brd. (R-1.3) R-13 ½" Bead Brd. (R-1.8)	.7	.4	1.0	.8	.4	1.4	1.1	.8	1.5	1.1	1.5	1.9	.075
F. R-11 1" Bead Brd. (R-3.6) R-11 ¾" Extr Poly Brd. (R-3.8) R-13 ½" Extr Poly Brd. (R-2.5) R-13 ¾" Bead Brd. (R-2.7)	.6	.4	1.0	.7	.4	1.3	1.0	.7	1.4	1.0	1.4	1.8	.070
G. R-13 ¾" Extr Poly Brd. (R-3.8) R-13 1" Bead Brd (R-3.6)	.6	.3	.9	.6	.3	1.2	1.0	.6	1.3	1.0	1.3	1.6	.065
H. R-11 1" Extr Brd. (R-5.0) R-13 1" Extr Poly Brd. (R-5.0) R-19 ½" Gypsum Brd. (R-0.5)	.5	.3	.8	.6	.3	1.1	.9	.6	1.2	.9	1.2	1.5	.060
I. R-19 ½" Asphalt Brd. (R-1.3) R-19 ½" Bead Brd. (R-1.8)	.5	.3	.8	.5	.3	1.0	.8	.5	1.1	.8	1.1	1.4	.055
J. R-11 R-8 Sheathing R-13 R-8 Sheathing R-19 ½" or ¾" Extr Poly R-19 ¾" or 1" Bead Brd.	.4	.2	.7	.5	.2	.9	.7	.5	1.0	.7	1.0	1.2	.050
K. R-19 1" Extr Poly Brd. (R-5.0)	.4	.2	.6	.4	.2	.9	.7	.4	.9	.7	.9	1.1	.045
L. R-19 R-8 Sheathing	.4	.2	.6	.4	.2	.8	.6	.4	.8	.6	.8	1.0	.040

No. 13 - Partitions Between Conditioned & Unconditioned Space. Brick or Brick Partitions	10		15			20			25		30	35	U
	L	M	L	M	H	L	M	H	M	H	H	H	
	HTM (Btuh per sq. ft.)												
M. 8" Brick, No Insul., Unfinished	1.3	0	3.8	1.8	0	6.4	4.3	1.8	6.9	4.3	6.9	9.4	.510
N. 8" Brick R-5	.4	0	1.1	.5	0	1.8	1.2	.5	1.9	1.2	1.9	2.7	.144
O. 8" Brick R-11	.2	0	.6	.3	0	1.0	.7	.3	1.0	.7	1.0	1.4	.077
P. 8" Brick R-19	.1	0	.4	.2	0	.6	.4	.2	.6	.4	.6	.9	.048
Q. 4" Brick 8" Block, No Insul.	1.0	0	3.0	1.4	0	5.0	3.4	1.4	5.4	3.4	5.4	7.4	.400
R. 4" Brick 8" Block R-5	.3	0	1.0	.5	0	1.7	1.1	.5	1.8	1.1	1.8	2.5	.133
S. 4" Brick 8" Block R-11	.2	0	.6	.3	0	.9	.6	.3	1.0	.6	1.0	1.4	.074
T. 4" Brick 8" Block R-19	.1	0	.4	.2	0	.6	.4	.2	.6	.4	.6	.9	.047

No. 14 - Masonry Walls, Block or Brick Finished or Unfinished - Above Grade	10		15			20			25		30	35	U
	L	M	L	M	H	L	M	H	M	H	H	H	
	HTM (Btuh per sq. ft.)												
A. 8" or 12" Block, No Insul., Unfinished	5.3	3.2	7.8	5.8	3.2	10.4	8.3	5.8	10.9	8.3	10.9	13.4	.510
B. 8" or 12" Block + R-5	1.5	.9	2.2	1.6	.9	2.9	2.3	1.6	3.1	2.3	3.1	3.8	.144
C. 8" or 12" Block + R-11	.8	.5	1.2	.9	.5	1.6	1.3	.9	1.6	1.3	1.6	2.0	.077
D. 8" or 12" Block + R-19	.5	.3	.7	.5	.3	1.0	.8	.5	1.0	.8	1.0	1.3	.048
E. 4" Brick + 8" Block, No Insul.	4.1	2.5	6.1	4.5	2.5	8.1	6.5	4.5	8.5	6.5	8.5	10.5	.400
F. 4" Brick + 8" Block + R-5	1.4	.8	2.0	1.5	.8	2.7	2.2	1.5	2.8	2.2	2.8	3.5	.133
G. 4" Brick + 8" Block + R-11	.8	.5	1.1	.8	.5	1.5	1.2	.8	1.6	1.2	1.6	1.9	.074
H. 4" Brick + 8" Block + R-19	.5	.3	.7	.5	.3	1.0	.8	.5	1.0	.8	1.0	1.2	.047

Reprinted with permission of ACCA, *Manual J,* 1986.

Note: The values for this table are listed in Btuh per square foot. To convert them to Watts per square meter, multiply the value given by 3.152. For example, an HTM of 6 Btuh per sq ft equals 18.9 watts per sq m (6 × 3.152).

FIGURE 12-6   (*continued*)

No. 15 - Masonry Walls, Block or Brick Below Grade -	Summer Temperature Difference and Daily Temperature Range												
	10		15			20			25		30	35	U
	L	M	L	M	H	L	M	H	M	H	H	H	
	HTM (Btuh per sq. ft.)												
All	0.0	0.0	0.0	0.0	0.0	0.0	0.0	0.0	0.0	0.0	0.0	0.0	

No. 16 - Ceilings Under a Ventilated Attic Space. Light Colored Roof													
	10		15			20			25		30	35	U
	L	M	L	M	H	L	M	H	M	H	H	H	
	HTM (Btuh per sq. ft.)												
A. No Insulation	13.1	11.4	15.3	13.5	11.4	17.5	15.7	13.5	17.9	15.7	17.9	20.1	.437
B. R-7 Insulation	3.4	2.9	3.9	3.5	2.9	4.5	4.0	3.5	4.6	4.0	4.6	5.2	.112
C. R-11 Insulation	2.5	2.2	2.9	2.6	2.2	3.3	3.0	2.6	3.4	3.0	3.4	3.8	.083
D. R-19 Insulation	1.6	1.4	1.9	1.6	1.4	2.1	1.9	1.6	2.2	1.9	2.2	2.4	.053
E. R-22 Insulation	1.4	1.2	1.7	1.5	1.2	1.9	1.7	1.5	2.0	1.7	2.0	2.2	.048
F. R-26 Insulation	1.1	1.0	1.3	1.2	1.0	1.5	1.4	1.2	1.6	1.4	1.6	1.7	.038
G. R-30 Insulation	1.0	.9	1.2	1.0	.9	1.3	1.2	1.0	1.4	1.2	1.4	1.5	.033
H. R-38 Insulation	.8	.7	.9	.8	.7	1.0	.9	.8	1.1	.9	1.1	1.2	.026
I. R-44 Insulation	.7	.6	.8	.7	.6	.9	.8	.7	.9	.8	.9	1.1	.023
J. R-57 Insulation	.5	.4	.6	.5	.4	.7	.6	.5	.7	.6	.7	.8	.017
K. Wood Decking, No Insulation	8.6	7.5	10.0	8.9	7.4	11.4	10.3	8.9	11.8	10.3	11.8	13.2	.287

No. 16 - Ceilings Under a Ventilated Attic (Btuh per sq. ft.) Dark Colored Roof													
	10		15			20			25		30	35	U
	L	M	L	M	H	L	M	H	M	H	H	H	
	HTM (Btuh per sq. ft.)												
A. No Insulation	16.6	14.9	18.8	17.0	14.9	21.0	19.2	17.0	21.4	19.2	21.4	23.6	.437
B. R-7 Insulation	4.3	3.8	4.8	4.4	3.8	5.4	4.9	4.4	5.5	4.9	5.5	6.0	.112
C. R-11 Insulation	3.2	2.8	3.6	3.2	2.8	4.0	3.7	3.2	4.1	3.7	4.1	4.5	.083
D. R-19 Insulation	2.0	1.8	2.3	2.1	1.8	2.5	2.3	2.1	2.6	2.3	2.6	2.9	.053
E. R-22 Insulation	1.8	1.6	2.1	1.9	1.6	2.3	2.1	1.9	2.4	2.1	2.4	2.6	.048
F. R-26 Insulation	1.4	1.3	1.6	1.5	1.3	1.8	1.7	1.5	1.9	1.7	1.9	2.1	.038
G. R-30 Insulation	1.3	1.1	1.4	1.3	1.1	1.6	1.5	1.3	1.6	1.5	1.6	1.8	.033
H. R-38 Insulation	1.0	.9	1.1	1.0	.9	1.2	1.1	1.0	1.3	1.1	1.3	1.4	.026
I. R-44 Insulation	.9	.8	1.0	.9	.8	1.1	1.0	.9	1.1	1.0	1.1	1.2	.023
J. R-57 Insulation	.6	.6	.7	.7	.6	.8	.7	.7	.8	.7	.8	.9	017
K. Wood Decking, No Insulation	10.9	9.8	12.3	11.2	9.8	13.8	12.6	11.2	14.0	12.6	14.0	15.5	.287

No. 17 - Roof on Exposed Beams or Rafters Light Colored Roof													
	10		15			20			25		30	35	U
	L	M	L	M	H	L	M	H	M	H	H	H	
	HTM (Btuh pr sq. ft.)												
A. 1½" Wood Decking, No Insul	8.8	7.6	10.3	9.1	7.6	11.8	10.6	9.1	12.1	10.6	12.1	13.5	.294
B. 1½" Wood Decking R-4	4.2	3.6	4.9	4.3	3.6	5.6	5.0	4.3	5.7	5.0	5.7	6.4	.140
C. 1½" Wood Decking R-5	3.6	3.1	4.2	3.7	3.1	4.8	4.3	3.7	4.9	4.3	4.9	5.5	.119
D. 1½" Wood Decking R-6	3.2	2.8	3.7	3.3	2.8	4.2	3.8	3.3	4.3	3.8	4.3	4.9	.106
E. 1½" Wood Decking R-8	2.6	2.3	3.1	2.7	2.3	3.5	3.2	2.7	3.6	3.2	3.6	4.0	.088
F. 2" Shredded Wood Planks	6.2	5.4	7.2	6.4	5.4	8.3	7.5	6.4	8.5	7.5	8.5	9.5	.207
G. 3" Shredded Wood Planks	4.6	4.0	5.4	4.3	4.0	6.2	5.5	4.8	6.3	5.5	6.3	7.1	.154
H. 1½" Fiber Board Insulation	5.1	4.4	5.9	5.2	4.4	6.8	6.1	5.2	6.9	6.1	6.9	7.8	.169
I. 2" Fiber Board Insulation	4.0	3.5	4.7	4.2	3.5	5.4	4.9	4.2	5.5	4.9	5.5	6.2	.135
J. 3" Fiber Board Insulation	2.9	2.5	3.4	3.0	2.5	3.9	3.5	3.0	4.0	3.5	4.0	4.5	.097
K. 1½" Wood Decking R-13	1.8	1.6	2.1	1.9	1.6	2.4	2.2	1.9	2.5	2.2	2.5	2.8	.060
L. 1½" Wood Decking R-19	1.2	1.1	1.4	1.3	1.1	1.6	1.5	1.3	1.7	1.5	1.7	1.9	.041

Reprinted with permission of ACCA, *Manual J*, 1986.

Note: The values for this table are listed in Btuh per square foot. To convert them to Watts per square meter, multiply the value given by 3.152. For example, an HTM of 6 Btuh per sq ft equals 18.9 watts per sq m (6 × 3.152).

FIGURE 12-6   (*continued*)

No. 17 - Roof on Exposed Beams or Rafters Dark Colored Roof	Summer Temperature Difference and Daily Temperature Range												
	10		15			20			25		30	35	U
	L	M	L	M	H	L	M	H	M	H	H	H	
	HTM (Btuh per sq. ft.)												
A. 1½" Wood Decking, No Insul.	11.2	10.0	12.6	11.5	10.0	14.1	12.9	11.5	14.4	12.9	14.4	15.9	.294
B. 1½" Wood Decking R-4	5.3	4.8	6.0	5.5	4.8	6.7	6.2	5.5	6.9	6.2	6.9	7.6	.140
C. 1½" Wood Decking R-5	4.5	4.0	5.1	4.6	4.0	5.7	5.2	4.6	5.8	5.2	5.8	6.4	.119
D. 1½" Wood Decking R-6	4.0	3.6	4.6	4.1	3.6	5.1	4.7	4.1	5.2	4.7	5.2	5.7	.106
E. 1½" Wood Decking R-8	3.3	3.0	3.8	3.4	3.0	4.2	3.9	3.4	4.3	3.9	4.3	4.8	.088
F. 2" Shredded Wood Planks	7.9	7.0	8.9	8.1	7.0	9.9	9.1	8.1	10.1	9.1	10.1	11.2	.207
G. 3" Shredded Wood Planks	5.9	5.2	6.6	6.0	5.2	7.4	6.8	6.0	7.5	6.8	7.5	8.3	.154
H. 1½" Fiber Board Insulation	6.4	5.7	7.3	6.6	5.7	8.1	7.4	6.6	8.3	7.4	8.3	9.1	.169
I. 2" Fiber Board Insulation	5.1	4.6	5.8	5.3	4.6	6.5	5.9	5.3	6.6	5.9	6.6	7.3	.136
J. 3" Fiber Board Insulation	3.7	3.3	4.2	3.8	3.3	4.7	4.3	3.8	4.8	4.3	4.8	5.2	.097
K. 1½" Wood Decking R-13	2.3	2.0	2.6	2.3	2.0	2.9	2.6	2.3	2.9	2.6	2.9	3.2	.060
L. 1½" Wood Decking R-19	1.6	1.4	1.8	1.6	1.4	2.0	1.8	1.6	2.0	1.8	2.0	2.2	.041

No. 18 - Roof-Ceiling Combination - Light Colored Roof	10		15			20			25		30	35	U
	L	M	L	M	H	L	M	H	M	H	H	H	
	HTM (Btuh per sq. ft.)												
A. No Insulation	8.6	7.5	10.0	8.9	7.5	11.5	10.3	8.9	11.8	10.3	11.8	13.2	.287
B. R-11 Batts	2.2	1.9	2.5	2.2	1.9	2.9	2.6	2.2	3.0	2.6	3.0	3.3	.072
C. R-19 Batts	1.5	1.3	1.7	1.5	1.3	2.0	1.8	1.5	2.0	1.8	2.0	2.3	.049
D. R-22 Batts (2" x 8" Rafters)	1.3	1.2	1.6	1.4	1.2	1.8	1.6	1.4	1.8	1.6	1.8	2.1	.045
E. R-26 Batts (2" x 8" Rafters)	1.2	1.0	1.4	1.2	1.0	1.6	1.4	1.2	1.6	1.4	1.6	1.8	.040
F. R-30 Batts (2" x 10" Rafters)	1.0	.9	1.2	1.1	.9	1.4	1.3	1.1	1.4	1.3	1.4	1.6	.035

No. 18 - Roof-Ceiling Combination Dark Colored Roof	10		15			20			25		30	35	U
	L	M	L	M	H	L	M	H	M	H	H	H	
	HTM (Btuh per sq. ft.)												
A. No Insulation	10.9	9.8	12.3	11.2	9.8	13.8	12.6	11.2	14.1	12.6	14.1	15.5	.287
B. R-11 Batts	2.7	2.4	3.1	2.8	2.4	3.5	3.2	2.8	3.5	3.2	3.5	3.9	.072
C. R-19 Batts	1.9	1.7	2.1	1.9	1.7	2.4	2.2	1.9	2.4	2.2	2.4	2.6	.049
D. R-22 Batts (2" x 8" Rafters)	1.7	1.5	1.9	1.8	1.5	2.2	2.0	1.8	2.2	2.0	2.2	2.4	.045
E. R-26 Batts (2" x 8" Rafters)	1.5	1.4	1.7	1.6	1.4	1.9	1.8	1.6	2.0	1.8	2.0	2.2	.040
F. R-30 Batts (2" x 10" Rafters)	1.3	1.2	1.5	1.4	1.2	1.7	1.5	1.4	1.7	1.5	1.7	1.9	.035

No. 19 - Floors Over a Basement or Enclosed Crawl Space	10		15			20			25		30	35	U
	L	M	L	M	H	L	M	H	M	H	H	H	
	HTM (Btuh per sq. ft.)												
All	0.0	0.0	0.0	0.0	0.0	0.0	0.0	0.0	0.0	0.0	0.0	0.0	

No. 20 - Floors Over an Open Crawl Space or Garage	10		15			20			25		30	35	U
	L	M	L	M	H	L	M	H	M	H	H	H	
	HTM (Btuh per sq. ft.)												
A. Hardwood Floor, No Insulation	3.5	1.9	5.4	3.9	1.9	7.3	5.8	3.9	7.7	5.8	7.7	9.6	.386
B. Hardwood Floor R-11	.8	.4	1.2	.8	.4	1.6	1.3	.8	1.7	1.3	1.7	2.1	.084
C. Hardwood Floor R-13	.7	.4	1.1	.8	.4	1.4	1.1	.8	1.5	1.1	1.5	1.9	.076
D. Hardwood Floor R-19	.5	.3	.8	.5	.3	1.0	.8	.5	1.1	.8	1.1	1.3	.054
E. Hardwood Floor R-30	3	.2	.5	.4	.2	.7	.6	.4	.7	.6	.7	.9	.037
F. Carpet Floor No Insulation	2.3	1.3	3.5	2.5	1.3	4.8	3.8	2.5	5.1	3.8	5.1	6.3	.253
G. Carpet Floor R-11	.7	.4	1.0	.8	.4	1.4	1.1	.8	1.5	1.1	1.5	1.9	.075
H. Carpet Floor R-13	.6	.3	1.0	.7	.3	1.3	1.0	.7	1.4	1.0	1.4	1.7	.068
I. Carpet Floor R-19	.4	.2	.7	.5	.2	.9	.7	.5	1.0	.7	1.0	1.2	.050
J. Carpet Floor R-30	.3	.2	.5	.4	.2	.7	.5	.4	.7	.5	.7	.9	.035

No. 21 - 23 Basement Floors, Concrete Slab on Grade	10		15			20			25		30	35	U
	L	M	L	M	H	L	M	H	M	H	H	H	
	HTM												
All	0.0	0.0	0.0	0.0	0.0	0.0	0.0	0.0	0.0	0.0	0.0	0.0	

Notes to Table — Heat Transfer Multipliers
1. The HTM shown in this table do not include credit for infiltration.
2. Wall U values include wood framing equal to 20% of the opaque wall area.
3. Ceiling U values include wood framing equal to 10% of the opaque ceiling area.
4. Floor U values include wood framing equal to 15% of the opaque floor area.
5. Summer HTM values include the effects of solar radiation and thermal mass!

Reprinted with permission of ACCA, *Manual J*, 1986.

Note: The values for this table are listed in Btuh per square foot. To convert them to Watts per square meter, multiply the value given by 3.152. For example, an HTM of 6 Btuh per sq ft equals 18.9 watts per sq m (6 × 3.152).

FIGURE 12-6 (continued)

Design Temperature, F	85	90	95	100	105	110
Infiltration, Btuh per sq ft of gross exposed wall area	0.7	1.1	1.5	1.9	2.2	2.6
Mechanical Ventilation, Btuh per cfm	11.0	16.0	22.0	27.0	32.0	38.0

Courtesy of the Hydronics Institute.

SI UNITS

Design Temperature, C	29	32	35	38	41	43
Infiltration, W per sq m of gross exposed wall area	2.2	3.5	4.7	6.0	6.9	8.2
Mechanical Ventilation, Watts per L/s	6.8	9.9	13.7	16.8	19.9	23.6

FIGURE 12-7   Infiltration values

Rates of Heat Gain from Occupants of Conditioned Spaces

Degree of Activity		Total Heat, Btu/h Adult Male	Adjusted, M/F[a]	Sensible Heat, Btu/h	Latent Heat, Btu/h	% Sensible Heat that is Radiant[b] Low V	High V
Seated at theater	Theater, matinee	390	330	225	105		
Seated at theater, night	Theater, night	390	350	245	105	60	27
Seated, very light work	Offices, hotels, apartments	450	400	245	155		
Moderately active office work	Offices, hotels, apartments	475	450	250	200		
Standing, light work; walking	Department store; retail store	550	450	250	200	58	38
Walking, standing	Drug store, bank	550	500	250	250		
Sedentary work	Restaurant[c]	490	550	275	275		
Light bench work	Factory	800	750	275	475		
Moderate dancing	Dance hall	900	850	305	545	49	35
Walking 3 mph; light machine work	Factory	1000	1000	375	625		
Bowling[d]	Bowling alley	1500	1450	580	870		
Heavy work	Factory	1500	1450	580	870	54	19
Heavy machine work; lifting	Factory	1600	1600	635	965		
Athletics	Gymnasium	2000	1800	710	1090		

Notes:
1. Tabulated values are based on 75 °F room dry-bulb temperature. For 80 °F room dry bulb, the total heat remains the same, but the sensible heat values should be decreased by approximately 20%, and the latent heat values increased accordingly.
2. Also refer to Table 4, Chapter 8, for additional rates of metabolic heat generation.
3. All values are rounded to nearest 5 Btu/h.
[a] Adjusted heat gain is based on normal percentage of men, women, and children for the application listed, with the postulate that the gain from an adult female is 85% of that for an adult male, and that the gain from a child is 75% of that for an adult male.
[b] Values approximated from data in Table 6, Chapter 8, where V is air velocity with limits shown in that table.
[c] Adjusted heat gain includes 60 Btu/h for food per individual (30 Btu/h sensible and 30 Btu/h latent).
[d] Figure one person per alley actually bowling, and all others as sitting (400 Btu/h) or standing or walking slowly (550 Btu/h).

SI UNITS   ### Rates of Heat Gain from Occupants of Conditioned Spaces

Degree of Activity		Total Heat, W Adult Male	Adjusted, M/F[a]	Sensible Heat, W	Latent Heat, W	% Sensible Heat that is Radiant[b] Low V	High V
Seated at theater	Theater, matinee	115	95	65	30		
Seated at theater, night	Theater, night	115	105	70	35	60	27
Seated, very light work	Offices, hotels, apartments	130	115	70	45		
Moderately active office work	Offices, hotels, apartments	140	130	75	55		
Standing, light work; walking	Department store; retail store	160	130	75	55	58	38
Walking, standing	Drug store, bank	160	145	75	70		
Sedentary work	Restaurant[c]	145	160	80	80		
Light bench work	Factory	235	220	80	140		
Moderate dancing	Dance hall	265	250	90	160	49	35
Walking 4.8 km/h; light machine work	Factory	295	295	110	185		
Bowling[d]	Bowling alley	440	425	170	255		
Heavy work	Factory	440	425	170	255	54	19
Heavy machine work; lifting	Factory	470	470	185	285		
Athletics	Gymnasium	585	525	210	315		

Notes:
1. Tabulated values are based on 24 °C room dry-bulb temperature. For 27 °C room dry bulb, the total heat remains the same, but the sensible heat values should be decreased by approximately 20%, and the latent heat values increased accordingly.
2. Also refer to Table 4, Chapter 8, for additional rates of metabolic heat generation.
3. All values are rounded to nearest 5 W.
[a] Adjusted heat gain is based on normal percentage of men, women, and children for the application listed, with the postulate that the gain from an adult female is 85% of that for an adult male, and that the gain from a child is 75% of that for an adult male.
[b] Values approximated from data in Table 6, Chapter 8, where V is air velocity with limits shown in that table.
[c] Adjusted heat gain includes 18 W for food per individual (9 W sensible and 9 W latent).
[d] Figure one person per alley actually bowling, and all others as sitting (117 W) or standing or walking slowly (231 W).

FIGURE 12-8   Heat gain from occupants

Design Cooling Load Factors through Glass (Btu/h·ft²)

Outdoor Design Temp.	Regular Single Glass						Regular Double Glass						Heat Absorbing Double Glass						Clear Triple Glass		
	85	90	95	100	105	110	85	90	95	100	105	110	85	90	95	100	105	110	85	90	95
No Awnings or Inside Shading																					
North	23	27	31	35	39	44	19	21	24	26	28	30	12	14	17	19	21	23	17	19	20
NE and NW	56	60	64	68	72	77	46	48	51	53	55	57	27	29	32	34	36	38	42	43	44
East and West	81	85	89	93	97	102	68	70	73	75	77	79	42	44	47	49	51	53	62	63	64
SE and SW	70	74	78	82	86	91	59	61	64	66	68	70	35	37	40	42	44	46	53	55	56
South	40	44	48	52	56	61	33	35	38	40	42	44	19	21	24	26	28	30	30	31	33
Horiz. Skylight	160	164	168	172	176	181	139	141	144	146	148	150	89	91	94	96	98	100	126	127	129
Draperies or Venetian Blinds																					
North	15	19	23	27	31	36	12	14	17	19	21	23	9	11	14	16	18	20	11	12	14
NE and NW	32	36	40	44	48	53	27	29	32	34	36	38	20	22	25	27	29	31	24	26	27
East and West	48	52	56	60	64	69	42	44	47	49	51	53	30	32	35	37	39	41	38	39	41
SE and SW	40	44	48	52	56	61	35	37	40	42	44	46	24	26	29	31	33	35	32	33	34
South	23	27	31	35	39	44	20	22	25	27	29	31	15	17	20	22	24	26	18	19	21
Roller Shades Half-Drawn																					
North	18	22	26	30	34	39	15	17	20	22	24	26	10	12	15	17	19	21	13	14	15
NE and NW	40	44	48	52	56	61	38	40	43	45	47	49	24	26	29	31	33	35	34	35	35
East and West	61	65	69	73	77	82	54	56	59	61	63	65	35	37	40	42	44	46	49	49	50
SE and SW	52	56	60	64	68	73	46	48	51	53	55	57	30	32	35	37	39	41	41	42	43
South	29	33	37	41	45	50	27	29	32	34	36	38	18	20	23	25	27	29	25	26	26
Awnings																					
North	20	24	28	32	36	41	13	15	18	20	22	24	10	12	15	17	19	21	11	12	13
NE and NW	21	25	29	·33	37	42	14	16	19	21	23	25	11	13	16	18	20	22	12	13	14
East and West	22	26	30	34	38	43	14	16	19	21	23	25	12	14	17	19	21	23	12	13	14
SE and SW	21	25	29	33	37	42	14	16	19	21	23	25	11	13	16	18	20	22	12	13	14
South	21	24	28	32	36	41	13	15	18	20	22	24	11	13	16	18	20	22	11	12	13

Courtesy of the Hydronics Institute.

SI UNITS — Window Glass Load Factors (GLF) for Single-Family Detached Residences[a]

Design Temperature, °C	Regular Single Glass						Regular Double Glass						Heat-Absorbing Double Glass						Clear Triple Glass		
	29	32	35	38	41	43	29	32	35	38	41	43	29	32	35	38	41	43	29	32	35
No inside shading																					
North	107	114	129	148	151	158	95	95	107	117	120	129	63	63	73	79	82	88	85	85	95
NE and NW	199	205	221	237	243	262	173	177	186	196	199	208	114	117	123	132	139	139	158	158	167
E and W	278	284	300	315	322	337	243	246	255	265	268	278	161	161	170	177	186	186	221	221	230
SE and SW[b]	249	255	271	287	290	309	218	221	230	240	243	252	142	145	155	161	170	170	196	199	205
South[b]	167	173	189	205	211	227	145	148	158	167	170	180	98	98	107	114	123	123	132	132	142
Horizontal skylight	492	492	508	524	527	539	432	435	442	451	454	464	284	287	293	300	303	309	391	394	401
Draperies, venetian blinds, translucent roller shades fully drawn																					
North	57	60	73	85	91	104	50	50	60	69	73	82	41	44	50	57	60	66	47	50	57
NE and NW	101	104	120	132	136	148	91	95	101	110	114	123	76	76	85	91	91	101	88	88	95
E and W	142	145	158	170	173	186	126	129	139	145	148	158	104	104	114	120	120	129	123	123	129
SE and SW[b]	126	129	145	155	161	173	114	117	123	132	136	145	91	95	101	107	110	117	110	114	120
South[b]	85	88	104	117	120	132	76	79	88	98	98	107	63	66	73	79	82	88	73	76	82
Horizontal skylight	246	249	262	271	274	284	224	224	233	240	243	249	183	186	192	199	199	205	218	218	224
Opaque roller shades, fully drawn																					
North	44	47	63	73	79	91	41	44	54	60	63	73	38	38	47	54	54	63	41	41	47
NE and NW	79	82	98	107	114	126	73	76	85	95	95	104	66	69	76	82	85	91	73	73	82
E and W	107	114	126	139	142	155	101	104	114	120	123	132	91	95	101	107	110	117	101	101	110
SE and SW[b]	98	101	114	126	132	145	91	95	104	110	114	123	82	85	91	98	101	107	91	91	98
South[b]	66	69	85	95	101	114	63	63	73	82	85	95	57	60	66	73	76	82	60	63	69
Horizontal skylight	189	192	202	214	218	227	180	180	189	196	199	205	164	164	173	180	180	186	177	180	186

[a]Glass load factors (GLFs) for single-family detached houses, duplexes, or multi-family, with both east and west exposed walls or only north and south exposed walls, W/m².

[b]Correct by +30% for latitude of 48° and by −30% for latitude of 32°. Use linear interpolation for latitude from 40 to 48° and from 40 to 32°.

To obtain GLF for other combinations of glass and/or inside shading: $GLF_a = (SC_a/SC_t)(GLF_t - U_t D_t) + U_a D_t$, where the subscripts a and t refer to the alternate and table values, respectively. SC_t and U_t are given in Table 5. $D_t = (t_a - 75)$, where $t_a = t_o - (DR/2)$; t_o is the outdoor design temperature and DR is the daily range.

FIGURE 12-9 Solar heat gain

Shade Line Factors

Direction Window Faces	Latitude, Degrees						
	25	30	35	40	45	50	55
E	0.8	0.8	0.8	0.8	0.8	0.8	0.8
SE	1.9	1.6	1.4	1.3	1.1	1.0	0.9
S	10.1	5.4	3.6	2.6	2.0	1.7	1.4
SW	1.9	1.6	1.4	1.3	1.1	1.0	0.9
W	0.8	0.8	0.8	0.8	0.8	0.8	0.8

Note: Distance shadow line falls below the edge of the overhang equals shade line factor multiplied by width of overhang. Values are averages for five hours of greatest solar intensity on August 1.

Reprinted with permission of ACCA, *Manual D,* 1984.

FIGURE 12-10 Shade line factors

Design Equivalent Temperature Differences

Design Temperature, F	85		90			95			100		105	110
Daily Temperature Range[a]	L	M	L	M	H	L	M	H	M	H	H	H
WALLS AND DOORS												
1. Frame and veneer-on-frame	17.6	13.6	22.6	18.6	13.6	27.6	23.6	18.6	28.6	23.6	28.6	33.6
2. Masonry walls, 8-in. block or brick	10.3	6.3	15.3	11.3	6.3	20.3	16.3	11.3	21.3	16.3	21.3	26.3
3. Partitions, frame	9.0	5.0	14.0	10.0	5.0	19.0	15.0	10.0	20.0	15.0	20.0	25.0
masonry	2.5	0	7.5	3.5	0	12.5	8.5	3.5	13.5	8.5	13.5	18.5
4. Wood doors	17.6	13.6	22.6	18.6	13.6	27.6	23.6	18.6	28.6	23.6	28.6	33.6
CEILINGS AND ROOFS[b]												
1. Ceilings under naturally vented attic or vented flat roof—dark	38.0	34.0	43.0	39.0	34.0	48.0	44.0	39.0	49.0	44.0	49.0	54.0
—light	30.0	26.0	35.0	31.0	26.0	40.0	36.0	31.0	41.0	36.0	41.0	46.0
2. Built-up roof, no ceiling—dark	38.0	34.0	43.0	39.0	34.0	48.0	44.0	39.0	49.0	44.0	49.0	54.0
—light	30.0	26.0	35.0	31.0	26.0	40.0	36.0	31.0	41.0	36.0	41.0	46.0
3. Ceilings under uncondi-tioned rooms	9.0	5.0	14.0	10.0	5.0	19.0	15.0	10.0	20.0	15.0	20.0	25.0
FLOORS												
1. Over unconditioned rooms	9.0	5.0	14.0	10.0	5.0	19.0	15.0	10.0	20.0	15.0	20.0	25.0
2. Over basement, enclosed crawl space or concrete slab on ground	0	0	0	0	0	0	0	0	0	0	0	0
3. Over open crawl space	9.0	5.0	14.0	10.0	5.0	19.0	15.0	10.0	20.0	15.0	20.0	25.0

[a] Daily Temperature Range
    L (Low) Calculation Value: 12   M (Medium) Calculation Value: 20   H (High) Calculation Value: 30
Applicable Range: Less than 15 deg.   Applicable Range: 15 to 25 deg.   Applicable Range: More than 25 deg.
[b] Ceilings and Roofs: For roofs in shade, eight-hour average = 11 deg temperature differential. At 90 F design and medium daily range, equivalent temperature differential for light-colored roof equals $11 + (0.71)(39 - 11) = 31$ deg.
Reprinted with permission of ACCA, *Manual J,* 1986.

FIGURE 12-11 Equivalent temperature differences

Recommended Rate of Heat Gain from Selected Restaurant Equipment[a]

| Appliance | Size | Input Rating, Btu/h | | Recommended Rate of Heat Gain, Btu/h | | | |
| | | | | Without Hood | | | With Hood |
		Max.	Standby[b]	Sens.	Latent	Total	Sensible
Electric, No Hood Required							
Blender, per quart of capacity	1 to 4 qt	1550		1000	520	1520	480
Cabinet (large hot holding)	16.2 to 17.3 ft^3	7100		610	340	960	290
Cabinet (small hot holding)	3.2 to 6.4 ft^3	3070		270	140	410	130
Coffee brewer	12 cups/2 brnrs	5660		3750	1910	5660	1810
Coffee brewing urn (large), per quart of capacity	23 to 40 qt	2130		1420	710	2230	680
Coffee heater, per warming burner	1 to 2 brnrs	340		230	110	340	110
Dishwasher (hood type chemical sanitizing), per 100 dishes/h	950 to 2000 dishes/h	1300		170	370	540	170
Dishwasher (conveyor type water sanitizing), per 100 dishes/h	5000 to 9000 dishes/h	1160		150	370	520	170
Display case (refrigerated), per ft^3 of interior	6 to 67 ft^3	154		62	0	62	0
Food warmer (infrared bulb), per lamp	1 to 6 bulbs	850		850	0	850	850
Food warmer (well type), per ft^3 of well	0.7 to 2.5 ft^3	3620		1200	610	1810	580
Freezer (large)	73 ft^3	4570		1840	0	1840	0
Griddle/grill (large), per ft^2 of cooking surface	4.6 to 11.8 ft^2	9200		620	340	960	340
Hot plate (high speed double burner)		16720		7810	5430	13240	6240
Ice maker (large)	220 lb/day	3720		9320	0	9320	0
Microwave oven (heavy duty commercial)	0.7 ft^3	8970		8970		8970	0
Mixer (large), per quart of capacity	80 qt	94		94	0	94	0
Refrigerator (large), per 100 ft^3 of space	25 to 74 ft^3	750		300	0	300	0
Rotisserie	300	10920		7200	3720	10920	3480
Serving cart (hot), per ft^3 of well	1.8 to 3.2 ft^3	2050		680	340	1020	330
Steam kettle (large), per quart of capacity	80 to 320 qt	300		23	16	40	13
Toaster (large pop-up)	10 slice	18080		9590	8500	18080	5800
Electric, Exhaust Hood Required							
Charbroiler, per ft^2 of cooking surface	1.5 to 4.6 ft^2	7320					3310
Fryer (deep fat), per lb of fat capacity	15 to 15 to 70 lb	1270					14
Fryer (pressurized), per lb of fat capacity	13 to 33	1570					59
Oven (large convection), per ft^3 of oven space	7 to 19 ft^3	4450					180
Oven (small convection), per ft^3 of oven space	1.4 to 5.3 ft^3	10340					150
Range (burners), per 2 burner section	2 to 10 burners	7170					2660
Gas, No Hood Required							
Broiler, per ft^2 of broiling area	2.7 ft^2	14770	61	5310	2860	8170	1220
Dishwasher (hood type chemical sanitizing), per 100 dishes/h	950 to 2000 dishes/h	1740	660[b]	510	200	710	230
Dishwasher (conveyor type water sanitizing), per 100 dishes/h	5000 to 9000 dishes/h	1370	660[b]	370	80	450	140
Griddle/grill (large), per ft^2 of cooking surface	4.6 to 11.8 ft^2	17000	330	1140	610	1750	460
Oven (pizza), per ft^2 of hearth	6.4 to 12.9 ft^2	4740	61[b]	620	220	840	84
Gas, Exhaust Hood Required							
Braising pan, per quart of capacity	105 to 140 qt	9840	620				2430
Charbroiler (large), per ft^2 of cooking area	4.6 to 11.8 ft^2	16440	510				790
Fryer (deep fat), per lb of fat capacity	11 to 70 lb	2270	300[b]				160
Oven (convection), per ft^3 of oven space	7.4 to 19.4 ft^3	8670	19[b]				250
Oven (pizza), per ft^2 of oven hearth	9.3 to 25.8 ft^2	7240	61[b]				130
Range (burners), per 2 burner section	2 to 10 burners	33600	1325				6590
Range (hot top/fry top), per ft^2 of cooking surface	3 to 8 ft^2	11800	330				3390
Steam							
Compartment steamer, per lb of food/h	46 to 450 lb	280		22	14	36	11
Dishwasher (hood type chemical sanitizing), per 100 dishes/h	950 to 2000 dishes/h	3150		880	380	1260	410
Dishwasher (conveyor water sanitizing), per 100 dishes/h	5000 to 9000 dishes/h	1180		150	370	520	170
Steam kettle, per quart capacity	13 to 32 qt	500		39	25	64	19

[a] In cases where heat gain is given per unit of capacity the heat gain is calculated by multiplying the capacity by the recommended heat gain per unit of capacity.
[b] Standby input rating is for the entire appliance regardless of size.

Reprinted with permission form ASHRAE, *FUNDAMENTALS HANDBOOK*, 1993.

FIGURE 12-12 Appliance heat gain

QUESTIONS

12-1. What factors must be considered when determining heat gain in buildings?

12-2. Define *sensible* and *latent heat*.

12-3. How does the orientation of the windows in the building affect the heat gain?

12-4. Why must the number of occupants and the type of activity be considered in a heat gain calculation?

12-5. What factors does a heat transfer multiplier take into account?

12-6. How is infiltration calculated for heat gain, and how does this compare with infiltration calculations in heat loss?

Design Exercises

12-7. Calculate the heat gain for the residence in Appendix D. Assume the residence will be built in Richmond, Virginia.

12-8. Calculate the heat gain for a top-floor apartment in Appendix A and one on a lower floor. Assume the apartments will be built in Seattle, Washington.

CHAPTER 13

Forced Air System Design (Air Conditioning)

13-1 FORCED AIR SYSTEMS (AIR CONDITIONING)

A motor-driven fan is used to circulate filtered air from a central unit through supply ducts to each of the rooms. When cold air is supplied, warm air is returned to be cooled. The ductwork supplies air into main trunks and then smaller ducts branch off these to feed the rooms, or individual ducts to each room may be used. The ducts may be circular or rectangular, and their size depends on the amount of heated or cooled air required to maintain the desired temperature of the room.

A variety of duct layout designs for the various duct systems is available, several of which are indicated in Fig. 13-1. These types of systems quickly deliver heating or cooling when required and are economical and easy to install. Filters are put in these systems to reduce the amount of dust in the air. Heating and cooling units may be located anywhere in the building, including basement, crawl space, utility room, attic, small closets, garage, and even on the roof (commonly done in commercial, institutional, and industrial buildings where flat roofs are used).

The location of the supply registers (also called *diffusers*) and return registers varies, depending on whether warm or cool air is the most important portion of the system. If heating is of prime importance, the air registers should be located below a window so that the heat will counteract the cold outside air which infiltrates and transmits the most at windows. In this case, the registers would be located on the floor or slightly up the wall. Be certain the registers will not be covered by furniture or drapes, or their usefulness will be seriously reduced. Where cold air is of prime importance, the suggested locations are in the ceiling and high in the wall.

Warm air furnaces may use oil, natural or bottled gas, or electricity as a fuel. The air-conditioning equipment may be a part of a heating and cooling package called a unitary system or a package which has part of the unit on the furnace and the condenser package outside (split system).

Forced air systems may have humidifiers which are controlled by a *humidistat* inside the building, and moisture is added to the air when required. Another option being used increasingly is an electronic air purifier which removes impurities such as dust from the air.

Ducts and fittings for forced air systems are discussed thoroughly in Sec. 10-2 and 10-3, and apply to ducts used for cooling and heating with one additional item: all insulation used should have a vapor-barrier covering to protect against any possibility of moisture condensation developing.

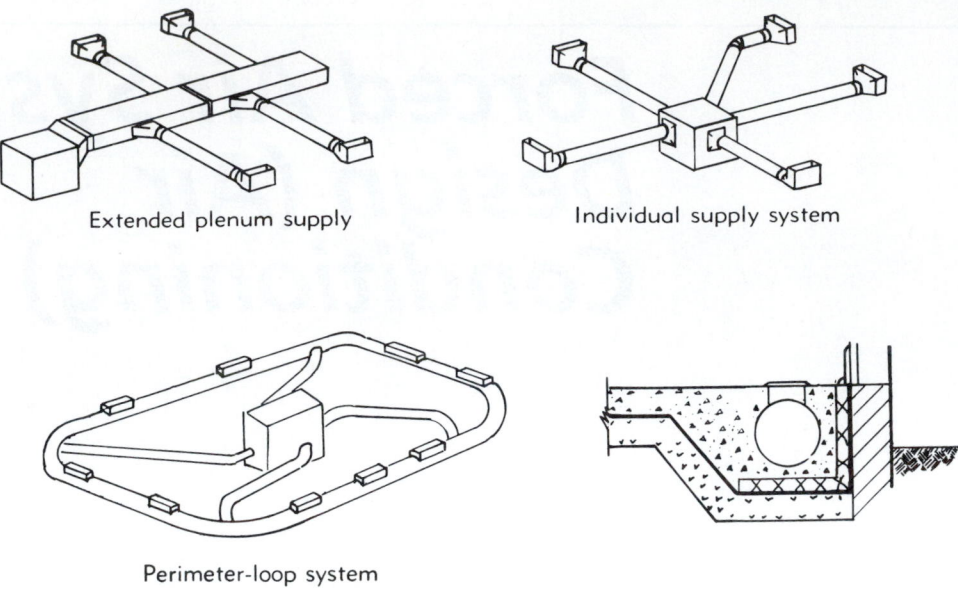

Extended plenum supply   Individual supply system

Perimeter-loop system

FIGURE 13-1   Duct layouts

13-2   EQUIPMENT

The forced air heating and cooling system may obtain its heat from a furnace, just as is used in forced air systems which supply only heat, with an added-on package of cooling coils placed next to the furnace bonnet and a condenser located outside the building (Fig. 13-2). The other unit which is becoming increasingly popular is the heat pump (Sec. 14-8), which produces both warm and cool air. Use of a heat pump means a furnace is not necessary.

In this section, the furnace with a separate cooling unit (commonly called an air conditioner) will be discussed. The size of the cooling unit is rated according to its cooling capacity in Btuh, often referred to as "tons" or in kilowatts. One ton is equal to 12,000 Btuh (so a 3-ton unit would have a capacity of 36,000 Btuh). The rating of the unit selected should be adequate to provide cooling Btuh equal to, or slightly more than, what the heat gain calculations call for. The wide range of sizes commonly available allows for the selection of a unit with the rated cooling capacity close to the required cooling calculated. Selection of a much larger unit will result in less efficient operation, and thus higher costs (due to the inefficiency of the on-off cycles, the time it takes to begin to cool, and the warming of the system as it is off, only to be cooled again as it is turned back on). But if a unit is too small, it may not be able to provide sufficient cool air. This becomes especially critical if any of the design assumptions (such as the amount of moisture, insulation, or the size or type glass used) varies, and the designer is not aware of the change. Also, the "tightness" of the construction, how well it is built, is assumed by the designer to be average. If it is not, there may be more heat gain than was calculated.

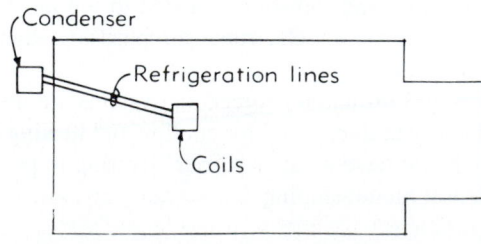

Condenser

Refrigeration lines

Coils

FIGURE 13-2   Split system

Cooling capacity (Btuh)	Watts	SEER (Btuh per watt)
24,000	2750	8.9
29,000	3100	9.4
36,000	4050	9.0
42,000	5000	8.4

High efficiency

Cooling capacity (Btuh)	Watts	SEER (Btuh per watt)
24,000	3800	6.3
30,000	4600	6.3
36,000	5800	6.2
42,000	7100	5.9

Standard

FIGURE 13-3   Seasonal energy efficiency ratios

The unit selected for use should also be checked for its energy efficiency. Most manufacturers have more than one type of unit available. It is important to get the most efficient model available. The tables in Fig. 13-3 give the manufacturer's specifications for "standard" and "high efficiency" models.

The efficiency of the models is checked by comparing the SEER (seasonal energy efficiency ratio) listed for each model. This SEER rating is obtained by dividing the total Btuh of the unit by its watts; the higher the number, the more efficient the unit. Note that the standard 24,000-Btuh model in Fig. 13-3 has an SEER of 6.3, while the highest-efficiency 24,000-Btuh model has an SEER of 8.9. This indicates that the highest-efficiency model is slightly more than 40% more efficient than the standard model. This means that the fuel bill for cooling will be about 40% less when using the highest-efficiency model, compared to the standard. High-efficiency units typically cost 50% more than the standard, but in terms of dollars, it may only be $200 to $300. The SEER ratings shown are typical, but each manufacturer must be checked since they will vary.

13-3   DUCT DESIGN

Step-by-Step Approach

1. Determine the heat loss and heat gain of each individual room, and tabulate them

    The heat loss (Sec. 8-13) and heat gain (Sec. 12-9) have already been calculated for the residence being designed, and they are tabulated as shown in Fig. 13-4.

2. The cooling unit size is selected based on the heat gain calculated. The unit selected should be as close as possible to the heat gain which was calculated.

    When the heat gain calculations are more than the capacity of an available unit, yet going to the next available unit would provide far too much capacity, the designer may want to review the calculations and suggest changes (perhaps in insulation, type of glass, or sunshields) that will reduce the load.

    From step 1, the heat gain calculations are 23,313 Btuh (2.1 tons). A 2½-ton unit is selected to provide enough capacity to allow for a pick-up allowance and for cooling lost in the ductwork.

3. Next, determine the location and number of supply outlets and return air intakes. The layout should allow:

    Heat loss: no more than 8,000 Btuh (2340 watts) per outlet.
    Heat gain: no more than 4,000 Btuh (1170 watts) per outlet.

    Extra outlets may be desirable in some rooms to provide the best air distribution. This is particularly true in large rooms.

SI UNITS

Supply Outlet	Heat Gain (BTUH)	Heat Loss (BTUH)	Supply Outlet	Heat Gain (Watts)	Heat Loss (Watts)
Bedroom #1	2,887	4,529	Bedroom #1	849	1324
Bedroom #2	1,330	2,772	Bedroom #2	403	807
Living	3,562	4,664	Living	981	1345
Living	3,562	4,663	Living	980	1345
Kitchen	2,341	2,080	Kitchen	706	603
Kitchen	2,341	2,080	Kitchen	706	602
Dining	2,175	3,056	Dining	628	898
Entry	1,038	2,031	Entry	313	644
Hall	1,037	2,030	Hall	312	643
Bedroom #3	2,180	3,483	Bedroom #3	666	1013
Bath	697	1,092	Bath	220	322
Totals	23,150	32,480	Totals	6764	9546

FIGURE 13-4   Tabulate heat loss and gain

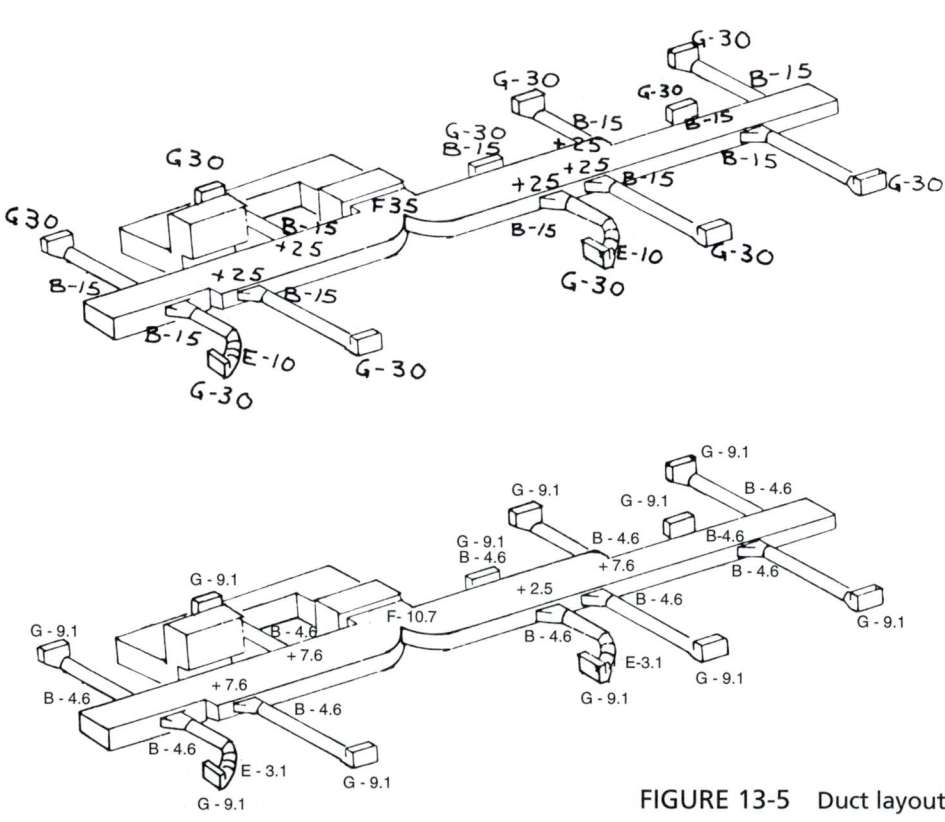

FIGURE 13-5   Duct layout

4. Note on the sketch the types of fittings which will be used and the actual length of each run from the furnace to the outlet. The fittings are the same as those used for heating (Chapter 10) and are also included at the end of this chapter. Fittings may be selected from Fig. 13-19, and the equivalent lengths are given for each type of fitting shown in the illustrations. Note the equivalent length of each fitting on the sketch.

The sketch in Fig. 13-5 shows the actual length of duct, the type of fitting, and each fitting's equivalent length.

5. Determine the total equivalent length of each run from bonnet to outlet by adding the actual length and the equivalent length of each fitting in the run, and tabulate the totals.

The tabulated total equivalent length of each run is calculated in Fig. 13-6 and Fig. 13-7.

SI UNITS

Outlet	Equivalent length	Actual length	Total equivalent length	Outlet	Equivalent length	Actual length	Total equivalent length
BEDROOM (1)	F 35 B 15 G 30 ――― 80	36	116	Bedroom (1)	F 10.7 B 4.6 G 9.1 ――― 24.4	11	35.4
BEDROOM (2)	F 35 + 25 B 15 G 30 ――― 105	26	131	Bedroom (2)	F 10.7 + 7.6 B 4.6 G 9.1 ――― 32	7.9	39.9
LIVING 1	F 35 B 15 + 25 G 30 ――― 105	33	138	Living 1	F 10.7 B 4.6 + 7.6 G 9.1 ――― 32	10	42
2	F 35 B 15 G 30 ――― 80	55	135	2	F 10.7 B 4.6 G 9.1 ――― 24.4	16.8	41.2
KITCHEN 1	F 35 B 15 G 30 ――― 80	46	126	Kitchen 1	F 10.7 B 4.6 G 9.1 ――― 24.4	14	38.4
KITCHEN 2	F 35 B 15 G 30 ――― 80	40	120	Kitchen 2	F 10.7 B 4.6 G 9.1 ――― 24.4	12	36.4
DINING	F 35 + 25 B 15 G 30 ――― 105	27	132	Dining	F 10.7 +7.6 B 4.6 G 9.1 ――― 32	8.2	40.2
ENTRY	F 35 + 25 B 15 E 10 G 30 ――― 115	16	131	Entry	F 10.7 + 7.6 B 4.6 E 3.1 G 9.1 ――― 35.1	5	40.1
HALL	F 35 B 15 G 30 ――― 80	10	90	Hall	F 10.7 B 4.6 G 9.1 ――― 24.4	3.1	27.5
BEDROOM (3)	F 35 + 25 B 15 G 30 ――― 105	30	135	Bedroom (3)	F 10.7 + 7.6 B 4.6 G 9.1 ――― 32	3.1	35.1
BATH	F 35 B 15 E 10 G 30 ――― 90	25	115	Bath	F 10.7 B 4.6 E 3.1 G 9.1 ――― 27.5	7.6	35.1

FIGURE 13-6   Equivalent length tabulation

Supply Outlet	Heat Gain (BTUH)	Heat Loss (BTUH)	Actual Length (ft)	Equivalent Length (ft)	Total Equiv. Length (ft)
Bedroom #1	2,887	4,529	36	80	116
Bedroom #2	1,330	2,772	26	105	131
Living	3,562	4,664	33	105	138
Living	3,562	4,663	55	80	135
Kitchen	2,341	2,080	46	80	126
Kitchen	2,341	2,080	40	80	120
Dining	2,175	3,056	27	105	132
Entry	1,038	2,031	16	115	131
Hall	1,037	2,030	10	80	90
Bedroom #3	2,180	3,483	30	105	135
Bath	697	1,092	25	90	115

SI UNITS

Supply Outlet	Heat Gain (Watts)	Heat Loss (Watts)	Actual Length	Equivalent Length	Total Equiv. Length
Bedroom #1	849	1324	11	24.4	35.4
Bedroom #2	403	807	7.9	32	39.9
Living	981	1345	10	32	42
Living	980	1345	16.8	24.4	41.2
Kitchen	706	603	12	24.4	36.4
Kitchen	706	602	14	24.4	38.4
Dining	628	898	8.2	32	40.2
Entry	313	644	5	35.1	40.1
Hall	312	643	3.1	24.4	27.5
Bedroom #3	666	1013	3.1	32	35.1
Bath	220	322	7.6	27.5	35.1

FIGURE 13-7 Duct lengths

6. Determine the total air volume required for the system. First determine the air volume required in cfm per ton or L/s per 3510 watts from Fig. 13-8. Using this information, the air volume for the system can be found as shown below.

In this design, the cfm for cooling is

$$400 \times \frac{23,150}{12,000} = 772 \text{ cfm}$$

The cfm for heating is

$$400 \times \frac{32,480}{12,000} = 1089 \text{ cfm (use 1200 cfm)}$$

Air Flow	cfm/ton	L/s per 3510 W
Typical*	400	190
Some Heat Pumps	450	215
Humid Areas	360	170
Dry Areas	429	204

*For normal residential cooling (70% sensible and 30% latent heat)
Reprinted with permission from ASHARE, *HVAC Handbook,* 1987.
Note: Metric added by author

FIGURE 13-8   Air flow volume

SI

In this design the L/s for cooling is

$$190 \text{ L/s} \times \frac{6764}{3510} = 366 \text{ L/s}$$

The L/s for heating is

$$190 \text{ L/s} \times \frac{9546}{3510} = 518 \text{ L/s (use 550)}$$

7. The heating requirement for each supply outlet is found by dividing the number of outlets in each room into the heat loss of each room.

   In this problem, the living room and kitchen have two outlets; they are noted twice in Fig. 13-4.

8. Calculate the air volume which will be delivered through each supply outlet. The air volume is calculated by dividing the Btuh for the outlet by the total Btuh or heat gain and then multiplying it times the total cfm.

   In Bedroom 1, the heat gain is 2,887. With a total heat gain of 23,313, the system is 1,200 cfm.

$$\frac{2,887}{23,150} \times 1,200 = 150 \text{ cfm}$$

SI

In bedroom 1 the heat gain is 849 W. With a total heat gain of 6764, the system is 550 L/s.

$$\frac{849}{6764} \times 550 = 69 \text{ L/s}$$

   This is repeated for each outlet, for both heat gain and heat loss, and tabulated as shown in Fig. 13-9.

9. The bonnet pressure is selected next. The approximate bonnet pressure required for a trunk (main duct) to carry the maximum volume of air (either heating or cooling cfm, whichever is larger) is found in Fig. 13-20.

   In this design, the total cfm (L/s) required is 1,200 cfm (550 L/s). From Fig. 13-20 for quiet operation in a residence (Type A), the suggested bonnet pressure is about 0.10 in. (24.8 Pa) of water; assume a system with .10 (24.8) is selected.

10. Next, the supply outlet size and its pressure loss are selected from the manufacturer's engineering data. A typical example of a manufacturer's data is shown in Fig. 13-10. The exact pressure loss will depend on variables such as the duct velocity and the an-

Supply Outlet	Heat Gain (BTUH)	Heat Loss (BTUH)	Actual Length (ft)	Equivalent Length (ft)	Total Equiv. Length (ft)	Cooling cfm	Heating cfm
Bedroom #1	2,887	4,529	36	80	116	150	167
Bedroom #2	1,330	2,772	26	105	131	69	103
Living	3,562	4,662	33	105	138	185	173
Living	3,562	4,663	55	80	135	185	173
Kitchen	2,341	2,080	46	80	126	121	77
Kitchen	2,341	2,080	40	80	120	121	77
Dining	2,175	3,056	27	105	132	113	113
Entry	1,038	2,031	16	115	131	54	75
Hall	1,037	2,030	10	80	90	54	75
Bedroom #3	2,180	3,483	30	105	135	113	129
Bath	697	1,092	25	90	115	36	40

SI UNITS

Supply Outlet	Heat Gain (Watts)	Heat Loss (Watts)	Actual Length	Equivalent Length	Total Equiv. Length	Cooling (L/s)	Heating (L/s)
Bedroom #1	849	1314	11	24.4	35.4	69	77
Bedroom #2	403	807	7.9	32	39.9	33	49
Living	981	1345	10	32	42	80	78
Living	980	1345	16.8	24.4	41.2	80	78
Kitchen	706	603	12	24.4	36.4	58	35
Kitchen	706	602	14	24.4	38.4	58	34
Dining	628	898	8.2	32	40.2	51	52
Entry	313	644	5	35.1	40.1	25	37
Hall	312	643	3.1	24.4	27.5	25	37
Bedroom #3	666	1013	3.1	32	35.1	54	59
Bath	220	322	7.6	27.5	35.1	18	19

FIGURE 13-9   Branch cfm

gles at which the register blades are set. Typically, the pressure loss will range from 0.01 to 0.02 in.; (2.4 to 4.8 L/s); many designers assume a 0.02-in. (4.8 L/s) loss and make a final selection later.

The selected supply outlet pressure losses are tabulated in Fig. 13-11.

11. The available bonnet pressure is divided between the supply and the return runs. It should be divided in proportion to the amount of supply and return runs on the project. A review of the sketch plan layout shows the approximate proportions of supply to return runs.

A review of the sketch plan layout (Fig. 13-6) shows that there are considerably more supply than return runs, so a division of 0.07 in. (17.36 Pa) for supply and 0.03 in. (7.44 Pa) for return is selected.

	300			400			500			600		
V = Duct Vel.												
Blade Set °	0	22½	45	0	22½	45	0	22½	45	0	22½	45
P_t	.01	.012	.027	.014	.021	.050	.023	.034	.082	.034	.051	.120
CFM	75			100			125			150		
T PWL-NC	8	L		10	L		13	L		16	L	
CFM	130			180			220			260		
T PWL-NC	9	L		13	L		17	L		20	L	
CFM	210			280			350			400		
T PWL-NC	13	L		17	L		21	L		24	28	
CFM	300			400			500			600		
T PWL-NC	15	L		20	L		25	26		29	32	

LISTED WIDTH: 6, 8, 10, 12, 14, 16, 18, 20, 22, 24, 26, 28, 30, 32, 34, 36

6 | 4
8 | 6 | 5 | 4
10 | 8 | 6 | 5 | 4
12 | 10 | 8 | 6 | 5 | 4

SYMBOLS:

- **V** = Duct velocity in fpm.
- **CFM** = Quantity of air in cubic ft./min.
- **NC** = Noise criteria (8 db room attenuation).
- **D** = Drop in feet.

- **Pt** = Total pressure inches H₂O
- **T** = Throw in Feet
- **PWL-NC INDEX** = A single number which expresses the PWL (sound power level) in relation to NC (noise criteria) curves.

FIGURE 13-10   Supply outlet data

Note: If later during the design the proportion selected does not provide satisfactory results, the available bonnet pressure can be reapportioned between supply and return.

12. Determine the pressure available for duct loss by subtracting the supply outlet loss (step 10) from the total pressure available for supply runs (step 11).

In this design, the pressure allowed for supply runs is 0.07 in. (17.36 Pa) (step 11). Beginning with the supply run in bedroom 1, the supply outlet loss is 0.02 in. (4.8 Pa) (step 11), and the pressure available for duct drop is 0.05 in. (12.56 Pa). Calculate the pressure available for duct loss for each of the supply runs, and tabulate the information as shown in Fig. 13-11.

13. In order to make a workable table of duct sizes, the pressure drop for duct loss must be calculated.

EXAMPLE   Since the various supply runs are all of different lengths, it is necessary to find the *allowable pressure drop per 100 ft* based on the *allowable duct loss* and the *total equivalent length* of the supply run. This allowable pressure drop may be calculated by using the equation:

$$\text{Allowable pressure drop per 100 ft} = \frac{\text{Allowable duct loss} \times 100}{\text{Total equivalent length}}$$

In this design, in bedroom 1 the supply run has an allowable duct loss of 0.05 in. (Step 12) and an equivalent length of 116 ft.

Using the formula, the allowable pressure drop per 100 ft is:

$$\text{Allowable pressure drop per 100 ft} = \frac{0.05 \text{ in. } (100)}{116 \text{ ft}} = 0.0431 \text{ in.}$$

Calculate the allowable pressure drop per 100 ft for each supply run, and tabulate the information as shown in Fig. 13-12.

SI

$$\text{Allowable pressure loss} = \frac{\text{Allowable duct loss}}{\text{Total equivalent length}}$$

Supply Outlet	Heat Gain (BTUH)	Heat Loss (BTUH)	Actual Length (ft)	Equivalent Length (ft)	Total Equiv. Length (ft)	Cooling cfm	Heating cfm	Supply Air Pressure	Supply Outlet Pressure Loss	Pressure Avail. for Duct Loss
Bedroom #1	2,887	4,529	36	80	116	150	167	0.07	0.02	0.05
Bedroom #2	1,330	2,772	26	105	131	69	103	0.07	0.02	0.05
Living	3,562	4,662	33	105	138	185	173	0.07	0.02	0.05
Living	3,562	4,663	55	80	135	185	173	0.07	0.02	0.05
Kitchen	2,341	2,080	46	80	126	121	77	0.07	0.02	0.05
Kitchen	2,341	2,080	40	80	120	121	77	0.07	0.02	0.05
Dining	2,175	3,056	27	105	132	113	113	0.07	0.02	0.05
Entry	1,038	2,031	16	115	131	54	75	0.07	0.02	0.05
Hall	1,037	2,030	10	80	90	54	75	0.07	0.02	0.05
Bedroom #3	2,180	3,483	30	105	135	113	129	0.07	0.02	0.05
Bath	697	1,092	25	90	115	36	40	0.07	0.02	0.05

SI UNITS

Supply Outlet	Heat Gain (Watts)	Heat Loss (Watts)	Actual Length	Equivalent Length	Total Equiv. Length	Cooling (L/s)	Heating (L/s)	Supply Air Pressure (Pa)	Supply Outlet Pressure Loss (Pa)	Pressure Avail. for Duct Loss
Bedroom #1	849	1314	11	24.4	35.4	69	77	17.36	4.8	12.56
Bedroom #2	403	807	7.9	32	39.9	33	49	17.36	4.8	12.56
Living	981	1345	10	32	42	80	78	17.36	4.8	12.56
Living	980	1345	16.8	24.4	41.2	80	78	17.36	4.8	12.56
Kitchen	706	603	12	24.4	36.4	58	35	17.36	4.8	12.56
Kitchen	706	602	14	24.4	38.4	58	34	17.36	4.8	12.56
Dining	628	898	8.2	32	40.2	51	52	17.36	4.8	12.56
Entry	313	644	5	35.1	40.1	25	37	17.36	4.8	12.56
Hall	312	643	3.1	24.4	27.5	25	37	17.36	4.8	12.56
Bedroom #3	666	1013	3.1	32	35.1	54	59	17.36	4.8	12.56
Bath	220	322	7.6	27.5	35.1	18	19	17.36	4.8	12.56

FIGURE 13-11   Pressure tabulation

In this design, Bedroom 1, the supply run has an allowable duct loss of 12.56 Pa (step 12) and an equivalent length of 35.4 m. Using the formula, the pressure drop per meter is:

$$\text{Allowable pressure loss} = \frac{12.56}{35.4} = 0.3548 \text{ (Use 0.35)}$$

Calculate the allowable pressure drop for each of the supply runs and tabulate the information as shown in Fig. 13-12.

Supply Outlet	Heat Gain (BTUH)	Heat Loss (BTUH)	Actual Length (ft)	Equivalent Length (ft)	Total Equiv. Length (ft)	Cooling cfm	Heating cfm	Supply Air Pressure	Supply Outlet Pressure Loss	Pressure Avail. for Duct Loss	Pressure Avail. Per 100 ft
Bedroom #1	2,887	4,529	36	80	116	150	167	0.07	0.02	0.05	0.043
Bedroom #2	1,330	2,772	26	105	131	69	103	0.07	0.02	0.05	0.038
Living	3,562	4,662	33	105	138	185	173	0.07	0.02	0.05	0.036
Living	3,562	4,663	55	80	135	185	173	0.07	0.02	0.05	0.037
Kitchen	2,341	2,080	46	80	126	121	77	0.07	0.02	0.05	0.040
Kitchen	2,341	2,080	40	80	120	121	77	0.07	0.02	0.05	0.042
Dining	2,175	3,056	27	105	132	113	113	0.07	0.02	0.05	0.038
Entry	1,038	2,031	16	115	131	54	75	0.07	0.02	0.05	0.038
Hall	1,037	2,030	10	80	90	54	75	0.07	0.02	0.05	0.056
Bedroom #3	2,180	3,483	30	105	135	113	129	0.07	0.02	0.05	0.037
Bath	697	1,092	25	90	115	36	40	0.07	0.02	0.05	0.043

SI UNITS

Supply Outlet	Heat Gain (Watts)	Heat Loss (Watts)	Actual Length	Equivalent Length	Total Equiv. Length	Cooling (L/s)	Heating (L/s)	Supply Air Pressure (Pa)	Supply Outlet Pressure Loss (Pa)	Pressure Avail. for Duct Loss	Pressure Avail. (Pa)
Bedroom #1	849	1314	11	24.4	35.4	69	77	17.36	4.8	12.56	0.35
Bedroom #2	403	807	7.9	32	39.9	33	49	17.36	4.8	12.56	0.31
Living	981	1345	10	32	42	80	78	17.36	4.8	12.56	0.30
Living	980	1345	16.8	24.4	41.2	80	78	17.36	4.8	12.56	0.30
Kitchen	706	603	12	24.4	36.4	58	35	17.36	4.8	12.56	0.35
Kitchen	706	602	14	24.4	38.4	58	34	17.36	4.8	12.56	0.31
Dining	628	898	8.2	32	40.2	51	52	17.36	4.8	12.56	0.31
Entry	313	644	5	35.1	40.1	25	37	17.36	4.8	12.56	0.31
Hall	312	643	3.1	24.4	27.5	25	37	17.36	4.8	12.56	0.46
Bedroom #3	666	1013	3.1	32	35.1	54	59	17.36	4.8	12.56	0.36
Bath	220	322	7.6	27.5	35.1	18	19	17.36	4.8	12.56	0.36

FIGURE 13-12   Available pressure tabulation

14. The round branch duct sizes may be determined by using the table in Fig. 13-21 or by using an air duct calculator if one is available. The table in Fig. 13-21 gives the size of round ducts based on the larger cfm required (either heating or cooling) and the allowable pressure drop (step 13). The size of the round duct may be converted to an equivalent rectangular duct size (Fig. 13-22) which will handle the required cfm within the allowable pressure drop.

In this design, bedroom 1 has an allowable duct loss of 0.043 in. per 100 ft (step 13). Using the table in Fig. 13-21, a 9.0-in. round duct is selected.

SI   In this design, Bedroom 1 has an allowable duct loss of 0.35 Pa per meter. Using the table in Fig. 13-21, a 250 mm round duct is selected.

a. Find the allowable pressure drop (along the bottom) (Friction loss).

b. Air quantity, ft^3/min (cfm), is found along the bottom.

c. Move horizontally from the friction loss and vertically from the air quantity until the two lines meet.

d. This intersection will occur on or near the diagonal lines which will give the duct diameters required (sizes on the line). If it falls right on the line, using that size duct will use the amount of friction loss available. If it falls between two lines, using the larger duct will use slightly less friction loss (the smaller duct would use more).

From the table in Fig. 13-21, determine the round duct sizes for each of the supply runs and tabulate the information as shown in Fig. 13-13.

The air duct calculator is used to obtain required round duct sizes by aligning the required air volume (cfm) and the allowable pressure drop (friction) per 100 ft of duct (0.043 in.) and reading the round duct diameter directly from the calculator (about 8.3 in., rounded off to 9.0 in.).

15. Round duct sizes selected in step 14 may be changed to equivalent rectangular duct sizes, carrying the same air volume required while maintaining the allowable pressure drops, by using the table in Fig. 13-22 or by using an air duct calculator:

Supply Outlet	Heat Gain (BTUH)	Heat Loss (BTUH)	Actual Length (ft)	Equivalent Length (ft)	Total Equiv. Length (ft)	Cooling cfm	Heating cfm	Supply Air Pressure	Supply Outlet Pressure Loss	Pressure Avail. for Duct Loss	Pressure Avail. Per 100 ft	Round Duct Size (in.)
Bedroom #1	2,887	4,529	36	80	116	150	167	0.07	0.02	0.05	0.043	9.0
Bedroom #2	1,330	2,772	26	105	131	69	103	0.07	0.02	0.05	0.038	8.0
Living	3,562	4,662	33	105	138	185	173	0.07	0.02	0.05	0.036	9.0
Living	3,562	4,663	55	80	135	185	173	0.07	0.02	0.05	0.037	9.0
Kitchen	2,341	2,080	46	80	126	121	77	0.07	0.02	0.05	0.040	8.0
Kitchen	2,341	2,080	40	80	120	121	77	0.07	0.02	0.05	0.042	8.0
Dining	2,175	3,056	27	105	132	113	113	0.07	0.02	0.05	0.038	8.0
Entry	1,038	2,031	16	115	131	54	75	0.07	0.02	0.05	0.038	7.0
Hall	1,037	2,030	10	80	90	54	75	0.07	0.02	0.05	0.056	6.0
Bedroom #3	2,180	3,483	30	105	135	113	129	0.07	0.02	0.05	0.037	8.0
Bath	697	1,092	25	90	115	36	40	0.07	0.02	0.05	0.043	6.0

SI UNITS

Supply Outlet	Heat Gain (Watts)	Heat Loss (Watts)	Actual Length	Equivalent Length	Total Equiv. Length	Cooling (L/s)	Heating (L/s)	Supply Air Pressure (Pa)	Supply Outlet Pressure Loss (Pa)	Pressure Avail. for Duct Loss	Pressure Avail. (Pa)	Round Duct Size (mm)
Bedroom #1	849	1314	11	24.4	35.4	69	77	17.36	4.8	12.56	0.35	250
Bedroom #2	403	807	7.9	32	39.9	33	49	17.36	4.8	12.56	0.31	200
Living	981	1345	10	32	42	80	78	17.36	4.8	12.56	0.30	250
Living	980	1345	16.8	24.4	41.2	80	78	17.36	4.8	12.56	0.30	250
Kitchen	706	603	9	24.4	36.4	58	35	17.36	4.8	12.56	0.35	200
Kitchen	706	602	14	24.4	38.4	58	34	17.36	4.8	12.56	0.31	200
Dining	628	898	8.2	32	40.2	51	52	17.36	4.8	12.56	0.31	200
Entry	313	644	5	35.1	40.1	25	37	17.36	4.8	12.56	0.31	160
Hall	312	643	3.1	24.4	27.5	25	37	17.36	4.8	12.56	0.46	160
Bedroom #3	666	1013	3.1	32	35.1	54	59	17.36	4.8	12.56	0.36	200
Bath	220	322	7.6	27.5	35.1	18	19	17.36	4.8	12.56	0.36	160

FIGURE 13-13 Round duct size tabulation

The round duct for bedroom 1 can be changed to a comparable rectangular duct by using Fig. 13-22. To use the table, the sizes of the rectangular duct are found with one dimension along the top and the other along the side. Their circular equivalent is the number in the middle of the page where the two sizes intersect.

To convert circular ducts to rectangular, find the circular equivalent in the center of the table and get the rectangular equivalents by reading up and to the side. There are many different sizes which can be used. The actual size selected depends to a large extent on the space available for installation of the duct.

From Fig. 13-22, determine a rectangular size for each supply run which could be used instead of the round duct previously selected. Tabulate the rectangular duct size as shown in Fig. 13-14.

The air duct calculator is used to find rectangular duct sizes by aligning the required cfm (148) and the allowable drop per 100 ft (0.043 in.) and reading off the rectangular duct dimensions from the calculator. The aligned sizes provide the various sizes. When using sizes that do not quite align, go to the next larger size.

16. Add the air volume of all the branch supply runs from each trunk duct. When there is more than one trunk duct (such as in Fig. 13-5), keep the totals for each trunk duct separate.

Supply Outlet	Heat Gain (BTUH)	Heat Loss (BTUH)	Actual Length (ft)	Equivalent Length (ft)	Total Equiv. Length (ft)	Cooling cfm	Heating cfm	Supply Air Pressure	Supply Outlet Pressure Loss	Pressure Avail. for Duct Loss	Pressure Avail. Per 100 ft	Round Duct Size (in.)	Rectangular Duct Size (in.)
Bedroom #1	2,887	4,529	36	80	116	150	167	0.07	0.02	0.05	0.043	9.0	7 × 10
Bedroom #2	1,330	2,772	26	105	131	69	103	0.07	0.02	0.05	0.038	8.0	6 × 9
Living	3,562	4,662	33	105	138	185	173	0.07	0.02	0.05	0.036	9.0	7 × 10
Living	3,562	4,663	55	80	135	185	173	0.07	0.02	0.05	0.037	9.0	7 × 10
Kitchen	2,341	2,080	46	80	126	121	77	0.07	0.02	0.05	0.040	8.0	6 × 9
Kitchen	2,341	2,080	40	80	120	121	77	0.07	0.02	0.05	0.042	8.0	6 × 9
Dining	2,175	3,056	27	105	132	113	113	0.07	0.02	0.05	0.038	8.0	6 × 9
Entry	1,038	2,031	16	115	131	54	75	0.07	0.02	0.05	0.038	7.0	6 × 7
Hall	1,037	2,030	10	80	90	54	75	0.07	0.02	0.05	0.056	6.0	4.5 × 7
Bedroom #3	2,180	3,483	30	105	135	113	129	0.07	0.02	0.05	0.037	8.0	6 × 9
Bath	697	1,092	25	90	115	36	40	0.07	0.02	0.05	0.043	6.0	4.5 × 7

SI UNITS

Supply Outlet	Heat Gain (Watts)	Heat Loss (Watts)	Actual Length	Equivalent Length	Total Equiv. Length	Cooling (L/s)	Heating (L/s)	Supply Air Pressure (Pa)	Supply Outlet Pressure Loss (Pa)	Pressure Avail. for Duct Loss	Pressure Avail. (Pa)	Round Duct Size (mm)	Rectangular Duct Size (mm)
Bedroom #1	883	1314	11	24.4	35.4	70	77	17.36	4.8	12.56	0.35	250	150 × 400
Bedroom #2	410	807	7.9	32	39.9	33	49	17.36	4.8	12.56	0.31	200	150 × 225
Living	1027	1345	10	32	42	82	78	17.36	4.8	12.56	0.30	250	150 × 400
Living	1027	1345	16.8	24.4	41.2	82	78	17.36	4.8	12.56	0.35	200	150 × 400
Kitchen	706	597	12	24.4	36.4	56	35	17.36	4.8	12.56	0.31	200	150 × 225
Kitchen	706	597	14	24.4	38.4	56	34	17.36	4.8	12.56	0.31	200	150 × 225
Dining	628	898	8.2	32	40.2	50	52	17.36	4.8	12.56	0.31	160	100 × 200
Entry	313	644	5	35.1	40.1	25	37	17.36	4.8	12.56	0.31	160	100 × 200
Hall	312	643	3.1	24.4	27.5	25	37	17.36	4.8	12.56	0.46	160	100 × 200
Bedroom #3	666	1013	3.1	32	35.1	53	59	17.36	4.8	12.56	0.36	200	150 × 225
Bath	229	322	7.6	27.5	35.1	18	19	17.36	4.8	12.56	0.36	160	100 × 200

FIGURE 13-14   Rectangular duct size tabulation

In this design, there are two trunk ducts (one toward the living room and kitchen, the other toward the bedrooms). Tabulate and total the air volume for each trunk duct.

The totals for both heating and cooling must be tabulated for each trunk and the largest value used. The trunk serving the living areas will have a higher cooling air volume requirement, while the trunk serving the bedroom areas will have a higher heating air volume requirement. Each trunk must be sized for the maximum air volume it will receive.

17. For best air flow distribution through the ducts, it is important that the friction losses be approximately equal in both trunk ducts.

Using the table in Fig. 13-21, select the trunk sizes required. When there is more than one trunk duct, size each separately using the required air volume (cfm) and the allowable duct friction (step 12).

The bedroom trunk must handle 443 cfm with an allowable pressure drop of 0.038 in. per 100 ft. Using Fig. 13-21, the bedroom trunk duct will be a 14-in. round duct.

The living space trunk duct has 757 cfm, and an allowable 0.036-in. pressure drop per 100 ft. Using Fig. 13-21, the living space trunk duct will be a 16.0-in. round duct.

Since the trunk ducts have been sized separately, based on the cfm and the allowable pressure drop, they are balanced.

The round duct sizes may be converted to rectangular duct sizes by use of Fig. 13-22. Once again, the space available for installation will have a large influence on the shape of the rectangular duct selected.

The bedroom trunk must handle 202 L/s with an allowable pressure drop of 0.31 Pa. Using Fig. 13-21, the bedroom trunk will be a 315 mm round duct.

Duct	cfm	Supply Static Pressure*	Total Equiv. Length (ft)	Supply Pressure Avail. Per 100 ft	Round Duct Size (in.)	Rectangular Duct Size (in.)
Main Branch	1,200	0.05	138	0.036	18.0	10 × 28
Bedroom Branch	439**	0.05	131	0.038	14.0	10 × 18
Living Branch	863***	0.05	138	0.036	16.0	10 × 20

*Allows 0.02 for supply outlet
**Based on heating requirements
***Based on cooling requirements

SI UNITS

Duct	L/s	Supply Static Pressure* (Pa)	Total Equiv. Length (m)	Supply Pressure Avail. Per branch	Round Duct Size (mm)	Rectangular Duct Size (mm)
Main Branch	550	12.56	42.1	0.30	500	200 × 1300
Bedroom Branch	202**	12.56	40	0.31	315	200 × 450
Living Branch	377***	12.56	42.1	0.30	400	200 × 750

*Allows 0.02 for supply outlet
**Based on heating requirement
***Based on cooling requirement

FIGURE 13-15   Tabulated main duct sizes

The living room trunk duct has 376 L/s, and an allowable pressure drop of 0.30 Pa. Using Fig. 13-21, the living space trunk duct will be a 400 mm round duct.

18. Review the supply trunk duct to determine if it is desirable to reduce its size as each branch duct leaves the trunk. Generally, such a reduction is suggested so that the velocity of air through the duct will not drop too low. Using the remaining air volume in the trunk duct, after each branch and the allowable pressure drop (step 9), select the reduced trunk sizes from Figs. 13-21 and 13-22.

Using the information in Fig. 13-16 and the tables in Figs. 13-21 and 13-22, determine the reduced trunk duct sizes for both the bedroom and the living space ducts, and tabulate the sizes as shown in Fig. 13-17.

Duct		cfm	Supply Static Pressure	Total Equiv. Length (ft)	Supply Pressure Avail. Per 100 ft	Round Duct Size (in.)	Rectangular Duct Size (in.)
Main Branch	A-B	1,200	0.05	138	0.036	18.0	10 × 28
Bedroom Branch	*B-C	439	0.05	131	0.038	14.0	10 × 18
	C-D	336					
	D-E	207					
	E-F	167					
Living Branch	*B-G	863	0.05	138	0.036	16.0	10 × 20
	G-H	788					
	H-I	713					
	I-J	600					
	J-K	427					
	K-L	306					
	L-M	185					

*Does not total the main because the maximum need of heating or air conditioning must be used for each length.

SI UNITS

Duct		L/s	Supply Static Pressure*	Total Equiv. Length	Supply Pressure Avail.	Round Duct Size	Rectangular Duct Size
Main	A-B	550	12.56	42.1	0.30	500	200 × 1300
Bedroom Branch	*B-C	204	12.56	40	0.31	315	200 × 450
	C-D	155					
	D-E	96					
	E-F	77					
Living Branch	*B-G	377	12.56	42.1	0.30	400	200 × 750
	G-H	297					
	H-I	239					
	I-J	181					
	J-K	101					
	K-L	50					
	L-M	25					

*Does not total the main because the maximum need of heating or air conditioning must be used for each length.

FIGURE 13-16   Reduced trunk sizes

Duct		cfm	Supply Static Pressure	Total Equiv. Length (ft)	Supply Pressure Avail. Per 100 ft	Round Duct Size (in.)	Rectangular Duct Size (in.)
Main	A-B	1,200	0.05	138	0.036	18.0	10 × 28
Bedroom Branch	B-C	439	0.05	131	0.038	14.0	10 × 18
	C-D	336				12.0	10 × 13
	D-E	207				9.0	10.0 × 11.7
	E-F	167				6.0	9.0 × 8.8
Living Branch	B-G	863	0.05	138	0.036	16.0	10 × 20
	G-H	788				14.0	10 × 18
	H-I	713				14.0	10 × 18
	I-J	600				14.0	10 × 18
	J-K	427				12.0	10 × 13
	K-L	306				10.0	10 × 8
	L-M	185					

SI UNITS

Duct		L/s	Supply Static Pressure	Total Equiv. Length	Supply Pressure Avail.	Round Duct Size	Rectangular Duct Size
Main	A-B	550	12.56	42.1	0.30	500	200 × 1300
Bedroom Branch	B-C	204	12.56	40	0.31	315	200 × 450
	C-D	155				315	200 × 450
	D-E	96				250	200 × 275
	E-F	77				250	200 × 275
Living Branch	B-G	377	12.56	42.1	0.30	400	200 × 750
	G-H	297				400	200 × 750
	H-I	239				400	200 × 750
	I-J	161				315	200 × 450
	J-K	101				250	200 × 275
	K-L	50				200	200 × 200
	L-M	25				160	200 × 160

FIGURE 13-17   Reduced trunk sizes

19. Next, the design turns to the return ducts. The first step is to select an allowable return air pressure drop. This was previously decided (step 11), but it is reviewed at this point as the designer reviews the supply trunk and branch duct sizes to see if it is desirable to increase the duct sizes (reducing the allowable air pressure drop) or perhaps to reduce the duct sizes (increasing the allowable air pressure drop for the supply runs). Remember that a change in allowable air pressure drop for the supply will affect the allowable air pressure drop for the return runs (step 11).

A review of the supply duct sizes is made, and it is decided that they will not be revised since the duct sizes selected so far seem reasonable. Therefore, the original decision, in step 11, to use 0.03 in. (7.44 Pa) as the allowable pressure drop for the return is unchanged, and the return design will be based on it.

20. The return trunk duct size is determined by the air volume it must handle and the allowable pressure drop. This allowable pressure drop must be converted to allowable pressure drop, just as was done for the supply runs. Using the air volume and the allowable pressure drop, the round return duct size may be selected from Fig. 13-21. The rectangular duct of equivalent size may be found in Fig. 13-22.

The return trunk duct will have to handle an air volume equal to what the supply trunk duct handles; from Fig. 13-15, this is 1,200 cfm. The allowable pressure drop selected for the return is 0.03 in. (steps 11 and 12). First, the allowable pressure drop must be converted into the allowable pressure drop *per* 100 ft (step 13). The equivalent length is shown in Fig. 13-18.

$$(0.03 \text{ in.} \times 100) \div 82 \text{ ft} = 0.036 \text{ in.}$$
allowable pressure drop per 100 ft

SI

The return duct will have to handle the air volume equal to that handled by the supply duct. From Fig. 13-15, this is 550 L/s. The allowable pressure drop calculated for the return is 0.30 (using the information from steps 11 and 12 and the total equivalent length).

Using the allowable pressure drop, select the round return duct size from the table in Fig. 13-21 and its equivalent rectangular size from Fig. 13-22. Tabulate the information as shown in Fig. 13-18.

If the return trunk sizes are too large, they may be reduced in size by:

a. Reapportioning the pressure drop available so there is less drop allowed for the supply runs and more pressure allowed for the return runs. This may require a revision of supply branch and trunk duct sizes.

b. Checking to see if the return air grilles can be located closer to the furnace, which would reduce the actual length of duct. This reduces the allowable pressure drop, resulting in a smaller duct size.

21. The size of the blower on the furnace is determined by the total air volume (the total cfm to be delivered) and the total static pressure requirements.

The total static pressure for furnace-blower combination units is the total of the actual pressure drops in the supply and return ducts. Most residential and small commercial buildings have furnace-blower combination units.

When the blowers are selected separately from the furnace, the total static pressure is the sum of the actual pressure losses in the supply and return ducts, filter loss, casing loss, and losses through any devices which are put on the system, such as air washers and purifiers.

Duct	Cfm	Actual length	Equivalent length	Total equivalent length	Loss per 100'	Round duct size (in)	Rectangular duct size (in)
Return	1200	12	70	82	0.036	18.0	12 × 24

Duct	L/s	Actual length	Equivalent length	Total equivalent length	Return pressure avail.	Round duct size	Rectangular duct size
Return	550	3.7	21.4	25.1	0.30	500	200 × 1300

FIGURE 13-18   Return duct sizes

In this design, the 0.10-in. pressure drop, on which the designs of the ducts are based, is selected for a standard blower-furnace combination. While the actual pressure drop is slightly less than 0.10 in., the actual drop is not calculated since, as each duct is selected, it is large enough to handle the air volume within the allowable pressure drop.

13-4   SYSTEM INSTALLATION

Forced air systems that provide both heating and cooling have the same basic system installation as forced air systems which just provide heating (Sec. 10-7). The primary difference is that they require a cooling unit that will have condensed moisture to dispose of. This is sometimes done by means of a pipe to the exterior or by draining to a floor drain or into a small sump pump that pumps the water into a drain line.

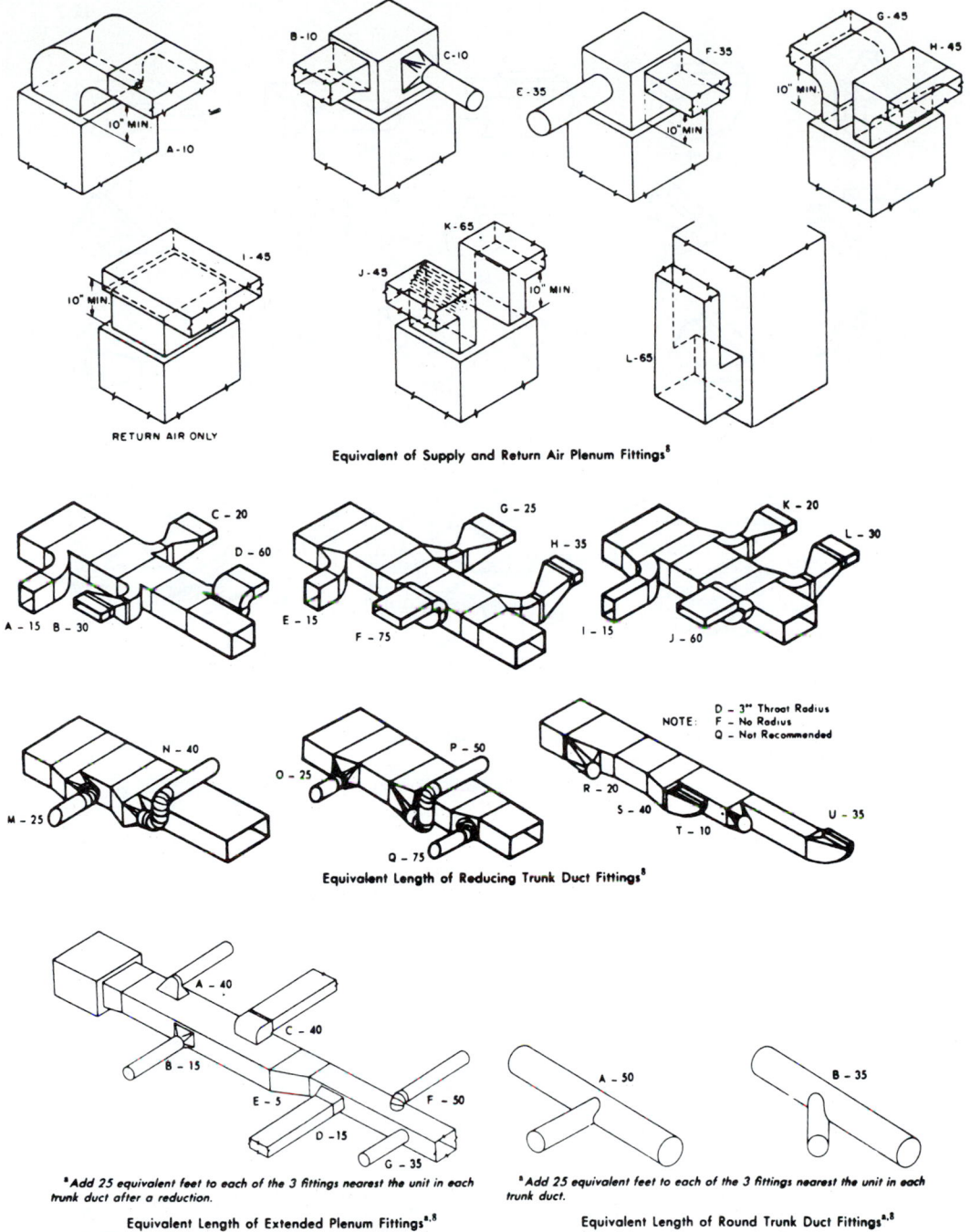

Reprinted with permission of ACCA, *Manual D*, 1984.

FIGURE 13-19   Equivalent fittings

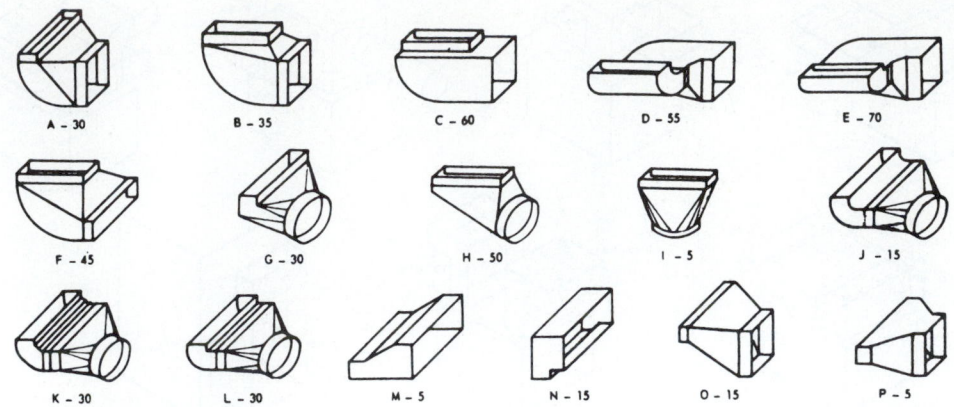

ᵃ *These values may also be used for floor diffuser boxes.*

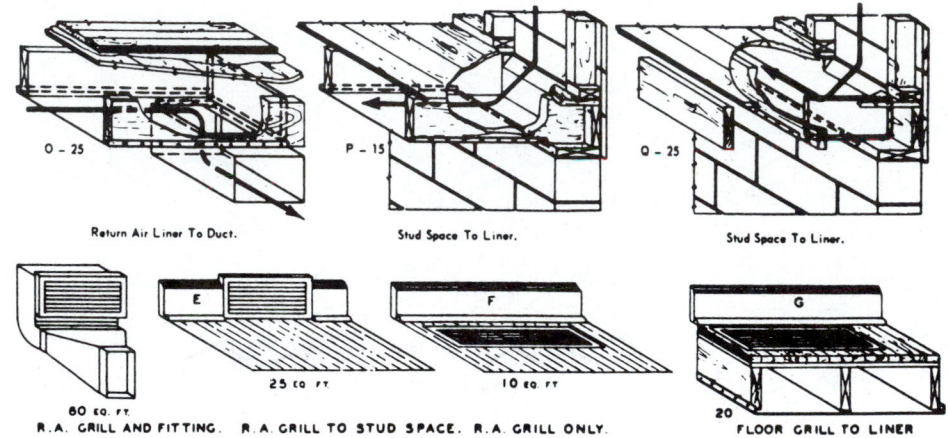

Return Air Liner To Duct.   Stud Space To Liner.   Stud Space To Liner.

R.A. GRILL AND FITTING.   R.A. GRILL TO STUD SPACE.   R.A. GRILL ONLY.   FLOOR GRILL TO LINER

ᵃ *Inside radius for A and B = 3 in., and for F and G = 5 in.*

Equivalent Length of Angles and Elbows for Individual and Branch Ductsᵃ·ᵇ

Reprinted with permission from ASHRAE, Systems Handbook, 1976

Equivalent Length of Boot Fittingsᵃ·ᵇ

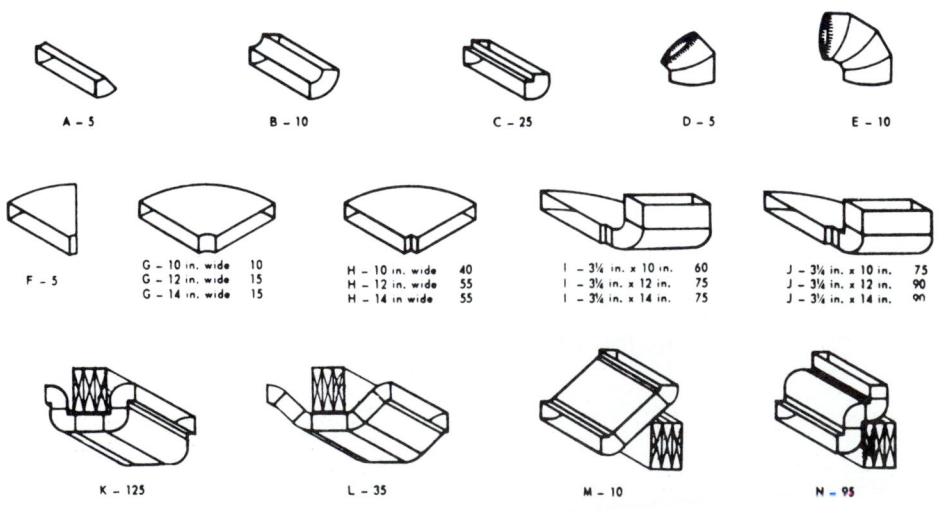

FIGURE 13-19   (continued)

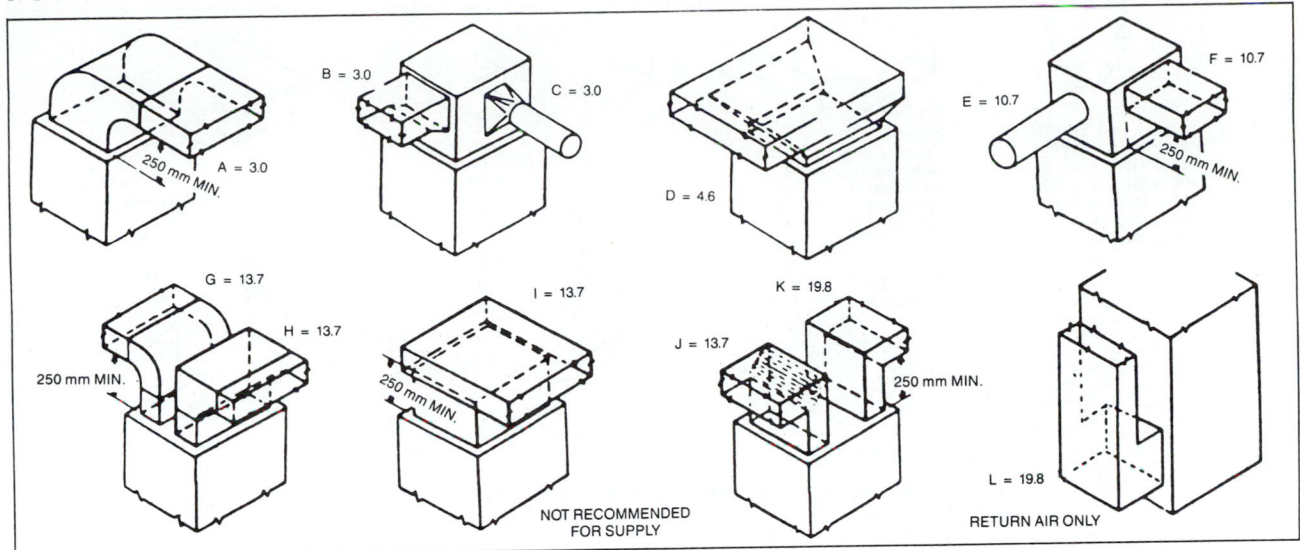

Equivalent Length in Metres of Supply and Return Air Plenum Fittings (ACCA 1984)

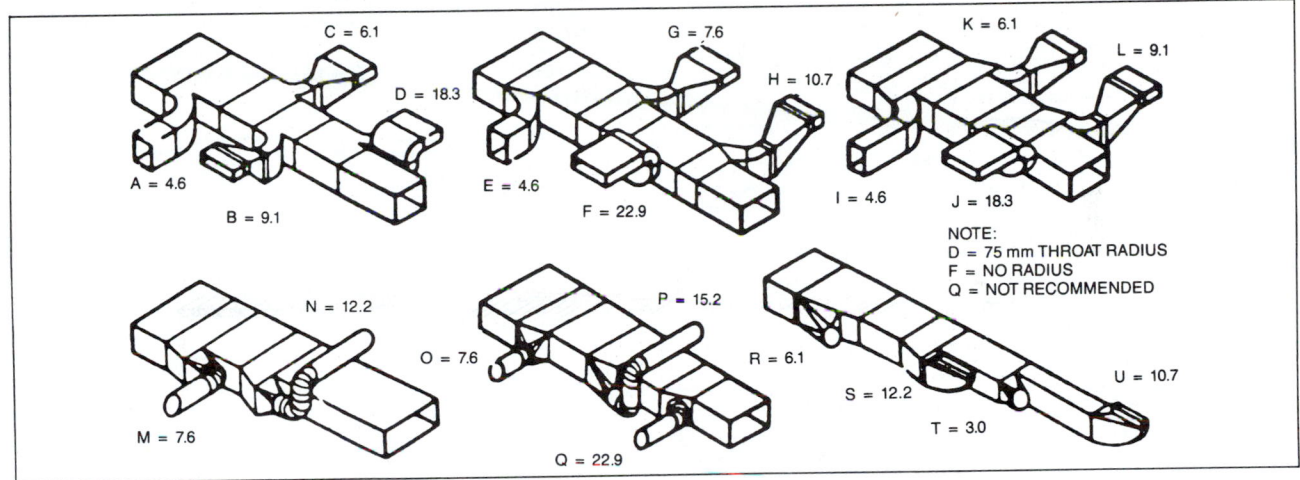

Equivalent Length in Metres of Reducing Trunk Duct Fittings (ACCA 1984)

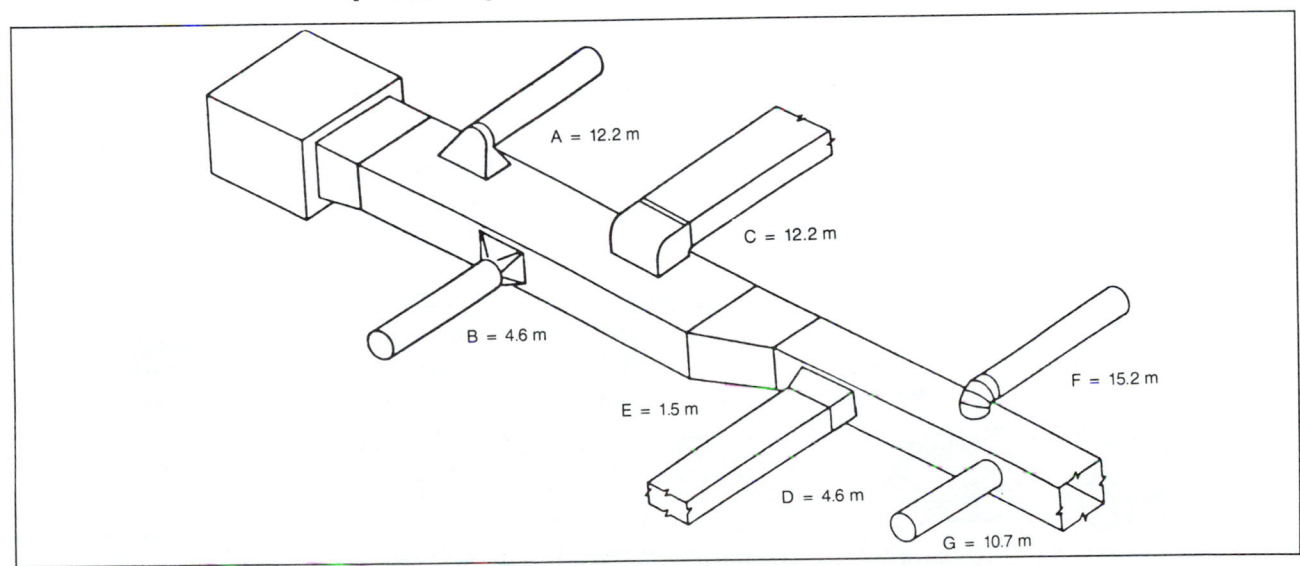

Equivalent Length in Metres of Extended Plenum Fittings (ACCA 1984)

FIGURE 13-19 (*continued*)

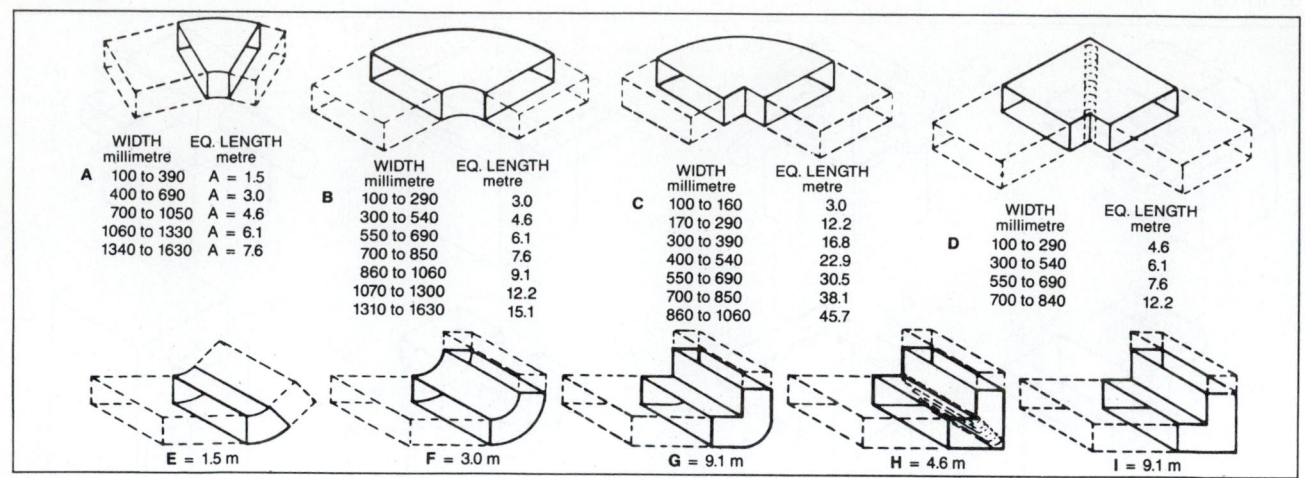

Equivalent Length of Angles and Elbows for Trunk Ducts (ACCA 1984)

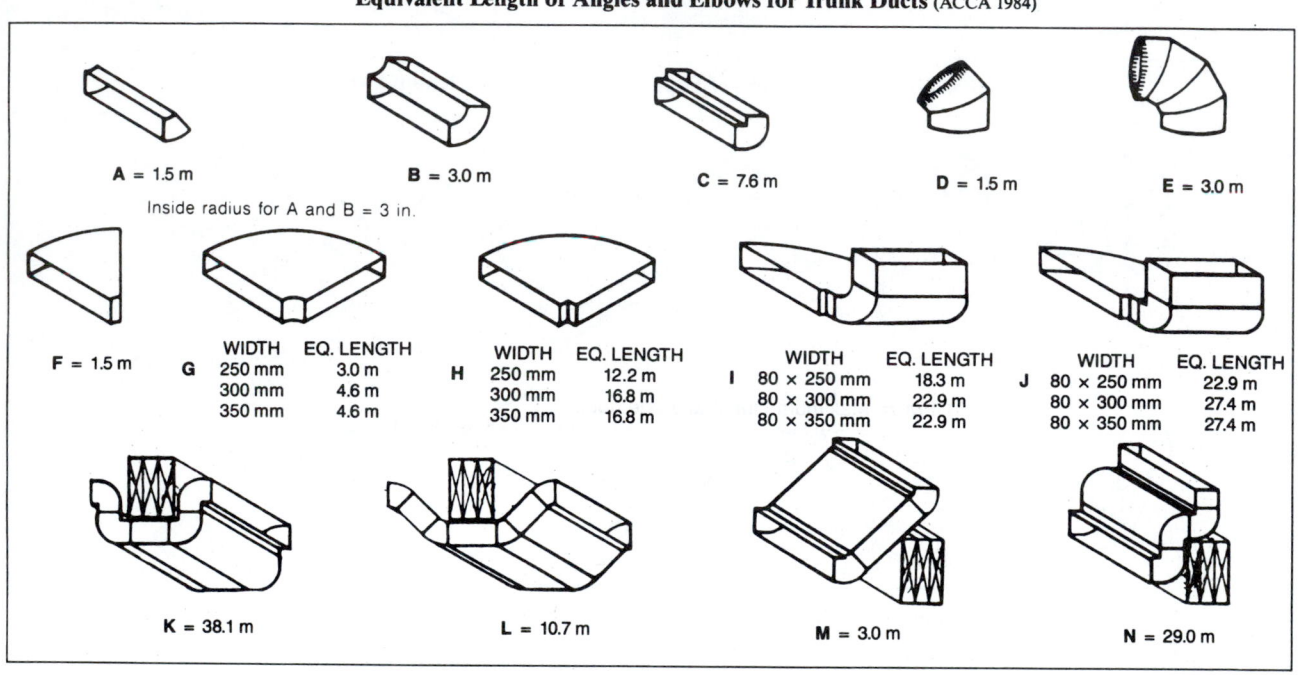

Equivalent Length of Angles and Elbows for Branch Ducts (ACCA 1984)

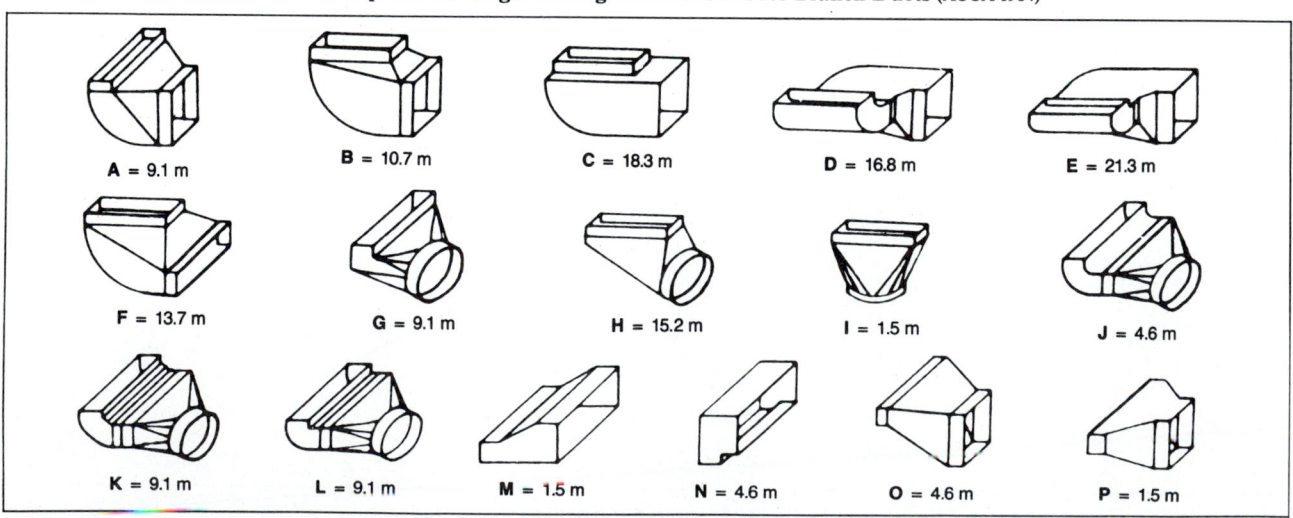

Equivalent Length of Boot Fittings

FIGURE 13-19   (*continued*)

CFM	Type A	Type B	Type C
1000	0.10	0.15	0.20
1500	.12	.17	.22
2000	.13	.18	.24
2500	.14	.20	.27
3000	.16	.22	.29
3500	.17	.24	.31
4000	.18	.26	.33
4500	0.20	0.28	0.36
5000	.21	.29	.38
5500	.22	.31	.40
6000	.24	.33	.42
6500	.25	.34	.45
7000	.27	.36	.47
7500	.28	.38	.49
8000	0.29	0.39	0.51
8500	.30	.41	.53
9000	.32	.43	.56
9500	.34	.45	.58
10,000	.34	.47	.60
10,500	.37	.48	.60
11,000	.39	.50	.60
11,500	0.40	0.50	0.60
12,000	.40	.50	.60
12,500	.40	.50	.60
13,000	.40	.50	.60
14,000	.40	.50	.60

CFM = Maximum cfm in any one duct.
Type A = Systems for quiet operation in residences, churches, concert halls, broadcasting studios, funeral homes, etc.
Type B = Systems for schools, theaters, public buildings, etc.
Type C = Systems for industrial buildings.

FIGURE 13-20   Suggested bonnet and return pressures (inches of water)

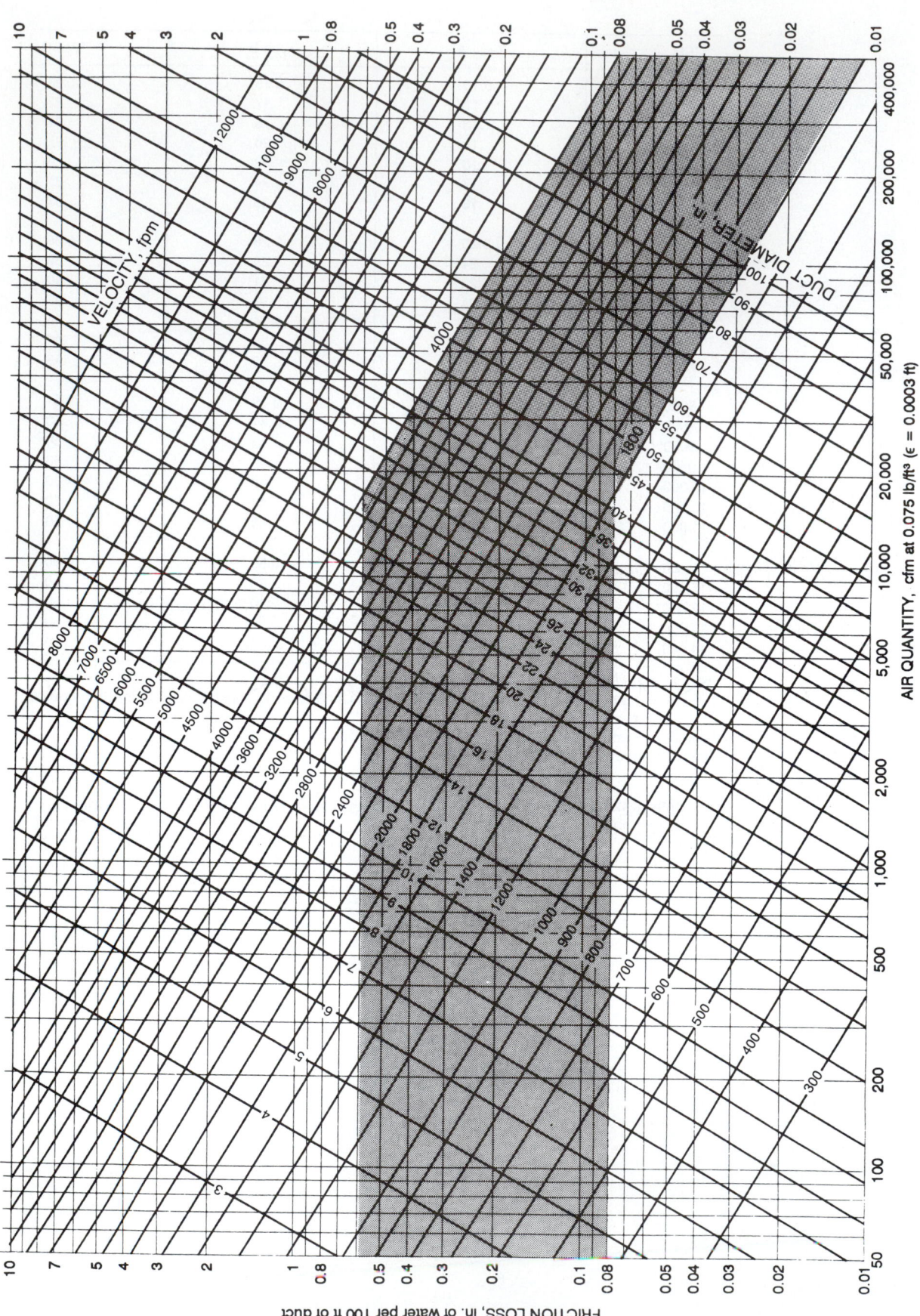

AIR QUANTITY, cfm at 0.075 lb/ft³ (ε = 0.0003 ft)

FIGURE 13-21   Friction chart

Reprinted with permission from ASHRAE, *Fundamentals Handbook*, 1993.

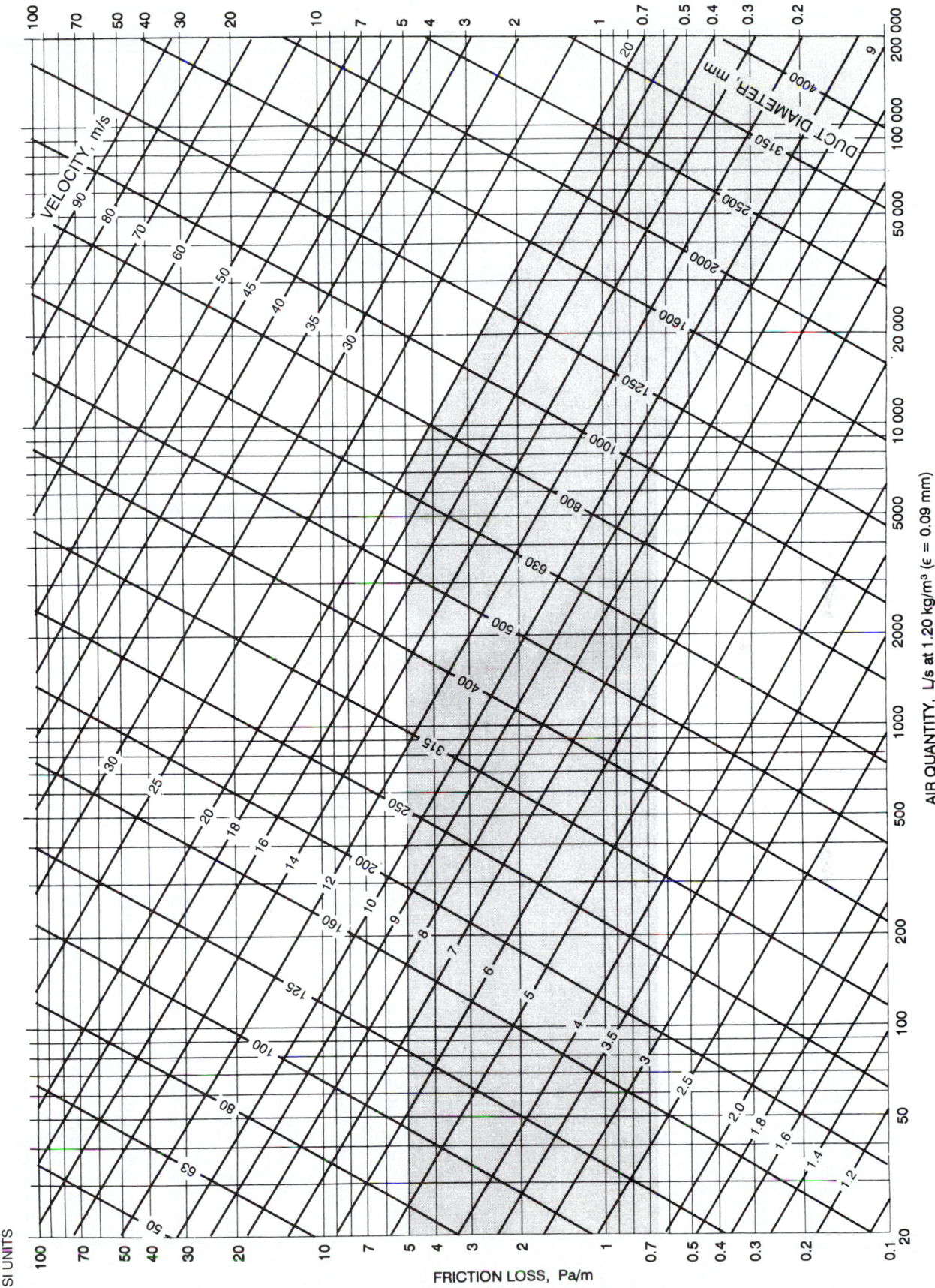

SI UNITS

FRICTION LOSS, Pa/m

AIR QUANTITY, L/s at 1.20 kg/m³ (ε = 0.09 mm)

FIGURE 13-21 Friction chart (*continued*)

349

Equivalent Rectangular Duct Dimension

Duct Diameter, in.	Rectangular Size, in.	Aspect Ratio														
		1.00	1.25	1.50	1.75	2.00	2.25	2.50	2.75	3.00	3.50	4.00	5.00	6.00	7.00	8.00
6	Width	—	6													
	Height	—	5													
7	Width	6	8													
	Height	6	6													
8	Width	7	9	9	11											
	Height	7	7	6	6											
9	Width	8	9	11	11	12	14									
	Height	8	7	7	6	6	6									
10	Width	9	10	12	12	14	14	15	17							
	Height	9	8	8	7	7	6	6	6							
11	Width	10	11	12	14	14	16	18	17	18	21					
	Height	10	9	8	8	7	7	7	6	6	6					
12	Width	11	13	14	14	16	16	18	19	21	21	24				
	Height	11	10	9	8	8	7	7	7	7	6	6				
13	Width	12	14	15	16	18	18	20	19	21	25	24	30			
	Height	12	11	10	9	9	8	8	7	7	7	6	6			
14	Width	13	14	17	18	18	20	20	22	24	25	28	30	36		
	Height	13	11	11	10	9	9	8	8	8	7	7	6	6		
15	Width	14	15	17	18	20	20	23	25	24	28	28	35	36	42	
	Height	14	12	11	10	10	9	9	9	8	8	7	7	6	6	
16	Width	15	16	18	19	20	23	23	25	27	28	32	35	42	42	48
	Height	15	13	12	11	10	10	9	9	9	8	8	7	7	6	6
17	Width	16	18	20	21	22	25	25	28	27	32	32	35	42	49	48
	Height	16	14	13	12	11	11	10	10	9	9	8	7	7	7	6
18	Width	16	19	21	23	24	25	28	28	30	32	36	40	42	49	56
	Height	16	15	14	13	12	11	11	10	10	9	9	8	7	7	7
19	Width	17	20	21	23	24	27	28	30	30	35	36	40	48	49	56
	Height	17	16	14	13	12	12	11	11	10	10	9	8	8	7	7
20	Width	18	20	23	25	26	27	30	30	33	35	40	45	48	56	56
	Height	18	16	15	14	13	12	12	11	11	10	10	9	8	8	7
21	Width	19	21	24	26	28	29	30	33	33	39	40	45	54	56	64
	Height	19	17	16	15	14	13	12	12	11	11	10	9	9	8	8
22	Width	20	23	26	26	28	32	33	36	36	39	44	50	54	56	64
	Height	20	18	17	15	14	14	13	13	12	11	11	10	9	8	8
23	Width	21	24	26	28	30	32	35	36	39	42	44	50	54	63	64
	Height	21	19	17	16	15	14	14	13	13	12	11	10	9	9	8
24	Width	22	25	27	30	32	34	35	39	39	42	48	55	60	63	72
	Height	22	20	18	17	16	15	14	14	13	12	12	11	10	9	9
25	Width	23	25	29	30	32	36	38	39	42	46	48	55	60	70	72
	Height	23	20	19	17	16	16	15	14	14	13	12	11	10	10	9
26	Width	24	26	30	32	34	36	38	41	42	46	52	55	66	70	72
	Height	24	21	20	18	17	16	15	15	14	13	13	11	11	10	9
27	Width	25	28	30	33	36	38	40	41	45	49	52	60	66	70	80
	Height	25	22	20	19	18	17	16	15	15	14	13	12	11	10	10
28	Width	26	29	32	35	36	38	43	44	45	49	56	60	66	77	80
	Height	26	23	21	20	18	17	17	16	15	14	14	12	11	11	10
29	Width	27	30	33	35	38	41	43	44	48	53	56	65	72	77	88
	Height	27	24	22	20	19	18	17	16	16	15	14	13	12	11	11
30	Width	27	31	35	37	40	43	45	47	48	53	60	65	72	77	88
	Height	27	25	23	21	20	19	18	17	16	15	15	13	12	11	11
31	Width	28	31	35	39	40	43	45	50	51	56	60	70	78	84	88
	Height	28	25	23	22	20	19	18	18	17	16	15	14	13	12	11
32	Width	29	33	36	39	42	45	48	50	54	56	60	70	78	84	96
	Height	29	26	24	22	21	20	19	18	18	16	15	14	13	12	12
33	Width	30	34	38	40	44	47	50	52	54	60	64	75	78	91	96
	Height	30	27	25	23	22	21	20	19	18	17	16	15	13	13	12
34	Width	31	35	39	42	44	47	50	52	57	60	64	75	84	91	96
	Height	31	28	26	24	22	21	20	19	19	17	16	15	14	13	12
35	Width	32	36	39	42	46	50	53	55	57	63	68	75	84	91	104
	Height	32	29	26	24	23	22	21	20	19	18	17	15	14	13	13
36	Width	33	36	41	44	48	50	53	55	60	63	68	80	90	98	104
	Height	33	29	27	25	24	22	21	20	20	18	17	16	15	14	13
38	Width	35	39	44	47	50	54	58	61	63	67	72	85	96	105	112
	Height	35	31	29	27	25	24	23	22	21	19	18	17	16	15	14

*Shaded area not recommended.

FIGURE 13-22   Rectangular duct chart

Equivalent Rectangular Duct Dimension (*Continued*).

Duct Diameter, in.	Rectangular Size, in.	Aspect Ratio														
		1.00	1.25	1.50	1.75	2.00	2.25	2.50	2.75	3.00	3.50	4.00	5.00	6.00	7.00	8.00
40	Width	37	41	45	49	52	56	60	63	66	70	76	90	96	105	120
	Height	37	33	30	28	26	25	24	23	22	20	19	18	16	15	15
42	Width	38	43	48	51	56	59	63	66	69	74	80	90	102	112	120
	Height	38	34	32	29	28	26	25	24	23	21	20	18	17	16	15
44	Width	40	45	50	54	58	61	65	69	72	81	84	95	108	119	128
	Height	40	36	33	31	29	27	26	25	24	23	21	19	18	17	16
46	Width	42	48	53	56	60	65	68	72	75	84	88	100	114	126	136
	Height	42	38	35	32	30	29	27	26	25	24	22	20	19	18	17
48	Width	44	49	54	60	62	68	70	74	78	88	92	105	120	126	136
	Height	44	39	36	34	31	30	28	27	26	25	23	21	20	18	17
50	Width	46	51	57	61	66	70	75	77	81	91	96	110	120	133	144
	Height	46	41	38	35	33	31	30	28	27	26	24	22	20	19	18
52	Width	48	54	59	63	68	72	78	83	84	95	100	115	126	140	152
	Height	48	43	39	36	34	32	31	30	28	27	25	23	21	20	19
54	Width	49	55	62	67	70	77	80	85	90	98	104	120	132	147	160
	Height	49	44	41	38	35	34	32	31	30	28	26	24	22	21	20
56	Width	51	58	63	68	74	79	83	88	93	102	108	125	138	147	160
	Height	51	46	42	39	37	35	33	32	31	29	27	25	23	21	20
58	Width	53	60	66	70	76	81	85	91	96	105	112	130	144	154	168
	Height	53	48	44	40	38	36	34	33	32	30	28	26	24	22	21
60	Width	55	61	68	74	78	83	90	94	99	109	116	130	144	161	
	Height	55	49	45	42	39	37	36	34	33	31	29	26	24	23	
62	Width	57	64	71	75	82	88	93	96	102	112	120	135	150	168	
	Height	57	51	47	43	41	39	37	35	34	32	30	27	25	24	
64	Width	59	65	72	79	84	90	95	99	105	116	124	140	156		
	Height	59	52	48	45	42	40	38	36	35	33	31	28	26		
66	Width	60	68	75	81	86	92	98	105	108	119	128	145	162		
	Height	60	54	50	46	43	41	39	38	36	34	32	29	27		
68	Width	62	70	77	82	90	95	100	107	111	123	132	150	168		
	Height	62	56	51	47	45	42	40	39	37	35	33	30	28		
70	Width	64	71	80	86	92	99	105	110	114	126	136	155			
	Height	64	57	53	49	46	44	42	40	38	36	34	31			
72	Width	66	74	81	88	94	101	108	113	117	130	140	160			
	Height	66	59	54	50	47	45	43	41	39	37	35	32			
74	Width	68	76	84	91	98	104	110	116	123	133	144	165			
	Height	68	61	56	52	49	46	44	42	41	38	36	33			
76	Width	70	78	86	93	100	106	113	118	126	137	148	165			
	Height	70	62	57	53	50	47	45	43	42	39	37	33			
78	Width	71	80	89	95	102	110	115	121	129	140	152				
	Height	71	64	59	54	51	49	46	44	43	40	38				
80	Width	73	83	90	98	104	113	118	124	132	144	156				
	Height	73	66	60	56	52	50	47	45	44	41	39				
82	Width	75	84	93	100	108	115	123	129	135	147	160				
	Height	75	67	62	57	54	51	49	47	45	42	40				
84	Width	77	86	95	103	110	117	125	132	138	151	164				
	Height	77	69	63	59	55	52	50	48	46	43	41				
86	Width	79	88	98	105	112	119	128	135	141	154	168				
	Height	79	70	65	60	56	53	51	49	47	44	42				
88	Width	80	90	99	107	116	124	130	138	144	158					
	Height	80	72	66	61	58	55	52	50	48	45					
90	Width	82	93	102	110	118	126	133	140	147	161					
	Height	82	74	68	63	59	56	53	51	49	46					
92	Width	84	94	104	112	120	128	138	143	150	165					
	Height	84	75	69	64	60	57	55	52	50	47					
94	Width	86	96	107	116	124	131	140	146	153	168					
	Height	86	77	71	66	62	58	56	53	51	48					
96	Width	88	99	108	117	126	135	143	151	159						
	Height	88	79	72	67	63	60	57	55	53						
98	Width	90	100	111	119	128	137	145	154	162						
	Height	90	80	74	68	64	61	58	56	54						
100	Width	91	103	113	123	132	140	148	157	165						
	Height	91	82	75	70	66	62	59	57	55						
102	Width	93	105	116	124	134	142	153	160	168						
	Height	93	84	77	71	67	63	61	58	56						
104	Width	95	106	117	128	136	146	155	162							
	Height	95	85	78	73	68	65	62	59							

'Shaded area not recommended.

FIGURE 13-22 (*continued*)

Equivalent Rectangular Duct Dimension (*Concluded*)

Duct Diameter, in.	Rectangular Size, in.	1.00	1.25	1.50	1.75	2.00	2.25	2.50	2.75	3.00	3.50	4.00	5.00	6.00	7.00	8.00
106	Width	97	109	120	130	140	149	158	165							
	Height	97	87	80	74	70	66	63	60							
108	Width	99	110	122	131	142	151	160	168							
	Height	99	88	81	75	71	67	64	61							
110	Width	101	113	125	135	144	153	163								
	Height	101	90	83	77	72	68	65								
112	Width	102	115	126	137	146	158	165								
	Height	102	92	84	78	73	70	66								
114	Width	104	116	129	140	150	160									
	Height	104	93	86	80	75	71									
116	Width	106	119	131	142	152	162									
	Height	106	95	87	81	76	72									
118	Width	108	121	134	144	154	164									
	Height	108	97	89	82	77	73									
120	Width	110	123	135	147	158										
	Height	110	98	90	84	79										

*Shaded area not recommended.

FIGURE 13-22   (*continued*)

Lgth Adj.[b]	Length of One Side of Rectangular Duct (a), mm																			
	100	125	150	175	200	225	250	275	300	350	400	450	500	550	600	650	700	750	800	900
100	109																			
125	122	137																		
150	133	150	164																	
175	143	161	177	191																
200	152	172	189	204	219															
225	161	181	200	216	232	246														
250	169	190	210	228	244	259	273													
275	176	199	220	238	256	272	287	301												
300	183	207	229	248	266	283	299	314	328											
350	195	222	245	267	286	305	322	339	354	383										
400	207	235	260	283	305	325	343	361	378	409	437									
450	217	247	274	299	321	343	363	382	400	433	464	492								
500	227	258	287	313	337	360	381	401	420	455	488	518	547							
550	236	269	299	326	352	375	398	419	439	477	511	543	573	601						
600	245	279	310	339	365	390	414	436	457	496	533	567	598	628	656					
650	253	289	321	351	378	404	429	452	474	515	553	589	622	653	683	711				
700	261	298	331	362	391	418	443	467	490	533	573	610	644	677	708	737	765			
750	268	306	341	373	402	430	457	482	506	550	592	630	666	700	732	763	792	820		
800	275	314	350	383	414	442	470	496	520	567	609	649	687	722	755	787	818	847	875	
900	289	330	367	402	435	465	494	522	548	597	643	686	726	763	799	833	866	897	927	984
1000	301	344	384	420	454	486	517	546	574	626	674	719	762	802	840	876	911	944	976	1037
1100	313	358	399	437	473	506	538	569	598	652	703	751	795	838	878	916	953	988	1022	1086
1200	324	370	413	453	490	525	558	590	620	677	731	780	827	872	914	954	993	1030	1066	1133
1300	334	382	426	468	506	543	577	610	642	701	757	808	857	904	948	990	1031	1069	1107	1177
1400	344	394	439	482	522	559	595	629	662	724	781	835	886	934	980	1024	1066	1107	1146	1220
1500	353	404	452	495	536	575	612	648	681	745	805	860	913	963	1011	1057	1100	1143	1183	1260
1600	362	415	463	508	551	591	629	665	700	766	827	885	939	991	1041	1088	1133	1177	1219	1298
1700	371	425	475	521	564	605	644	682	718	785	849	908	964	1018	1069	1118	1164	1209	1253	1335
1800	379	434	485	533	577	619	660	698	735	804	869	930	988	1043	1096	1146	1195	1241	1286	1371
1900	387	444	496	544	590	663	674	713	751	823	889	952	1012	1068	1122	1174	1224	1271	1318	1405
2000	395	453	506	555	602	646	688	728	767	840	908	973	1034	1092	1147	1200	1252	1301	1348	1438
2100	402	461	516	566	614	659	702	743	782	857	927	993	1055	1115	1172	1226	1279	1329	1378	1470
2200	410	470	525	577	625	671	715	757	797	874	945	1013	1076	1137	1195	1251	1305	1356	1406	1501
2300	417	478	534	587	636	683	728	771	812	890	963	1031	1097	1159	1218	1275	1330	1383	1434	1532
2400	424	486	543	597	647	695	740	784	826	905	980	1050	1116	1180	1241	1299	1355	1409	1461	1561
2500	430	494	552	606	658	706	753	797	840	920	996	1068	1136	1200	1262	1322	1379	1434	1488	1589
2600	437	501	560	616	668	717	764	810	853	935	1012	1085	1154	1220	1283	1344	1402	1459	1513	1617
2700	443	509	569	625	678	728	776	822	866	950	1028	1102	1173	1240	1304	1366	1425	1483	1538	1644
2800	450	516	577	634	688	738	787	834	879	964	1043	1119	1190	1259	1324	1387	1447	1506	1562	1670
2900	456	523	585	643	697	749	798	845	891	977	1058	1135	1208	1277	1344	1408	1469	1529	1586	1696

Lgth Adj.[b]	Length of One Side of Rectangular Duct (a), mm																			
	1000	1100	1200	1300	1400	1500	1600	1700	1800	1900	2000	2100	2200	2300	2400	2500	2600	2700	2800	2900
1000	1093																			
1100	1146	1202																		
1200	1196	1256	1312																	
1300	1244	1306	1365	1421																
1400	1289	1354	1416	1475	1530															
1500	1332	1400	1464	1526	1584	1640														
1600	1373	1444	1511	1574	1635	1693	1749													
1700	1413	1486	1555	1621	1684	1745	1803	1858												
1800	1451	1527	1598	1667	1732	1794	1854	1912	1968											
1900	1488	1566	1640	1710	1778	1842	1904	1964	2021	2077										
2000	1523	1604	1680	1753	1822	1889	1952	2014	2073	2131	2186									
2100	1558	1640	1719	1793	1865	1933	1999	2063	2124	2183	2240	2296								
2200	1591	1676	1756	1833	1906	1977	2044	2110	2173	2233	2292	2350	2405							
2300	1623	1710	1793	1871	1947	2019	2088	2155	2220	2283	2343	2402	2459	2514						
2400	1655	1744	1828	1909	1986	2060	2131	2200	2266	2330	2393	2453	2511	2568	2624					
2500	1685	1776	1862	1945	2024	2100	2173	2243	2311	2377	2441	2502	2562	2621	2678	2733				
2600	1715	1808	1896	1980	2061	2139	2213	2285	2355	2422	2487	2551	2612	2672	2730	2787	2842			
2700	1744	1839	1929	2015	2097	2177	2253	2327	2398	2466	2533	2598	2661	2722	2782	2840	2896	2952		
2800	1772	1869	1961	2048	2133	2214	2292	2367	2439	2510	2578	2644	2708	2771	2832	2891	2949	3006	3061	
2900	1800	1898	1992	2081	2167	2250	2329	2406	2480	2552	2621	2689	2755	2819	2881	2941	3001	3058	3115	3170

[a] Table based on $D_e = 1.30 (ab)^{0.625}/(a + b)^{0.25}$.
[b] Length of adjacent side of rectangular duct (b), mm.

FIGURE 13-22 *(continued)*

QUESTIONS

13-1. What term is commonly used to refer to the cooling capacity of a cooling unit, and how does it relate to Btuh?

13-2. What does the term SEER mean, and why is it important to check the SEER of the cooling unit?

13-3. What type of equipment may be used when both heating and air conditioning are required?

13-4. What factors determine where the supplies and returns are located?

Design Exercises

13-5. Design a forced air heating and cooling system for the residence in Appendix D, based on the heat loss calculations in Exercise 8-15 and the heat gain calculations in Exercise 12-7.

13-6. Design a forced air heating and cooling system for one of the top-floor apartments and one of the lower-floor apartments in Appendix A. Base the design on the heat loss calculations in Exercises 8-17 and the heat gain calculations in Exercise 12-8.

CHAPTER 14

Solar Energy and Heat Pumps

14-1 OVERVIEW

The idea of harnessing the sun's energy for use in homes and factories is nothing new. Work began on solar furnaces more than two centuries ago—but with little success. In the United States, solar hot water heaters were used in the early 1900s in Florida, Arizona, and California. The introduction of mass-produced hot water heaters which were low in cost and which used inexpensive oil, natural gas, and electricity all but stopped the further development of solar hot water heaters. The higher initial cost of the solar unit made it uneconomical except where the price of oil or electricity was high.

Since the 1930s, limited experimentation continued in the application of solar energy, and interest increased markedly over World War II. But by 1960, the basic obstacle which had to be overcome was that the solar units were not "economically feasible" when compared with the low cost of other fuels. Not being "economically feasible" means that the amount of money saved on fuel costs is not sufficient to pay the increased cost of a solar system over a set period (say, 10 years).

14-2 USES

Solar energy is being used to:

1. Heat hot water for use in buildings.
2. Provide heating for buildings.
3. Provide cooling for buildings.
4. Provide heating and cooling for buildings.
5. Provide heating and hot water for buildings.
6. Provide heating, cooling, and hot water for buildings (Fig. 14-1).

As outlined above, quite often the solar energy package will provide more than one service, usually at very little extra cost.

In this chapter, we will review some solar systems available, how they work, and what they have to offer.

14-3 COLLECTORS AND STORAGE

All systems are made up of two basic components—collectors and storage areas.

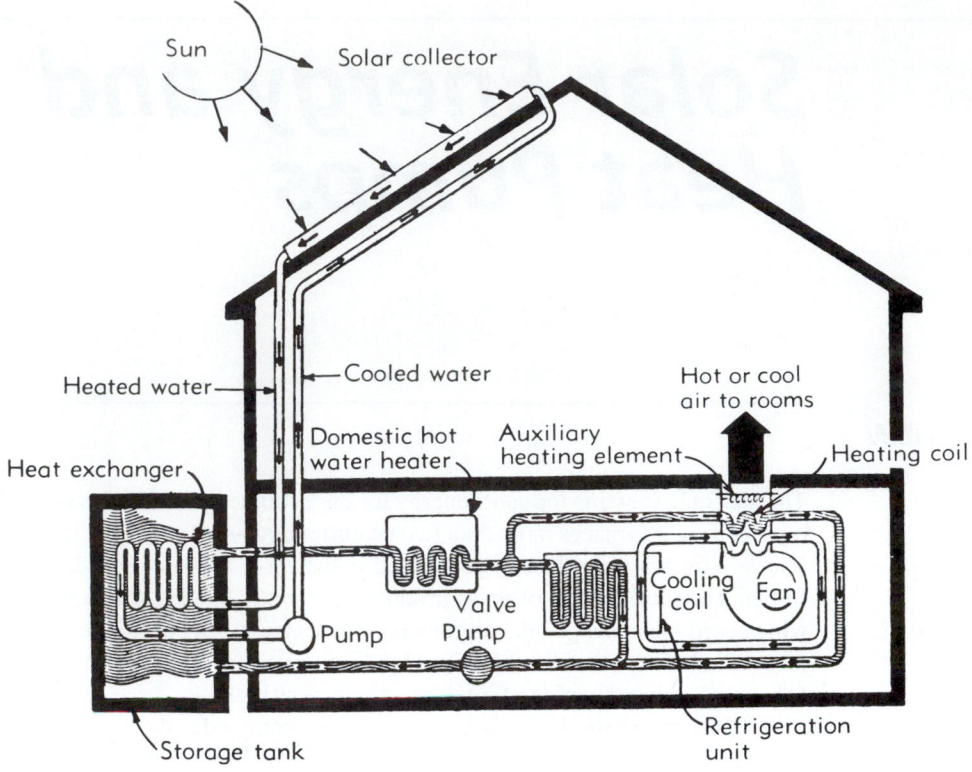

FIGURE 14-1   Solar heating and cooling system

Collectors

First, all systems must have some type of collector. The flat-plate collector (shown in Fig. 14-2) is the collector most commonly used to gather in the heat from the sun. Basically, all collectors are similar in that they have a casing (or frame), insulation, plate, heat transfer medium (liquid or gas), and glazing (Fig. 14-3). There is no one design that is used, however, and several different collector designs are shown in Fig. 14-4. But it is important to note there are two basic types of heat transfer mediums—*liquid and air*—and this makes for two basic types of collectors—liquid and air.

The collector is located outside and is angled to receive the maximum amount of sunshine possible (Fig. 14-5). To increase the amount of the sun's rays which hit the collector,

FIGURE 14-2   Flat plate collector installation

FIGURE 14-3   Flat plate collector section

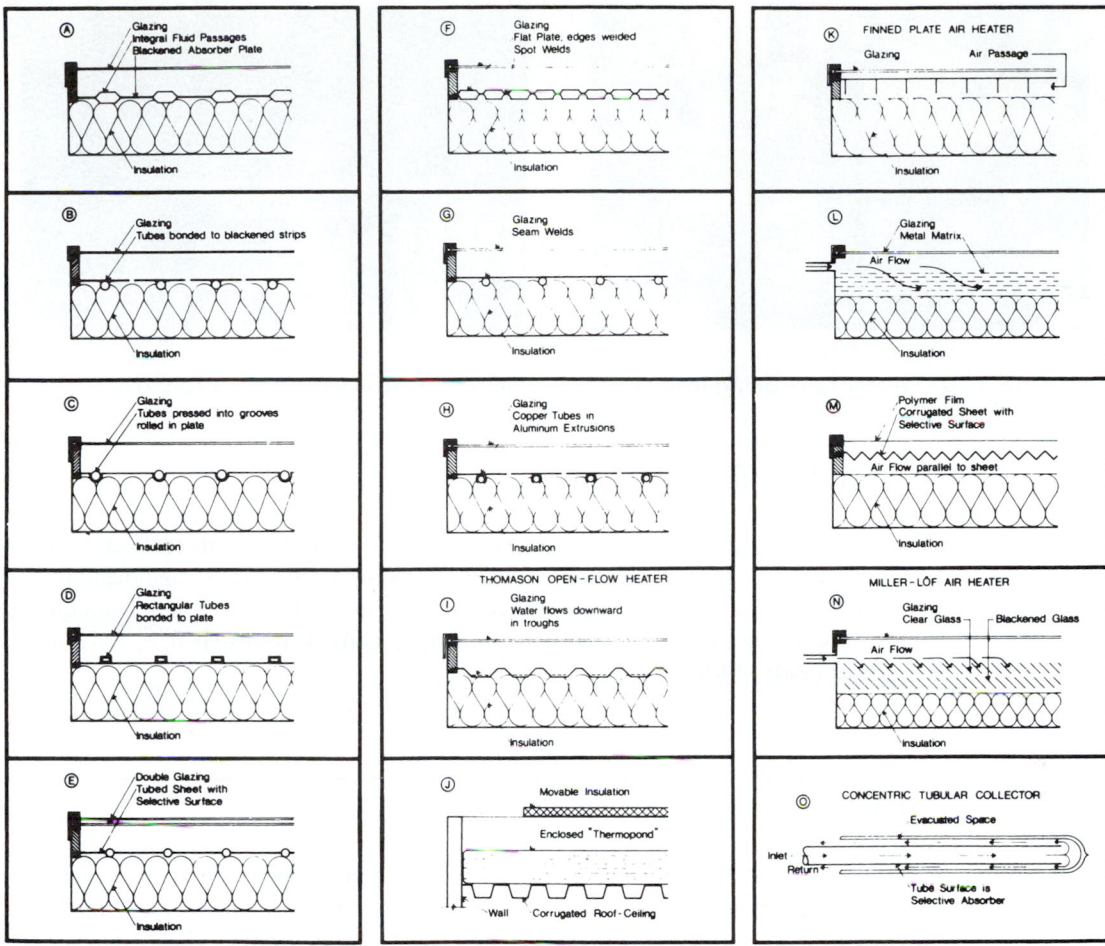

FIGURE 14-4   Variations of solar water and air heaters

FIGURE 14-5   Collectors angled to receive sun

FIGURE 14-6   Reflective surface

some systems even use a reflective surface in front of the collector (Fig. 14-6), with the reflective panel placed either on an adjoining flat roof or in front of the collector. The collector is commonly located on the roof, and many units are also designed to be placed on the ground.

FIGURE 14-7   Water tank being installed   FIGURE 14-8   Solar hot air collector

Storage Media

Once the heat exchange medium (liquid or air) is warmed in the collector, the heat it absorbs is transferred to a storage medium for future use. The storage mediums most commonly used are coarse aggregates (clean, washed rock), water, and a combination of the coarse aggregates and water. A water storage tank, well insulated, is about to be covered over in Fig. 14-7.

14-4   HEATING, COOLING, AND HOT WATER

The system shown in Fig. 14-1 utilizes a liquid heat transfer medium (water or water and anti-freeze), a collector on the roof, and water as the heat storage medium. Such a system is the most versatile, providing the most benefits to the owner. The system shown has auxiliary units for heating, cooling, and hot water for the times when the solar system cannot provide sufficient heat (or cooling).

Heating

The heat transfer medium is heated in the collector and is then circulated through the storage tank, transferring its heat to the storage medium—in this case, water. This heated water then circulates through the hot water tank to heat all or part of the hot water required for the building. The heated water then passes a fan which forces air past the hot water coils. The air passing the coils is warmed and circulates through the building.

There are many variations on the design; in Fig. 14-1, the water from the storage tank passes first through the fan and forced air and then through the hot water tank. A hot water heating system is shown in Fig. 14-8 and discussed in Sec. 14-5.

Cooling

To provide cooling for the building, the system is, in effect, reversed. While the collectors used to gather the heat ideally face south, the radiators used to give off heat to the exterior face north. (Although some systems utilize the collectors already on the south, it is not as effective.)

In order to use the system for cooling, the storage medium must be made as low in temperature (cool) as possible; then it circulates past the refrigerant in the refrigeration unit, which then circulates past the fan to blow cool air into the building. One method used is to have the heat pump transfer heat from the house to the heat storage medium and into the storage tank, where the temperature of the entire storage medium slowly increases. Then at night, the radiation system turns on and the heat transfer medium cools the storage medium by absorbing the heat as it passes through the tank. Circulating it up through the night radiator collectors facing north, the collectors radiate the heat to the surrounding atmosphere, and then the medium goes back to the storage tank to constantly repeat the process of lowering the water temperature in the storage tank so it can be used for cooling.

Another method used with a system of this type is to connect to a well; then if the water in the storage tank reaches a certain temperature (about 65°F), the system will automatically switch to the well as its source of cool water. Since many of the same controls (valves and pumps) are utilized in both the heating and the cooling operations, it is often this dual feature that makes the system economically feasible.

Another method used converts the sun's energy into the energy source used to power the cooling equipment. In this type of installation, storage batteries are charged during the daytime hours to operate a heat pump.

The use of solar energy for cooling is not nearly as advanced as its use for heating. A great deal more research is required before economical, dependable solar cooling units are available.

14-5   HOT WATER HEATING

Hot water solar systems are similar to the system described (in Sec. 14-4) except that the storage medium (water) is used to heat the hot water, which is then circulated through the system. An auxiliary heat source is still required to supplement the solar system. This system could also be used for cooling (similar to Fig. 14-1) if the devices in the building are selected to effectively handle chilled water. Realistically, this is rarely the case in residential work. Chilled water systems are discussed in Sec. 7–1. Solar hot water is also discussed in Sec. 2–18.

14-6   HOT AIR SYSTEM (HEATING)

A hot air heating system uses air as the heat transfer medium and rocks as the heat storage medium. As Fig. 14-8 shows, the collector may be located on the ground. The air in the collector is heated as it is pulled across the vanes in the collector, and it is then pushed over the rocks in the storage medium, transferring much of the heat collected to the rocks. This continuous flow of air is constantly warmed in the collector and then cooled as it transfers its heat to the rocks.

When the thermostat in the building calls for heat, the distribution fans come on and circulate the cooler air from the building over the rocks in the storage area, and this warmed air is then circulated back through the building.

When both the collection and the distribution are on, the air flow from the building passes through the collector to provide heat directly from the collector. This commonly occurs when the building thermostat calls for heat during the daytime while the collector is also activated to gather the heat from the sun. With a system of this type, it is usually recommended that there be continuous air circulation in the building. A solar system of this type is designed to continuously circulate air in the building, air that is relatively cooler than that from the forced air system using a furnace. In the solar system, the air will usually feel just slightly warm, or perhaps even a little cool to the touch since the air temperature will be lower than your body temperature. Studies of systems using continuous air circulation show that less heat is required when such a system is used.

Also important in this system is the fact that it is easily adaptable to existing forced air systems. In such an installation, the collector could be set on the ground.

14-7 SUPPLEMENTARY HEAT

Almost all solar systems are more expensive to install than conventional boiler or furnace systems. As discussed in Chapter 8, the conventional system is sized to deliver all of the heat required down to a particular design temperature ($-10°F$ in Rutland, Vt., and 20°F in Raleigh, N.C.). It is quite inexpensive to size a boiler or furnace to provide all of the heat required to meet those demands, even with an extra 20% pickup load added on. But keep in mind that during most of the heating season, the temperatures are well above the design temperature used. (Many times it might reach the low during the night, but it warms up 15° to 25°F during the day.)

Yet providing all of the heat required to keep a house warm in a northern city on *the coldest day ever recorded there* would require a solar unit about four times as large as one that would provide 90% of the heat required during the entire heating season.

For these reasons, many of the solar systems have auxiliary or supplemental heating systems along with the solar system. In this situation, the solar system would provide a certain percentage of the heat required (perhaps 40%, 50%, 60%, or more), and the supplemental heating boiler or furnace would provide the rest. In this manner, the most economically feasible installation can be obtained, and the problem of allocating huge areas to contain the storage medium is overcome.

A solar system with supplemental heat can be designed to save a significant amount of the total heat bill, but it is important to realize that it is the *total annual fuel bill* being discussed, that the savings will be highest in the milder months, and that the heat bills will increase in the colder months. So it is the "average" being discussed, as illustrated in Fig. 14-9.

While this type of system may not satisfy the "purist" in terms of total solar systems, it provides a very "cost-effective" system which will save large amounts of fuel. In addition, it makes solar systems easily adaptable to existing installations where the existing system may be used as the supplemental system. It is most important not only that fuels be conserved in new construction but also that systems be developed that are adaptable to existing heating systems (hot water and air).

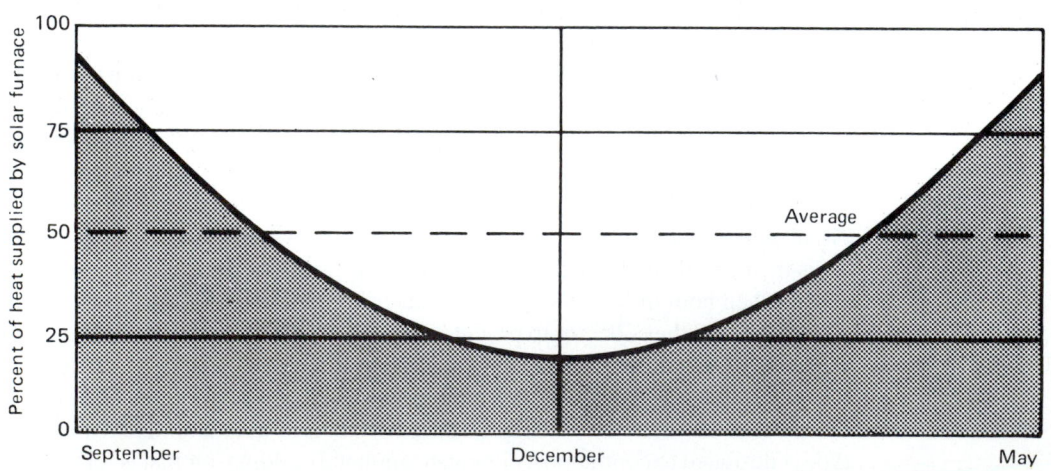

FIGURE 14-9   Yearly heat supplied

14-8   HEAT PUMPS

Use of refrigeration to provide cooling is discussed in Sec. 7–4, as are the basic principles which apply to such use. The use of refrigeration to provide both heating and cooling for a space is accomplished with a device called a *heat pump.* This increasingly popular form of heating and cooling a space requires explanation as to the principles involved and its advantages and disadvantages.

First, the discussion of the cooling principles of refrigerants in Sec. 7–4 must be carefully read and understood. In effect, it states that the refrigerant can take heat from one place (air or water) and move it to another by use of an evaporator, compressor, and condenser.

Now, using those basic cooling principles and a valve control, it is possible to reverse the cycle, causing the refrigerant to absorb heat from an outside surrounding medium (usually air, but water is sometimes used) and to release this heat inside the building (in the air or water being used to heat the building).

The reversing valve control allows the refrigerant to be used to provide cooling or heating, depending on the direction of flow after it leaves the compressor. If the refrigerant flows to the condenser, it provides cooling; if it flows toward the evaporator section, it provides heating.

The most commonly used surrounding mediums are the outside air and the forced air in the ductwork.

Outside air is the surrounding medium most likely to be used to draw heat from in the winter for heating and to release heat to in the summer for cooling. The temperature of this outside air affects the efficiency of the unit in providing heating and cooling Btuh. Typically, in the winter as the outside air temperature goes down, the amount of heat which the heat pump will produce goes down also. In effect, the more heat needed in the space, the less Btuh the unit will provide.

At this point, it is important to realize that a heat pump is not electric heat. Actually, it is a type of solar heating which draws heat from the surrounding air. It requires electrical power to run the equipment, but it is more efficient than electric heat. The efficiency of the heat pump can be determined from the manufacturer's engineering data. While there are many sizes and capacities of units available, we will review the typical data for a 3-ton (36,000-Btuh) cooling capacity unit. Since each manufacturer's units vary, so will their data; and as more efficient units are developed, the result will be higher values from the heat pump.

Note in Fig. 14-10 that at a 47°F (8°C) outside temperature, the unit will provide 36,000 Btuh (11,300 W) for heating. As the temperature goes down, the unit will supply less heat, and yet more heat is required inside the space. When it gets cold—say, 17°F (−8°C)—outside, the unit will provide 19,600 Btuh. It is obvious that as the temperature goes down, so does the ability of the heat pump to produce heat.

It is when the heat pump alone is producing the heat that the unit is economical to operate. The coefficient of performance [(Btuh/watts) ÷ 3.14] is not always given by the manufacturer, but it has been calculated for the 3-ton unit in Fig. 14-11. This coefficient of performance offers a comparison of the heat pump and electric resistance heat. The table shows a COP of 3.03 at 47°F (8°C), dropping to 1.85 at 17°F (−8°C). Due to this fluctuation, it is difficult to give an actual comparison with other fuels and systems. It depends very much on the geographic area it is being used in and how many hours it operates at the different temperatures.

For comparison, electric heat has a COP of 1.0. As shown in Sec. 8–19, electric heat tends to cost about twice as much as oil or natural gas. So, for the heat pump to be competitive, it needs to operate at a COP of about 2.0, and at 40°F (4°C). But at lower temperatures, it becomes less and less efficient.

Typical 3-ton (36,000 Btuh) heat pump
Cooling capacity, 62° outside wet bulb,
85° air temperature entering evaporator.

| Outside air temperature | | Cooling | | Operating | |
db°F	db°C	Btuh	Watts	Watts	SEER
85	29	38,600	11,300	4200	9.19
95	35	36,400	10,655	4430	8.21
105	41	32,000	9,370	4750	6.73

Heating capacity

| Outside air temperature | Heating | | Operating | |
	Btuh	Watts	Watts	COP
47°F/8°C	36,600	10,540	3842	3.03
17°F/−8°C	19,600	5,730	3364	1.85

FIGURE 14-10   Typical three-ton heat pump

Since the heat pump puts out 19,600 Btuh (5,730 W) at 17°F (−8°C), the rest of the heat is supplied by electric heat resistance units, which mount inside the evaporator blower discharge area; or else electric duct heaters, which are placed right in the ductwork, are used to supplement the heat pump. These supplemental units run on electricity and provide the standard 3,413 Btuh per 1,000 W (watts) (1 k W). These electric resistance units are generally available in increments of 3, 5, 10, 15, and 20 kW, depending on the manufacturer. Many of the smaller heat pump units have space to install only the smaller sizes, perhaps to a limit of 10 kW, and the manufacturer's information must be checked. The large units, such as the 3-ton unit being discussed here, will take the 15-kW unit.

The 15-kW unit would be made up of three 5-kW elements which would operate at three different stages set to provide the required heat to the space. The first stage is controlled by the room thermostat, and it typically operates on a 2°F (1°C) differential from the setting on the room thermostat. This means that if the room thermostat is set, say, at 70°F (21°C), as the thermostat calls for heat, the heat pump goes on and begins to send heat to the space. If the temperature of the room falls below 68°F (20°C), then the first 5-kW unit would come on and begin to provide additional heat to supplement the heat pump. The second and third stages of electrical resistance heat are activated by outdoor thermostat settings which are adjustable (from 50° to 0°F or 10°C to −18°C).

This differential of 2°F (1°C) between the thermostat setting and the room temperature, which activates the first stage, may mean that in cold weather the room temperature may stabilize at 68°F (20°C). This is because 68°F (20°C) is the temperature at which the first stage shuts off; to get a 70°F (21°C) reading on those days, it may be necessary to raise the thermostat reading to about 72°F (22°C).

The cooling efficiency of the heat pump should also be checked. Many times, the less expensive units have much lower efficiency ratings than the individual cooling units discussed in Sec. 13–2. It is important that the most efficient model be selected. For example, the 3-ton unit being discussed in this section has its engineering data shown in Fig. 14-10. As the temperature goes up, the cooling Btuh provided goes down; at 85°F (29°C) it produces 38,600 Btuh (11,300 W) for cooling using 4,200 W, while at 95°F (35°C) it produces 36,400 Btuh (10,655 W) using 4,430 W. Its SEER (seasonal energy efficiency ratings) ranges from 9.19 to 6.73 as listed in the table. The SEER most commonly used for comparison of cooling units is based on its performance at a 95°F temperature; based on that, this unit's SEER would be 8.21.

In reviewing the operation and efficiency of the heat pump:

1. The heat pump is more efficient in warmer climates than in cooler areas. In Florida and southern California, for example, heat pumps are reasonably efficient. As you approach "mid-South" states, those states in a band from North Carolina and Virginia on the east to Northern California on the west, the efficiency begins to fall off. In northern states, these units would depend on the electric resistance heat for so much of the time that efficiency is minimal.

2. Technology will continue to improve the efficiency of the heat pump. It is part of the designer's responsibility to keep abreast of developments that will provide increased efficiency.

3. A comparison of the heat pump with the other fuels and systems must be made for each geographical area, and its complexity suggests that a computer be used. This type of analysis is available in many areas now. However, the designer should be aware that it is being made available by the same people who are selling heat pumps and perhaps by those who sell the electricity to run the unit. This is not to infer that it would not be "technically accurate"; only that the utmost caution should always be used when using calculations, analyses, claims, and the like from any interested party (such as the company selling the unit). The best idea is to check actual installations and compare the costs with those of comparable installations. But be certain that they are comparable installations, that the thermostats are set at about the same temperatures, and that the actual electrical and fuel bills are available (don't rely on word of mouth).

4. While heat pumps are most efficient than electric resistance heat, in northern and mid-South states actual use has shown that it is not less expensive to heat with a heat pump than with oil and natural gas fuels.

5. While the currently available heat pump is a solar unit, it will probably soon be available with a solar collector panel which can be attached to it to provide greater efficiency. This may be one of the first practical, small and mid-sized solar solar system projects. Currently, at least one manufacturer is working on its development.

6. The shortage of fuel oil due to the embargo in the early 1970s, the shortage of natural gas available to the user, and the general uncertainty over future supplies and prices of these fuels have caused many owners and designers to consider using the heat pump in many areas where it might not ordinarily be considered. In effect, most people feel that, one way or another, at least with the heat pump, the fuel (the electricity) will be available to run the unit and provide the heating and cooling required.

7. One large unit—instead of two smaller units—is efficient and generally less expensive, especially during the heating season, unless one section of the house (or building) will be closed off. For example, based on the data in Fig. 14-11, if a building requires a 4-ton cooling unit as 17°F (-8°C), the heat pump will provide 29,000 Btuh

4-ton heat pump, heating capacity

Outside air temperature		Heating		Operating
db°F	db°C	Btuh	Watts	Watts
47	8	50,000	10,540	5293
17	−8	29,000	8,490	4315

4-ton heat pump, heating capacity

Outside air temperature		Heating		Operating
db°F	db°C	Btuh	Watts	Watts
47	8	24,600	7,220	3250
17	−8	14,000	4,100	2665

FIGURE 14-11   4-ton and 2-ton heat pumps

(8490 W)for heating while using 4315 W. Using two 2-ton units, at 17°F (-8°C) each unit will provide 14,000 Btuh (4100 W) using 2665 W, for a total of 28,000 Btuh (8200 W) using 5330 watts (about 23% more). Now, if part of the building were to be closed off—say, a large bedroom wing—with the temperature set quite low, it might be more economical to have two units—in this case, one unit to serve the bedrooms and one for the rest of the building.

QUESTIONS

14-1. What factors have made solar energy more attractive in recent years?

14-2. What may solar energy be used for in a building?

14-3. Briefly describe what a *collector* is and how it works.

14-4. What are the two basic types of heat transfer mediums?

14-5. What are the basic components of a collector?

14-6. What are the commonly used storage mediums, and what is the function of a storage medium in the solar energy system?

14-7. Sketch a schematic of a solar energy system providing heating and hot water only; label all the parts.

14-8. Sketch a schematic of a solar energy system providing heating, cooling, and hot water; label all the parts.

14-9. Why is supplementary heat often *required* in solar heating systems?

14-10. Why might it be *desirable* to have supplemental heat with a solar heating system?

14-11. Describe briefly how a heat pump operates.

14-12. What happens to the heating and cooling outputs of a heat pump as the temperature varies?

CHAPTER 15

Electrical Systems

15-1   CODES

All buildings require electrical systems to provide power for the lights and to run various appliances and equipment. The safety of the system is of prime importance, and minimum requirements are included in building codes. Most applicable codes have separate electrical sections, or else completely separate electrical codes are prepared. In addition, many local codes make reference to the *National Electric Code* (NEC) or are based on the national code. The designer must first determine what code is applicable to the locale in which the building will be built, and then be certain that the electrical design is in accordance with the code. The NEC is used as a basis for this portion of the text. Generally, these codes place limitations on the type and size of the wiring to be used, the circuit size, the outlet spacings, the conduit requirements, and the like. The tables used in this text are from the 1993 NEC. Be certain that you always have a copy of the latest edition available for your use.

15-2   UNDERWRITERS LABORATORIES (UL)

UL is an independent organization which tests various electrical fixtures and devices to determine if they meet minimum specifications as set up by UL. The device to be tested is furnished by the manufacturer, and if the test shows that it meets the minimum specifications, it will be put on the UL official published list, referred to as "listed by Underwriters Laboratories, Inc." The approved device may then have a UL label on it. A typical UL seal is shown in Fig. 15-1, and many consumers will not buy any electrical device which does not have a UL label.

FIGURE 15-1   Typical Underwriters Laboratory seal

15-3   LICENSES

Most municipalities have laws which require that any person who wishes to engage in the business of installing electrical systems must be licensed (usually by the state or the province) to do so. This generally means that the person must have a minimum number of years of experience working with a licensed electrician and must pass a written test which deals primarily with the electrical code being used and with methods of installation.

By requiring a license, it is assured that the electrician knows, at a minimum, the code requirements and the installation procedures. There are areas where no laws require that only licensed electricians may install electrical systems, and, in effect, there is no protection for the consumer against an unskilled electrician. Always insist on licensed electricians for all installations.

15-4   PERMITS

Many municipalities require a permit before any electrical installations may be made on the project. Depending on the municipality, a complete electrical drawing may be required and may even be reviewed by the municipality before installation may begin, while others may require no drawings at all. In general, most municipalities that require electrical permits also require licensed electricians.

In addition, these municipalities will probably have electrical inspectors, trained personnel who check the project during regularly scheduled visits. Typically, they will want to inspect the installation after the rough wiring is in and before it is concealed in the construction, and again when all of the fixtures and devices are installed and wired back to the panel and the service and meter installed.

On large projects, it may be necessary for many electrical inspections since the work may be done in stages. For example, conduit which will be encased in concrete may have to be checked before the concrete is poured, and conduit to be built into the masonry walls will have to be checked before the walls are begun. These types of covering up will occur throughout the project. Be certain that the installer and the designer are aware of when inspections are required and of what will be inspected. Also, it is important that close coordination and cooperation be maintained with the inspector, since the inspector could slow down the progress of the work if inspection is not made promptly. Whenever possible, the inspector will need to know as early as possible when inspections will be necessary.

15-5   TERMINOLOGY

Circuit: Two or more wires which carry electricity from the source to an electrical device and back.

Circuit Breaker: A switch which automatically stops the flow of electricity in a particular circuit when the circuit is overloaded.

Conductor: The wire used to carry electricity.

Conduit: A channel or tube which is designed to carry the conductors in locations where the conductors need protection.

Convenience Outlet: An outlet which receives the plugs of electrical devices, such as lamps, radios, clocks, etc. Also referred to as a *receptacle.*

Fixtures: The lighting fixtures used. They may be wall- or ceiling-mounted, recessed, or surface-mounted. Also included are table and floor lighting fixtures.

Ground: To minimize injuries from shock and possible damage from lightning, the electrical system should be equipped with a wire (called a ground) that connects to the earth.

Service Entrance: The wires, fittings, and equipment which bring the electricity into the building.

Service Panel: The main panel which receives the electricity at the service entrance, breaks it down, and distributes it through the various circuits.

Switch: The control used to turn the flow of electricity on or off to the electrical device to which it is connected.

15-6   SERVICE ENTRANCE

The service entrance furnishes electricity to a house; a three-wire service bringing in 120/240 volt, single-phase power is generally standard. These three wires attach to the house at a mast (Fig. 15-2), are installed underground (Fig. 15-3), and run through a metal conduit through the meter and into the service panel. The meter and service panel should be placed as close to the mast as possible. This will place the main breaker as close to the meter as possible.

The location of the service entrance is generally controlled by economics. Often, the service is located near the point of greatest power usage (in a house, this is usually the kitchen) since larger wires are needed where high usage is indicated, and the closer the entrance to the high-usage area, the less of the larger, more expensive wiring that is required (Fig. 15-4).

Electric meters are weatherproof and should be located on the outside of the house so readings can be taken by the power company without disturbing anyone or when no one is at home.

In many locales, service is brought to the buildings in electrical wires below ground (Fig. 15-3). In this manner, unsightly wires, masts, and fittings are concealed. When the en-

FIGURE 15-2   Overhead service

FIGURE 15-3   Underground service

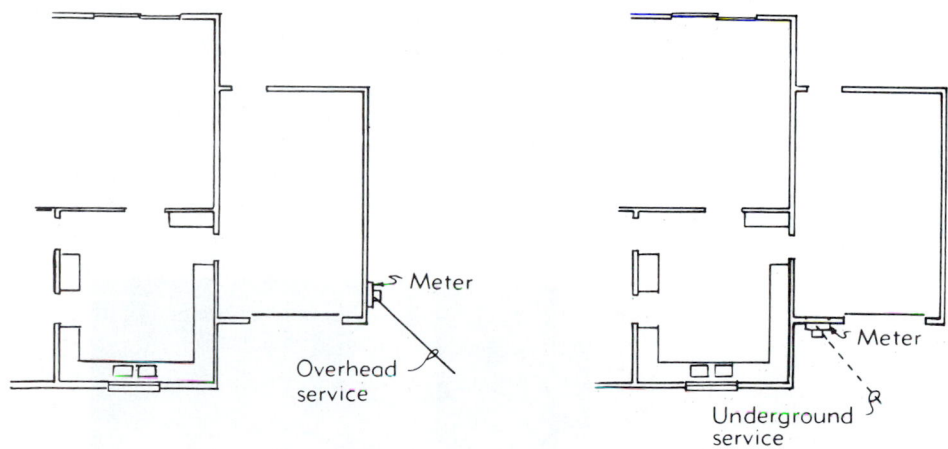

FIGURE 15-4   Service location

tire community is served by underground service, the power poles and lines are never missed. However, this is more expensive than running services above ground.

The sizing of the service entrance, based on the amount of power it must supply, should be guided by the following information:

100 amperes: Adequate power to provide general-purpose circuits, water heater, electric laundry, and cooking.

150 amperes: Adequate power to provide general-purpose circuits, water heater, electric laundry, cooking, and for a small house, the air conditioning and heating.

200 amperes: Adequate power to provide general-purpose circuits, water heater, electric laundry and cooking, air conditioning and heating.

Once all of the circuits, fixtures, appliances, and type of heating are determined, the size of the service entrance may be calculated by experienced personnel. It is best to anticipate a little high when selecting the size of the service entrance. It is much more expensive to increase the size later than when the building is being built.

15-7   SERVICE PANEL

This distribution box is the main panel (Fig. 15-5) that receives the service electricity, breaks it down, and distributes it through branch circuits to the places where the electricity is needed. Inside the panel is a main disconnect switch which cuts off power to the entire building and the circuit breakers (or fuses) which control the power to the individual circuits serving the house.

The circuit breakers (or fuses) are protective devices which will automatically cut off power to any circuit which is overloaded or short-circuited. The circuit breaker may be turned back on by flipping its surface-mounted switch; if the circuit is still overloaded, the switch will immediately flip off again. Fuses burn out when the circuit is overloaded and must be replaced to activate the circuit again. Circuit breakers and fuses are sized as to the amperage they will carry, and when replaced, the replacement should have the same rating.

The service panel is sized to match the service coming in, commonly 100, 150, or 200 amps. The number of circuits required in the building will determine the size of the panel. Panels are rated in amperage and poles, the poles being how many circuits it will handle. All 120-volt circuits, as required for most lighting, convenience outlets, and appliances, require one pole each. All appliances such as ranges, clothes dryers, hot water heaters, large air conditioners, and many motors require 240 volts, and two poles are required. The panel selected should have extra poles so that additional circuits may be run if and when they may be desired.

The panel may be surface-mounted or recessed and should be conveniently located for easy servicing and resetting of circuit breakers and fuses as required. It should not be located anywhere there is a possibility of water being on the panel or on the floor around it. Typically, it is located in a garage, corridor, basement, or utility room of the building.

FIGURE 15-5   Service panel

15-8 FEEDER CIRCUITS

On large buildings where the wiring for circuits would have long runs, a feeder circuit may be run from the service panel to a subdistribution panel (Fig. 15-6 and Fig. 15-7). Locating the subdistribution panel conveniently to service the larger feeder conductors (wires) will allow a minimal voltage drop when compared with the excessive voltage drop that occurs when branch circuits are in excess of 75 to 100 ft long.

15-9 BRANCH CIRCUITS

The branch circuit connects the service panel to the electrical device it supplies (Fig. 15-8). It may supply power to a single device, such as a water heater, range, or air conditioner, or it may service a group or series of devices, such as convenience outlets and lights. The branch circuit may have a variety of capacities, such as 15, 20, 30, 40, or 50 amps, depending on the requirements of the electrical devices serviced. The general-purpose circuits for lights and convenience outlets are usually sized at 15 or 20 amps. One general-purpose branch circuit will provide a maximum of 20 amps × 120 volts = 2,400 watts, of which the NEC allows 80% (1920 watts). There is a tendency for people to put higher wattage bulbs (lamps) in lighting fixtures (luminaires) as they need replacing; a group of light fixtures totalling 1,200 to 1,600 watts may be placed on a circuit and still allow for additional

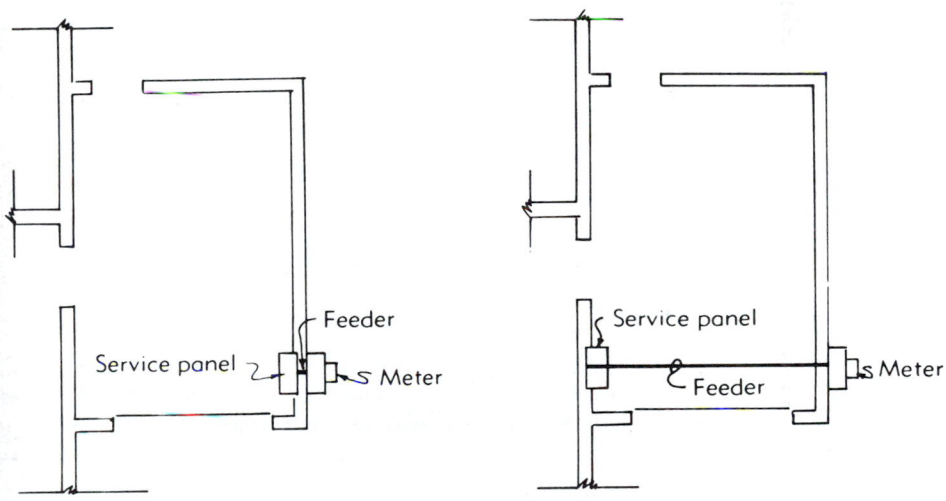

FIGURE 15-6   Feeder circuit

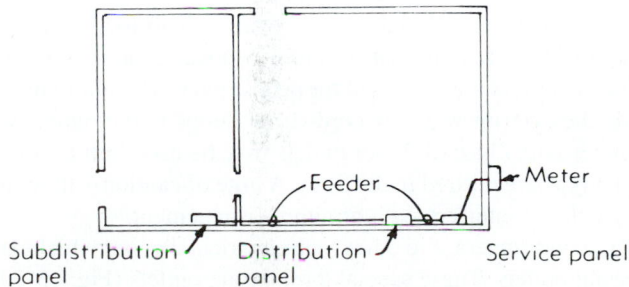

FIGURE 15-7   Feeder circuit

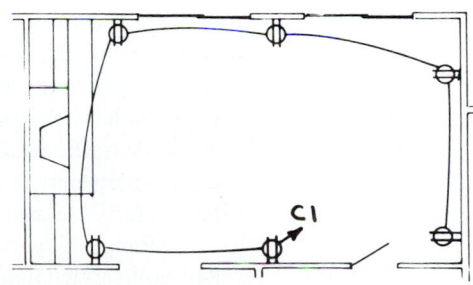

FIGURE 15-8   Typical branch circuit

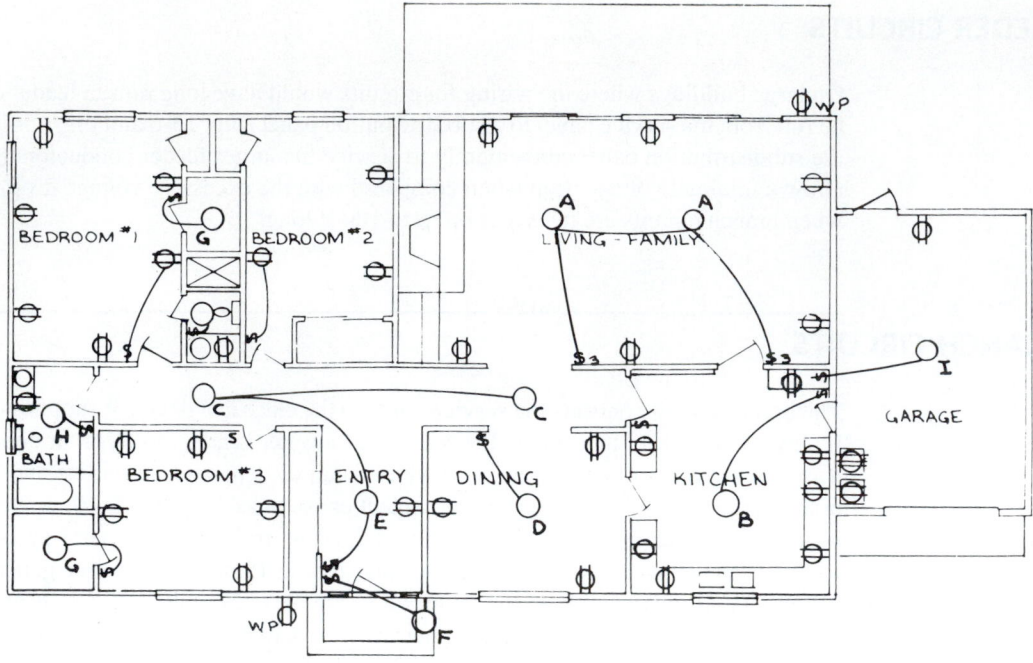

FIGURE 15-9 Circuit labeling (only switch circuits shown)

future usage. The same is true with convenience outlets; generally no more than six should be on a circuit.

The wiring layout for most buildings shows the branch circuit arrangements (Fig. 15-9). If not shown on the layout, it will be left to the electrical contractor to decide on groupings.

15-10 RECEPTACLES

Receptacles, also referred to as convenience outlets (Fig. 15-10), are used to plug in lights and small appliances around the house. Each room should be laid out with no less than one outlet per wall and with outlets no more than 10 ft apart. The amount, location, and type will vary with the room, depending on both the design of the room and the furniture layout. Duplex outlets (two receptacles) are most commonly used, but single and triple receptacles are also used. Also used are strips which allow movement of the receptacle to any desired location; these strips are available in 3-ft and 6-ft lengths and may even be used around the entire room. When specified, one of the receptacles may be controlled by a wall switch. This is particularly desirable in rooms where portable fixtures are used. Typical room layouts are shown in Fig. 15-11. Some room designs make it difficult to locate the outlets on the walls. Plans such as Fig 15-12 require that furniture be located out from the wall and the space between the furniture and the wall used for pedestrian traffic out to the deck. If the outlets were placed on the exterior wall, the cords from lamps to the outlets would cross the traffic area, creating a safety hazard. Floor outlets may be used in this situation; these outlets may be located anywhere desired in the floor. A note of caution—they should be carefully planned so they will not interfere with furniture arrangement.

Ranges, dryers, large air conditioners, and other such electrical devices which require 240-volt service require special outlets. These special three-prong outlets (Fig. 15-10) are designed so that conventional 120-volt devices cannot be plugged into them. The symbol

FIGURE 15-10   Typical receptacles

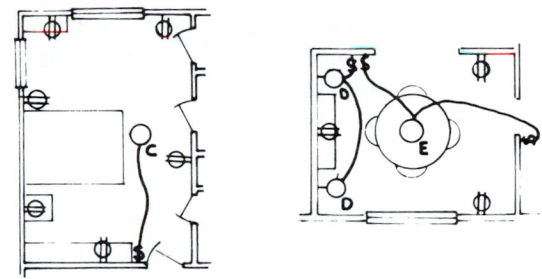

FIGURE 15-11   Typical room layouts

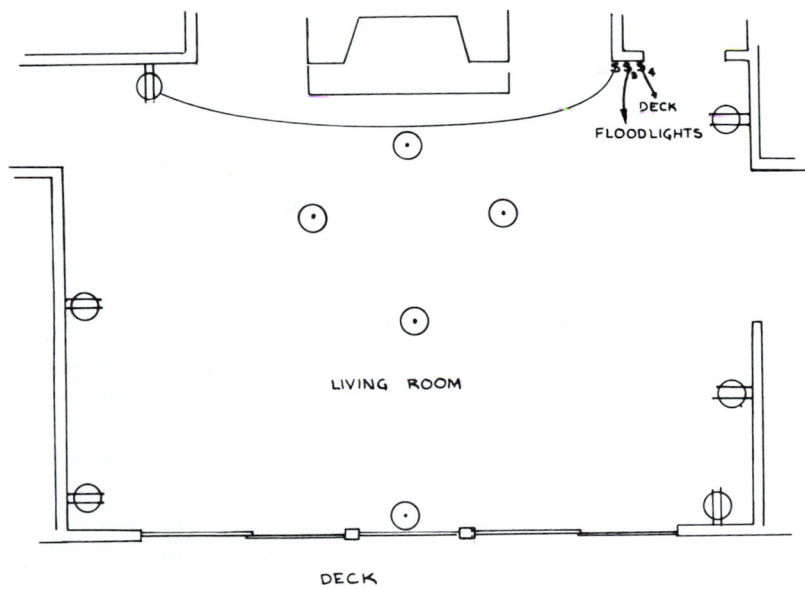

FIGURE 15-12   Floor receptacles

used will vary only slightly from that used for 120-volt convenience outlets in that there are three straight lines through the circle instead of two. Typically, each one of these special outlets is on a circuit by itself.

Three commonly used specialty outlets are the split-wired outlet grounded fault circuit interrupt (GFCI) and the weatherproof outlet. A split-wired outlet (Fig. 15-13) is any outlet which has the top outlet on a different circuit than the bottom or which has one outlet that may be switched off and on from a wall switch. The grounded fault circuit interrupt is

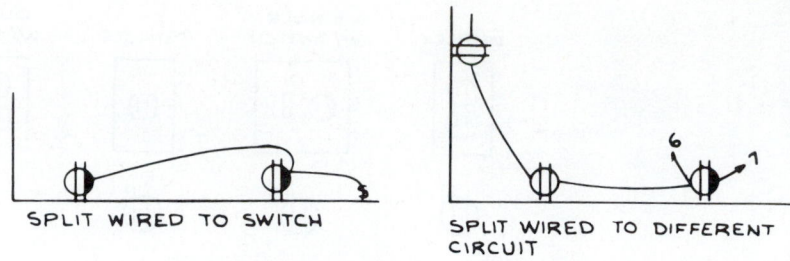

FIGURE 15-13   Split-wired outlets

connected to the building ground, which deenergizes a circuit should an overload occur. The weatherproof outlet is used in all exterior locations because it resists damage from weather. It is noted on the plan by use of the standard outlet symbol and the letters WP next to it.

15-11   LUMINAIRES (LIGHTING FIXTURES)

The luminaires used in the project must be carefully selected with the client. Lighting of some type will be required throughout the house, and quite often exterior lighting will also be required. The size, type, and location of fixtures throughout must be coordinated with the style of the house and the client's preferences. The lighting may come from built-in and surface-mounted ceiling and wall luminaires; floor and table lamps may be used as a supplement, or they may be preferred throughout.

Fluorescent lighting (tubes) is more efficient than incandescent lighting (bulbs), and it is effective as indirect lighting in valances and coves around rooms and also where high lighting levels are desired such as in bathrooms, work areas, and workshop areas. Objections to fluorescent lighting include: its higher initial cost, slow starting, and a tendency to flicker.

Many clients prefer a permanent ceiling fixture that can be turned on from the wall switch, while others prefer that the wall switch activate a convenience outlet, and they use a table or floor lamp in that outlet to light the room. Small rooms frequently have one central ceiling fixture activated by a wall switch. This provides the general illumination required and may be supplemented by wall, floor, and table fixtures. Shallow closets generally don't require fixtures if there is average room illumination. However, many clients want fixtures in the closets, and, if so, a ceiling fixture may be used. The closet fixture may be operated by a pull chain, but, again, many clients prefer either a wall switch just outside the door or a door-operated switch which turns the light on when the door is opened and shuts it off when the door is closed. The lighting symbols used are shown in Fig. 15-14. For those fixtures activated by a switch, a line must be shown connecting the switch and the fixtures it controls.

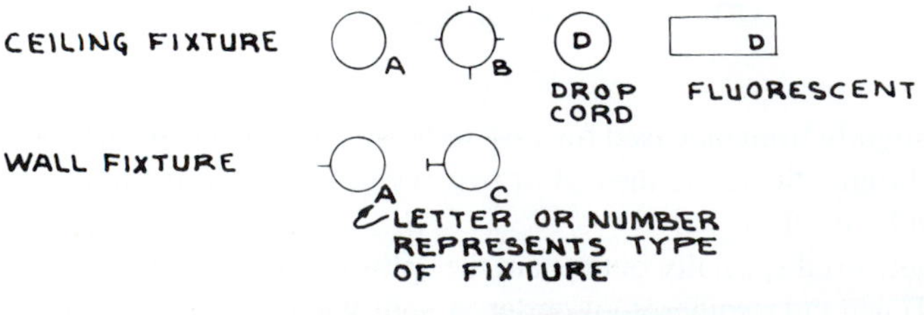

FIGURE 15-14   Fixture symbols

15-12 SWITCHES

Wall switches are used to control lighting fixtures in the various rooms of the house, and they may also be used to control convenience outlets. The switch most commonly used is the *toggle switch,* which has a small arm which is pushed up and down. Also used is the *mercury switch,* a completely noiseless switch which has mercury in a sealed tube; when the switch is turned on, the mercury completes the circuit.

Wall switches are located 4 ft above the floor and a few inches in from the door frame on the latch side (doorknob side) of the entry door into a room. Generally, they are located just inside the room.

Rooms with two entrances often have two switches controlling the fixture, referred to as three-way switches. When the fixture is to be controlled from three locations, it will require the use of two 3-way switches and one 4-way switch (Fig. 15-15).

A dimmer control is used when it is desirable to vary the intensity of the light on incandescent luminaires being given off, from very low to bright. A delayed action switch is used when it is desirable to have the lights go out about a minute after the switch is turned off.

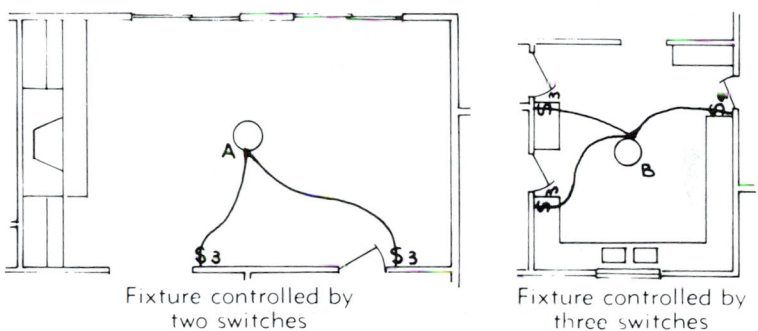

Fixture controlled by two switches

Fixture controlled by three switches

FIGURE 15-15   Switches

Low-Voltage Switching

The flexibility desired in home lighting systems has led to the development of low-voltage wiring, which allows flexibility in the control of fixtures through the use of centrally located remote-control switches which control any or all of the lights in the house.

Low-voltage wiring (about the size of wiring used to wire a door chime) connects the wall switch to a relay center. When the switch is turned on, it sends a low voltage (usually 24 V) to the relay center; this triggers a relay switch which activates the line voltage (120 V) at the light. The savings in wiring costs, the convenience of a centrally located remote-control selector switch, and the elimination of any possible electric shocks at the switch make this increasingly popular in homes.

15-13 CONDUCTORS

The wires used to supply electricity throughout the system are called *conductors.* Copper has traditionally been used as the conductor material, and the wiring practices developed over the years have been based on its use. Aluminum conductors are also available and are allowed in almost all codes. One of the biggest reasons for using aluminum conductors is that their costs range from one-third to one-half those of copper conductors. Aluminum conductors must be larger than copper conductors that carry the same amperage. This means that for installations requiring that the conductors be placed in conduit, the larger aluminum conductors may require larger conduits. (Conduits are discussed in Sec. 15-14.)

The various types of insulation which are placed on the conductor wires have been standardized and are listed in Fig. 15-16. The types of conductors are referred to by the type letter assigned to the insulation used. For example, RHW conductor has a moisture- and heat-resistant rubber insulation, with a maximum operating temperature of 75°C (167°F), and it may be used in both dry and wet locations. The maximum operating temperature has an effect on the allowable ampacities which the NEC will allow for the conductor (Fig. 16-24); the higher the maximum operating temperature, the higher the allowable ampacity. All individual conductors must be protected by raceways. Individual conductors in raceways are used extensively in commercial and industrial installations where changes in the electrical requirements are likely to occur.

When two or more conductors, each insulated separately, are grouped together in one common covering, they are referred to as *cables* (Fig. 15-17). These cables are used extensively in electrical wiring, particularly in residences. They are designated according to type of insulation used and where they are used; generally, they do not have to be protected by raceways.

All conductor sizes are given by an AWG or MCM number. Standard available conductor sizes are given in Fig. 16-24. Those conductor sizes based on the American Wire Gauge (from No. 16 to No. 4/0) are listed as AWG—for example, No. 12 AWG. The larger the AWG number, the smaller the conductor. The cross-sectional area of any conductor is listed in *circular* mils, with a circular mil defined as the area of a circle one mil in diameter; then the area is the square of the diameter (in mils). All conductors larger than 4/0 AWG are sized in direct relation to the circular mil and are labeled MCM or one thousand (M) circular mils. A wire with an area of 500,000 circular mils would be called 500 MCM.

15-14   RACEWAYS

A channel which is designed exclusively, and used solely, for holding wires, cables, or bus bars is called a *raceway*. The raceway may be made of metal or insulating material. Metal raceways include rigid and flexible metal conduits, electric metallic tubing (EMT), and cellular metal raceways, and all may be concealed in the construction or exposed in the space. The most popular insulating raceways are made of cement, asbestos, or impregnated fiber, all of which are commonly used in underground exposures since they are corrosion-resistant.

Types

Rigid metal conduit is usually steel (ferrous) or aluminum (nonferrous). The steel conduit has its outside surface coated in zinc to galvanize it, making it resistant to rusting, and zinc is applied to the interior surface of the conduit to make it easier to pull the conductors through it. Steel conduit is also available dipped in a clear, elastic enamel. The enamel coating makes it easier to pull the conductors and provides some resistance to corrosion, but it should not be used outside or where severe corrosive conditions may be present. Rigid metal conduit is extensively used throughout the building and may be placed in the walls, ceiling, and floors of the construction. In addition, corrosion-resistant metal conduit may also be used outside.

Trade Name	Type Letter	Max. Operating Temp.	Application Provisions	Insulation	AWG or kcmil — Thickness of Insulation — Mils	Outer Covering****
Moisture- and Heat-Resistant Rubber	RHW-2	90°C 194°F	Dry and wet locations.	Moisture- and Heat-Resistant Rubber	14-10 .... 45 8-2 .... 60 1-4/0 .... 80 213-500 .... 95 501-1000 ....110 1001-2000 ....125 For 601-2000 volts, see Table 310-62.	*Moisture-resistant, flame-retardant, non-metallic covering
Silicone-Asbestos	SA	90°C 194°F 125°C 257°F	Dry and damp locations. For special application.†	Silicone Rubber	14-10 .... 45 8-2 .... 60 1-4/0 .... 80 213-500 .... 95 501-1000 ....110 1001-2000 125	Asbestos, glass or other suitable braid material
Synthetic Heat-Resistant	SIS††	90°C 194°F	Switchboard wiring only.	Heat-Resistant Rubber	14-10 .... 30 8 .... 45 6-2 .... 60 1-4/0 80	None
Thermoplastic and Asbestos	TA	90°C 194°F	Switchboard wiring only.	Thermo-plastic and Asbestos	Th'pl'. Asb. 14-8 .... 20 20 6-2 .... 30 25 1-4/0 .... 40 30	Flame-retardant, nonmetallic covering
Thermoplastic and Fibrous Outer Braid	TBS	90°C 194°F	Switchboard wiring only.	Thermo-plastic	14-10 .... 30 8 .... 45 6-2 .... 60 1-4/0 80	Flame-retardant, nonmetallic covering

Trade Name	Type Letter	Max. Operating Temp.	Application Provisions	Insulation	AWG or kcmil — Thickness of Insulation — Mils	Outer Covering****
Perfluoro-alkoxy	PFA	90°C 194°F 200°C 392°F	Dry and damp locations. Dry locations — special applications.†	Perfluoro-alkoxy	14-10 .... 20 8-2 .... 30 1-4/0 45	None
Perfluoro-alkoxy	PFAH	250°C 482°F	Dry locations only. Only for leads within apparatus or within raceways connected to apparatus. (Nickel or nickel-coated copper only.)	Perfluoro-alkoxy	14-10 .... 20 8-2 .... 30 1-4/0 45	None
Heat-Resistant Rubber	RH	75°C 167°F	Dry and damp locations.	Heat-Resistant Rubber	**14-12 .... 30 10 .... 45 8-2 .... 60 1-4/0 .... 80 213-500 .... 95 501-1000 ....110 1001-2000 ....125 For 601-2000 volts, see Table 310-62.	*Moisture-resistant, flame-retardant, non-metallic covering
Heat-Resistant Rubber	RHH††	90°C 194°F	Dry and damp locations.			
Moisture- and Heat-Resistant Rubber	RHW††, †††	75°C 167°F	Dry and wet locations. For over 2000 volts insulation shall be ozone-resistant.	Moisture- and Heat-Resistant Rubber	14-10 .... 45 8-2 .... 60 1-4/0 .... 80 213-500 .... 95 501-1000 ....110 1001-2000 ....125 For 601-2000 volts, see Table 310-62.	*Moisture-resistant, flame-retardant, non-metallic covering

* Some rubber insulations do not require an outer covering.

** For 14-12 sizes RHH shall be 45 mils thickness insulation.

**** Some insulations do not require an outer covering.

† Where environmental conditions require maximum conductor operating temperatures above 90° C.

†† Insulation and outer coverings that meet the requirements of flame-retardant limited smoke and are so listed shall be permitted to be designated limited smoke with the suffix /LS after the Code type designation.

††† Listed wire types designated with the suffix –2 such as RHW–2 shall be permitted to be used at a 90°C operating temperature wet or dry.

FIGURE 15-16   Conductor application and insulations

Trade Name	Type Letter	Max. Operating Temp.	Application Provisions	Insulation	AWG or kcmil	Thickness of Insulation	Mils	Outer Covering****
Moisture- and Heat-Resistant Thermoplastic	THWN ††, †††	75°C 167°F	Dry and wet locations.	Flame-Retardant, Moisture- and Heat-Resistant Thermoplastic	14-12 .... 10 .... 8-6 .... 4-2 .... 1-4/0 .... 250-500 .... 501-1000		15 20 30 40 50 60 70	Nylon jacket or equivalent
Moisture-Resistant Thermoplastic	TW††	60°C 140°F	Dry and wet locations.	Flame-Retardant, Moisture-Resistant Thermoplastic	14-10 .... 8 .... 6-2 .... 1-4/0 .... 213-500 .... 501-1000 .... 1001-2000		30 45 60 80 95 110 125	None
Underground Feeder & Branch-Circuit Cable-Single Conductor. (For Type UF cable employing more than one conductor, see Article 339.)	UF	60°C 140°F	See Article 339.	Moisture-Resistant	14-10 .... 8-2 .... 1-4/0		*60 *80 *95	Integral with insulation
		**75°C 167°F		Moisture- and Heat-Resistant				

* Some rubber insulations do not require an outer covering.

** For 14-12 sizes RHH shall be 45 mils thickness insulation.

**** Some insulations do not require an outer covering.

† Where environmental conditions require maximum conductor operating temperatures above 90° C.

†† Insulation and outer coverings that meet the requirements of flame-retardant limited smoke and are so listed shall be permitted to be designated limited smoke with the suffix /LS after the Code type designation.

††† Listed wire types designated with the suffix –2 such as RHW–2 shall be permitted to be used at a 90°C operating temperature wet or dry.

FIGURE 15-16   *(continued)*

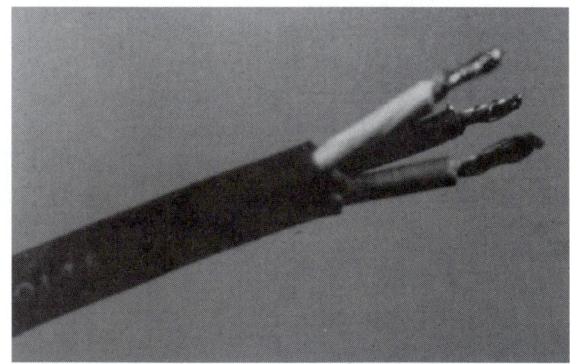

FIGURE 15-17   Cable

Flexible metal conduit is made of a continuous length of galvanized, spirally wound steel strip. It is used primarily when it is necessary to make connections from a junction box to machinery and to go around obstructions. Only the waterproof type may be used in wet locations.

Use of liquid-tight flexible metal conduit is restricted to connections of motors or portable equipment when the connections must be flexible. Further, its use is not allowed where it may be subjected to physical damage, where it may be in contact with rapidly moving parts, where temperatures exceed 60°C (140°F), and in hazardous locations.

Plastic-coated galvanized steel conduit is available for locations where the conduit will be subjected to the highly corrosive actions of fumes, gases, or chemicals. Polyvinyl chloride is extruded over the conduit, and all connections are taped with a vinyl, pressure-sensitive electrical insulating tape.

Also available for installations requiring high corrosion-resistance and high strength is an alloy of a mild carbon steel with 2% nickel and 1% copper. This provides a highly corrosion-resistant conduit which is available galvanized or enameled.

Electrical metallic tubing (EMT) is a thin-walled metallic conduit which weighs about one-third less than rigid metal conduits. It may be used anywhere except where it will be subject to severe physical damage during or after installation, in cinder concrete or fill underground, or in any hazardous locations. This type of conduit connects to its fittings with set screws, saving the time often required to put screw threads on rigid metal conduits.

Aluminum conduit is lightweight and sturdy. When used where it will be in contact with cement products, it must be mastic coated.

Surface metal raceways (Fig. 15-18) must be installed in dry locations and must not be concealed in the construction. This type of raceway is used extensively with metal partitions (Fig. 15-19) where the raceway has a backplate which attaches to the metal stud and a snap-on cover. The NEC limits conductor sizes used in this type of raceway to No. 6 AWG and smaller. The number of conductors allowed must be taken from the raceway manufacturer's data sheet and is based on the amount which Underwriters Laboratories will allow. This type of installation is used extensively when remodeling, adding to existing systems, and building new additions where room arrangements will be subject to changes.

Underfloor raceways, made of metal or fiber, are commonly used in any building where periodic remodeling or a change in the occupancy may occur. This includes most office buildings and many institutional and industrial buildings. While the size and number of conductors used will depend on the size of the raceway and is controlled by UL, it may also be found in the manufacturer's data. Some typical underfloor raceways are shown in Fig. 15-20. The NEC states that for raceways less than 4 in. wide, a minimum of ¾ in. of wood or concrete must cover the raceway. Raceways 4 to 8 in. wide and at least 1 in. apart (Fig. 15-21) must have at least a 1-in concrete cover. When they are less than 1 in. apart, raceways must have a 1½-in. concrete cover. Also available are flush raceways with removable covers (Fig. 15-22).

Cellular metal floor raceways are installed when the hollow spaces in cellular metal floors are used for the distribution of conductors through the building (Fig. 15-23). The NEC limits conductor size to No. 1 AWG or less, and the number of conductors which may be put in an individual cell (single, enclosed tubular space) is limited to 40% of the cross-sectional cell area, except that the limit does not apply to type AC metal-clad cable or to nonmetallic sheathed cable.

Wireways are sheet-metal troughs with hinged or removable covers. Wireways are used primarily for exposed inside work; if they are used outside, they must be of rain-tight construction. Conductor size is limited to 500 MCM, and the number of conductors is lim-

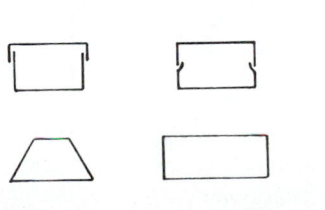

FIGURE 15-18   Surface runway shapes

FIGURE 15-19   Surface runway installation

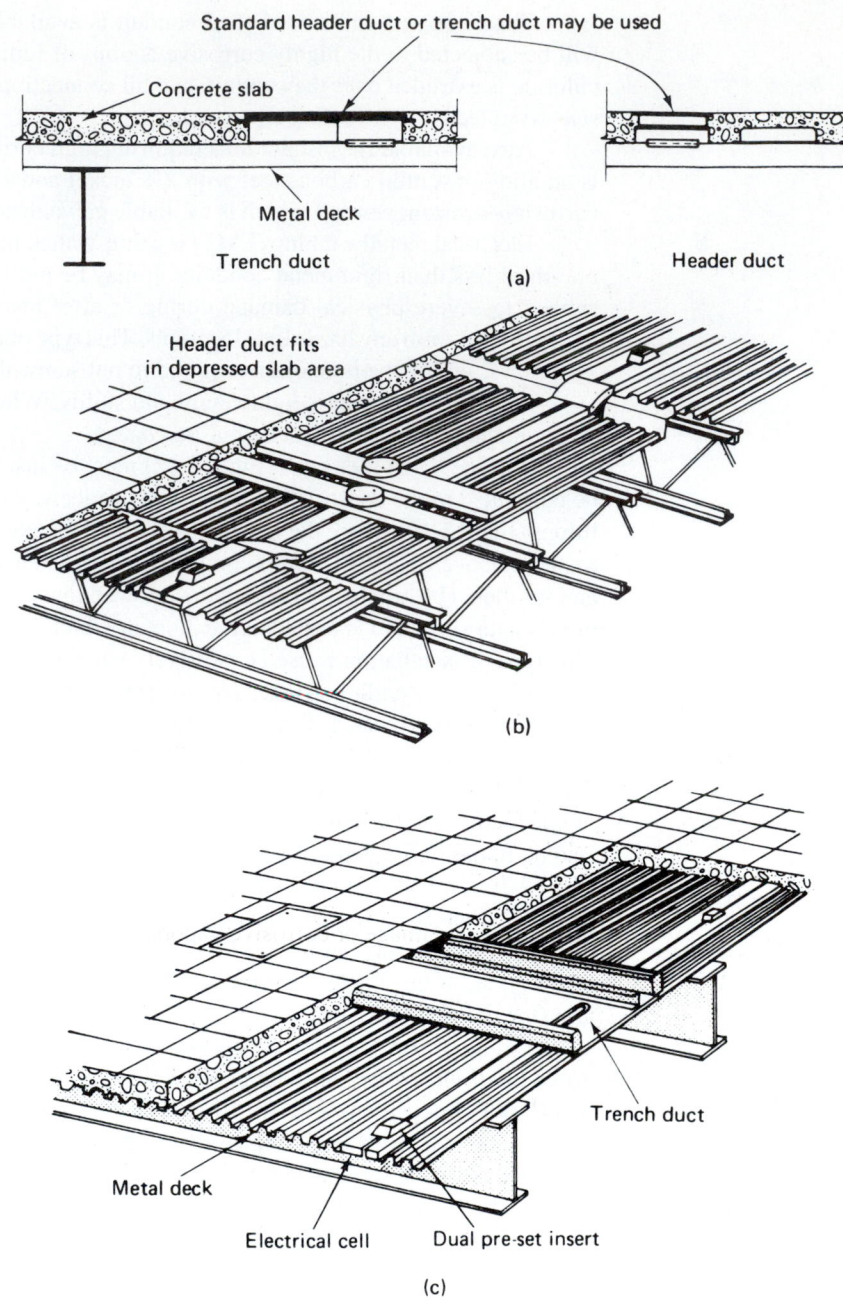

FIGURE 15-20   Raceways

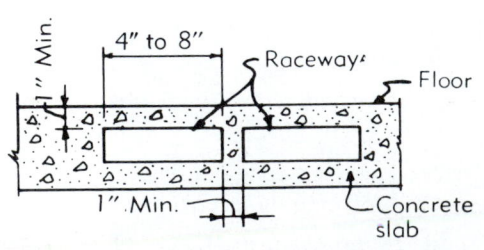

FIGURE 15-21   Raceway spacing

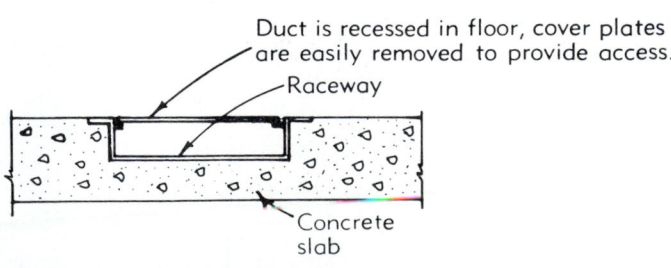

FIGURE 15-22   Raceway

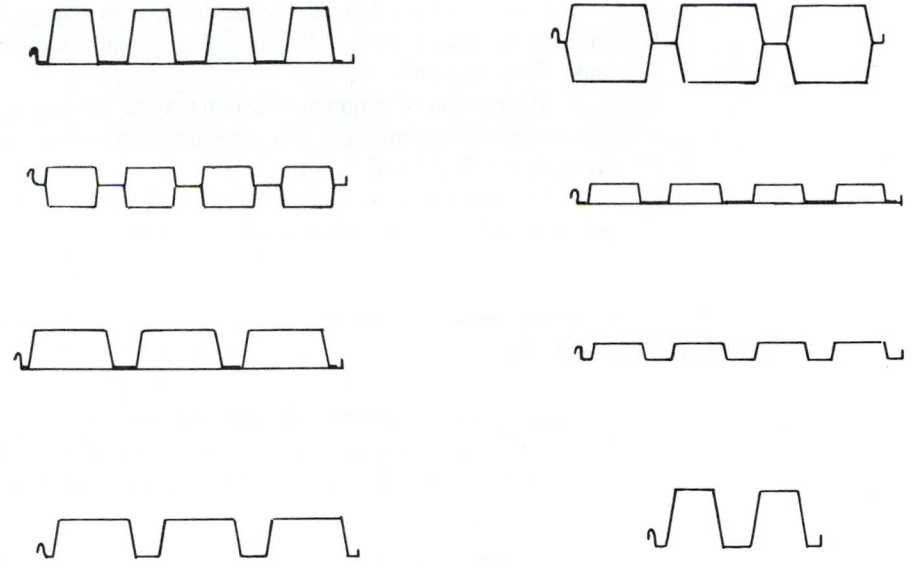

FIGURE 15-23   Cellular metal raceways

ited to a maximum of 30. The total area of all conductors installed in the wireway is limited to 20% of the interior cross-sectional area of the wireway.

Busways (also called bus ducts) are factory-assembled conductors mounted in a steel housing. They have high current-carrying capacities and are often used as service-entrance conductors, feeders and subfeeders, and even as branch circuits where large service loads are required. The NEC allows the use of busways only where they will be exposed. They may not be concealed in the construction or installed where they may be subjected to severe physical damage or corrosive vapors.

Installation

Rigid nometallic conduit used below ground must be moisture-resistant and, when used above ground, also flame-retardant and resistant to impact. When used for direct burial, not encased in concrete, the material must be strong enough to withstand any loads which may be placed on it after installation. Typical materials used below ground are fiber, cement, soapstone, rigid polyvinyl chloride, and high-density polyethylene. Rigid polyvinyl chloride is used above ground. This type of conduit should *not* be used:

1. Less than 8 ft above the ground, outdoors, unless protected from any possible physical damage.
2. In combustible construction where it is concealed (built into the walls, floors, ceilings).
3. In any location where it might be subject to physical damage.
4. In any location where the temperature may be higher than that which the conduit has been tested for.
5. Where the conductor's maximum insulation temperature is higher than that of the conduit being used.
6. If there is any possibility that the electrical service running through it may exceed 600 V, unless encased in a minimum of 2 in. (50 mm) of concrete.
7. In sunlight, unless the conduit being used has been tested and approved for such use.
8. In hazardous locations.

Rigid metal conduits are connected to their fittings by screwing the threaded fittings onto the threaded conduits. Whenever the conduit has been cut, it must be rethreaded to make the connection.

The number of conductors which may be put in a rigid metal conduit depends on the conductor and conduit sizes. The sizes are selected from Fig. 16-25, and typical examples are shown in Sec. 16-4.

The number of conductors which may be put in electrical metallic tubing again depends on the conductor and conduit sizes. The sizes are also selected from Fig. 16-25.

15-15 AMPS, OHMS, VOLTS

The design of electrical systems in a building requires that the designer have a "working familiarity" with amperes, ohms, and volts. These three electrical terms are often used by the designer to determine the total electrical load requirements of the building, and they are all related.

Ampere (Amp, Amperage—A): A unit (or measure) of the flow of electrons passing through a circuit (current).

Volt (Voltage—V): The unit of electrical pressure required to push the amperage through the circuit.

Ohm (Ω): The unit of electrical resistance which resists the flow of electrons through the circuit. Ohm's law states that the current in an electrical circuit is equal to the pressure divided by the resistance.

$$\text{Amps} = \frac{\text{Volts}}{\text{Ohms}}$$

Watt (W): The unit of electrical energy or electrical power. It indicates how much power has been used.

$$\text{Watts} = \text{Amps} \times \text{Volts}$$

Kilowatt (kW): 1,000 watts; for example, 9,500 W equals 9.5 kW.

QUESTIONS

15-1. Briefly describe what the Underwriters Laboratories does.

15-2. What does it mean when an electric fixture has a UL tag on it?

15-3. Define the following terms:

   a. Service entrance

   b. Service panel

   c. Feeder circuits

   d. Branch circuits

   e. Receptacles

   f. Luminaires

15-4. When is a *split-wired outlet* used?

15-5. What is the difference between conductors and cable?

15-6. When are raceways used, and what are three commonly used types?

15-7. Show, in a formula, the relationship between amps, volts, and ohms; then briefly describe each.

CHAPTER 16

Electrical System Design

16-1 PRELIMINARY INFORMATION

Ideally, the electrical designer should be involved from the very beginning in the design of the project. It would be best, in some situations, if the designer could even be involved in the selection of the site for the project. On a large project, it may be necessary to extend high-voltage lines to the project site. This will take time and the owner may have to pay part of the cost. The designer is the person who could best discuss the situation with the power company. All of the utilities (whether sewer, water storm sewer, natural gas lines, etc.) must be checked to determine if they are near the property and whether they can be brought to the property economically or not. Such information is needed early in the design stage.

Before actually beginning the design layout of the project, the designer will need to accumulate certain information:

1. Determine whether electrical service is available. If it is not, arrangements must be made with the power company to extend service to the building site. Large projects may require more voltage or more wattage than existing service can supply. Each of these situations requires coordination with the power company as early in the design stage as possible. Costs which may have to be paid by the owner should be thoroughly discussed, written, and given to the owner.

2. Obtain a list from the owner of all the types of equipment, appliances, etc., to be used in the building which will require electricity. While the electrical designer will know the electrical requirements of much of the equipment, it may be necessary to find the manufacturer's specifications for certain items, such as motor sizes and power required.

3. Working with the architectural designer, locate all of the equipment and appliances on the floor plan. In commercial projects, this sometimes takes many meetings with the architects, owners, and manufacturer's representatives. There are times when the type of equipment used and its location must be approved by governmental agencies.

4. Review with the architect where the basic mechanical equipment, such as the service entrance, the power and lighting panels, and the conduit or cable, will be located.

5. Discuss with the owners what future plans are as far as adding to the building, remodeling, constructing other buildings, increasing future equipment requirements, and anything else that could affect the size and location of the electrical service. Many times, the service entrance must be sized to anticipate future expansion as well as present building plans. Once the basic information has been gathered, the designer can begin to design the system itself.

16.2 ELECTRICAL SYSTEM DESIGN

The system is designed by first drawing, or having a print made, of the floor plan of the building.

Step-by-Step Approach

In this problem, the floor plan of the residence shown in Appendix C will be used to illustrate the design process:

1. The receptacles are located first, as discussed in Sec. 15-10. The various types of receptacles available are shown in Fig. 15-10, along with the symbols used to represent each. All switching, sizing of wire, and circuit layout will be done later.

   Locate the receptacles and switches on the floor plan as shown in Fig. 16-1.

2. Locate on the floor plan all connections to be made for appliances and equipment. This calls for close coordination with the architectural designer since each appliance and piece of equipment must be known in order to locate the receptacles properly. Many times, the plans do not show everything which must be connected. For example, a garbage disposal must be connected to power and may even have a switch on the wall to turn it on and off; yet it is seldom shown on the floor plan. Specifically, the electrical designer should ask the architectural designer to list all appliances and equipment in a letter or on the drawing for complete coordination. A checklist of typical appliances that require electrical connections is listed in Fig. 16-2. In addition, the electrical designer will need to know the voltage and amperage required. Typical requirements for various appliances are shown in Fig. 16-3, but these may vary among different manufacturers and should be checked.

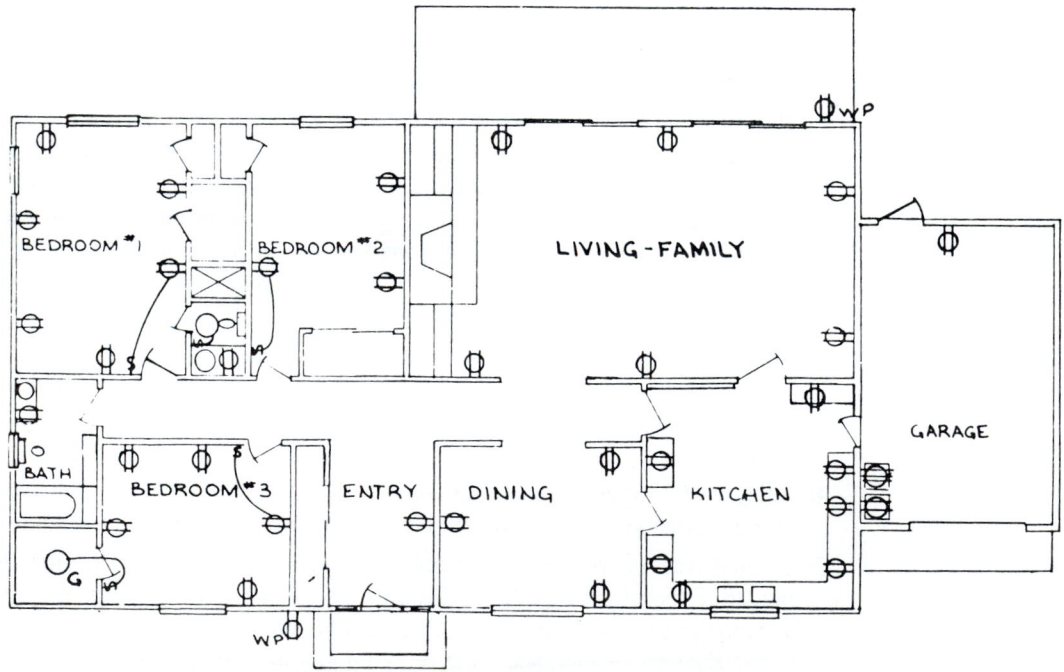

FIGURE 16-1   Locate switches and receptacles

Device		Device	Watts
Air conditioner, central		Air conditioner, central	3,000 to 5,000
Air conditioner, room		Air conditioner, room	800 to 1,500
Clothes dryer		Clothes dryer	4,000 to 8,000
Garbage disposal		Garbage disposal	300 to 500
Heat pump		Heat pump	3,000 to 6,000
Humidifier		Humidifier	80 to 200
Iron, hand		Iron, hand	600 to 1,200
Lamp, incandescent		Lamp, incandescent	10 to 250
Lamp, fluorescent		Lamp, fluorescent	15 to 60
Radio		Radio	40 to 150
Range		Range	8,000 to 14,000
Range, oven		Range, oven	4,000 to 6,000
Range, top		Range, top	4,000 to 8,000
Television		Television	200 to 400
Water heaters		Water heaters	2,000 to 5,000

FIGURE 16-2   Typical residential checklist

FIGURE 16-3   Residential appliance checklist

Also, check as to the number of appliances or equipment. For example, quite often two water heaters are used in residences and both will need connections. Also, in some residences and many commercial buildings, more than one heating and/or cooling unit may be used.

Using the appropriate symbols for the various receptacles required (Fig. 16-19), lay out the electrical requirements for appliances and equipment as shown in Fig. 16-4.

3. Locate the lighting fixtures.

As shown in Fig. 16-5, lay out on the floor plan all lighting fixtures, using the appropriate symbols from Fig. 15-14 for all wall and ceiling fixtures.

At the same time, the designer must make a list of what types of fixtures will be used throughout and how many watts each will use. This list will be used later when grouping circuits, and it must also be included in the specifications or on the drawings as a "Fixture Schedule" (Fig. 16-6). The list is also used by the electrical estimator when determining the cost to be charged for the work, then by the electrical purchasing agent when the material is ordered for the project, and then by the electrician who actually installs the work. Many times, the architectural designer decides what fixtures are to be used, often in coordination with the electrical designer.

When a minimum amount of illumination is required, the electrical designer may have to calculate the number and types of lighting fixtures which should be used. A complete discussion of this is given in Chapter 17.

4. The circuit layout may be an *individual branch circuit* which feeds only one receptacle, light, appliance, or piece of equipment, or a *branch circuit* which supplies power to two or more receptacles, lights, appliances, or pieces of equipment. An example of an individual branch circuit would be the circuit to a clothes dryer (240 V) and back to the electrical panel. The NEC states that an individual branch circuit should supply any load required to service that single item.

15- and 20-ampere branch circuits are used for receptacles, light fixtures, and small appliances. They are limited according to what will be connected to them.

a. When the circuit serves fixed appliances and light fixtures or portable appliances, the total of the fixed appliances shall be no more than 50% of the branch circuit rating. Assuming a 15-amp, 120-V branch circuit, it would have a maximum rating of 15 amps × 120 V = 1,800 W (refer to Sec. 15-15 for formula explanation). In

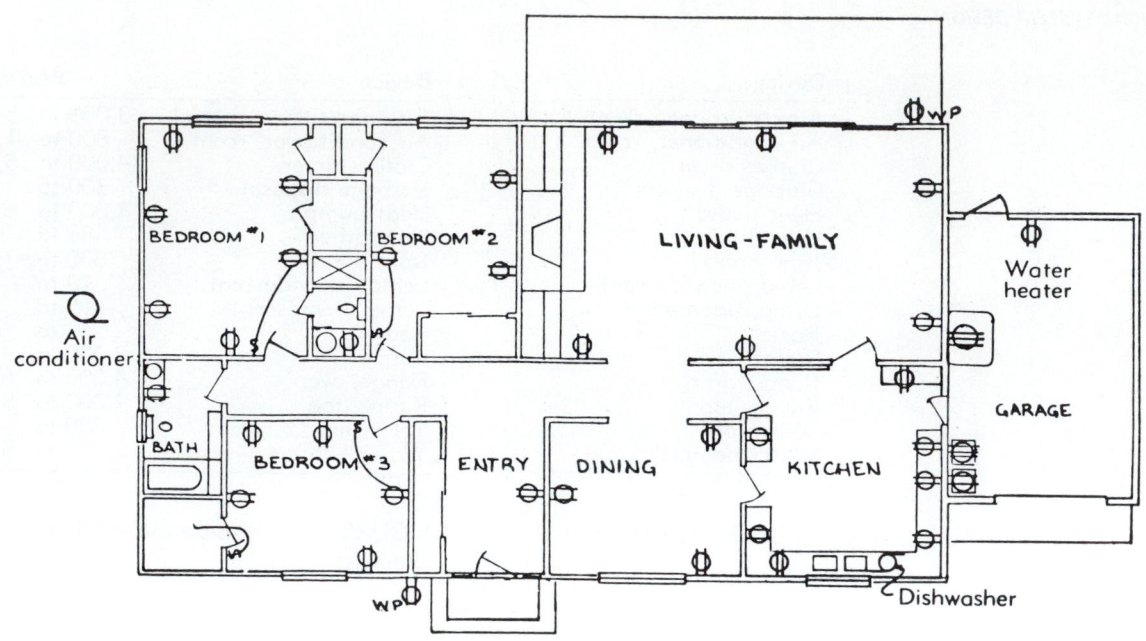

FIGURE 16-4 Lay out power requirements

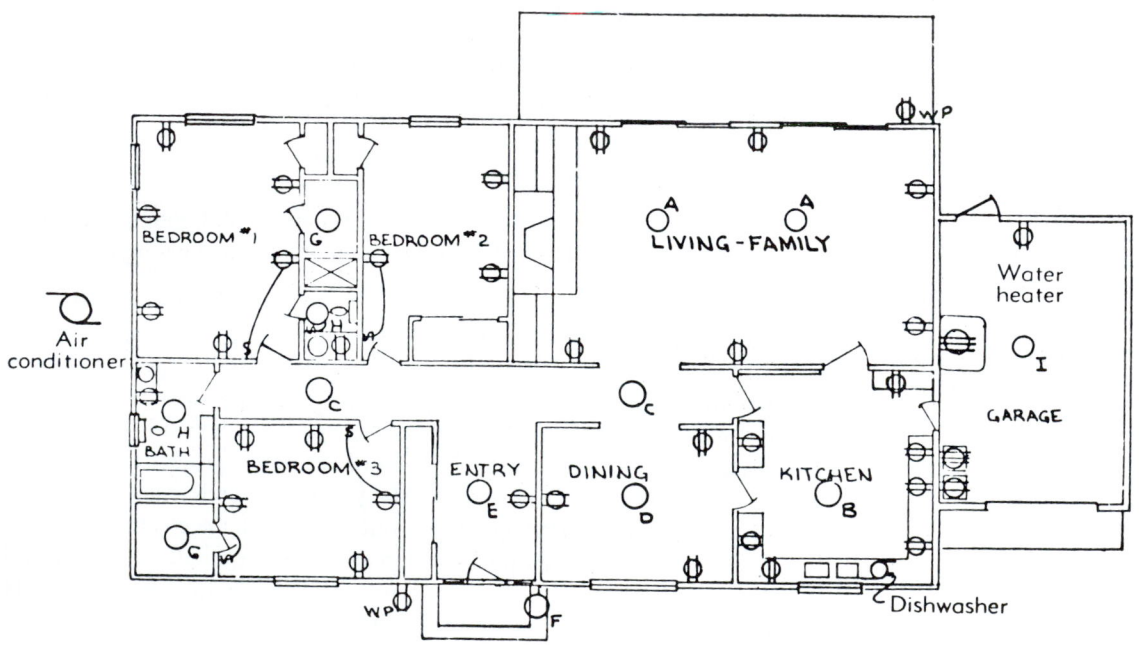

FIGURE 16-5 Lay out fixtures (luminaires)

FIXTURE SCHEDULE		
NO.	TYPE	REMARKS

Number or letter used on drawing ⟶

Fixture manufacturer and fixture number

FIGURE 16-6 Fixture schedule

this case, the fixed appliances would be limited to 900 W, leaving the other 900 W available to supply the light fixtures or portable appliances also served by the branch circuit. A 20-amp, 120-V branch circuit would have a maximum of 2,400 W.

b. When the load on the circuit will be a continuous operating load, such as for store lights, the total load should not exceed 80% of the circuit rating. The lighting load must include any ballasts, transformers, or autotransformers which are part of the lighting system. Since a 15-amp branch has a full rating of 1,800 W, the limit would be 80% or 12 amps and 1,440 W. A 20-amp, 2,400-W branch would be limited to 16 amps and 1,920 W of connected load.

c. When portable appliances will be used on a circuit, the limit for any one portable appliance is 80% of the branch circuit rating.

d. Receptacles are computed at a load of 1½ amps each and limited to 80% of the rating. This limits a branch circuit serving only receptacles to its rating divided by 1½ amps. For example, a 15-amp circuit is limited to a maximum of 8 outlets and a 20-amp circuit to 10 outlets.

e. A minimum of two 20-amp circuits is required for small appliances in the kitchen, laundry, dining room, family room, and breakfast or dinette area. These are in addition to the other receptacles required, and no lights or fixed appliances should be connected to these circuits. A typical layout of the two circuits is shown in Fig. 16-7.

f. A minimum of one 20-amp circuit is required as an individual branch to the laundry room receptacle. Any other special requirements, such as an electric clothes dryer requiring 230-V service, must also be added.

The designer must be certain not to exceed the code requirements. In actual practice, the designer tends to be a little more conservative, generally limiting a 15-amp branch to 1,000 to 1,200 W and a 20-amp branch to 1,300 to 1,600 W. Receptacles are generally limited to about six on a circuit. This allows the circuit to take additional loads, such as when higher wattage bulbs are used to replace those originally installed and calculated. In addition, more and more small appliances and equipment are being purchased and connected to receptacles (for example, air purifiers, humidifiers, stereos, and the like). The designer must also try to anticipate any future requirements. Such a layout allows for the extension of a circuit if it is necessary to add a light or a receptacle instead of adding a whole new circuit from the panel. If it is possible that the occupant will want to install individual air conditioners, an individual branch circuit to each receptacle required may be desired.

Thirty-, 40-, 50- and 60-ampere branch circuits will be used for fixed appliances, equipment, and heavy-duty lampholders (in other than residential occupancies). Generally, the electrical requirements of the connected load must be determined, and the total load connected to the branch circuit should be limited to 80% of the branch circuit rating.

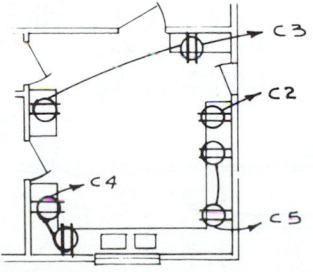

FIGURE 16-7 Typical kitchen circuit layout

The code states that any branch circuit serving a single motor shall have an ampacity (amperage rating) of not less than 125% of the motor's full-load current rating (this is the same as saying that no motor can exceed 80% of the branch circuit rating).

For example, if the motor to operate the air conditioner requires 22 amps, it will require a minimum branch circuit of 22 amps × 1.25% = 27.5 amperes, and a 30-ampere branch circuit will be used. This means that the electrical designer will need the manufacturers' data for all equipment and appliances selected to be certain the proper branch circuit sizes are used. Many times, the fuse sizes required for individual branch circuits are listed in the manufacturers' data.

A typical circuit layout is shown in Fig. 16-8. Later, each circuit group will have an arrow at the end of with a circuit number to be assigned later (Fig. 16-9).

5. Lay out the switches required to control the lights, appliances, equipment, and any desired receptacles. The discussion of switches in Sec. 15-12 generally outlines where they are most commonly used and the symbols used.

The plan in Fig. 16-9 shows the switches and circuits added to the floor plan.

6. Calculate the electrical load, the total of all general lighting, appliance, and equipment loads in the building.

a. *The general lighting load* is calculated for all types of occupancies (Fig. 16-20) based on the unit load given in the table (in watts) times the square footage of the building.

The square footage shall be determined using the outside dimensions of the building involved and the number of stories. For dwellings, do not include any open porches, garages, or carports. Any unfinished or unused spaces do not have to be included in the square footage *unless* they are adaptable for future use.

For the residence being designed, the lighting load is taken as 3 W per sq ft (Fig. 16-20). Since its outside dimensions (excluding garage) are 61 ft 6 in. × 36 ft, there are 2,214 sq ft.

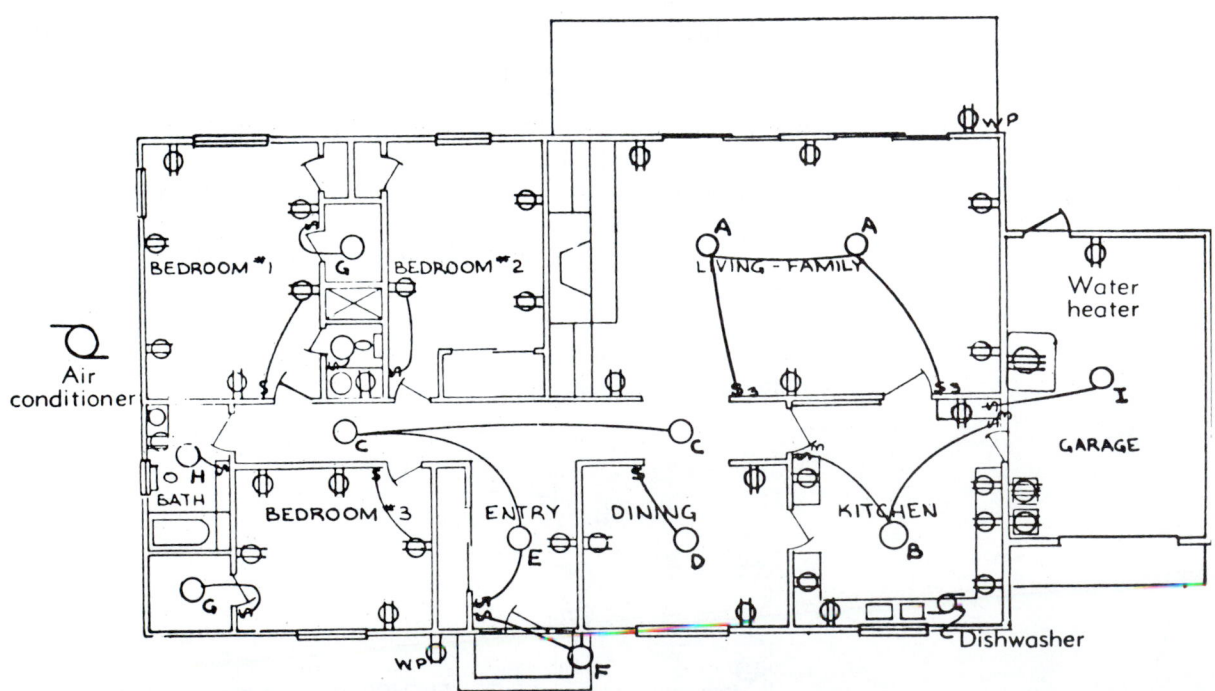

FIGURE 16-8   Typical branch circuit

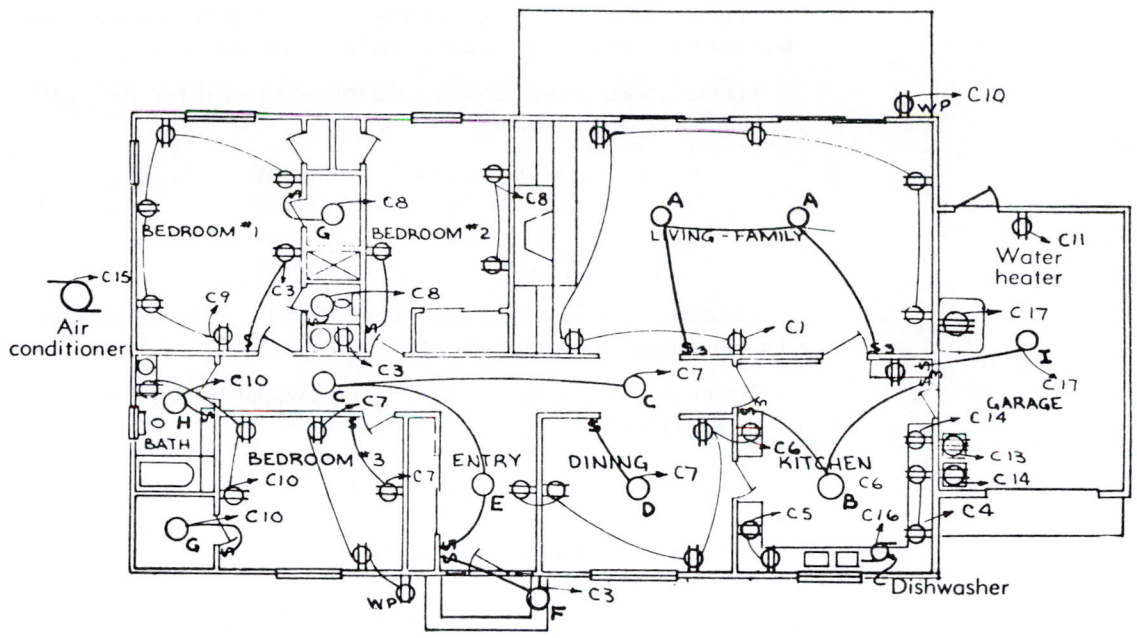

FIGURE 16-9   Add light switches and circuits (circuits shown and numbered)

Minimum general lighting load = 3 W/sq ft × 2,214 sq ft = 6,642 W

b. *The appliance and laundry circuit load* is calculated next.

Since the code requires two 20-amp branch circuits, the load would be based on 1,500 W (from the code) for each branch circuit. In addition, one 20-amp circuit is required for laundry room appliances. This results in a total of three 20-amp branch circuits for appliances.

Appliance and laundry load = 3 circuits × 1,500 W = 4,500 W

c. *Subtotal* the general lighting, appliance, and laundry branch circuit loads.

In this example:

General lighting	6,642 W
Appliance & laundry circuits	4,500 W
Subtotal	11,142 W

d. The *demand load* allowed by the code takes into account the fact that all of the electrical connections will not be in use at one time. While there are limits to this reduction for certain types of occupancies, in a dwelling the first 3,000 W are taken as 100%, and from 3,000 to 120,000 W, only 35% of the load is calculated (from Fig. 16-21).

For this example, the load subtotal is 11,142 W.

First 3,000 W at 100%	3,000 W
Remaining 8,142 W at 35%	2,850 W
Total demand load	5,850 W

The loads of all other appliances and equipment (motors) must be added to this demand load in order to determine the total service load on the system.

e. To determine the *appliance and equipment load,* all appliances and equipment which will not be on the lines discussed above must be listed along with their electrical requirements. While typical ratings are given in Fig. 16-4, it is most important that the manufacturer's data be used in the design.

For the residence being designed, the following is a list of fixed appliances and equipment and their ratings:

Water heater	3,800 W
Clothes dryer	4,400 W
Dishwasher	1,000 W
Range	11,700 W

The demand load for an *electric range,* consisting of an oven and a countertop cooking unit, is taken from Fig. 16-22.

For an electric range with a rating of 11.7 kW, the demand load would be 8 kW (or 8,000 W).

Electric range demand load = 8,000 W

The demand load for a *clothes dryer* is the total amount of power required according to the manufacturer's data.
In this case, the full 4,400 W must be used in the calculation.

Clothes dryer demand load = 4,400 W

The demand for *fixed appliances* (other than the range, clothes dryer, and air conditioning and space heating equipment) is taken as 100% of the total amount they require; *except* that when there are four or more of these fixed appliances (other than those omitted), the demand load can be taken as 75% of the fixed appliance load.
In this design, there are only two of these fixed appliances, the water heater at 3,800 W and the dishwasher at 1,000 W. The total of the ratings is 4,800 W, also the demand load.

Fixed appliances demand load = 4,800 W

Motors, such as those used in *central air conditioners,* have their demand loads calculated as 125% of the motor rating.
In this case, the air conditioner is rated at 9,000 watts. The total demand load will be 9,000 W × 1.25% = 11,250 W.

Air conditioner demand load = 11,250 W

The demand load for all of the lighting and appliances has now been calculated, and it should be tabulated as shown

General lighting } Appliances & laundry }	5,850
Electric range	8,000
Clothes dryer	4,400
Fixed appliances	4,800
Air conditioner	11,250
	34,300 W

7. Size the minimum service entrance based on the demand load from step 6. Service entrances and the typical sizes used for residential work are discussed in Sec. 15-6. The service entrance size is found by dividing the demand load for the building by the voltage serving the building. Most commonly 240-V service is used.

In this design, the total demand load of 34,300 W is divided by the 240-V service for a minimum service entrance of 143 amps.

$$\text{Minimum service entrance} = 34{,}300 \text{ W} \div 240 \text{ V} = 142.9 \text{ or } 143 \text{ amps}$$

8. Sizing the feeder (the circuit conductors between the service equipment and the branch overcurrent device—circuit breaker or fuse) is the next step. The feeder size is based on the total demand load calculated. The size is then selected from the tables in Fig. 16-24. The tables list the size of the conductor in the left column and the different insulation temperature ratings across the top. Generally, conductors with insulation ratings of 60°, 75°, and 95°C (140°, 167°, and 185°F) are most commonly used.

   For example, Fig. 16-24, a No. 6 copper conductor, 140°F (60°C) insulation, will carry 55 amps without subjecting the insulation to damaging heat. The same No. 6 conductor with 167°F (75°C) insulation can safely carry 65 amps. These values must be adjusted using a correction factor from the lower portion of the table if the room temperatures are within certain ranges. For example, if the No. 6 conductor with 140°F insulation will carry 55 amps and if the room temperature is 131°F (45°C), the load-carrying capacity would be corrected by a factor of 0.71, 55 amps × 0.71 = 39 amps, the maximum safe allowable amperage.

   To size the feeder, assuming RHW conductor, copper (75°C, 167°F) and a 150-amp demand load, as calculated. From Fig. 16-24, a 1/0 AWG is selected as the feeder size for 150-amp, 240-V service.

   150-amp feeder demand load, 1/0 AWG, RHW, copper

   Most designers would review their design before using a feeder conductor size so close to its capacity. In this case, the calculated demand load is 143 amps, while the conductor's capacity is listed as 150 amps. It is probably a better design solution to use 2/0 AWG copper conductors with a rated capacity of 175 amps. This allows for future addition of appliances and other equipment by the occupant.

   The size of the *neutral feeder conductor* may be determined as 70% of the demand load calculated for the range (step 6) plus all other demand loads on the system.

   In this design, the neutral feeder demand load would be:

Range load (8,000 W × 70%)	5,600 W
All other demand loads	26,300 W
Neutral demand load	31,900 W

   Neutral net computed load = 31,900 W/240 V = 133 amps

   Select the size of the neutral feeder from Fig. 16-24:

   133-amp neutral feeder demand load, 1/0 AWG, RHW, copper

   There are times when this calculation will result in smaller neutral feeder sizes, while at other times it will not change.

   Now size the feeder conduit. The minimum conduit size is determined for new work by using the table in Fig. 16-25. The conductor sizes are listed in the left column, and conduit (or tubing) sizes across the top. The numbers within the body of the table indicate the maximum number of conductors of any given size which can be put in the conduit in accordance with the NEC. First, find the conductor size being used in the left column. Then move to the right until the number shown is the same as, or larger than, the number of conductors to be placed in the conduit. Then move vertically upward and read the conduit size.

In this design, the three feeders are, two No. 2/0 and One 1/0 AWG copper conductors (use 3-2/0 to select conduit). Using Fig. 16-24, find the 2/0 in the left column, and move horizontally to the right to the number 3 and then vertically up to the 2 in. The code requires a 2-in. conduit to carry three No. 2/0 AWG copper conductors.

9. Determine the minimum number of lighting circuits by dividing the general lighting load by the voltage, finding the amperage required and dividing the amperage into circuits.

In this design, the general lighting load is calculated in step 6 as 6,642 W, and the voltage used for the lighting is 120 V.

$$6{,}642 \text{ W} \div 120 \text{ V} = 56 \text{ amps}$$

Since the Code limits the branch circuit size to 80% of the rating, this may be broken up into four 20-amp branches for a total of 64 amps, or into five 15-amp branches for a total of 60 amps. Remember, this is the minimum number of branch circuits. In laying out the circuits, almost all designs will have more circuits than the minimum. This is because, as discussed in Sec. 15-9 and 16-2, step 4, most designers will limit each circuit to five or six receptacles, lights or a combination of receptacles and lights.

10. Lay out all branch circuits on the drawing. The branch circuits for receptacles and switching are discussed in Sec. 15-9 and 16-2. Remember that all of these general-use receptacles and all lighting will use 120-V service.

A typical circuit layout is shown in Fig. 16-9. Note that each circuit is numbered, beginning with 1. It will be necessary to know the total number of 120- and 240-V circuits required.

In this design, there is a total of 14 circuits of which 11 are 120-V and 3 are 240-V.

11. Select the lighting panel based on the number of circuits and the required amperage. Be certain that all the pole space is not taken up so there is room for expansion. Remember that each 120-V circuit takes up one pole, while each 240-V circuit takes up two poles.

In this design, there are:

11 120-V circuits	11 poles
3 240-V circuits (3 × 2)	6 poles
Total poles required	17 poles

For this design, a minimum 150-amp, 24-pole lighting panel would be selected. Selection of a 200-amp panel would allow for expansion.

12. Lay out the panel circuits, either on the drawing or in tabulated form as shown in Fig. 16-10. In large designs, with more than one panel, this provides the electrician with a schedule of what circuits will be served from what box. It is also used to note the size of the conductors used for circuits (step 8).

While a lighting panel layout is not often done for a residence, it is helpful to both the electrician and the designer if one is included. For commercial projects, a panel layout is almost always included. A typical commercial application is shown in

200 AMP, 32 Pole 120/208V		
20A – 120V	2 12→	Receptacles C-1 thru C-8
20A 120V	2 12→	Lights C-9 thru C-15
20A 120V	2 12→	Dishwasher C-16
30A 208V	3 10→	Clothes Dryer C-17
20A 120V		Spares

FIGURE 16-10   Circuits ands panels

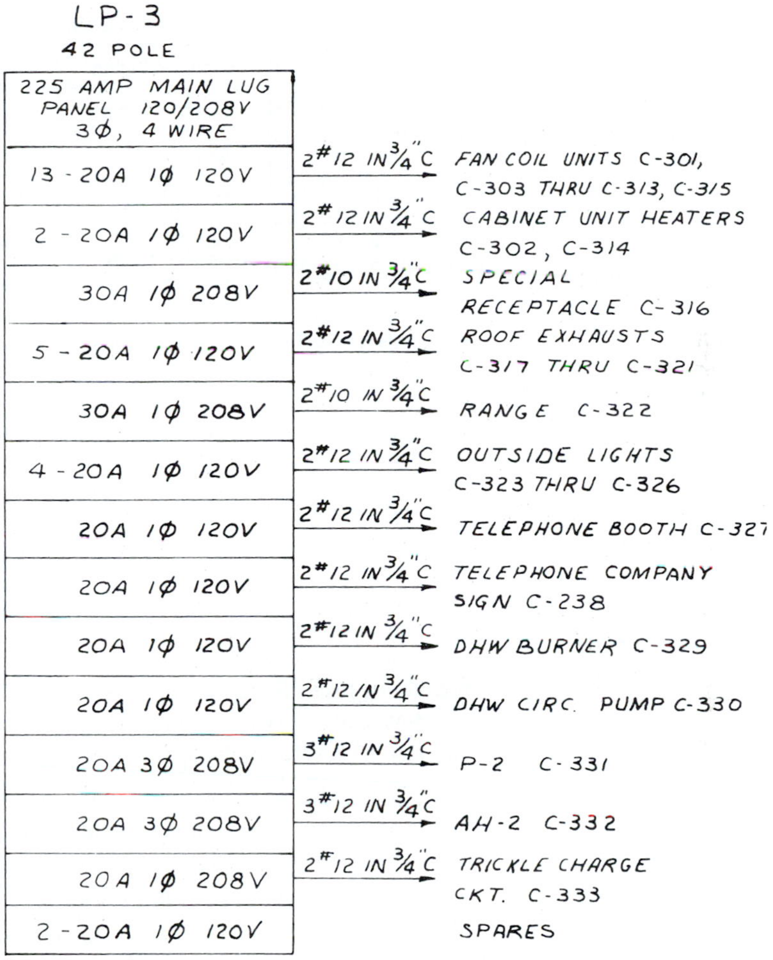

LP-3

42 POLE

225 AMP MAIN LUG PANEL 120/208V 3∅, 4 WIRE		
13 - 20A 1∅ 120V	2#12 IN ¾"C	FAN COIL UNITS C-301, C-303 THRU C-313, C-315
2 - 20A 1∅ 120V	2#12 IN ¾"C	CABINET UNIT HEATERS C-302, C-314
30A 1∅ 208V	2#10 IN ¾"C	SPECIAL RECEPTACLE C-316
5 - 20A 1∅ 120V	2#12 IN ¾"C	ROOF EXHAUSTS C-317 THRU C-321
30A 1∅ 208V	2#10 IN ¾"C	RANGE C-322
4 - 20A 1∅ 120V	2#12 IN ¾"C	OUTSIDE LIGHTS C-323 THRU C-326
20A 1∅ 120V	2#12 IN ¾"C	TELEPHONE BOOTH C-327
20A 1∅ 120V	2#12 IN ¾"C	TELEPHONE COMPANY SIGN C-238
20A 1∅ 120V	2#12 IN ¾"C	DHW BURNER C-329
20A 1∅ 120V	2#12 IN ¾"C	DHW CIRC. PUMP C-330
20A 3∅ 208V	3#12 IN ¾"C	P-2 C-331
20A 3∅ 208V	3#12 IN ¾"C	AH-2 C-332
20A 1∅ 208V	2#12 IN ¾"C	TRICKLE CHARGE CKT. C-333
2 - 20A 1∅ 120V		SPARES

FIGURE 16-11   Commercial panels and circuits

Fig. 16-11. The panel drawing for the residence is shown in Fig. 16-13 with each of the branch circuits, and the circuit numbers, noted.

13. Size the conductors for all of the branch circuits, and note them on the panel drawing or in the tabulation. The conductors are sized just as the feeder conductor was sized in step 8.

Branch circuit conductors to general purpose receptacles and light fixtures must be a minimum of No. 14 AWG when used with a 15-amp overcurrent device (circuit

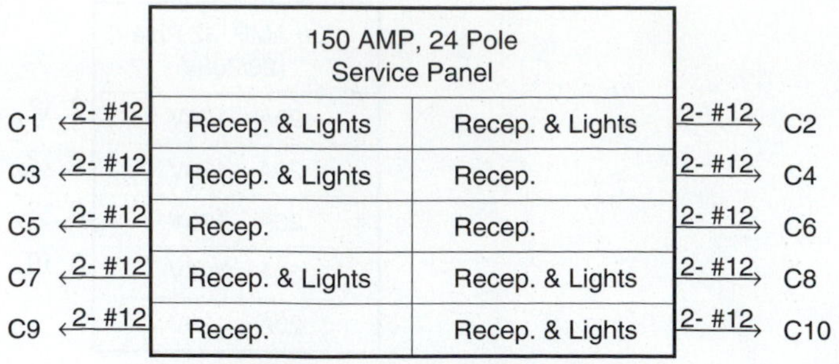

FIGURE 16-12   Conductor notation

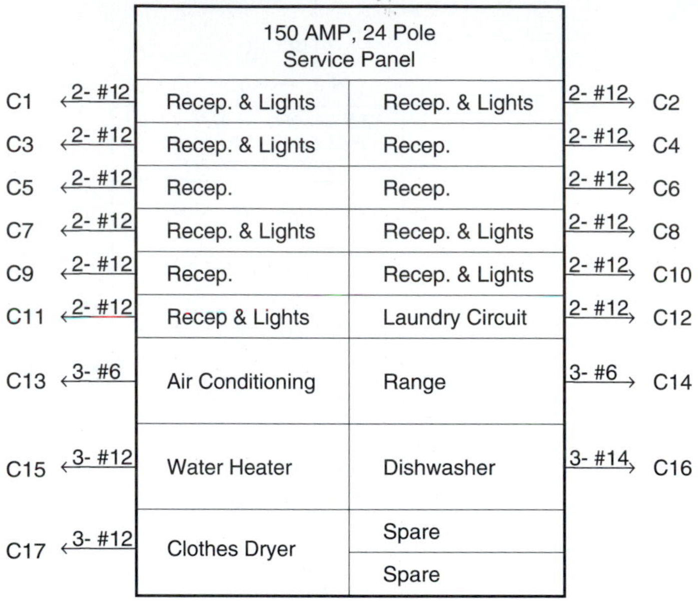

FIGURE 16-13   Conductor notation

breaker or fuse). Because No. 14 AWG conductors are limited to a maximum load of 1,725 W and a maximum circuit length of 30 ft, most designers use 20-amp breakers and No. 12 AWG conductors.

In this design the conductors for the 20-amp branch circuits serving the receptacles and light fixtures are sized first. The conductor size selected from the table in Fig. 16-24 is No. 12 AWG, RHW, copper (see note at bottom of table). This is noted on the panel drawing as shown in Fig. 16-12.

Next, the conductor for each branch circuit is sized, based on the amperage required and the type of conductor being used. The NEC requires limiting of such loads to 80% of the ratings. To allow for this, the calculated amperage is multiplied by 1.25.

Air conditioner—11,250 W:

$$11,250 \text{ W} \div 240 \text{ V} = 46.9 \text{ amps}$$
$$46.9 \times 1.25 = 58.7 \text{ amps}$$

From Fig. 16-24, use No. 6 AWG, RHW, copper
Water heater—3,800 W:

$$3,800 \text{ W} \div 240 \text{ V} = 15.8 \text{ amps}$$
$$15.8 \times 1.25 = 19.75 \text{ amps}$$

From Fig. 16-24, use No. 12 AWG, RHW, copper.
Clothes dryer—4,400 W:

$$4,400 \text{ W} \div 240 \text{ V} = 18.3 \text{ amps}$$
$$18.3 \times 1.25 = 22.9 \text{ amps}$$

From Fig. 16-24, use No. 10 AWG, RHW, copper.
Dishwasher—1,000 W:

$$1,000 \div 120 \text{ V} = 8.3 \text{ amps}$$
(a 20-amp breaker is used)

From Fig. 16-24, use No. 12 AWG, RHW, copper.
Range—11,700 W (nameplate rating):

$$11,700 \text{ W} \div 240 \text{ V} = 48.8 \text{ amps}$$
$$48.8 \times 1.25 = 61 \text{ amps}$$

From Fig. 16-24, use No. 6 AWG, RHW, copper.

Note the conductor sizes for each branch circuit on the panel drawing as shown in Fig. 16-13.

Whenever there will be more than three conductors in a raceway, the allowable ampacity of each conductor must be reduced. The amount of reduction required is shown in Fig. 16-24, note 8.

For example, if six No. 10 AWG copper conductors are placed in a 1-in. conduit, according to the reduction values in Fig. 16-24, the allowable ampacity will be reduced to 80% of the allowable values given in Fig. 16-24. The allowable ampacity for No. 10 AWG copper conductors is 30. Since this must be reduced by 80%:

$$30 \times 0.80 = 24 \text{ amps (allowed ampacity)}$$

The design of the electrical system for the residence in Appendix C is complete. The electrical system selected is 150-amp service with a 24-pole lighting panel. The feeders selected are three No. 2/0 AWG, RHW, copper conductors. The panel layout is shown in Fig. 16-13, with all branch circuits and their conductor sizes noted.

14. On commercial, industrial, and institutional projects, the electrical supply feeds into several panels, oftentimes broken up into *lighting panels* to serve light fixtures and general-purpose receptacles and *power panels* to supply the electricity to all appliances and equipment in the project. Quite often, there are several of each type of panel.

In such a design, the circuits are labeled with a circuit number and a panel number. For example, C-3, LP-1 would be circuit 3 from lighting panel 1. Typical examples of this type of notation are shown in Fig. 16-14.

Also, in such a design, it is necessary to have a feeder from the service entrance to the main distribution panel, which is, in turn, connected to the lighting and power panels by feeders (Fig. 16-15). Each of the feeders must be sized in accordance with the code.

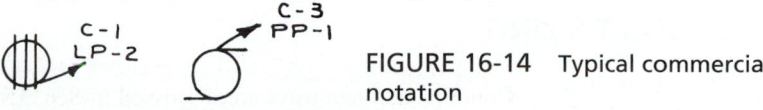

FIGURE 16-14   Typical commercial notation

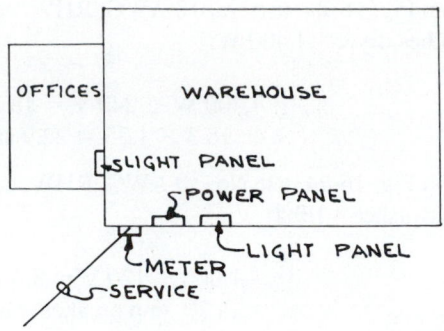

FIGURE 16-15   Typical commercial panel layout

**Optional Calculation for Dwelling Unit
Load in kVA**

Largest of the following four selections.

(1) 100 percent of the nameplate rating(s) of the air conditioning and cooling, including heat pump compressors.

(2) 100 percent of the nameplate ratings of electric thermal storage and other heating systems where the usual load is expected to be continuous at the full nameplate value. Systems qualifying under this selection shall not be figured under any other selection in this table.

(3) 65 percent of the nameplate rating(s) of the central electric space heating including integral supplemental heating in heat pumps.

(4) 65 percent of the nameplate rating(s) of electric space heating if less than four separately controlled units.

(5) 40 percent of the nameplate rating(s) of electric space heating of four or more separately controlled units.

Plus: 100 percent of the first 10 kVA of all other load. 40 percent of the remainder of all other load.

FIGURE 16-16   Optional residential demand load (load in kVA)

16-3   OPTIONAL CALCULATION FOR A ONE-FAMILY RESIDENCE

There is an optional method of calculating the demand load for a residence, based on percentages of the loads listed in Fig. 16-16.

Fig. 16-17 is a list of loads calculated for the residence in Appendix C.

Under the optional method of calculating the demand load for a one-family residence, a much smaller service is required (117.6 amps compared with 143 amps), and a 150-amp service would still be used.

Based on amps, the feeder required (from Fig. 16-24) is a No. 1 AWG, RHW, copper conductor.

Neutral feeder calculations are the same as for the feeder *except* for the reduced range load.

16-4   CONDUIT SIZING

Conduits and raceways are discussed in Sec. 15-14, and in this section the sizing of rigid metal conduit and electrical metallic tubing is covered. The size of the conduit is directly related to the number and size of the conductors which will be placed in it.

	Watts
Air conditioner	9,000
General lighting 2213 x 3	6,642
2-20 amp appliance circuits	
2 @ 1500 each	3,000
1-20 amp laundry circuit	1,500
Water heater	3,800
Clothes dryer	4,400
Dishwasher	1,000
Range	11,700
Total	41,042 = 41.04 kw

Optional demand load	
Air conditioner @ 100%	= 9,000
First 10kw @ 100%	10,000
Balance @ 40% (23.04 x 0.4)	9,216
	28,216 watts

Service load
28,216 watts ÷ 240 volts = 117.6 amps

FIGURE 16-17   Optional demand load calculations

All of the tables to be used are based on the NEC's maximum percentages of the conduit that can be filled with conductors. Based on these percentages, the table in Fig. 16-25 gives the maximum number of conductors which can be placed in the trade (standard) sizes of conduit or tubing listed.

For example:

1. What size conduit is required to run three No. 3 AWG conductors in?

   From Fig. 16-25, 1¼-in. conduit or tubing is required for three No. 3 AWG conductors.

2. How many No. 1 AWG conductors may be placed in 2-in. conduit or tubing?

   A maximum of five No. 1 AWG conductors may be placed in 2-in. conduit or tubing.

Remember that when more than three conductors are placed in any raceway (including conduit or tubing), the allowable ampacity of each conductor is derated (reduced) in accordance with note 8 for Fig. 16-24.

16-5   VOLTAGE DROP

The NEC limits the amount of voltage drop (the loss of voltage due to resistance in the conductors) for power, heating, and lighting, or any combination of these loads, to 3%. In addition, the maximum total voltage drop for feeders and branch circuits should be no more than 5%, leaving 2% for branch circuit loss.

In most residences, and small buildings, the voltage drop in the feeder is small because the length of the conductor is short.

The voltage drop can be calculated by using the formula:

$$\text{Voltage drop} = \frac{I \times L \times R}{1,000}$$

where

I = current carried in the conductor (amps)
L = length of current-carrying conductor (ft)

R = resistance of conductor (ohms per 1,000 ft)

1,000 = constant (converting R per 1,000 ft to R per ft)

Example *Given:* The commercial building shown in Fig. 16-18.

Problem: Determine the total voltage drop from the feeder to panel B.

The feeder to panel B has an I of 93 amps, and the conductor is No. 3 AWG copper. The length of the conductor from the main distribution panel to panel B is about 85 ft. Resistances for the various sizes of copper and aluminum conductors are given in Fig. 16-25. This table lists the DC resistances in ohms per M (1,000) ft at 25°C (77°F), which are then converted to AC resistances by multiplying by the factors also given in Fig. 16-26. For a No. 3 conductor, there is no revision; but note that as the conductors get larger, so do the factors. From Fig. 16-26, the resistance for a No. 3 conductor is 0.254 ohms per 1,000 ft.

The total voltage drop for the feeder to panel B is:

$$\frac{93 \text{ amps} \times 85 \text{ ft} \times 0.254 \text{ ohms}}{1,000} = 2.01 \text{ V}$$

The percentage of voltage drop, in this case, would be 2.01 V (voltage drop) divided by 240 (voltage), or less than 1%.

Example *Given:* The residence in Appendix C.

Problem: Check the voltage drop in the branch circuit to bedroom 1 since it is the one farthest from the panel.

The circuit has four receptacles on it, and its load is about 1,440 (4 duplex outlets at 180 W each = 4 × 2 × 180 = 1,440 W). This 1,440 W gives an amperage of 1,440 W ÷ 120 V = 12 amps. The approximate length of the branch circuit is 90 ft, and the resistance of a No. 12 AWG conductor (unstranded) is 2.01 ohms per 1,000 ft.

$$\frac{12 \text{ amps} \times 90 \text{ ft} \times 2.01 \text{ ohms}}{1,000} = 2.17 \text{ V}$$

The percentage of voltage drop, in this case, would be 2.17 V (voltage drop) divided by 120 (voltage) or about 1.8%, within the 2% drop that is considered good design practice.

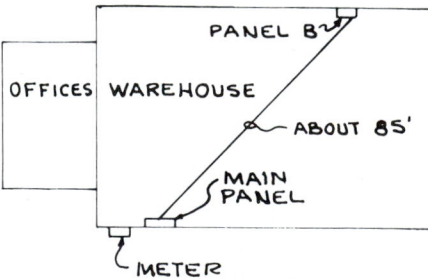

FIGURE 16-18   Voltage drop

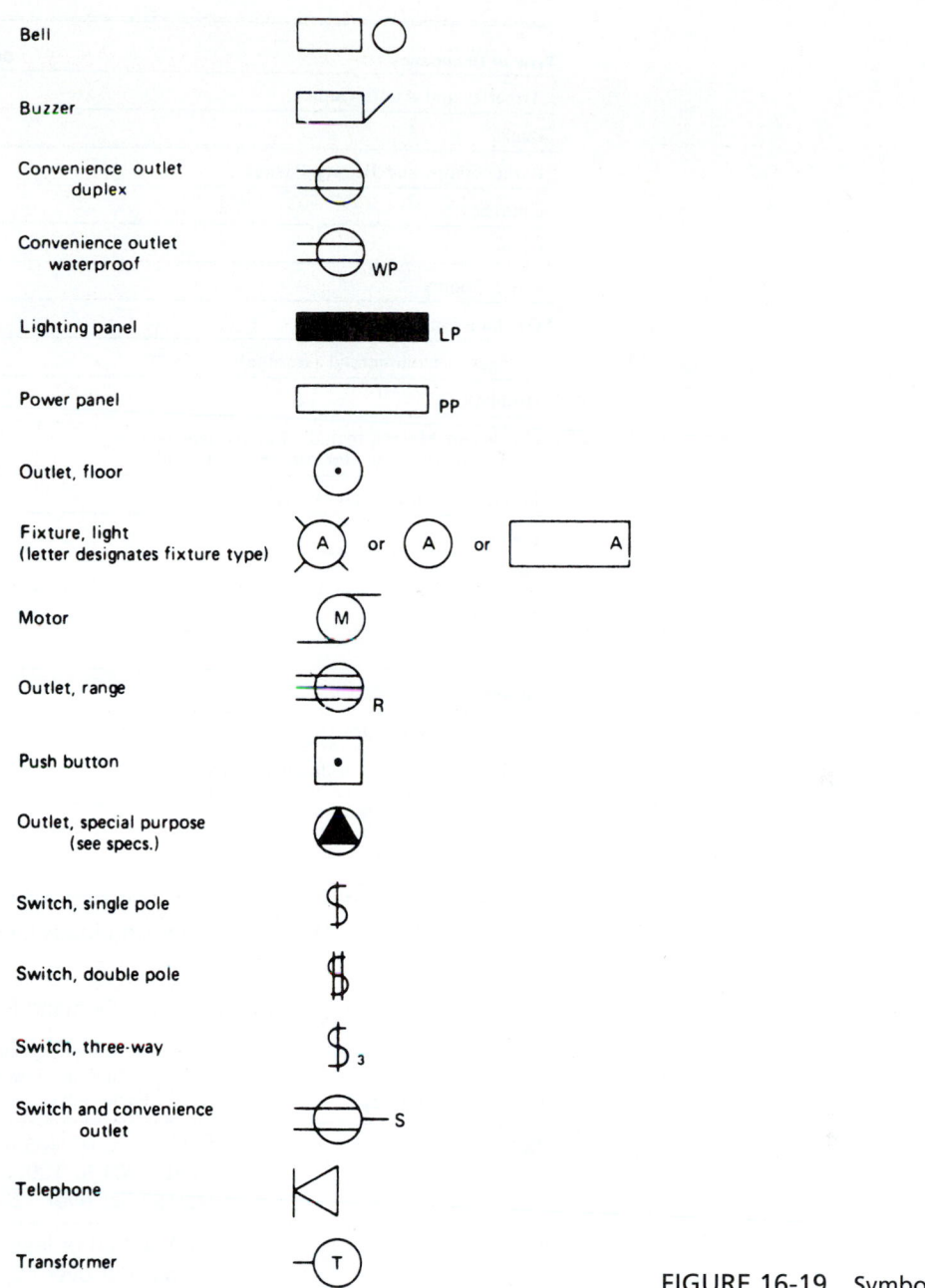

Bell	
Buzzer	
Convenience outlet duplex	
Convenience outlet waterproof	WP
Lighting panel	LP
Power panel	PP
Outlet, floor	
Fixture, light (letter designates fixture type)	A or A or A
Motor	M
Outlet, range	R
Push button	
Outlet, special purpose (see specs.)	
Switch, single pole	
Switch, double pole	
Switch, three-way	
Switch and convenience outlet	S
Telephone	
Transformer	T

FIGURE 16-19   Symbols

Type of Occupancy	Unit Load per Sq. Ft. (Volt-Amperes)
Armories and Auditoriums	1
Banks	3½**
Barber Shops and Beauty Parlors	3
Churches	1
Clubs	2
Court Rooms	2
*Dwelling Units	3
Garages — Commercial (storage)	½
Hospitals	2
*Hotels and Motels, including apartment houses without provisions for cooking by tenants	2
Industrial Commercial (Loft) Buildings	2
Lodge Rooms	1½
Office Buildings	3½**
Restaurants	2
Schools	3
Stores	3
Warehouses (storage)	¼
In any of the above occupancies except one-family dwellings and individual dwelling units of two-family and multifamily dwellings:	
Assembly Halls and Auditoriums	1
Halls, Corridors, Closets, Stairways	½
Storage Spaces	¼

FIGURE 16-20   General lighting loads by occupancies

Lighting Load Feeder Demand Factors

Type of Occupancy	Portion of Lighting Load to Which Demand Factor Applies (wattage)	Demand Factor Percent
Dwelling Units	First 3,000 or less at	100
	Next 3,001 to 120,000 at	35
	Remainder over 120,000 at	25
*Hospitals	First 50,000 or less at	40
	Remainder over 50,000 at	20
*Hotels and Motels—Including Apartment Houses without Provision for Cooking by Tenants	First 20,000 or less at	50
	Next 20,001 to 100,000 at	40
	Remainder over 100,000 at	30
Warehouses (Storage)	First 12,500 or less at	100
	Remainder over 12,500 at	50
All Others	Total Wattage	100

* The demand factors of this table shall not apply to the computed load of feeders to areas in hospitals, hotels, and motels where the entire lighting is likely to be used at one time: as in operating rooms, ballrooms, or dining rooms.
Reprinted with permission from NFPA 70-1993, the *National Electrical Code*®, Copyright © 1992, National Fire Protection Association, Quincy, MA 02269. This reprinted material is not the official position of the National Fire Protection Association, which is represented only by the standard in its entirety.

FIGURE 16-21   Demand factors

Conductor properties

Size AWG/kcmil	Area Cir. Mils	Conductors Stranding Quantity	Conductors Stranding Diam. In.	Conductors Overall Diam. In.	Conductors Overall Area In.²	DC Resistance at 75°C (167°F) Copper Uncoated ohm/MFT	DC Resistance at 75°C (167°F) Copper Coated ohm/MFT	DC Resistance at 75°C (167°F) Aluminum ohm/MFT
18	1620	1	—	0.040	0.001	7.77	8.08	12.8
18	1620	7	0.015	0.046	0.002	7.95	8.45	13.1
16	2580	1	—	0.051	0.002	4.89	5.08	8.05
16	2580	7	0.019	0.058	0.003	4.99	5.29	8.21
14	4110	1	—	0.064	0.003	3.07	3.19	5.06
14	4110	7	0.024	0.073	0.004	3.14	3.26	5.17
12	6530	1	—	0.081	0.005	1.93	2.01	3.18
12	6530	7	0.030	0.092	0.006	1.98	2.05	3.25
10	10380	1	—	0.102	0.008	1.21	1.26	2.00
10	10380	7	0.038	0.116	0.011	1.24	1.29	2.04
8	16510	1	—	0.128	0.013	0.764	0.786	1.26
8	16510	7	0.049	0.146	0.017	0.778	0.809	1.28
6	26240	7	0.061	0.184	0.027	0.491	0.510	0.808
4	41740	7	0.077	0.232	0.042	0.308	0.321	0.508
3	52620	7	0.087	0.260	0.053	0.245	0.254	0.403
2	66360	7	0.097	0.292	0.067	0.194	0.201	0.319
1	83690	19	0.066	0.332	0.087	0.154	0.160	0.253
1/0	105600	19	0.074	0.373	0.109	0.122	0.127	0.201
2/0	133100	19	0.084	0.419	0.138	0.0967	0.101	0.159
3/0	167800	19	0.094	0.470	0.173	0.0766	0.0797	0.126
4/0	211600	19	0.106	0.528	0.219	0.0608	0.0626	0.100
250	—	37	0.082	0.575	0.260	0.0515	0.0535	0.0847
300	—	37	0.090	0.630	0.312	0.0429	0.0446	0.0707
350	—	37	0.097	0.681	0.364	0.0367	0.0382	0.0605
400	—	37	0.104	0.728	0.416	0.0321	0.0331	0.0529
500	—	37	0.116	0.813	0.519	0.0258	0.0265	0.0424
600	—	61	0.099	0.893	0.626	0.0214	0.0223	0.0353
700	—	61	0.107	0.964	0.730	0.0184	0.0189	0.0303
750	—	61	0.111	0.998	0.782	0.0171	0.0176	0.0282
800	—	61	0.114	1.03	0.834	0.0161	0.0166	0.0265
900	—	61	0.122	1.09	0.940	0.0143	0.0147	0.0235
1000	—	61	0.128	1.15	1.04	0.0129	0.0132	0.0212
1250	—	91	0.117	1.29	1.30	0.0103	0.0106	0.0169
1500	—	91	0.128	1.41	1.57	0.00858	0.00883	0.0141
1750	—	127	0.117	1.52	1.83	0.00735	0.00756	0.0121
2000	—	127	0.126	1.63	2.09	0.00643	0.00662	0.0106

These resistance values are valid ONLY for the parameters as given. Using conductors having coated strands, different stranding type, and, especially, other temperatures, change the resistance.

Formula for temperature change: $R_2 = R_1 [1+\alpha(T_2-75)]$ where: $\alpha_{cu} = 0.00323$, $\alpha_{AL} = 0.00330$.

Conductors with compact and compressed stranding have about 9 percent and 3 percent, respectively, smaller bare conductor diameters than those shown. See Table 5A for actual compact cable dimensions.

The IACS conductivities used: bare copper = 100%, aluminum = 61%.

Class B stranding is listed as well as solid for some sizes. Its overall diameter and area is that of its circumscribing circle.

(FPN): The construction information is per NEMA WC8-1976 (Rev 5-1980). The resistance is calculated per National Bureau of Standards Handbook 100, dated 1966, and Handbook 109, dated 1972.

FIGURE 16-23   Conductor properties

Demand load for electric ranges

Demand Loads for Household Electric Ranges, Wall-Mounted Ovens, Counter-Mounted Cooking Units, and Other Household Cooking Appliances over 1¾ kW Rating, Column A to be used in all cases except as otherwise permitted in Note 3 below.

NUMBER OF APPLIANCES	Maximum Demand (See Notes) COLUMN A (Not over 12 kW Rating)	Demand Factors Percent (See Note 3) COLUMN B (Less than 3½ kW Rating)	Demand Factors Percent (See Note 3) COLUMN C (3½ kW to 8½ kW Rating)
1	8kW	80%	80%
2	11kW	75%	65%
3	14kW	70%	55%
4	17kW	66%	50%
5	20kW	62%	45%
6	21kW	59%	43%
7	22kW	56%	40%
8	23kW	53%	36%
9	24kW	51%	35%
10	25kW	49%	34%
11	26kW	47%	32%
12	27kW	45%	32%
13	28kW	43%	32%
14	29kW	41%	32%
15	30kW	40%	32%
16	31kW	39%	28%
17	32kW	38%	28%
18	33kW	37%	28%
19	34kW	36%	28%
20	35kW	35%	28%
21	36kW	34%	26%
22	37kW	33%	26%
23	38kW	32%	26%
24	39kW	31%	26%
25	40kW	30%	26%
26–30	15 kW plus 1kW for each range	30%	24%
31–40	15 kW plus 1kW for each range	30%	22%
41–50	25 kW plus ½ kW for each range	30%	20%
51–60	25 kW plus ½ kW for each range	30%	18%
61 & over	25 kW plus ½ kW for each range	30%	16%

Note 1. Over 12 kW through 27 kW ranges all of same rating. For ranges individually rated more than 12 kW but not more than 27 kW, the maximum demand in Column A shall be increased 5 percent for each additional kW of rating or major fraction thereof by which the rating of individual ranges exceeds 12 kW.

Note 2. Over 12 kW through 27 kW ranges of unequal ratings. For ranges individually rated more than 12 kW and of different ratings but none exceeding 27 kW an average value of rating shall be computed by adding together the ratings of all ranges to obtain the total connected load (using 12 kW for any range rated less than 12 kW) and dividing by the total number of ranges: and then the maximum demand in Column A shall be increased 5 percent for each kW or major fraction thereof by which this average value exceeds 12 kW.

Note 3. Over 1¾ kW through 8¼ kW. In lieu of the method provided in Column A, it shall be permissible to add the nameplate ratings of all ranges rated more than 1¾ kW but not more than 8¼ kW and multiply the sum by the demand factors specified in Column B or C for the given number of appliances.

Note 4. Branch-Circuit Load. It shall be permissible to compute the branch-circuit load for one range in accordance with Table 220-19. The branch-circuit load for one wall-mounted oven or one counter-mounted cooking unit shall be the nameplate rating of the appliance. The branch-circuit load for a counter-mounted cooking unit and not more than two wall-mounted ovens, all supplied from a single branch circuit and located in the same room, shall be computed by adding the nameplate rating of the individual appliances and treating this total as equivalent to one range.

Note 5. This table also applies to household cooking appliances rated over 1¾ kW and used in instructional programs.

See Table 220-20 for commercial cooking equipment.

FIGURE 16-22   Demand load for electric ranges

Allowable Ampacities of Insulated Conductors
Rated 0-2000 Volts, 60° to 90°C (140° to 194°F)
Not More Than Three Conductors in Raceway or Cable or Earth
(Directly Buried), Based on Ambient Temperature of 30°C (86°F)

Size	Temperature Rating of Conductor. See Table 310-13.						Size
	60°C (140°F)	75°C (167°F)	90°C (194°F)	60°C (140°F)	75°C (167°F)	90°C (194°F)	
AWG kcmil	TYPES TW†, UF†	TYPES FEPW†, RH†, RHW†, THHW†, THW†, THWN†, XHHW† USE†, ZW†	TYPES TA, TBS, SA SIS, FEP†, FEPB†, MI RHH†, RHW-2, THHN†, THHW†, THW-2, THWN-2, USE-2, XHH, XHHW† XHHW-2, ZW-2	TYPES TW†, UF†	TYPES RH†, RHW†, THHW†, THW†, THWN†, XHHW† USE†	TYPES TA, TBS, SA, SIS, THHN†, THHW†, THW-2, THWN-2, RHH†, RHW-2 USE-2 XHH, XHHW XHHW-2, ZW-2	AWG kcmil
	COPPER			ALUMINUM OR COPPER-CLAD ALUMINUM			
18	. . . .	. . . .	14	. . . .	. . . .	. . . .	. . . .
16	. . . .	. . . .	18	. . . .	. . . .	. . . .	. . . .
14	20†	20†	25†	. . . .	. . . .	. . . .	. . . .
12	25†	25†	30†	20†	20†	25†	12
10	30	35†	40†	25	30†	35†	10
8	40	50	55	30	40	45	8
6	55	65	75	40	50	60	6
4	70	85	95	55	65	75	4
3	85	100	110	65	75	85	3
2	95	115	130	75	90	100	2
1	110	130	150	85	100	115	1
1/0	125	150	170	100	120	135	1/0
2/0	145	175	195	115	135	150	2/0
3/0	165	200	225	130	155	175	3/0
4/0	195	230	260	150	180	205	4/0
250	215	255	290	170	205	230	250
300	240	285	320	190	230	255	300
350	260	310	350	210	250	280	350
400	280	335	380	225	270	305	400
500	320	380	430	260	310	350	500
600	355	420	475	285	340	385	600
700	385	460	520	310	375	420	700
750	400	475	535	320	385	435	750
800	410	490	555	330	395	450	800
900	435	520	585	355	425	480	900
1000	455	545	615	375	445	500	1000
1250	495	590	665	405	485	545	1250
1500	520	625	705	435	520	585	1500
1750	545	650	735	455	545	615	1750
2000	560	665	750	470	560	630	2000

CORRECTION FACTORS

Ambient Temp. °C	For ambient temperatures other than 30°C (86°F), multiply the allowable ampacities shown above by the appropriate factor shown below.						Ambient Temp. °F
21-25	1.08	1.05	1.04	1.08	1.05	1.04	70-77
26-30	1.00	1.00	1.00	1.00	1.00	1.00	78-86
31-35	.91	.94	.96	.91	.94	.96	87-95
36-40	.82	.88	.91	.82	.88	.91	96-104
41-45	.71	.82	.87	.71	.82	.87	105-113
46-50	.58	.75	.82	.58	.75	.82	114-122
51-55	.41	.67	.76	.41	.67	.76	123-131
56-60	. . . .	.58	.71	. . . .	.58	.71	132-140
61-70	. . . .	.33	.58	. . . .	.33	.58	141-158
71-80	. . . .	. . . .	.41	. . . .	. . . .	.41	159-176

†Unless otherwise specifically permitted elsewhere in this Code, the overcurrent protection for conductor types marked with an obelisk (†) shall not exceed 15 amperes for No. 14, 20 amperes for No. 12, and 30 amperes for No. 10 copper; or 15 amperes for No. 12 and 25 amperes for No. 10 aluminum and copper-clad aluminum after any correction factors for ambient temperature and number of conductors have been applied.

FIGURE 16-24

Allowable Ampacities of Single Insulated Conductors,
Rated 0 through 2000 Volts, In Free Air
Based on Ambient Air Temperature of 30°C (86°F)

Size	Temperature Rating of Conductor. See Table 310-13.						Size
	60°C (140°F)	75°C (167°F)	90°C (194°F)	60°C (140°F)	75°C (167°F)	90°C (194°F)	
AWG kcmil	TYPES TW†, UF†	TYPES FEPW†, RH†, RHW†, THHW†, THW†, THWN†, XHHW†, ZW†	TYPES TA, TBS, SA SIS, FEP†, FEPB†, MI, RHH†, RHW-2, THHN†, THHW†, THW-2, THWN-2, USE-2, XHH, XHHW†, XHHW-2, ZW-2	TYPES TW†, UF†	TYPES RH†, RHW†, THHW†, THW†, THWN†, XHHW†	TYPES TA, TBS, SA, SIS, THHN†, THHW†, THW-2, THWN-2, RHH†, RHW-2, USE-2, XHH, XHHW†, XHHW-2, ZW-2	AWG kcmil
	COPPER			ALUMINUM OR COPPER-CLAD ALUMINUM			
18	. . . .	. . . .	18	. . . .	. . . .	. . . .	. . . .
16	. . . .	. . . .	24	. . . .	. . . .	. . . .	. . . .
14	25†	30†	35†	. . . .	. . . .	. . . .	. . . .
12	30†	35†	40†	25†	30†	35†	12
10	40†	50†	55†	35†	40†	40†	10
8	60	70	80	45	55	60	8
6	80	95	105	60	75	80	6
4	105	125	140	80	100	110	4
3	120	145	165	95	115	130	3
2	140	170	190	110	135	150	2
1	165	195	220	130	155	175	1
1/0	195	230	260	150	180	205	1/0
2/0	225	265	300	175	210	235	2/0
3/0	260	310	350	200	240	275	3/0
4/0	300	360	405	235	280	315	4/0
250	340	405	455	265	315	355	250
300	375	445	505	290	350	395	300
350	420	505	570	330	395	445	350
400	455	545	615	355	425	480	400
500	515	620	700	405	485	545	500
600	575	690	780	455	540	615	600
700	630	755	855	500	595	675	700
750	655	785	885	515	620	700	750
800	680	815	920	535	645	725	800
900	730	870	985	580	700	785	900
1000	780	935	1055	625	750	845	1000
1250	890	1065	1200	710	855	960	1250
1500	980	1175	1325	795	950	1075	1500
1750	1070	1280	1445	875	1050	1185	1750
2000	1155	1385	1560	960	1150	1335	2000

CORRECTION FACTORS							
Ambient Temp. °C	For ambient temperatures other than 30°C (86°F), multiply the allowable ampacities shown above by the appropriate factor shown below.						Ambient Temp. °F
21-25	1.08	1.05	1.04	1.08	1.05	1.04	70-77
26-30	1.00	1.00	1.00	1.00	1.00	1.00	78-86
31-35	.91	.94	.96	.91	.94	.96	87-95
36-40	.82	.88	.91	.82	.88	.91	96-104
41-45	.71	.82	.87	.71	.82	.87	105-113
46-50	.58	.75	.82	.58	.75	.82	114-122
51-55	.41	.67	.76	.41	.67	.76	123-131
56-60	. . . .	.58	.71	. . . .	.58	.71	132-140
61-70	. . . .	.33	.58	. . . .	.33	.58	141-158
71-80	. . . .	. . . .	.41	. . . .	. . . .	.41	159-176

†Unless otherwise specifically permitted elsewhere in this Code, the overcurrent protection for conductor types marked with an obelisk (†) shall not exceed 15 amperes for No. 14, 20 amperes for No. 12, and 30 amperes for No. 10 copper; or 15 amperes for No. 12 and 25 amperes for No. 10 aluminum and copper-clad aluminum.

FIGURE 16-24   (continued)

Notes to Ampacity Tables of 0 to 2000 Volts

1. Explanation of Tables. For explanation of type letters and for recognized size of conductors for the various conductor insulations, see Section 310-13. For installation requirements, see Sections 310-1 through 310-10 and the various articles of this Code. For flexible cords, see Tables 400-4, 400-5(A), and 400-5(B).

3. 120/240 Volts, 3-Wire, Single-Phase Dwelling Services and Feeders. For dwelling units, conductors, as listed below, shall be permitted to be utilized as 120/240-volt, 3-wire, single-phase service-entrance conductors, service lateral conductors, and feeder conductors that supply the total load to a dwelling unit and installed in raceway or cable with or without an equipment grounding conductor. The grounded conductor shall be permitted to be not more than two AWG sizes smaller than the ungrounded conductors for application of this note, provided the requirements of Sections 215-2, 220-22, and 230-42 are met.

Conductor Types and Sizes
RH-RHH-RHW-THHW-THW-THWN-THHN-XHHW-USE

Copper	Aluminum or Copper-Clad AL	Rating in Amps
AWG	AWG	
4	2	100
3	1	110
2	1/0	125
1	2/0	150
1/0	3/0	175
2/0	4/0	200
3/0	250 kcmil	225
4/0	300 kcmil	250
250 kcmil	350 kcmil	300
350 kcmil	500 kcmil	350
400 kcmil	600 kcmil	400

5. Bare Conductors. Where bare conductors are used with insulated conductors, their allowable ampacities shall be limited to those permitted for the adjacent insulated conductors.

6. Mineral-Insulated, Metal-Sheathed Cable. The temperature limitation on which the ampacities of mineral-insulated, metal-sheathed cable are based is determined by the insulating materials used in the end seal. Termination fittings incorporating unimpregnated, organic, insulating materials are limited to 90°C (194°F) operation.

7. Type MTW Machine Tool Wire.

(FPN): For the allowable ampacities of Type MTW wire, see Table 13-5(a) in the Electrical Standard for Industrial Machinery, NFPA 79-1991.

8. Adjustment Factors.

(a) More than Three Current-Carrying Conductors in a Raceway or Cable. Where the number of current-carrying conductors in a raceway or cable exceeds three, the allowable ampacities shall be reduced as shown in the following table:

Number of Current-Carrying Conductors	Percent of Values in Tables as Adjusted for Ambient Temperature if Necessary
4 through 6	80
7 through 9	70
10 through 20	50
21 through 30	45
31 through 40	40
41 and above	35

Where single conductors or multiconductor cables are stacked or bundled longer than 24 inches (610 mm) without maintaining spacing and are not installed in raceways, the allowable ampacity of each conductor shall be reduced as shown in the above table.

Exception No. 1: Where conductors of different systems, as provided in Section 300-3, are installed in a common raceway or cable, the derating factors shown above shall apply to the number of power and lighting (Articles 210, 215, 220, and 230) conductors only.

Exception No. 2: For conductors installed in cable trays, the provisions of Section 318-11 shall apply.

Exception No. 3: Derating factors shall not apply to conductors in nipples having a length not exceeding 24 inches (610 mm).

Exception No. 4: Derating factors shall not apply to underground conductors entering or leaving an outdoor trench if those conductors have physical protection in the form of rigid metal conduit, intermediate metal conduit, or rigid nonmetallic conduit having a length not exceeding 10 feet (3.05 m) above grade and the number of conductors does not exceed four.

Exception No. 5: For other loading conditions, adjustment factors and ampacities shall be permitted to be calculated under Section 310-15(b).

(FPN): See Appendix B, Table B-310-11 for adjustment factors for more than three current-carrying conductors in a raceway or cable with load diversity.

(b) More than One Conduit, Tube, or Raceway. Spacing between conduits, tubing, or raceways shall be maintained.

9. Overcurrent Protection. Where the standard ratings and settings of overcurrent devices do not correspond with the ratings and settings allowed for conductors, the next higher standard rating and setting shall be permitted.

Exception: As limited in Section 240-3.

FIGURE 16-24

Conduit or Tubing Trade Size (inches)		½	¾	1	1¼	1½	2	2½	3	3½	4	5	6
Type Letters	Conductor Size AWG/kcmil												
RHW,	14	3	6	10	18	25	41	58	90	121	155		
	12	3	5	9	15	21	35	50	77	103	132		
	10	2	4	7	13	18	29	41	64	86	110		
	8	1	2	4	7	9	16	22	35	47	60	94	137
RHH	6	1	1	2	5	6	11	15	24	32	41	64	93
	4	1	1	1	3	5	8	12	18	24	31	50	72
(with	3	1	1	1	3	4	7	10	16	22	28	44	63
outer	2		1	1	3	4	6	9	14	19	24	38	56
covering)	1		1	1	1	3	5	7	11	14	18	29	42
	1/0		1	1	1	2	4	6	9	12	16	25	37
	2/0			1	1	1	3	5	8	11	14	22	32
	3/0			1	1	1	3	4	7	9	12	19	28
	4/0			1	1	1	2	4	6	8	10	16	24
	250				1	1	1	3	5	6	8	13	19
	300				1	1	1	3	4	5	7	11	17
	350				1	1	1	2	4	5	6	10	15
	400				1	1	1	1	3	4	6	9	14
	500				1	1	1	1	3	4	5	8	11
	600					1	1	1	2	3	4	6	9
	700					1	1	1	1	3	3	6	8
	750						1	1	1	3	3	5	8

Note 1. This table is for concentric stranded conductors only. For cables with compact conductors, the dimensions in Table 5A shall be used.
Note 2. Conduit fill for conductors with a -2 suffix is the same as for those types without the suffix.

FIGURE 16-25

Multiplying Factors for Converting DC Resistance to 60-Hertz AC Resistance

	Multiplying Factor			
Size	For Nonmetallic-Sheathed Cables in Air or Nonmetallic Conduit		For Metallic-Sheathed Cables or all Cables in Metallic Raceways	
	Copper	Aluminum	Copper	Aluminum
Up to 3 AWG	1.	1.	1.	1.
2	1.	1.	1.01	1.00
1	1.	1.	1.01	1.00
0	1.001	1.000	1.02	1.00
00	1.001	1.001	1.03	1.00
000	1.002	1.001	1.04	1.01
0000	1.004	1.002	1.05	1.01
250 MCM	1.005	1.002	1.06	1.02
300 MCM	1.006	1.003	1.07	1.02
350 MCM	1.009	1.004	1.08	1.03
400 MCM	1.011	1.005	1.10	1.04
500 MCM	1.018	1.007	1.13	1.06
600 MCM	1.025	1.010	1.16	1.08
700 MCM	1.034	1.013	1.19	1.11
750 MCM	1.039	1.015	1.21	1.12
800 MCM	1.044	1.017	1.22	1.14
1000 MCM	1.067	1.026	1.30	1.19
1250 MCM	1.102	1.040	1.41	1.27
1500 MCM	1.142	1.058	1.53	1.36
1750 MCM	1.185	1.079	1.67	1.46
2000 MCM	1.233	1.100	1.82	1.56

Reprinted with permission from NFPA, 70-1993, the *National Electrical Code®,* Copyright © 1992, National Fire Protection Association, Quincy, MA 02269. This reprinted material is not the official position of the National Fire protection Association, which is represented only by the standard in its entirety.

FIGURE 16-26   Multiplying factors

QUESTIONS

16-1. What type of information will the designer need to accumulate before actually beginning to design the electrical system?

16-2. What is an *individual branch circuit,* and when is it used?

16-3. How many circuits are required to accommodate kitchen appliances?

16-4. What is meant by the *demand load* in reference to the electrical load of the building?

16-5. When selecting the panel size, what considerations for the future should be taken into account?

16-6. How is the general lighting load for a building determined?

16-7. How is the minimum service entrance determined?

16-8. In selecting the service entrance size, what should be considered?

Design Exercises

16-9. Design the electrical service entrance system for the residence in Appendix D.

> Water heater 3,800 watts
> Clothes dryer 4,400 watts
> Dishwasher 1,000 watts
> Range 11,700 watts
> Air conditioner 9,000 watts

16-10 Design the electrical service entrance system for one of the apartments in Appendix A. Calculate the total load and service for the apartment building.

> sq. ft., 38′ × 28′ (sq. m., 11.6 m × 8.55 m)
> Air conditioner, 2,000 watts
> Electric range, 10,500 watts
> Water heater, 3,500 watts
> Dishwasher, 1,000 watts
> No clothes dryer

CHAPTER 17

Lighting

17-1 GENERAL

The basic reasons for providing light in a space are to make the objects in the space visible and to allow the conduct of activities which must take place in the space. Part of the light required may be provided by windows and skylights, which allow sunlight to come into the space. Many times, this sunlight is sufficient to provide all the light required on sunny days. However, as the sun moves, it shines first on the east portion of the building, and it moves toward the west as the day progresses. This means that windows facing the east are subjected to the strong rays of the sun in the morning, but must count on reflected light later in the day. Windows facing north get no direct sun, while windows facing south get a considerable amount. And on cloudy days, no direct sunlight is available. In effect, while natural light from windows and skylights can be counted on to provide some general light, it does not provide a controlled light source by which to perform activities.

Therefore, in most projects, at least some electric lighting must be introduced to provide a constant, controlled amount of light so that the space can be used whenever necessary, whether sunny or cloudy, day or night. This chapter discusses the terminology and procedures used to arrive at a lighting design that will provide the required amount of light.

Since lighting should be an integral part of the building, the lighting designer must work closely with the architect to achieve a lighting solution which blends with and becomes a part of the architectural design.

17-2 TERMINOLOGY

Light: That radiant energy which has a wavelength that the human eye can see. Light from the sun may be *sunlight* or it may be *skylight.* Light which is produced electrically should be referred to as *electric light* and not artificial light (as it is commonly called). This is because, even though it is there because of our technology, it is not "artificial."

Luminous Flux: The term used to refer to that portion of the radiant energy which the human eye can see (it produces the sensation of light in the human eye).

Lumen (lm): A quantity of light used to express the total output of a light source. It is the unit of measure for luminous flux and for light. One lumen is the amount of light which falls on 1 sq ft of surface area when all points on the surface are 1 ft from a standard candle (Fig. 17-1).

Footlambert (fl): The unit of measure for brightness for any perfectly diffused (spread-out) surface which emits or reflects light. Footlamberts are expressed in lumens per square foot; 1 footlambert equals 1 lumen per sq ft. When lumens are measured from a light source (such as a light fixture or a reflecting wall or ceiling), they are referred to as footlamberts. (See "Footcandle" below.)

Reflectance: The ability of a surface to reflect light. It is expressed as the ratio of the light reflected by a surface to the amount of light which strikes the surface.

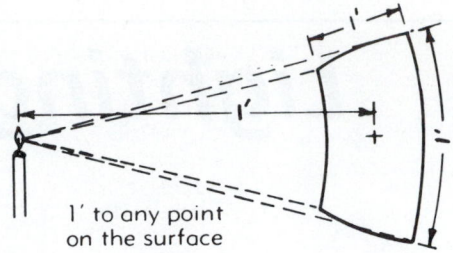

One lumen is the amount of light which falls on 1 sq. ft. of surface area which is located 1′ from a standard candle.

FIGURE 17-1   Lumen

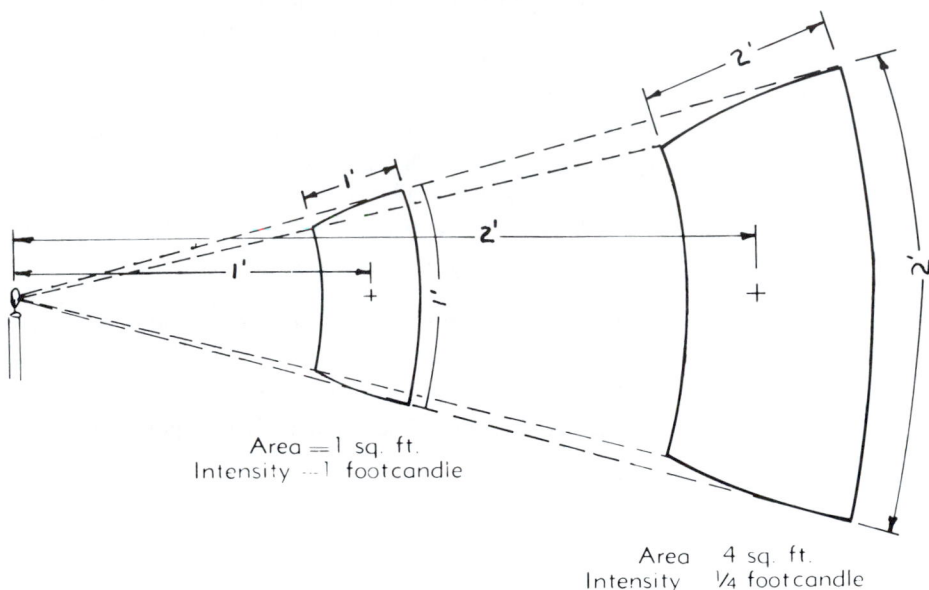

Area = 1 sq. ft.
Intensity = 1 footcandle

Area   4 sq. ft.
Intensity   ¼ footcandle

FIGURE 17-2   Footcandle

Transmission Factor: The ability of a surface or object to allow the light to transmit (pass through it). It is expressed as the ratio of the light transmitted through a body to the amount of light which strikes the surface.

Illumination:. The density of light (luminous flux) on a surface. The amount of illumination is found by dividing the amount of luminous flux (in lumens) by the area over which the flux is distributed. The unit of measure for illumination is the footcandle.

Footcandle (fc): The unit of measure for illumination, expressed in lumens per square foot, indicating how much light is actually on a surface. When lumens are measured at the surface where they will be "used," such as a desk top, they are referred to as footcandles. (See "Footlambert" above.) The number of footcandles for a given surface is found by dividing the number of lumens on a surface by the area of surface in square feet. One footcandle is the equivalent of 1 lumen distributed uniformly over 1 sq ft of surface located 1 ft from a standard candle (Fig. 17-2). The number of footcandles required varies considerably with the use of the space or area being lit; some typical footcandle requirements are given in Fig. 17-3.

Brightness: The property of a surface. For example, the surface of a wall either absorbs or reflects light, exhibiting a certain amount of brightness. Technically, it is "the luminous in-

Schools		Industrial	
corridors	20	general lighting	30-50
auditorium	20-30	laundries	30-50
study halls	70-80	locker rooms	30
classrooms	70	machines, rough	40-50
chalkboard area	150	machines, fine	80-100
drafting rooms	80-100	intricate work	150-400
laboratories	80	assembly areas	50-100
art	100	Theaters	
sewing	120-150	entrance, lobby	30-40
shop	100	foyer	30-40
gymnasium	50	auditorium	10
Stores		Post office	
circulation area	30	lobby	30
marchandise	100-150	mail sorting	100
showcases	150-200	corridors, storage	30

FIGURE 17-3 Typical footcandle requirements

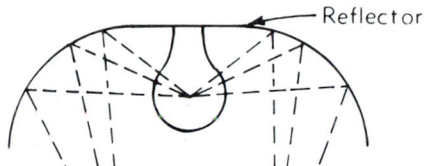

Reflector

FIGURE 17-4 Reflector

tensity of any surface in a given direction per unit of projected area of the surface viewed from that direction."

Candlepower (cp): A measurement of the luminous intensity of the source of light, stated in terms of candles. The amount of candlepower from any light source varies with the direction from the source in which the measurement is taken. This is because different types of light sources give off their light in different directions, some aiming up to the ceiling and others toward the floor. Manufacturers have available a candlepower distribution curve which shows in what direction the light will go and how much. The typical candlepower distribution of various luminaires is shown in Fig. 17-19.

Lamp: A man-made source of light, commonly referred to as a bulb (for incandescent lamps) and a tube (fluorescent lamps).

Reflector: A device which redirects the light of a lamp by reflecting it to the desired location. As the light leaves the source, it hits the reflector (Fig. 17-4) and is directed in the direction that the reflector is aimed.

Globe: An enclosing device which covers all or part of the lamp. It may protect the lamp, diffuse the light, redirect the light, or modify the color of the light.

Luminaire: A complete lighting unit consisting of a light source, globe, reflector, housing, wiring, and any supporting brackets that are a part of the housing.

17-3 ELECTRIC LAMPS

The available electric light sources are grouped into two classes:

1. Incandescent light sources (which use a tungsten filament).

2. Gaseous discharge light sources (such as fluorescent, mercury vapor, sodium, and neon).

For general lighting purposes inside residences and commercial, industrial, and institutional projects, the incandescent and fluorescent lamps are used. Mercury vapor, quartz, and sodium lamps are predominantly used outside to light large areas, such as parking lots, and to serve as flood lights.

Incandescent lamps have three main parts: the filament, the base, and the hollow-glass bulb which encloses the filament. The filament is made of tungsten and is connected by lead-in wires to the base. When the base is screwed into a socket and the power turned on, the circuit is complete. As the tungsten filament heats, the lamp gives off light.

The bulb bases most commonly used are the *medium* (usually up to 200 W) and the *mogul* (over 200 W). But other bases are also used, depending on the type and the size of the lamp.

The glass bulb is available with clear or frosted glass, coated inside or outside with a diffusing or reflecting material, or etched on the inside.

The typical luminous characteristics of an incandescent lamp with a frosted bulb are listed on Fig. 17-5. Among the advantages of incandescent lamps are:

1. They operate directly from the standard electrical distribution circuits.
2. A wide range of lighting distribution characteristics is available.
3. The standard sockets make them easily interchangeable.
4. Surrounding air temperature does not affect their performance.
5. Their radiation characteristics are good within the luminous range.
6. They can be easily equipped with a dimmer that can control the amount of light the lamp gives off, from none to full outlet.

The incandescent lamp is used extensively in residences in ceiling, wall, floor, and table luminaires. In commercial applications, it is primarily for specialty usage, such as spotlighting.

Fluorescent lamps are devices which convert invisible ultraviolet radiation into visible radiation. This conversion is made by the chemical phosphorus which is distributed over the inside of the tube.

The fluorescent lamp is used extensively in commercial, industrial, and institutional projects. The typical fluorescent lamp has a cylindrical, sealed, glass tube which contains a mixture of low-pressure mercury vapor and an inert gas—usually argon (and phosphorus on the inside of the tube). Each end of the fluorescent tube has a cathode built in to supply the electrons which start and maintain the mercury arc. This mercury arc is absorbed by the phosphorus which re-radiates it as visible radiation.

The typical fluorescent lamp is available with straight tubes, for most commercial uses, and with circular, square, or U-shaped tubes for specialty uses. Lamp lengths are given in inches. A range of lighting "colors" is available from the lamps; they may be referred to as "day light" and "white," and some manufacturers use terms such as "cool white," "soft white," and "warm white." The "deluxe" lamp produces a warmer, more lifelike light for flesh tones by adding more red to the phosphor. This deluxe type of lamp is used extensively in retail stores.

The cathodes commonly used in fluorescent lamps are:

1. Instant start (Fig. 17-20).
2. Rapid start (Fig. 17-6).

The instant-start fluorescent lamp, of which the slimline lamp is the best known, makes use of a high-voltage transformer which strikes an arc without any of the cathode preheating that was necessary in early lamp designs. This type of lamp has only one pin at each end of the tube. While more expensive and slightly less efficient than the rapid-start lamp, it is manufactured in some lengths and currents which are not available with the rapid-start. Since it will operate at lower temperatures (below 50°F) than the rapid-start lamp, it is

Incandescent General-Service Lamps for 120-, 125-, and 130-Volt Circuits (Will Operate in any Position but Lumen Maintenance is Best for 40 to 1500 Watts when Burned Base Up)

Watts	Bulb and other description	Base	Filament	Rated average life (h)	Maximum Overall length (mm)	(in.)	Average light center length (mm)	(in.)	Approximate color temperature (K)	Maximum bare bulb temperature (°C)*	Base temperature (°C)†	Approximate initial lumens	Initial lumens per watt‡	Lamp lumen depreciation (%)§
10	S-14 inside frosted	Medium	C-9	1500	89	3 1/2	64	2 1/2	2420	46.75	41	80	8.0	89
15	A-15 inside frosted	Medium	C-9	2500	89	3 1/2	60	2 3/8	—	—	—	126	8.4	83
25	A-19 inside frosted	Medium	C-9	2500	98	4 1/8	64	2 1/2	2550	43	42	230	9.2	79
25[d]	T-19 white ‖	Medium	CC-6	2500	113	4 7/16	79	3 1/8	—	—	—	235	9.2	80
34[a,b]	A-19 inside frosted	Medium	CC-6 or CC-8	1500	113	4 7/16	79	3 1/8	2550	—	—	410	12.1	—
40	A-19 inside frosted or white ‖	Medium	C-9	1500	113	4 7/16	79	3 1/8	2650	127	105	474	11.9	88
40[d]	T-19 white ‖	Medium	CC-6	1000	113	4 7/16	79	3 1/8	—	—	—	430	12.3	88
50	A-19 inside frosted	Medium	CC-6	1000	113	4 7/16	79	3 1/8	—	—	—	680	13.6	—
52[a,c]	A-19 inside frosted or clear	Medium	CC-8	1000	113	4 7/16	79	3 1/8	—	—	—	800	15.4	—
55[a]	A-19 clear or white ‖	Medium	CC-8	750	113	4 7/16	79	3 1/8	—	—	—	638	11.6	—
60	A-19 inside frosted or white ‖	Medium	CC-6	1000	113	4 7/16	79	3 1/8	2790	124	93	1060	14.4	93
60[d]	T-19 white ‖	Medium	CC-8	1000	113	4 7/16	79	3 1/8	—	—	—	860	14.3	92
67[a,d]	A-19 inside frosted or clear	Medium	CC-8	750	113	4 7/16	79	3 1/8	—	—	—	1130	16.9	—
70[a]	A-19 clear or white ‖	Medium	CC-8	750	113	4 7/16	79	3 1/8	—	—	—	1173	16.8	—
75	A-19 inside frosted or white ‖	Medium	CC-6	750	113	4 7/16	79	3 1/8	2840	135	96	1190	15.8	92
90[a]	A-19 inside frosted or clear	Medium	CC-8	750	113	4 7/16	79	3 1/8	—	—	—	1620	18.0	—
100	A-19 inside frosted or white ‖	Medium	CC-8	750	113	4 7/16	79	3 1/8	2905	149	98	1740	17.4	91
95[d]		Medium	CC-8		113	4 7/16	79	3 1/8	—	—	—	1710	18.0	90
100[d]	A-19 inside frosted or white ‖	Medium	CC-8	750	113	4 7/16	79	3 1/8	—	—	—	1683	16.8	—
100	T-19 white ‖	Medium	CC-8	750	113	4 7/16	79	3 1/8	—	—	—	1710	17.1	91
100	A-21 inside frosted	Medium	CC-6	750	133	5 1/4	98	3 7/8	2880	127	90	1688	16.9	90
135[a]	A-21 inside frosted or clear	Medium	CC-87	750	139	5 1/2	103	4 1/16	—	—	—	2580	19.1	—
150[d]	T-19 white ‖	Medium	CC-8	1000	135	5 5/16	98	3 7/8	—	—	—	2650	17.7	87
150	A-21 inside frosted	Medium	CC-8	750	139	5 1/2	103	4 1/16	2960	—	—	2873	19.2	89

FIGURE 17-5 Incandescent lamp characteristics

Nominal length (millimeters) (inches) Bulb	1050 42 T-6	1600 64 T-6	1800 72 T-8	2400 96 T-8	1200 48 T-12	1500 60 T-17
Base	Single Pin	Single Pin	Single Pin	Single Pin	Med. Bipin	Mog. Bipin
Approx. lamp current (amperes)	0.200^f	0.200^f	0.200^f	0.200^f	0.425	0.425
Approx. lamp volts	150	233	220	295	104	107
Approx. lamp watts	25.5	38.5	38	51	40.5	42
Rated life (hours)^b	7500	7500	7500	7500	7500–12000	7500–9000
Lamp lumen depreciation (LLD)^c	76	77	83	89	83	89
Initial lumens^d						
Cool White	1835	3000	3030	4265	3100	2900
Deluxe Cool White	1265	2100	2100	2910		2020
Warm White	1875	3050	3015	4215	3150	2940
Deluxe Warm White	1275	2100				1990
White	1900	2945	3050	4225	3150	2940
Daylight	1605	2600	2650	3525	2565	2410
Factors for Calculating Luminance^e						
Candelas per Square Meter	6.1	3.93	2.55	1.87	2.65	1.5

^{a} The life and light output ratings of fluorescent lamps are based on their use with ballasts that provide proper operating characteristics. Ballasts that do not provide proper electrical values may substantially reduce either lamp life or light output, or both.
^{b} Rated life under specified test conditions at three hours per start. At longer burning intervals per start, longer life can be expected.
^{c} Per cent of initial light output at 70 per cent rated life at three hours per start. Average for cool white lamps. Approximate values.
^{d} At 100 hours. Where lamp is made by more than one manufacturer, light output is the average of manufacturers. For the lumen output of other colors of fluorescent lamps, multiply the cool white lumens by the relative light output value from the table below Fig. 8–116.

Reprinted with permission from IES, *IES Lighting Handbook, 1984 Reference Handbook.*

FIGURE 17-6   Typical hot-cathode fluorescent lamps (rapid starting)

often used in outdoor situations or in any location where the temperature may be low (such as in many manufacturing facilities, outdoor loading docks, and covered parking areas).

The rapid-start fluorescent lamp is the more popular of the two since it has a higher light output per foot and is slightly less expensive than the instant-start. This type of lamp has about a 1-sec starting time and operates on a cathode current which flows continuously from a separate ballast; no starter is needed.

Typical fluorescent lamp data are given in Fig. 17-6.

The advantages of fluorescent lamps are as follows:

1. They provide more light (lumens) for less operating cost than incandescent lamps.

2. Since they provide more light (lumens) per watt, less wattage is required, resulting in fewer branch circuits.

3. Less heat is given off by fluorescent lamps.

17-4   LIGHTING SYSTEMS

The luminaire may project *direct, indirect,* or *semi-indirect* light on the surface or in the area to be lit. *Direct* lighting (Fig. 17-7) projects light directly onto the surface or object, and there is little reflection of the light rays. *Indirect* lighting projects all of its light onto a surface (usually a ceiling), and the light is reflected from that surface onto the object (Fig. 17-8). *Semi-indirect* lighting projects part of the light onto a surface to be reflected onto the object, and part of the light travels directly to the object (Fig. 17-9). Even when a direct lighting system is used, the light that falls onto an object or surface is usually a combination of light reflected from the walls, ceiling, and even floors as well as the direct light.

The use of diffusers below fluorescent lamps makes possible an even distribution of lighting without causing any glare. Thus, the tremendous selection and availability of such diffusers has made the direct lighting system the most popular in commercial, industrial, and institutional projects.

Nominal length (millimeters) (inches) Bulb	Circline				U-Shaped		Lightly Loaded Lamps						
(millimeters)	165 dia.	210 dia.	300 dia.	400 dia.	600	600	900	1200	1200	1200	900	1200	1500
(inches)	6½ dia.	8¼ dia.	12 dia.	16 dia.	24	24	36	48	48	48	36	48	60
Bulb	T-9	T-9	T-10	T-10	T-12	T-12	T-12	T-12	T-12	T-10	T-8	T-8	T-8
Base	4-Pin	4-Pin	4-Pin	4-Pin	Med. Bipin	Med. Bipin	Med. Bipin	Med. Bipin	Med. Bipin	Med. Bipin	Med. Bipin	Med. Bipin	Med. Bipin
Leg spacing					92 mm (3⅝ in)	152 mm (6 in)							
Approx. lamp current (amperes)	0.38	0.37	0.425	0.415	0.42	0.43	0.43	0.43	0.43	0.42	0.265	0.265	0.265
Approx. lamp volts	48	61	81	108	103	100	81	101	101	104	100	135	172
Approx. lamp watts[f]	20	22.5	33	41.5	41	40.5	32.4	41	41	41	25	32	40
Rated life (hours)[b]	12000	12000	12000	12000	12000	12000	18000	20000	15000	24000	20000	20000	20000
Lamp lumen depreciation (LLD)[c]		72	82	77	84	84	81	84					
Initial lumens[d]													
Cool White	800	1065	1870	2580	2900	2935	2210	3150	3250	3200	*	*	*
Deluxe Cool White		875	1425	2000	2020	2065	1555	2200		2270			
Warm White	825	1065	1835	2550	2850	2965	2235	3175	3250	3250			
Deluxe Warm White	630	800	1375	1950	1980	2040	1505	2165					
White		1100	1870	2650	2850	2965	2255	3185	3250	3250			
Daylight		906	1550	2165			1900	2615	2650	2700			
Factors for Calculating Luminance[e]													
Candelas per Square Meter	11.5	8.4	4.6	3.3	2.65	2.65	3.58	2.65	2.65	3.18	5.42	4.03	3.18
Candelas per Square Foot	1.07	0.78	0.43	0.31	0.25	0.25	0.33	0.25	0.25	0.30	0.50	0.37	0.30

Medium Loaded Lamps

Nominal length (millimeters) (inches) Bulb										
(millimeters)	600	750	900	1050	1200	1500	1600	1800	2100	2400
(inches)	24	30	36	42	48	60	64	72	84	96
Bulb	T-12	T-12	T-12	T-12	T-12	T-12	T-12	T-12	T-12	T-12
Base	Recess D.C.	Recess D.C.	Recess D.C.	Recess D.C.	Recess D.C.	Recess D.C.	Recess D.C.	Recess D.C.	Recess D.C.	Recess D.C.
Approx. lamp current (amperes)	0.8	0.8	0.8	0.8	0.8	0.8	0.8	0.78	0.8	0.79
Approx. lamp volts	41		59		78	98		117	135	153
Approx. lamp watts[f]	37		50		63	75.5		87	100	113
Rated life (hours)[b]	9000	9000	9000	9000	12000	12000	12000	12000	12000	12000
Lamp lumen depreciation (LLD)[c]	77		77	77	82	82	82	82	82	82
Initial lumens[d]										
Cool White	1700	2290	2885	3516	4300	5400	5800	6650	7800	9150
Deluxe Cool White					3050			4550		6533
Warm White	1700				4300	5500		6500		9200
Deluxe Warm White										6475
White								6475		9200
Daylight	1400		2476	3100	4300	4650	4900	5600	6867	7800
Factors for Calculating Luminance[e]										
Candelas per Square Meter	5.54	4.35	3.58	3.04	2.65	2.1	1.95	1.7	1.44	1.24
Candelas per Square Foot	0.51	0.40	0.33	0.28	0.25	0.20	0.18	0.16	0.13	0.12

Highly Loaded Lamps

Nominal length (millimeters) (inches) Bulb										
(millimeters)	1200	1800	2400	1200	1500	1800	2400	1200	1800	2400
(inches)	48	72	96	48	60	72	96	48	72	96
Bulb	T-10[g]	T-10[g]	T-10[g]	T-12[i]	T-12	T-12[i]	T-12[i]	PG-17	PG-17	PG-17
Base	Recess D.C.	Recess D.C.	Recess D.C.	Recess D.C.	Recess D.C.	Recess D.C.	Recess D.C.	Recess D.C.	Recess D.C.	Recess D.C.
Approx. lamp current (amperes)	1.5	1.5	1.5	1.5	1.5	1.52	1.5	1.5	1.5	1.5
Approx. lamp volts	80	120	160	84		125	163	84	125	163
Approx. lamp watts[f]	105[h]	150[h]	205[h]	116		168	215	116	168	215
Rated life (hours)[b]	9000	9000	9000	9000	9000	9000	9000	9000	9000	9000
Lamp lumen depreciation (LLD)[c]	66	66	66	69		72	72	69	69	69
Initial lumens[d]										
Cool White	6700[h]	10000[h]	14000[h]	6900	8950	10640	15250	7450	11500	16000
Deluxe Cool White	4690[h]	7000[h]	9800[h]	4900		7400	10750	5200		11200
Warm White				6700		10500	14650	7000		15000
White						10500	15000			
Daylight				5700		9300	12650	6000	9300	13300
Factors for Calculating Luminance[e]										
Candelas per Square Meter	3.18	2.04	1.49	2.65	2.1	1.7	1.24	1.91[j]	1.2[j]	0.89[j]
Candelas per Square Foot	0.30	0.19	0.14	0.25	0.20	0.16	0.12	0.18[j]	0.11[j]	0.08[j]

[a] The life and light output ratings of fluorescent lamps are based on their use with ballasts that provide proper operating characteristics. Ballasts that do not provide proper electrical values may substantially reduce either lamp life or light output, or both.

[b] Rated life under specified test conditions at three hours per start. At longer burning intervals per start, longer life can be expected.

[c] Per cent of initial light output at 70 per cent rated life at three hours per start. Average for cool white lamps. Approximate values.

[d] At 100 hours. Where lamp is made by more than one manufacturer, light output is the average of manufacturers. For the lumen output of other colors of fluorescent lamps, multiply the cool white lumens by the relative light output value from the table below.

[e] To calculate approximate lamp luminance, multiply the lamp lumens of the lamp color desired by the appropriate factor. Factors derived using method by E. A. Linsday, "Brightness of Cylindrical Fluorescent Sources", Illuminating Engineering, Vol. XXXIX, January, 1944, p. 23.

[f] Includes watts for cathode heat.

[g] A jacketed T-10 design is also available for use in applications where lamps are directly exposed to cold temperatures.

[h] Peak value. At 25 °C (77 °F) lumen and wattage values are lower.

[i] These lamps available in several variations (outdoor, low temperature, jacketed) with the same, or slightly different, ratings.

[j] Average luminance for center section of lamp. Parts of surface of lamp will have higher luminance.

Note: All electrical and lumen values apply only under standard photometric conditions.

[k] Initial Lumens for 3100 K and 4100 K color lamps (GTE Sylvania) are: 25-W, 2150; 32-W, 2900; and 40-W, 3650.

Reprinted with permission from IES, *IES Lighting Handbook, 1993 Lighting Handbook*.

FIGURE 17-6 (*continued*)

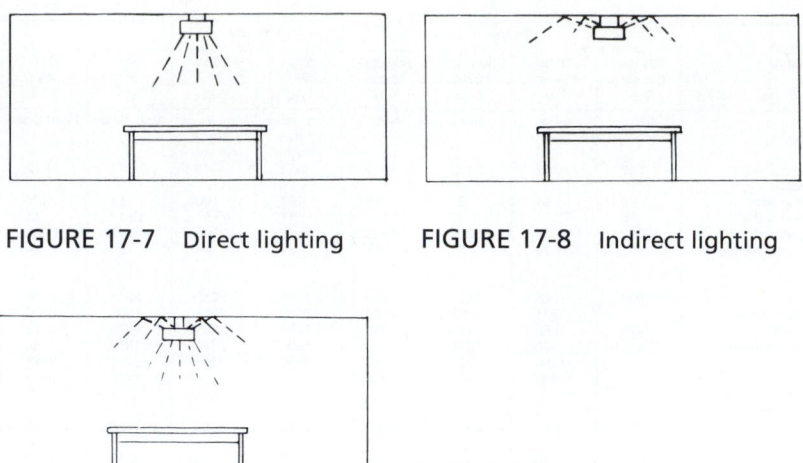

FIGURE 17-7   Direct lighting   FIGURE 17-8   Indirect lighting

FIGURE 17-9   Semi-indirect lighting

17-5   METHODS OF LIGHTING

The three methods of providing light to any space are localized, general, and a combination of localized and general.

Localized lighting (Fig. 17-10) provides light only at the point where it is needed. This is accomplished primarily by providing individual luminaires in specific locations. For example, to provide the localized lighting necessary for reading a paper in a chair, a floor, ceiling, or table luminaire would probably be used. Localized lighting generally produces pools or areas of light among larger areas of shadow. It is used primarily in residences and apartments and, to some extent, in industrial plants.

General lighting (Fig. 17-11) provides a uniform level of lighting over an entire area. In this type of design, the luminaires are evenly distributed within the space. The luminaires may use various types of diffusing covers to spread the light. General lighting is often used in stores, classrooms, markets, factories, and other such general-use spaces.

Combination lighting (Fig. 17-12) provides enough general illumination for the entire space, with additional localized lighting in areas where such things as desks, showcases, drafting tables, or machines are located. With this type of lighting, the general lighting level would be lower than when only general lighting is used. Combination lighting is used in almost all types of buildings.

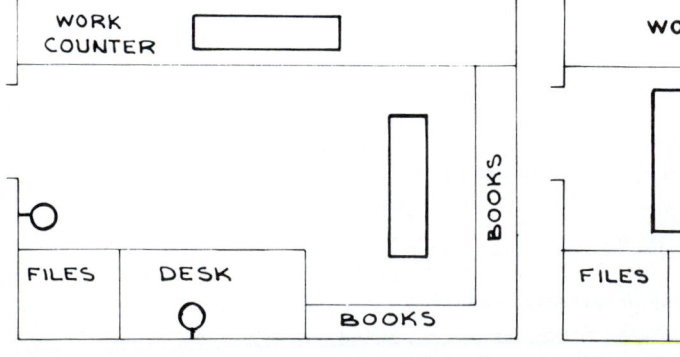

FIGURE 17-10   Lighting combinations

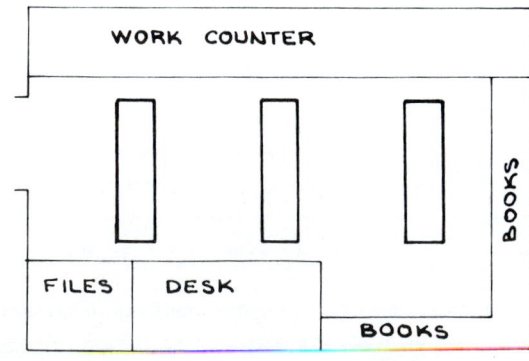

FIGURE 17-11   General and local lighting

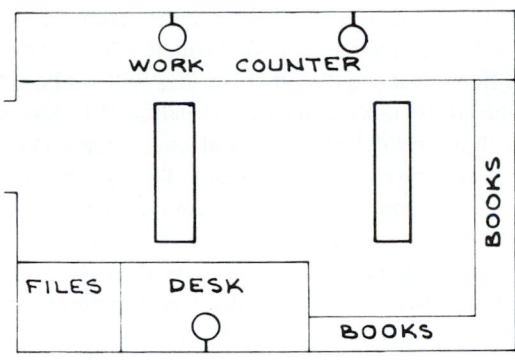

FIGURE 17-12 Combined lighting

17-6 PRINCIPLES OF DESIGN

There are several basic principles of lighting design which should be discussed before the actual design is begun.

General lighting *requirements* must be determined for the surface; that is, the number of footcandles (lumens per square foot) must be determined for the use of the surface. Next, the designer will have to take into account the room area, the light loss due to room proportions, the colors of the walls, the coefficient of utilization, and the maintenance factor.

The number of *footcandles* required for a variety of common uses is shown in Fig. 17-3. Selecting the required number of footcandles is among the first steps in the lighting design.

Next, the *room area* in square feet must be determined, taking the room dimensions from the drawings.

The *coefficient of utilization* is a measure of the amount of the total lumens put out by a luminaire that will actually reach the working plane, or the surface being lit. The higher the percentage of lumens that reaches the surface, the higher the coefficient. The coefficient is a function of the room dimensions, the reflection factors of walls and ceilings, and the characteristics of any particular luminaire. For any particular luminaire (such as those shown in Fig. 17-19), the two variables become the room ratio and the reflection factors of walls and ceilings.

Reflection factors of the floor, ceiling, and walls must also be considered. The reflection factors of the surface finish materials, such as paints, ceiling tile, carpet, and wood, are easily found in the manufacturers' data sheets.

Ceiling reflection factors are usually taken as the reflection factor of the ceiling finish. When the ceiling has more than one finish, each finish must be considered in relation to the area it covers. Most ceiling factors are listed in manufacturers' brochures as 50%, 70%, and 80%.

Wall reflection factors must take into consideration not only the finish on the wall but also the other vertical reflection surfaces, including doors, curtains (or other window coverings), windows (when they are not covered), cabinets, and any other items which may be on the walls, such as a chalkboard or bulletin board. The wall factors listed in the manufacturers' material are 10%, 30%, and 50%. Average rooms have wall reflection factors of 30%, and it would be unlikely that any room would have a factor exceeding 50%.

Floor reflection factors also must be based on the reflection factor not only of the floor but also of all horizontal surfaces, including furniture (desks, chairs, couches, files, and so on). A conservative factor of 10% is often used, but the increased use of lighter floor colors has caused many designers to use a floor value of 30%. Many manufacturers list both 10% and 30% floor reflection factors. As a compromise value, an effective floor cavity reflectance (p_{FC}) of 20% is used in the CU tables in Fig. 17-19. When the actual value of the floor cavity reflectance is more or less than the 20%, the CU values in Fig. 17-19 are adjusted by the appropriate factor from Fig. 17-18.

Since the lighting design is usually done before all the room finishes are selected, the designer must have a "feeling" for what will be done in the space. Many times, paint colors, ceilings and floor tiles, and the other finishes will not be actually selected (as to color and style) until the project is well under construction. The designer will have to check carefully as to what each space will be used for and what colors and finishes the architect, owner, and decorator may use. From this information, the lighting designer must make an "educated guess," and when in doubt, the designer should use lower values to be on the safe side.

The *maintenance factor* takes into consideration all of the reasons that would cause a luminaire (with its lamp) to operate at less than 100% of its design and test capabilities after it is installed. Reasons for such decreased performance include:

1. The output of the luminaire decreases with age.
2. The voltage of the luminaire is generally less than the voltage at which it is rated.
3. Longtime chemical changes and discolorations affect the reflecting surface and diffusing grilles.
4. Dust and dirt affect the lamp, its reflecting surface and its diffusing grilles.

These loss of performance factors are grouped into:

LLD = Lamp Lumen Depreciation Factor
LDD = Luminaire Dirt Depreciation Factor

These two factors are combined (multiplied by each other) to obtain the maintenance factor.

This maintenance factor may be kept to a minimum by regularly cleaning the entire luminaire and periodically replacing the lamp, but the factor cannot be eliminated.

Values of lamp lumen depreciation (LLD) for various lamps are given in Figs. 17-5 and 17-17. The values are listed as the percentage of initial lumens produced at 70% of life for the various types of lamps in common use.

The luminaire dirt depreciation (LDD) categories range from I through VI (Fig. 17-16) and are listed in Fig. 17-17. Each category applies to certain types of luminaire and lamp situations. The appropriate LDD category classification to be used for each type is listed in the typical luminaire listings in Fig. 17-19.

The selection for cleanliness (very clean through very dirty) is a judgment factor based on the designer's anticipation of the quality of maintenance and housekeeping to which the luminaire will be subjected.

LDD values are listed as the percentage of initial lumens produced by the specific luminaire type.

> *Caution:* Care must be used in selecting both the surface reflection factors and the maintenance factor. Many designers (and architects/engineers) have a tendency to use values which are too high so they will have to use fewer luminaires, saving money on the number of luminaires bought, the wiring, the circuits, and so on. This becomes false economy since the lighting system will not provide the light required in the space.

The breakdown of a space into the room cavities used to determine the coefficient of utilization is shown in Fig. 17-13.

The cavity ratios are used to determine how the proportions of a room will affect the characteristics of light flux. For example, a room that is narrow in relation to its height will allow less flux to reach the working plane (surface) than one that is wide.

The ceiling and floor cavity ratios (CCR and FCR, respectively) are used in conjunction with the wall, ceiling, and floor reflectances. The ratios are used to combine separate wall and ceiling reflectances into a single effective ceiling cavity reflectance, p_{CC}, or to combine separate wall and floor reflectances into a single effective floor cavity reflectance, p_{FC}.

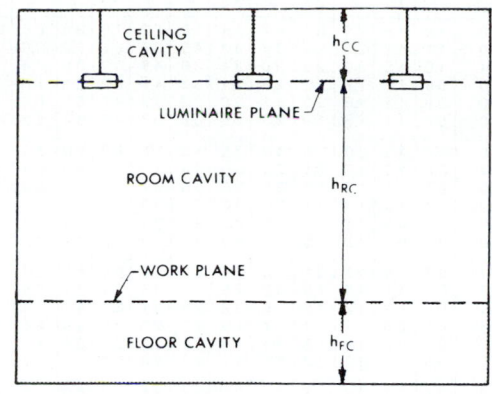

Reprinted with permission from IES, *IES Lighting Handbook, 1993 Lighting Handbook.*

FIGURE 17-13 Cavities

The RCR, CCR, and FCR may be calculated by use of the equation:

$$\text{Cavity ratio} = \frac{5h \,(\text{Room length} + \text{Room width})}{\text{Room length} \times \text{Room width}}$$

Where:

$$h = h_{RC} \text{ for the room cavity ratio (RCR)}$$
$$= h_{CC} \text{ for the ceiling cavity ratio (CCR)}$$
$$= h_{FC} \text{ for the floor cavity ratio (FCR)}$$

When the luminaire is surface-mounted or recessed, the CCR = 0 and the ceiling reflectance is used as the p_{CC}.

Many cavity ratios are given in Fig. 17-14.

To determine the coefficient of utilization (CU) for a particular room size and luminaire, it is necessary to:

1. Determine the RCR, CCR, and FCR, either by calculating the ratio or by using Fig. 17-14.

2. Determine the effective ceiling cavity reflectance p_{CC} and the effective floor cavity reflectance p_{FC} (Fig. 17-15).

3. Determine the coefficient of utilization (CU) based on the RCR, p_{CC}, p_{FC}, and wall reflectance (referred to as p_{WC}). The CUs for many typical luminaires are shown in Fig. 17-19. When the exact ratios and reflectances are not given in the tables, they may be interpolated (averaged from the values given).

4. When the effective floor cavity reflectance varies greatly from the 20% *value* used in Fig. 17-19, a multiplying factor is found in Fig. 17-18 by which the CU is adjusted.

Example *Given:*

Room: 20 ft × 30 ft (6.1m × 9.15m)

Fixture mounting height; 6 ft (1.83m) above working plane

Working plane: 2 ft 6 in. above finished floor (0.76m)

Footcandles desired: 100 fc (store, merchandise)

Ceiling reflectance: 80%

Room Dimensions		Cavity Depth																			
Width	Length	1	1.5	2	2.5	3	3.5	4	5	6	7	8	9	10	11	12	14	16	20	25	30
8	8	1.2	1.9	2.5	3.1	3.7	4.4	5.0	6.2	7.5	8.8	10.0	11.2	12.5	—	—	—	—	—	—	—
	10	1.1	1.7	2.2	2.8	3.4	3.9	4.5	5.6	6.7	7.9	9.0	10.1	11.3	12.4	—	—	—	—	—	—
	14	1.0	1.5	2.0	2.5	3.0	3.4	3.9	4.9	5.9	6.9	7.8	8.8	9.7	10.7	11.7	—	—	—	—	—
	20	0.9	1.3	1.7	2.2	2.6	3.1	3.5	4.4	5.2	6.1	7.0	7.9	8.8	9.6	10.5	12.2	—	—	—	—
	30	0.8	1.2	1.6	2.0	2.4	2.8	3.2	4.0	4.7	5.5	6.3	7.1	7.9	8.7	9.5	11.0	—	—	—	—
	40	0.7	1.1	1.5	1.9	2.3	2.6	3.0	3.7	4.5	5.3	5.9	6.5	7.4	8.1	8.8	10.3	11.8	—	—	—
10	10	1.0	1.5	2.0	2.5	3.0	3.5	4.0	5.0	6.0	7.0	8.0	9.0	10.0	11.0	12.0	—	—	—	—	—
	14	0.9	1.3	1.7	2.1	2.6	3.0	3.4	4.3	5.1	6.0	6.9	7.8	8.6	9.5	10.4	12.0	—	—	—	—
	20	0.7	1.1	1.5	1.9	2.3	2.6	3.0	3.7	4.5	5.3	6.0	6.8	7.5	8.3	9.0	10.5	12.0	—	—	—
	30	0.7	1.0	1.3	1.7	2.0	2.3	2.7	3.3	4.0	4.7	5.3	6.0	6.6	7.3	8.0	9.4	10.6	—	—	—
	40	0.6	0.9	1.2	1.6	1.9	2.2	2.5	3.1	3.7	4.4	5.0	5.6	6.2	6.9	7.5	8.7	10.0	12.5	—	—
	60	0.6	0.9	1.2	1.5	1.7	2.0	2.3	2.9	3.5	4.1	4.7	5.3	5.9	6.5	7.1	8.2	9.4	11.7	—	—
12	12	0.8	1.2	1.7	2.1	2.5	2.9	3.3	4.2	5.0	5.8	6.7	7.5	8.4	9.2	10.0	11.7	—	—	—	—
	16	0.7	1.1	1.5	1.8	2.2	2.5	2.9	3.6	4.4	5.1	5.8	6.6	7.2	8.0	8.7	10.2	11.6	—	—	—
	24	0.6	0.9	1.2	1.6	1.9	2.2	2.5	3.1	3.7	4.4	5.0	5.6	6.2	6.9	7.5	8.7	10.0	12.5	—	—
	36	0.6	0.8	1.1	1.4	1.7	1.9	2.2	2.8	3.3	3.9	4.4	5.0	5.5	6.0	6.6	7.8	8.8	11.0	—	—
	50	0.5	0.8	1.0	1.3	1.5	1.8	2.1	2.6	3.1	3.6	4.1	4.6	5.1	5.6	6.2	7.2	8.2	10.2	—	—
	70	0.5	0.7	1.0	1.2	1.5	1.7	2.0	2.4	2.9	3.4	3.9	4.4	4.9	5.4	5.8	6.8	7.8	9.7	12.2	—
14	14	0.7	1.1	1.4	1.8	2.1	2.5	2.9	3.6	4.3	5.0	5.7	6.4	7.1	7.8	8.5	10.0	11.4	—	—	—
	20	0.6	0.9	1.2	1.5	1.8	2.1	2.4	3.0	3.6	4.2	4.9	5.5	6.1	6.7	7.3	8.6	9.8	12.3	—	—
	30	0.5	0.8	1.0	1.3	1.6	1.8	2.1	2.6	3.1	3.7	4.2	4.7	5.2	5.8	6.3	7.3	8.4	10.5	—	—
	42	0.5	0.7	1.0	1.2	1.4	1.7	1.9	2.4	2.9	3.3	3.8	4.3	4.7	5.2	5.7	6.7	7.6	9.5	11.9	—
	60	0.4	0.7	0.9	1.1	1.3	1.5	1.8	2.2	2.6	3.1	3.5	3.9	4.4	4.8	5.2	6.1	7.0	8.8	10.9	—
	90	0.4	0.6	0.8	1.0	1.2	1.4	1.6	2.0	2.5	2.9	3.3	3.7	4.1	4.5	5.0	5.8	6.6	8.3	10.3	12.4
17	17	0.6	0.9	1.2	1.5	1.8	2.1	2.3	2.9	3.5	4.1	4.7	5.3	5.9	6.5	7.0	8.2	9.4	11.7	—	—
	25	0.5	0.7	1.0	1.2	1.5	1.7	2.0	2.5	3.0	3.5	4.0	4.5	5.0	5.5	6.0	7.0	8.0	10.0	12.5	—
	35	0.4	0.7	0.9	1.1	1.3	1.5	1.7	2.2	2.6	3.1	3.5	3.9	4.4	4.8	5.2	6.1	7.0	8.7	10.9	—
	50	0.4	0.6	0.8	1.0	1.2	1.4	1.6	2.0	2.4	2.8	3.1	3.5	3.9	4.3	4.5	5.4	6.2	7.7	9.7	11.6
	80	0.4	0.5	0.7	0.9	1.1	1.2	1.4	1.8	2.1	2.5	2.9	3.3	3.6	4.0	4.3	5.1	5.8	7.2	9.0	10.9
	120	0.3	0.5	0.7	0.8	1.0	1.2	1.3	1.7	2.0	2.3	2.7	3.0	3.4	3.7	4.0	4.7	5.4	6.7	8.4	10.1
20	20	0.5	0.7	1.0	1.2	1.5	1.7	2.0	2.5	3.0	3.5	4.0	4.5	5.0	5.5	6.0	7.0	8.0	10.0	12.5	—
	30	0.4	0.6	0.8	1.0	1.2	1.5	1.7	2.1	2.5	2.9	3.3	3.7	4.1	4.5	4.9	5.8	6.6	8.2	10.3	12.4
	45	0.4	0.5	0.7	0.9	1.1	1.3	1.4	1.8	2.2	2.5	2.9	3.3	3.6	4.0	4.3	5.1	5.8	7.2	9.1	10.9
	60	0.3	0.5	0.7	0.8	1.0	1.2	1.3	1.7	2.0	2.3	2.7	3.0	3.4	3.7	4.0	4.7	5.4	6.7	8.4	10.1
	90	0.3	0.5	0.6	0.8	0.9	1.1	1.2	1.5	1.8	2.1	2.4	2.7	3.0	3.3	3.6	4.2	4.8	6.0	7.5	9.0
	150	0.3	0.4	0.6	0.7	0.8	1.0	1.1	1.4	1.7	2.0	2.3	2.6	2.9	3.2	3.4	4.0	4.6	5.7	7.2	8.6
24	24	0.4	0.6	0.8	1.0	1.2	1.5	1.7	2.1	2.5	2.9	3.3	3.7	4.1	4.5	5.0	5.8	6.7	8.2	10.3	12.4
	32	0.4	0.5	0.7	0.9	1.1	1.3	1.5	1.8	2.2	2.6	2.9	3.3	3.6	4.0	4.3	5.1	5.8	7.2	9.0	11.0
	50	0.3	0.5	0.6	0.8	0.9	1.1	1.2	1.5	1.8	2.2	2.5	2.8	3.1	3.4	3.7	4.4	5.0	6.2	7.8	9.4
	70	0.3	0.4	0.6	0.7	0.8	1.0	1.1	1.4	1.7	2.0	2.2	2.5	2.8	3.0	3.3	3.8	4.4	5.5	6.8	8.2
	100	0.3	0.4	0.5	0.6	0.8	0.9	1.0	1.3	1.6	1.8	2.1	2.4	2.6	2.9	3.1	3.7	4.2	5.2	6.5	7.9
	160	0.2	0.4	0.5	0.6	0.7	0.8	1.0	1.2	1.4	1.7	1.9	2.1	2.4	2.6	2.8	3.3	3.8	4.7	5.9	7.1
30	30	0.3	0.5	0.7	0.8	1.0	1.2	1.3	1.7	2.0	2.3	2.7	3.0	3.3	3.7	4.0	4.7	5.4	6.7	8.4	10.0
	45	0.3	0.4	0.6	0.7	0.8	1.0	1.1	1.4	1.7	1.9	2.2	2.5	2.7	3.0	3.3	3.8	4.4	5.5	6.9	8.2
	60	0.3	0.4	0.5	0.6	0.7	0.9	1.0	1.2	1.5	1.7	2.0	2.2	2.5	2.7	3.0	3.5	4.0	5.0	6.2	7.4
	90	0.2	0.3	0.4	0.6	0.7	0.8	0.9	1.1	1.3	1.6	1.8	2.0	2.2	2.5	2.7	3.1	3.6	4.5	5.6	6.7
	150	0.2	0.3	0.4	0.5	0.6	0.7	0.8	1.0	1.2	1.4	1.6	1.8	2.0	2.2	2.4	2.8	3.2	4.0	5.0	5.9
	200	0.2	0.3	0.4	0.5	0.6	0.7	0.8	1.0	1.1	1.3	1.5	1.7	1.9	2.0	2.2	2.6	3.0	3.7	4.7	5.6
36	36	0.3	0.4	0.6	0.7	0.8	1.0	1.1	1.4	1.7	1.9	2.2	2.5	2.8	3.0	3.3	3.9	4.4	5.5	6.9	8.3
	50	0.2	0.4	0.5	0.6	0.7	0.8	1.0	1.2	1.4	1.7	1.9	2.1	2.5	2.6	2.9	3.3	3.8	4.8	5.9	7.2
	75	0.2	0.3	0.4	0.5	0.6	0.7	0.8	1.0	1.2	1.4	1.6	1.8	2.0	2.3	2.5	2.9	3.3	4.1	5.1	6.1
	100	0.2	0.3	0.4	0.5	0.6	0.7	0.8	0.9	1.1	1.3	1.5	1.7	1.9	2.1	2.3	2.6	3.0	3.8	4.7	5.7
	150	0.2	0.3	0.3	0.4	0.5	0.6	0.7	0.9	1.0	1.2	1.4	1.6	1.7	1.9	2.1	2.4	2.8	3.5	4.3	5.2
	200	0.2	0.2	0.3	0.4	0.5	0.6	0.7	0.8	1.0	1.1	1.3	1.5	1.6	1.8	2.0	2.3	2.6	3.3	4.1	4.9
42	42	0.2	0.4	0.5	0.6	0.7	0.8	1.0	1.2	1.4	1.6	1.9	2.1	2.4	2.6	2.8	3.3	3.8	4.7	5.9	7.1
	60	0.2	0.3	0.4	0.5	0.6	0.7	0.8	1.0	1.2	1.4	1.6	1.8	2.0	2.2	2.4	2.8	3.2	4.0	5.0	6.0
	90	0.2	0.3	0.3	0.4	0.5	0.6	0.7	0.9	1.0	1.2	1.4	1.6	1.7	1.9	2.1	2.4	2.8	3.5	4.4	5.2
	140	0.2	0.2	0.3	0.4	0.5	0.5	0.6	0.8	0.9	1.1	1.2	1.4	1.5	1.7	1.9	2.2	2.5	3.1	3.9	4.6
	200	0.1	0.2	0.3	0.4	0.4	0.5	0.6	0.7	0.9	1.0	1.1	1.3	1.4	1.6	1.7	2.0	2.3	2.9	3.6	4.3
	300	0.1	0.2	0.3	0.3	0.4	0.5	0.5	0.7	0.8	0.9	1.1	1.3	1.4	1.5	1.7	1.9	2.2	2.8	3.5	4.2
50	50	0.2	0.3	0.4	0.5	0.6	0.7	0.8	1.0	1.2	1.4	1.6	1.8	2.0	2.2	2.4	2.8	3.2	4.0	5.0	6.0
	70	0.2	0.3	0.3	0.4	0.5	0.6	0.7	0.9	1.0	1.2	1.4	1.5	1.7	1.9	2.0	2.4	2.7	3.4	4.3	5.1
	100	0.1	0.2	0.3	0.4	0.4	0.5	0.6	0.7	0.9	1.0	1.2	1.3	1.5	1.6	1.8	2.1	2.4	3.0	3.7	4.5
	150	0.1	0.2	0.3	0.3	0.4	0.5	0.5	0.7	0.8	0.9	1.1	1.2	1.3	1.5	1.6	1.9	2.1	2.7	3.3	4.0
	300	0.1	0.2	0.2	0.3	0.3	0.4	0.5	0.6	0.7	0.8	0.9	1.0	1.1	1.3	1.4	1.6	1.9	2.3	2.9	3.5
60	60	0.2	0.2	0.3	0.4	0.5	0.6	0.7	0.8	1.0	1.2	1.3	1.5	1.7	1.8	2.0	2.3	2.7	3.3	4.2	5.0
	100	0.1	0.2	0.3	0.3	0.4	0.5	0.5	0.7	0.8	0.9	1.1	1.2	1.3	1.5	1.6	1.9	2.1	2.7	3.3	4.0
	150	0.1	0.2	0.2	0.3	0.3	0.4	0.5	0.6	0.7	0.8	0.9	1.0	1.2	1.3	1.4	1.6	1.9	2.3	2.9	3.5
	300	0.1	0.1	0.2	0.2	0.3	0.3	0.4	0.5	0.6	0.7	0.8	0.9	1.0	1.1	1.2	1.4	1.6	2.0	2.5	3.0
75	75	0.1	0.2	0.3	0.3	0.4	0.5	0.5	0.7	0.8	0.9	1.1	1.2	1.3	1.5	1.6	1.9	2.1	2.7	3.3	4.0
	120	0.1	0.2	0.2	0.3	0.3	0.4	0.5	0.6	0.7	0.8	0.9	1.0	1.1	1.2	1.3	1.5	1.7	2.2	2.7	3.3
	200	0.1	0.1	0.2	0.2	0.3	0.3	0.4	0.5	0.5	0.6	0.7	0.8	0.9	1.0	1.1	1.3	1.5	1.8	2.3	2.7
	300	0.1	0.1	0.2	0.2	0.2	0.3	0.3	0.4	0.5	0.6	0.7	0.7	0.8	0.9	1.0	1.2	1.3	1.7	2.1	2.5
100	100	0.1	0.1	0.2	0.2	0.3	0.3	0.4	0.5	0.6	0.7	0.8	0.9	1.0	1.1	1.2	1.4	1.6	2.0	2.5	3.0
	200	0.1	0.1	0.1	0.2	0.2	0.3	0.3	0.4	0.4	0.5	0.6	0.7	0.7	0.8	0.9	1.0	1.2	1.5	1.9	2.2
	300	0.1	0.1	0.1	0.2	0.2	0.2	0.3	0.3	0.4	0.5	0.5	0.6	0.7	0.7	0.8	0.9	1.1	1.3	1.7	2.0
150	150	0.1	0.1	0.1	0.2	0.2	0.2	0.3	0.3	0.4	0.5	0.5	0.6	0.7	0.7	0.8	0.9	1.1	1.3	1.7	2.0
	300	—	0.1	0.1	0.1	0.1	0.2	0.2	0.2	0.3	0.3	0.4	0.5	0.5	0.6	0.6	0.7	0.8	1.0	1.2	1.5
200	200	—	0.1	0.1	0.1	0.1	0.2	0.2	0.2	0.3	0.3	0.4	0.5	0.5	0.6	0.6	0.7	0.8	1.0	1.2	1.5
	300	—	0.1	0.1	0.1	0.1	0.1	0.2	0.2	0.2	0.3	0.3	0.4	0.4	0.5	0.5	0.6	0.7	0.8	1.0	1.2
300	300	—	—	0.1	0.1	0.1	0.1	0.1	0.2	0.2	0.2	0.3	0.3	0.3	0.4	0.4	0.5	0.5	0.6	0.7	0.8
500	500	—	—	—	—	0.1	0.1	0.1	0.1	0.1	0.1	0.2	0.2	0.2	0.2	0.2	0.3	0.3	0.4	0.5	0.6

FIGURE 17-14   Cavity ratios

Per Cent Effective Ceiling or Floor Cavity Reflectances for Various Reflectance Combinations

Per Cent Base† Reflectance	90										80										70										60										50									
Per Cent Wall Reflectance / Cavity Ratio	90	80	70	60	50	40	30	20	10	0	90	80	70	60	50	40	30	20	10	0	90	80	70	60	50	40	30	20	10	0	90	80	70	60	50	40	30	20	10	0	90	80	70	60	50	40	30	20	10	0
0.2	89	88	88	87	87	86	85	84	84	82	79	78	78	77	77	76	76	75	74	72	70	69	68	68	67	67	66	66	65	64	60	59	59	59	58	58	57	56	55	53	50	50	49	49	48	48	47	46	46	44
0.4	88	87	86	85	84	83	81	80	79	76	79	77	76	75	74	73	72	71	70	68	69	68	67	66	65	64	63	62	61	58	60	59	59	58	57	55	54	53	52	50	50	49	48	48	47	46	45	45	44	42
0.6	87	86	84	82	80	79	77	76	74	73	78	76	75	73	71	70	68	66	65	63	69	67	65	64	63	61	59	58	57	54	60	58	57	56	55	53	51	51	50	46	50	48	47	46	45	44	43	42	42	38
0.8	87	85	82	80	77	75	73	71	69	67	78	75	73	71	69	67	65	63	61	57	69	67	64	62	60	58	56	55	53	50	59	58	57	55	54	52	49	48	47	43	50	48	47	45	44	42	40	38	38	36
1.0	86	83	80	77	75	72	69	66	64	62	77	74	72	69	67	65	62	60	57	55	68	65	62	60	58	55	52	50	50	47	59	57	56	55	51	48	46	44	43	41	50	48	46	44	43	41	38	36	36	34
1.2	85	82	78	75	72	69	66	63	60	57	76	73	70	67	64	61	58	55	53	51	67	64	61	59	57	54	50	48	46	44	59	56	54	51	49	46	44	42	40	38	50	47	45	43	41	39	36	35	34	29
1.4	85	80	77	73	69	65	62	59	57	52	76	72	68	65	62	59	55	53	50	48	67	63	60	58	55	51	47	45	44	41	59	56	53	49	47	44	41	39	38	36	50	47	45	42	40	38	35	34	32	27
1.6	84	79	75	71	67	63	59	56	53	50	75	71	67	63	60	57	53	50	47	44	67	62	59	56	53	47	45	43	41	38	58	55	52	48	45	42	39	37	35	33	50	47	44	41	39	36	33	32	30	26
1.8	83	78	73	69	64	60	56	53	50	48	75	70	66	62	58	54	50	47	44	41	66	61	58	54	51	46	42	40	38	35	58	55	51	47	44	40	37	35	33	31	50	46	43	40	38	35	31	30	28	25
2.0	83	77	72	67	62	58	53	50	47	43	74	69	64	60	56	52	48	45	41	38	66	60	56	52	49	45	42	40	38	33	57	54	50	46	43	39	35	33	31	29	50	46	43	40	37	34	30	28	26	24
2.2	82	76	70	65	59	54	50	47	44	40	74	68	63	58	54	49	45	42	38	35	66	60	55	51	48	43	38	36	34	32	58	53	49	45	42	37	34	31	29	28	50	46	42	38	36	33	29	27	24	22
2.4	82	75	69	64	58	53	48	45	41	37	73	67	61	56	52	47	43	39	36	33	65	60	54	50	46	41	37	35	32	30	58	53	48	44	41	36	32	30	27	26	50	46	42	38	35	31	27	25	23	21
2.6	81	74	67	62	56	51	46	42	38	35	73	66	60	55	50	45	41	37	34	31	65	59	54	49	45	40	35	33	30	28	58	53	47	43	39	35	31	28	25	24	50	46	41	37	34	30	26	23	21	20
2.8	81	73	66	60	54	49	44	40	36	34	73	65	59	53	48	43	39	35	32	29	65	59	53	48	43	38	33	30	28	26	58	53	46	42	38	34	30	27	24	22	50	46	41	36	33	29	25	22	20	19
3.0	80	72	64	58	52	47	42	38	34	30	72	65	58	52	47	42	37	34	30	27	64	58	52	47	42	37	32	29	27	24	57	52	46	42	38	32	28	25	23	20	50	45	40	36	32	28	24	21	19	17
3.2	79	71	63	56	50	45	40	36	32	28	72	65	57	51	45	40	35	33	28	25	64	58	51	46	40	36	31	28	25	23	57	51	45	40	36	31	27	23	22	18	50	44	39	35	31	27	23	20	18	16
3.4	79	70	62	54	48	43	38	34	30	27	71	64	56	49	44	39	34	32	27	24	64	57	50	45	39	35	29	27	24	22	57	51	45	39	35	30	26	23	20	17	50	44	39	35	30	26	22	19	17	15
3.6	78	69	61	53	47	42	36	32	28	25	71	63	54	48	43	38	32	30	25	23	63	56	49	44	38	33	28	25	22	20	57	50	44	39	34	29	25	22	19	16	50	44	39	34	29	25	21	18	16	14
3.8	78	69	60	51	45	40	35	31	27	23	70	62	53	47	41	36	31	28	24	22	63	56	49	43	37	32	27	24	21	19	57	50	43	38	33	29	24	21	19	15	50	44	38	34	29	25	21	17	15	13
4.0	77	69	58	51	44	39	33	29	25	22	70	61	53	46	40	35	30	26	22	20	63	55	48	42	36	31	26	23	20	17	57	50	42	37	32	28	23	20	18	14	50	44	38	33	28	24	20	17	15	12
4.2	77	62	57	50	43	37	32	28	24	21	69	60	52	45	39	34	29	25	21	18	62	55	47	41	35	30	25	22	19	16	56	49	42	36	32	27	22	19	17	14	50	43	37	32	28	24	20	17	14	12
4.4	76	61	56	49	42	36	31	27	23	20	69	60	51	44	38	33	28	24	20	17	62	54	46	40	34	29	24	21	18	15	56	49	41	36	31	27	22	19	16	13	50	43	37	32	27	23	19	16	13	11
4.6	76	60	55	47	40	35	30	26	21	19	69	59	50	43	37	32	27	23	19	15	62	53	45	39	33	28	24	21	17	14	56	48	41	35	30	26	21	18	16	13	50	43	36	31	26	22	18	15	13	10
4.8	75	59	54	46	39	34	28	24	20	17	68	58	49	42	36	31	26	21	18	14	62	53	45	38	32	27	23	20	16	13	56	48	40	34	29	25	21	17	15	12	50	43	36	31	26	21	17	15	12	09
5.0	75	59	53	45	38	33	28	24	20	16	68	58	48	41	35	30	25	21	18	14	61	52	44	38	31	26	22	19	16	12	55	48	40	34	28	24	20	17	14	11	47	43	35	30	25	21	17	14	12	09
6.0	73	61	49	41	34	29	24	20	16	11	66	55	44	38	31	27	22	19	16	10	60	51	41	35	28	24	19	16	13	09	55	45	37	31	25	21	17	14	11	07	50	42	34	29	23	19	15	13	10	06
7.0	70	58	45	38	30	27	21	18	14	08	64	53	41	35	28	24	19	16	13	07	58	48	38	32	26	22	17	14	14	06	54	43	35	28	24	20	15	12	09	05	49	41	32	27	21	18	14	11	08	05
8.0	68	55	42	35	27	23	18	15	12	06	62	50	38	32	25	21	17	14	11	05	57	46	35	29	23	19	15	13	11	05	53	42	33	28	22	18	14	11	08	04	49	40	30	25	19	16	13	10	07	03
9.0	66	52	38	31	25	21	16	14	11	05	61	49	36	30	23	19	15	12	10	04	56	45	33	27	21	18	14	12	10	04	52	40	31	26	20	16	12	10	07	03	48	39	29	24	18	15	11	09	07	03
10.0	65	51	36	29	22	19	15	11	09	04	59	46	33	27	21	18	14	11	08	03	55	43	31	25	19	16	12	10	08	03	51	39	29	24	18	15	11	09	07	02	47	37	27	22	17	14	10	08	06	02

* Values in this table are based on a length to width ratio of 1.6.
† Ceiling, floor or floor of cavity.

FIGURE 17-15 Cavity reflectances

Per Cent Base† Reflectance table

Per Cent Base† Reflectance	40										30										20										10										0									
Per Cent Wall Reflectance	90	80	70	60	50	40	30	20	10	0	90	80	70	60	50	40	30	20	10	0	90	80	70	60	50	40	30	20	10	0	90	80	70	60	50	40	30	20	10	0	90	80	70	60	50	40	30	20	10	0
Cavity Ratio																																																		
0.2	40	40	39	39	39	38	38	37	36	36	31	31	30	30	29	29	28	28	27	27	21	20	20	20	20	20	19	19	19	17	11	11	11	10	10	10	10	09	09	09	02	02	01	01	01	01	01	00	00	0
0.4	41	40	39	39	38	37	36	35	34	34	31	31	30	30	29	28	28	27	26	25	22	21	20	20	20	19	19	18	18	16	12	12	11	11	11	10	10	09	09	08	04	03	03	02	02	02	01	01	00	0
0.6	41	40	39	38	37	36	34	33	32	31	32	31	30	29	28	27	26	25	24	23	23	21	21	20	19	19	18	17	16	15	13	13	12	11	11	10	10	09	08	08	05	05	04	03	03	02	02	01	01	0
0.8	41	40	38	37	36	35	33	32	31	29	32	31	30	29	28	26	25	24	23	21	24	22	21	20	19	18	17	16	16	14	15	13	13	12	11	11	10	09	08	08	07	06	05	04	03	03	02	02	01	0
1.0	42	40	38	37	35	33	31	29	29	27	33	32	30	29	27	26	24	23	22	20	25	23	22	21	19	18	17	16	15	13	16	14	13	12	12	11	10	09	08	07	08	07	06	05	04	03	03	02	01	0
1.2	42	40	38	36	34	32	30	29	27	25	33	32	30	28	27	25	23	22	21	19	25	23	22	20	19	17	16	15	14	12	17	15	14	13	12	11	10	09	07	06	10	08	07	06	05	04	03	02	01	0
1.4	42	39	37	35	33	31	29	27	25	23	34	32	30	28	26	24	22	21	19	18	26	24	22	20	18	17	16	15	13	12	18	16	14	13	12	11	10	09	07	06	11	09	08	06	05	04	03	02	01	0
1.6	42	39	37	35	32	30	27	25	23	22	34	33	29	27	25	23	22	20	18	17	26	24	22	20	18	16	15	14	13	11	19	17	15	14	12	11	10	08	07	06	12	10	09	07	06	05	04	03	02	0
1.8	42	39	36	34	31	28	26	24	22	21	35	33	29	27	25	23	21	19	17	16	27	25	23	20	17	16	15	14	12	10	19	17	15	14	13	11	10	08	06	05	13	11	09	08	07	05	04	03	02	0
2.0	42	39	36	34	31	28	25	23	21	19	35	33	29	26	24	22	20	18	16	14	28	25	23	21	18	16	15	13	11	09	20	18	16	14	13	11	09	08	06	05	14	12	10	09	07	06	05	03	02	0
2.2	42	39	36	33	30	27	24	22	19	18	36	32	29	26	24	22	19	17	15	13	28	25	23	20	18	16	14	12	10	09	21	19	16	14	13	11	09	07	06	05	15	13	11	09	07	06	04	03	01	0
2.4	43	39	35	33	29	27	24	21	18	17	36	32	29	26	24	22	19	16	14	12	29	26	23	20	18	16	14	12	10	08	22	19	17	15	13	11	09	07	06	05	16	13	11	09	08	06	05	03	01	0
2.6	43	39	35	32	29	26	23	20	17	15	36	33	29	25	23	21	18	16	14	12	29	26	23	20	18	16	14	11	09	08	23	20	17	15	13	11	09	07	06	04	17	14	12	10	08	06	05	03	02	0
2.8	43	39	35	32	28	25	22	19	16	14	37	33	29	25	23	20	17	15	13	11	30	27	23	20	17	15	13	11	09	07	23	20	18	16	14	11	09	07	05	03	17	15	13	10	08	07	05	03	02	0
3.0	43	39	35	31	27	24	21	18	16	13	37	33	29	25	22	20	17	15	12	11	30	27	23	20	17	15	13	11	09	07	24	21	18	16	13	11	09	07	05	03	18	16	13	11	09	07	05	04	02	0
3.2	43	39	35	31	27	23	20	17	15	13	37	33	29	25	22	19	16	14	12	10	31	27	23	20	17	15	12	11	09	06	25	21	18	16	13	11	09	07	05	03	19	16	14	11	09	07	05	03	02	0
3.4	43	39	34	30	26	23	20	17	14	12	37	33	29	24	22	19	16	14	11	09	31	27	23	20	17	15	12	10	08	06	26	22	18	16	13	11	09	07	05	03	20	17	14	12	09	07	06	04	02	0
3.6	44	38	34	30	26	22	19	16	14	11	38	33	28	24	21	18	15	13	10	09	32	27	23	20	17	15	12	10	08	05	26	22	19	16	13	11	09	06	04	03	20	17	15	12	10	08	06	04	02	0
3.8	44	38	33	29	25	22	18	15	13	10	38	33	28	24	21	18	15	13	10	08	32	28	23	20	17	14	12	10	07	05	27	23	19	17	14	11	08	06	04	02	21	18	15	13	10	08	06	04	02	0
4.0	44	38	33	29	25	21	18	15	12	10	38	33	28	24	20	18	14	12	09	08	33	28	23	20	17	14	12	09	07	05	27	23	20	17	14	11	08	06	04	02	22	18	16	13	11	08	06	04	02	0
4.2	44	38	33	29	24	21	17	15	12	10	38	33	28	24	20	17	14	12	09	07	33	28	23	20	17	14	11	09	07	04	28	24	20	17	14	11	09	06	04	02	22	19	16	13	10	08	06	04	02	0
4.4	44	38	33	28	24	20	17	14	11	09	39	33	28	24	20	17	14	11	09	06	34	28	24	20	17	14	11	09	07	04	28	24	20	17	14	11	09	06	04	02	23	19	16	13	11	08	06	04	02	0
4.6	44	38	32	28	23	19	16	14	11	08	39	33	28	23	20	17	13	10	08	06	34	29	24	20	16	14	11	09	07	04	29	25	20	17	14	11	08	06	04	02	23	20	17	14	11	08	06	04	02	0
4.8	44	38	32	27	22	19	16	13	10	08	39	33	28	23	19	16	13	10	08	05	35	29	24	20	16	13	11	08	06	04	30	25	21	17	14	11	08	06	04	02	24	20	17	14	11	08	06	04	02	0
5.0	45	38	31	27	22	19	15	13	10	07	39	33	28	23	19	16	13	10	08	05	35	29	24	20	16	13	10	08	06	04	30	25	21	17	14	11	08	06	04	02	25	21	17	14	11	08	06	04	02	0
6.0	44	37	30	25	20	17	13	11	08	05	39	33	27	23	18	15	11	09	06	04	36	30	24	20	16	13	10	08	05	02	31	26	21	18	14	11	08	06	03	01	27	23	18	15	12	09	06	04	02	0
7.0	44	36	29	24	19	16	12	10	07	04	40	33	26	22	17	14	10	08	05	03	36	30	23	19	15	12	09	07	04	02	32	27	21	17	13	10	08	05	03	01	28	24	19	15	12	09	06	03	01	0
8.0	44	35	28	23	18	15	11	09	06	03	40	33	26	21	16	13	09	07	04	02	37	30	23	19	15	12	08	06	03	01	33	27	21	17	13	10	07	05	03	01	30	25	20	15	13	09	06	03	01	0
9.0	44	34	26	21	16	13	10	08	05	02	40	32	25	20	15	12	09	07	04	02	37	29	23	19	14	11	08	06	03	01	34	28	21	17	13	10	07	05	02	01	31	25	20	15	12	09	06	04	02	0
10.0	43	34	25	20	15	12	09	07	05	02	40	32	24	19	14	11	08	06	03	01	37	29	22	18	13	10	07	05	03	01	34	28	21	17	12	10	07	05	02	01	31	25	20	15	12	09	06	04	02	0

* Values in this table are based on a length to width ratio of 1.6.
† Ceiling, floor or floor of cavity.

Reprinted with permission from IES, *IES Lighting Handbook*, *1993 Lighting Handbook*.

FIGURE 17-15 (continued)

Wall reflectance: 50%

Floor reflectance: 20%

Maintenance factor: Fair

Luminaires: Fluorescent, 40-W, instant-start, cool white (#40, Fig. 17-19) mounted 1 ft 0 in. from the ceiling

Problem: Find the total lumens and the number of luminaires required and the number of footcandles they provide.

1. Determine the heights (h) of the zonal cavities.

$$h_{CC} = 1.0$$
$$h_{RC} = 6.0$$
$$h_{FC} = 2.5$$

2. Determine the cavity ratios, using the formula or taking them directly from Fig. 17-14 if the room size is given.

$$CCR = 0.4$$
$$RCR = 2.5$$
$$FCR = 1.0$$

3. Determine the effective ceiling and floor cavity reflectances from Fig. 17-15 by use of the given reflectances and the calculated cavity ratios.

$$p_{CC} = 74\%$$
$$p_{WC} = 50\% \text{ (the given value)}$$
$$p_{FC} = 20\% \text{ (the given value)}$$

4. Determine the coefficient of utilization (CU) from the manufacturer's specifications or from Fig. 17-19. The RCR, p_{FC}, and p_{CC} must be interpolated (averaged) from the values given. From Fig. 17-19, the values are:

Pcc	80	70
RCR = 2.0	0.55	0.54
3.0	0.48	0.47

You will need to do a two-way average of RCR 2.5 and p_{CC} 74%. First RCR: RCR 2.5 is midway between 2.0 and 3.0, so the interpolation to RCR 2.5 at p_{CC} 80 and 70 is:

At 80:

2.0   0.55

3.0   0.48

difference = 0.07

RCR 2.5 = 0.55 minus (0.5) times the difference

= 0.515

At 70:

2.0   0.54

3.0   0.47

difference = 0.07

RCR 2.5 = 0.54 minus (0.5) times the difference

= 0.505

$p_{CC} = 74\%$:

Mainte-nance Cat-egory	Top Enclosure	Bottom Enclosure
I	1. None.	1. None
II	1. None 2. Transparent with 15 per cent or more uplight through apertures. 3. Translucent with 15 per cent or more uplight through apertures. 4. Opaque with 15 per cent or more uplight through apertures.	1. None 2. Louvers or baffles
III	1. Transparent with less than 15 per cent upward light through apertures. 2. Translucent with less than 15 per cent upward light through apertures. 3. Opaque with less than 15 per cent uplight through apertures.	1. None 2. Louvers or baffles
IV	1. Transparent unapertured. 2. Translucent unapertured. 3. Opaque unapertured.	1. None 2. Louvers
V	1. Transparent unapertured. 2. Translucent unapertured. 3. Opaque unapertured.	1. Transparent unapertured 2. Translucent unapertured
VI	1. None 2. Transparent unapertured. 3. Translucent unapertured. 4. Opaque unapertured.	1. Transparent unapertured 2. Translucent unapertured 3. Opaque unapertured

Reprinted with permission from IES, *1993 Lighting Handbook.*

FIGURE 17-16   Lamp maintenance category

Using the values calculated at RCR, the p_{CC} can be interpolated to determine the coefficient of utilization.

RCR 2.5

Pcc	80	70
CU	0.515	0.505

The coefficient of utilization is calculated by interpolating between the P_{CC} values.

$$CH = \text{(if subtracting from 80)}$$
$$0.515 - [(0.515 - 0.505) \times 0.6] = 0.51$$

5. Determine the lamp lumen depreciation (LLD) from the table in Fig. 17-20. Based on fluorescent lamps, 48″, 40 watt cool white:

$$LLD = 83\%$$

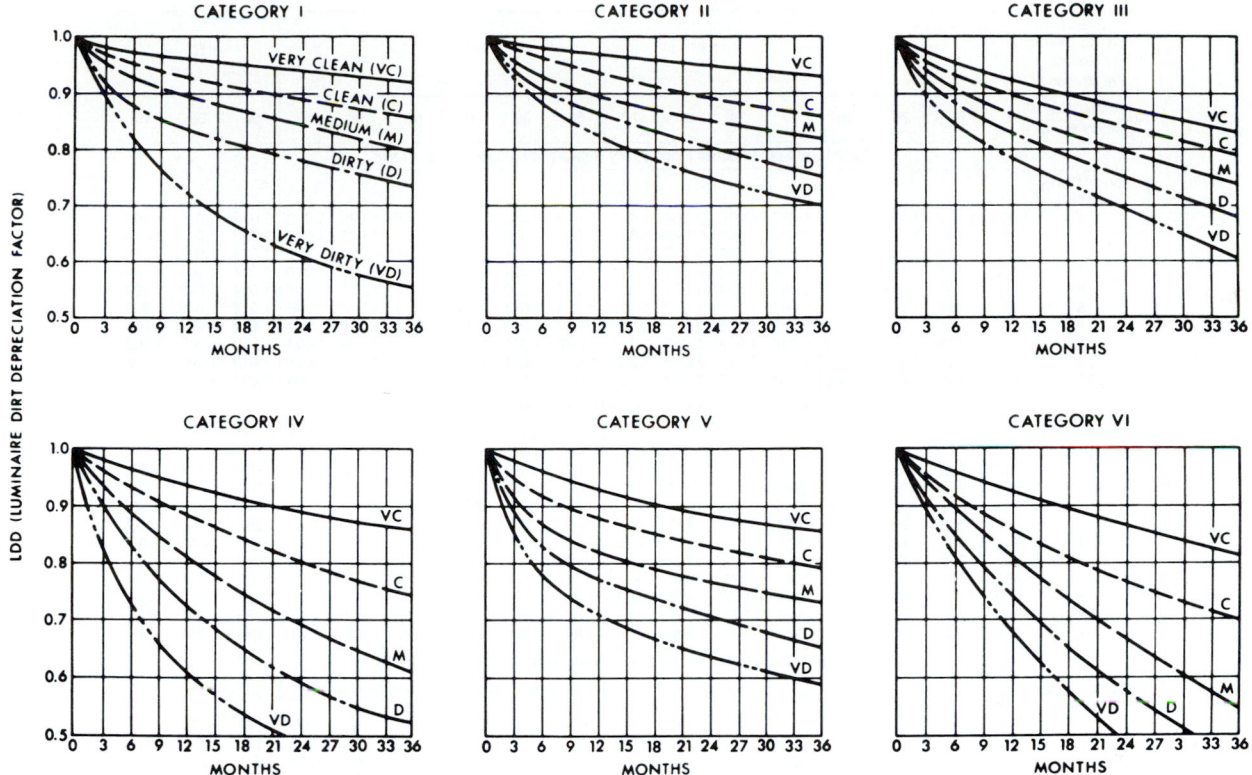

Reprinted with permission from IES, *1993 Lighting Handbook.*

FIGURE 17-17   Luminaire dirt depreciation factors

6. Determine the luminaire dirt depreciation (LDD) from Fig. 17-17. The luminaire se-
    lected (type 40, Fig. 17-19) lists an LDD maintenance category of V. Assuming a
    dirty maintenance (D), with a 24-month interval, the LDD from Fig. 17-17 is 0.7.

7. Determine the total lumens required for the space:

$$\text{Total lumens} = \frac{\text{Footcandles} \times \text{Room area}}{\text{CU} \times \text{LLD} \times \text{LDD}}$$

$$\text{Total lumens} = \frac{100\,(20 \times 30)}{0.51 \times 0.83 \times 0.70} = 202{,}491 \text{ lumens}$$

8. Determine the minimum number of lamps required by dividing the number of lumens
    per lamp into the total amount of lumens required. Since 40-W, instant start, cool
    white fluorescent lamps are used, From Fig. 17-20 the lumens per lamp are 3,100.

9. Determine the number of lamps required by dividing the total number of lumens re-
    quired by the lumens per lamp:

$$\text{Lamps} = \frac{202{,}491}{3{,}100} = 65.3 = 66 \text{ lamps}$$

10. Determine the number of luminaires required by dividing the number of lamps re-
    quired by the number of lamps in the luminaire selected.

% Effective Ceiling Cavity Reflectance, ρ_{CC}	80				70				50			30			10		
% Wall Reflectance, ρ_W	70	50	30	10	70	50	30	10	50	30	10	50	30	10	50	30	10
For 30 Per Cent Effective Floor Cavity Reflectance (20 Per Cent = 1.00)																	
Room Cavity Ratio																	
1	1.092	1.082	1.075	1.068	1.077	1.070	1.064	1.059	1.049	1.044	1.040	1.028	1.026	1.023	1.012	1.010	1.008
2	1.079	1.066	1.055	1.047	1.068	1.057	1.048	1.039	1.041	1.033	1.027	1.026	1.021	1.017	1.013	1.010	1.006
3	1.070	1.054	1.042	1.033	1.061	1.048	1.037	1.028	1.034	1.027	1.020	1.024	1.017	1.012	1.014	1.009	1.005
4	1.062	1.045	1.033	1.024	1.055	1.040	1.029	1.021	1.030	1.022	1.015	1.022	1.015	1.010	1.014	1.009	1.004
5	1.056	1.038	1.026	1.018	1.050	1.034	1.024	1.015	1.027	1.018	1.012	1.020	1.013	1.008	1.014	1.009	1.004
6	1.052	1.033	1.021	1.014	1.047	1.030	1.020	1.012	1.024	1.015	1.009	1.019	1.012	1.006	1.014	1.008	1.003
7	1.047	1.029	1.018	1.011	1.043	1.026	1.017	1.009	1.022	1.013	1.007	1.018	1.010	1.005	1.014	1.008	1.003
8	1.044	1.026	1.015	1.009	1.040	1.024	1.015	1.007	1.020	1.012	1.006	1.017	1.009	1.004	1.013	1.007	1.003
9	1.040	1.024	1.014	1.007	1.037	1.022	1.014	1.006	1.019	1.011	1.005	1.016	1.009	1.004	1.013	1.007	1.002
10	1.037	1.022	1.012	1.006	1.034	1.020	1.012	1.005	1.017	1.010	1.004	1.015	1.009	1.003	1.013	1.007	1.002
For 10 Per Cent Effective Floor Cavity Reflectance (20 Per Cent = 1.00)																	
Room Cavity Ratio																	
1	.923	.929	.935	.940	.933	.939	.943	.948	.956	.960	.963	.973	.976	.979	.989	.991	.993
2	.931	.942	.950	.958	.940	.949	.957	.963	.962	.968	.974	.976	.980	.985	.988	.991	.995
3	.939	.951	.961	.969	.945	.957	.966	.973	.967	.975	.981	.978	.983	.988	.988	.992	.996
4	.944	.958	.969	.978	.950	.963	.973	.980	.972	.980	.986	.980	.986	.991	.987	.992	.996
5	.949	.964	.976	.983	.954	.968	.978	.985	.975	.983	.989	.981	.988	.993	.987	.992	.997
6	.953	.969	.980	.986	.958	.972	.982	.989	.977	.985	.992	.982	.989	.995	.987	.993	.997
7	.957	.973	.983	.991	.961	.975	.985	.991	.979	.987	.994	.983	.990	.996	.987	.993	.998
8	.960	.976	.986	.993	.963	.977	.987	.993	.981	.988	.995	.984	.991	.997	.987	.994	.998
9	.963	.978	.987	.994	.965	.979	.989	.994	.983	.990	.996	.985	.992	.998	.988	.994	.999
10	.965	.980	.989	.995	.967	.981	.990	.995	.984	.991	.997	.986	.993	.998	.988	.994	.999
For 0 Per Cent Effective Floor Cavity Reflectance (20 Per Cent = 1.00)																	
Room Cavity Ratio																	
1	.859	.870	.879	.886	.873	.884	.893	.901	.916	.923	.929	.948	.954	.960	.979	.983	.987
2	.871	.887	.903	.919	.886	.902	.916	.928	.926	.938	.949	.954	.963	.971	.978	.983	.991
3	.882	.904	.915	.942	.898	.918	.934	.947	.936	.950	.964	.958	.969	.979	.976	.984	.993
4	.893	.919	.941	.958	.908	.930	.948	.961	.945	.961	.974	.961	.974	.984	.975	.985	.994
5	.903	.931	.953	.969	.914	.939	.958	.970	.951	.967	.980	.964	.977	.988	.975	.985	.995
6	.911	.940	.961	.976	.920	.945	.965	.977	.955	.972	.985	.966	.979	.991	.975	.986	.996
7	.917	.947	.967	.981	.924	.950	.970	.982	.959	.975	.988	.968	.981	.993	.975	.987	.997
8	.922	.953	.971	.985	.929	.955	.975	.986	.963	.978	.991	.970	.983	.995	.976	.988	.998
9	.928	.958	.975	.988	.933	.959	.980	.989	.966	.980	.993	.971	.985	.996	.976	.988	.998
10	.933	.962	.979	.991	.937	.963	.983	.992	.969	.982	.995	.973	.987	.997	.977	.989	.999

Reprinted with permission from IES, *1993 Lighting Handbook.*

FIGURE 17-18 Multiplying factors for other than 20 percent effective floor cavity reflectance

$$\text{Luminaires} = \frac{66}{2} = 33 \text{ luminaires}$$

The designer has several choices in luminaire placement. Using a 4-ft luminaire, a maximum of 7 luminaires could be placed end to end in a row. To install the luminaires required in this manner would require 4.7 rows. The luminaires could also be placed in 5 alternate rows of 6 and 5 luminaires for a total of 33.

The spacing used should be checked against spacing-mounting-height ratios in Fig. 17-19. In this design, the mounting height above the working plane is 6 ft. Using Fig. 17-19, the maximum spacing (S) should be no more than the mounting height times 1.0:

$$\frac{\text{MAX S}}{MH_{WP}} = 1.0$$

or

$$\text{Max } S = MH_{WP} \times 1.0$$

In this design, the maximum spacing should be 6.0 ft.

To calculate the number of footcandles, the original formula

$$\text{Lumens} = \frac{\text{Footcandles} \times \text{Room area}}{\text{Coefficient of utilization (cu)} \times \text{Maintenance factor (mf)}}$$

is rewritten:

$$\text{Footcandles} = \frac{\text{Lumens} \times \text{cu} \times \text{mf}}{\text{Room area}}$$

$$= \frac{3{,}100(66) \times 0.51 \times 0.83 \times 0.70}{600}$$

$$= 101 \text{ fc}$$

Since 100 fc are desired, the design is complete.

Typical Luminaire	Maint. Cat.	SC	RCR ↓	ρCC→ 80, ρW 50	30	10	70, 50	30	10	50, 50	30	10	30, 50	30	10	10, 50	30	10	0	WDRC
1 — Pendant diffusing sphere with incandescent lamp (35½% up, 45% down)	V	1.5	0	.87	.87	.87	.81	.81	.81	.70	.70	.70	.59	.59	.59	.49	.49	.49	.45	
			1	.71	.66	.62	.65	.61	.58	.55	.52	.49	.46	.44	.42	.38	.36	.34	.30	.368
			2	.60	.53	.48	.55	.50	.45	.47	.42	.38	.39	.35	.32	.31	.29	.26	.23	.279
			3	.52	.44	.38	.48	.41	.36	.40	.35	.31	.33	.29	.26	.27	.24	.21	.18	.227
			4	.45	.37	.32	.42	.35	.30	.35	.30	.25	.29	.25	.21	.23	.20	.17	.14	.192
			5	.40	.32	.27	.37	.30	.25	.31	.25	.21	.26	.21	.18	.21	.17	.14	.12	.166
			6	.35	.28	.23	.33	.26	.21	.28	.22	.18	.23	.19	.15	.19	.15	.12	.10	.146
			7	.32	.25	.19	.29	.23	.18	.25	.20	.16	.21	.16	.13	.17	.13	.11	.09	.130
			8	.29	.22	.17	.27	.20	.16	.23	.17	.14	.19	.15	.12	.15	.12	.09	.07	.117
			9	.26	.19	.15	.24	.18	.14	.21	.16	.12	.17	.13	.10	.14	.11	.08	.07	.107
			10	.24	.17	.13	.22	.16	.12	.19	.14	.11	.16	.12	.09	.13	.10	.08	.06	.098
2 — Concentric ring unit with incandescent silvered-bowl lamp (83% up, 3½% down)	II	N.A.	0	.83	.83	.83	.72	.72	.72	.50	.50	.50	.30	.30	.30	.12	.12	.12	.03	
			1	.72	.69	.66	.62	.60	.57	.43	.42	.40	.26	.25	.25	.10	.10	.10	.03	.018
			2	.63	.58	.54	.54	.50	.47	.38	.35	.33	.23	.22	.20	.09	.09	.08	.02	.015
			3	.55	.49	.45	.47	.43	.39	.33	.30	.28	.20	.19	.17	.08	.07	.07	.02	.013
			4	.48	.42	.37	.42	.37	.33	.29	.26	.23	.18	.16	.15	.07	.06	.06	.02	.012
			5	.43	.36	.32	.37	.32	.28	.26	.23	.20	.16	.14	.12	.06	.06	.05	.01	.011
			6	.38	.32	.27	.33	.28	.24	.23	.20	.17	.14	.12	.11	.06	.05	.04	.01	.010
			7	.34	.28	.23	.30	.24	.21	.21	.17	.15	.13	.11	.09	.05	.04	.04	.01	.009
			8	.31	.25	.20	.27	.21	.18	.19	.15	.13	.12	.10	.08	.05	.04	.03	.01	.008
			9	.28	.22	.18	.24	.19	.16	.17	.14	.11	.10	.09	.07	.04	.03	.03	.01	.008
			10	.25	.20	.16	.22	.17	.14	.16	.12	.10	.10	.08	.06	.04	.03	.03	.01	.007
29 — Metal or dense diffusing sides with 45°CW × 45°LW shielding (39% up, 32% down)	II	1.1	0	.75	.75	.75	.69	.69	.69	.57	.57	.57	.46	.46	.46	.37	.37	.37	.32	
			1	.66	.64	.62	.61	.59	.57	.51	.50	.48	.42	.41	.40	.33	.33	.32	.28	.094
			2	.59	.55	.52	.54	.51	.48	.46	.43	.41	.38	.36	.34	.30	.29	.28	.25	.091
			3	.52	.48	.44	.48	.44	.41	.41	.38	.35	.34	.32	.30	.27	.26	.25	.22	.085
			4	.47	.42	.38	.43	.39	.35	.37	.33	.31	.31	.28	.26	.25	.23	.22	.19	.079
			5	.42	.37	.33	.39	.34	.31	.33	.30	.27	.28	.25	.23	.23	.21	.20	.17	.073
			6	.38	.33	.29	.35	.31	.27	.30	.27	.24	.25	.23	.21	.21	.19	.18	.16	.068
			7	.35	.29	.26	.32	.28	.24	.28	.24	.21	.23	.21	.19	.19	.17	.16	.14	.063
			8	.32	.26	.23	.29	.25	.22	.25	.22	.19	.22	.19	.17	.18	.16	.15	.13	.059
			9	.29	.24	.21	.27	.23	.20	.23	.20	.17	.20	.17	.15	.17	.15	.13	.12	.056
			10	.27	.22	.19	.25	.21	.18	.22	.18	.16	.19	.16	.14	.16	.14	.12	.11	.052
34 — Prismatic bottom and sides, open top, 4-lamp suspended unit—see note 7 (33% up, 50% down)	VI	1.4/1.2	0	.91	.91	.91	.85	.85	.85	.74	.74	.74	.64	.64	.64	.54	.54	.54	.50	
			1	.80	.77	.74	.75	.72	.70	.65	.63	.61	.57	.55	.54	.49	.47	.47	.43	.179
			2	.70	.65	.61	.66	.62	.58	.58	.54	.52	.50	.48	.46	.43	.42	.40	.37	.166
			3	.62	.56	.51	.58	.53	.49	.51	.47	.44	.45	.42	.39	.39	.37	.35	.32	.153
			4	.55	.49	.44	.52	.46	.42	.46	.41	.38	.40	.37	.34	.35	.32	.30	.27	.140
			5	.50	.43	.38	.47	.41	.36	.41	.37	.33	.36	.33	.30	.32	.29	.26	.24	.129
			6	.45	.38	.33	.42	.36	.32	.37	.33	.29	.33	.29	.26	.29	.26	.23	.21	.119
			7	.40	.34	.29	.38	.32	.28	.34	.29	.26	.30	.26	.23	.26	.23	.21	.19	.111
			8	.37	.30	.26	.35	.29	.25	.31	.26	.23	.28	.24	.21	.24	.21	.19	.17	.103
			9	.34	.27	.23	.32	.26	.22	.29	.24	.21	.25	.22	.19	.22	.19	.17	.15	.096
			10	.31	.25	.21	.29	.24	.20	.26	.22	.19	.23	.20	.17	.21	.18	.15	.14	.090
35 — 2-lamp prismatic wraparound—see note 7 (11½ up, 58½ down)	V	1.5/1.2	0	.81	.81	.81	.78	.78	.78	.72	.72	.72	.66	.66	.66	.61	.61	.61	.59	
			1	.71	.68	.66	.68	.66	.63	.63	.61	.59	.58	.57	.56	.54	.53	.52	.50	.223
			2	.63	.58	.55	.60	.56	.53	.56	.53	.50	.52	.50	.47	.48	.46	.45	.43	.201
			3	.56	.50	.46	.54	.49	.45	.50	.46	.43	.47	.43	.41	.43	.41	.39	.37	.183
			4	.50	.44	.40	.48	.43	.39	.45	.40	.37	.42	.38	.35	.39	.36	.34	.32	.167
			5	.45	.39	.34	.43	.38	.34	.40	.36	.32	.38	.34	.31	.35	.32	.30	.28	.153
			6	.40	.34	.30	.39	.34	.30	.37	.32	.28	.34	.30	.27	.32	.29	.26	.25	.142
			7	.37	.31	.27	.35	.30	.26	.33	.29	.25	.31	.27	.24	.30	.26	.23	.22	.131
			8	.33	.28	.24	.32	.27	.23	.30	.26	.23	.29	.25	.22	.27	.24	.21	.20	.122
			9	.31	.25	.21	.30	.25	.21	.28	.24	.20	.26	.23	.20	.25	.22	.19	.18	.114
			10	.28	.23	.19	.27	.22	.19	.26	.21	.18	.24	.21	.18	.23	.20	.17	.16	.107
40 — Fluorescent unit dropped diffuser, 4-lamp 610 mm (2') wide—see note 7 (1% up, 60½% down)	V	1.2	0	.73	.73	.73	.71	.71	.71	.68	.68	.68	.65	.65	.65	.62	.62	.62	.60	
			1	.63	.60	.58	.62	.59	.57	.59	.57	.55	.56	.55	.53	.54	.53	.51	.50	.259
			2	.55	.51	.47	.54	.50	.46	.51	.48	.45	.49	.46	.44	.47	.45	.43	.42	.236
			3	.48	.43	.39	.47	.42	.39	.45	.41	.38	.43	.40	.37	.42	.39	.36	.35	.212
			4	.43	.37	.33	.42	.37	.33	.40	.36	.32	.39	.35	.32	.37	.34	.31	.30	.191
			5	.38	.33	.29	.37	.32	.28	.36	.31	.28	.35	.31	.28	.33	.30	.27	.26	.173
			6	.34	.29	.25	.34	.29	.25	.33	.28	.24	.31	.27	.24	.30	.27	.24	.23	.158
			7	.31	.26	.22	.31	.26	.22	.30	.25	.22	.29	.25	.21	.28	.24	.21	.20	.144
			8	.28	.23	.20	.28	.23	.20	.27	.23	.19	.26	.22	.19	.25	.22	.19	.18	.133
			9	.26	.21	.18	.26	.21	.18	.25	.21	.17	.24	.20	.17	.24	.20	.17	.16	.123
			10	.24	.19	.16	.24	.19	.16	.23	.19	.16	.22	.19	.16	.22	.18	.16	.15	.115

Note: Coefficients of Utilization for 20 Per Cent Effective Floor Cavity Reflectance (ρFC = 20).

Reprinted with permission from IES, *1993 Lighting Handbook.*

FIGURE 17-19   Coefficients of utilization

Typical Luminaire	Typical Intensity Distribution and Per Cent Lamp Lumens		ρcc →	80			70			50			30			10			0	WDRC	ρcc →	
			ρw →	50	30	10	50	30	10	50	30	10	50	30	10	50	30	10	0		ρw →	
	Maint. Cat.	SC	RCR ↓	Coefficients of Utilization for 20 Per Cent Effective Floor Cavity Reflectance (ρFC = 20)																		RCR ↓
49	I	1.4/1.2	0	1.13	1.13	1.13	1.09	1.09	1.09	1.01	1.01	1.01	.94	.94	.94	.88	.88	.88	.85		
			1	.95	.90	.86	.92	.87	.83	.85	.82	.78	.79	.76	.74	.74	.72	.69	.66	.464	1
	12; ▲		2	.82	.74	.68	.79	.72	.66	.73	.68	.63	.68	.64	.60	.63	.60	.56	.53	.394	2
			3	.71	.62	.55	.69	.61	.54	.64	.57	.52	.59	.54	.49	.55	.51	.47	.44	.342	3
			4	.62	.53	.46	.60	.52	.45	.56	.49	.43	.52	.46	.41	.49	.44	.40	.37	.300	4
	85 ▼		5	.55	.46	.39	.54	.45	.39	.50	.43	.37	.47	.40	.36	.44	.38	.34	.32	.267	5
			6	.50	.41	.34	.48	.40	.33	.45	.38	.32	.42	.36	.31	.39	.34	.30	.27	.240	6
			7	.45	.36	.30	.43	.35	.29	.41	.34	.28	.38	.32	.27	.36	.30	.26	.24	.218	7
2-lamp fluorescent strip unit with 235° reflector fluorescent lamps			8	.41	.32	.26	.40	.32	.26	.37	.30	.25	.35	.29	.24	.33	.27	.23	.21	.199	8
			9	.37	.29	.24	.36	.28	.23	.34	.27	.22	.32	.26	.22	.30	.25	.21	.19	.183	9
			10	.34	.26	.21	.33	.26	.21	.32	.25	.20	.30	.24	.20	.28	.23	.19	.17	.170	10

Typical Luminaires	ρcc →	80			70			50			30			10			0
	ρw →	50	30	10	50	30	10	50	30	10	50	30	10	50	30	10	0
	RCR ↓	Coefficients of utilization for 20 Per Cent Effective Floor Cavity Reflectance, ρFC															
50	1	.42	.40	.39	.36	.35	.33	.25	.24	.23	Coves are not recommended for lighting areas having low reflectances.						
	2	.37	.34	.32	.32	.29	.27	.22	.20	.19							
	3	.32	.29	.26	.28	.25	.23	.19	.17	.16							
	4	.29	.25	.22	.25	.22	.19	.17	.15	.13							
	5	.25	.21	.18	.22	.19	.16	.15	.13	.11							
	6	.23	.19	.16	.20	.16	.14	.14	.12	.10							
	7	.20	.17	.14	.17	.14	.12	.12	.10	.09							
	8	.18	.15	.12	.16	.13	.10	.11	.09	.08							
Single row fluorescent lamp cove without reflector, mult. by 0.93 for 2 rows and by 0.85 for 3 rows.	9	.17	.13	.10	.15	.11	.09	.10	.08	.07							
	10	.15	.12	.09	.13	.10	.08	.09	.07	.06							
51 ρcc from below ~65%	1				.60	.58	.56	.58	.56	.54							
	2				.53	.49	.45	.51	.47	.43							
	3				.47	.42	.37	.45	.41	.36							
	4				.41	.36	.32	.39	.35	.31							
	5				.37	.31	.27	.35	.30	.26							
	6				.33	.27	.23	.31	.26	.23							
Diffusing plastic or glass	7				.29	.24	.20	.28	.23	.20							
1) Ceiling efficiency ~60%; diffuser transmittance ~50%; diffuser reflectance ~40%. Cavity with minimum obstructions and painted with 80% reflectance paint—use ρc = 70.	8				.26	.21	.18	.25	.20	.17							
	9				.23	.19	.15	.23	.18	.15							
2) For lower reflectance paint or obstructions—use ρc = 50.	10				.21	.17	.13	.21	.16	.13							
52 ρcc from below ~60%	1				.71	.68	.66	.67	.66	.65	.65	.64	.62				
	2				.63	.60	.57	.61	.58	.55	.59	.56	.54				
	3				.57	.53	.49	.55	.52	.48	.54	.50	.47				
	4				.52	.47	.43	.50	.45	.42	.48	.44	.42				
	5				.46	.41	.37	.44	.40	.37	.43	.40	.36				
	6				.42	.37	.33	.41	.36	.32	.40	.35	.32				
Prismatic plastic or glass.	7				.38	.32	.29	.37	.31	.28	.36	.31	.28				
1) Ceiling efficiency ~67%; prismatic transmittance ~72%; prismatic reflectance ~18%. Cavity with minimum obstructions and painted with 80% reflectance paint—use ρc = 70.	8				.34	.28	.25	.33	.28	.25	.32	.28	.25				
	9				.30	.25	.22	.30	.25	.21	.29	.25	.21				
2) For lower reflectance paint or obstructions—use ρc = 50.	10				.27	.23	.19	.27	.22	.19	.26	.22	.19				
53 ρcc from below ~45%	1							.51	.49	.48				.47	.46	.45	
	2							.46	.44	.42				.43	.42	.40	
	3							.42	.39	.37				.39	.38	.36	
	4							.38	.35	.33				.36	.34	.32	
	5							.35	.32	.29				.33	.31	.29	
Louvered ceiling.	6							.32	.29	.26				.30	.28	.26	
1) Ceiling efficiency ~50%; 45° shielding opaque louvers of 80% reflectance. Cavity with minimum obstructions and painted with 80% reflectance paint—use ρc = 50.	7							.29	.26	.23				.28	.25	.23	
	8							.27	.23	.21				.26	.23	.21	
	9							.24	.21	.19				.24	.21	.19	
2) For other conditions refer to Fig. 6–18.	10							.22	.19	.17				.22	.19	.17	

Reprinted with permission from IES, *1993 Lighting Handbook.*

FIGURE 17-19   (*continued*)

Lamp description	Lamp watts	Lamp lumen depreciation†	Lamp life (h)‡	Nominal length (mm)	Nominal length (in.)	Base (end caps)	Nominal lumens§ 3000K RE70	3500K RE70	4100K RE70	3000K RE80	3500K RE80	4100K RE80	5000K RE80
F42T6 ‖	24	76	7500	1067	42	Single Pin							
F64T6 ‖	38	77	7500	1626	64	Single Pin							
F72T8 ‖	37	83	7500	1854	73	Single Pin							
F96T8 ‖	50	89	7500	2438	96	Single Pin							
F48T12	40	83	8250	1219	48	Med. Bipin							
F60T17	40	89	7500	1524	60	Mog. Bipin							
F24T12	20	81	7500	610	24	Single Pin							
F36T12	30	82	7500	914	36	Single Pin							
F42T12	35	80	7500	1067	42	Single Pin							
F48T12	39	82	9000	1219	48	Single Pin	3000[a]	3017		3075	3050[a]	3150[c]	
F60T12	50	78	12,000	1524	60	Single Pin							
F64T12	51	78	12,000	1626	64	Single Pin	4700	4700	4700[a]	4817	4817	4825	4800[c]
F72T12	55	89	12,000	1829	72	Single Pin	6450	6450	6450	6617	6617	6575	6500[c]
F84T12	67	91	12,000	2134	84	Single Pin							
F96T12	75	89	12,000	2438	96	Single Pin							

Lamp description	Lamp watts	Base (end caps)	Nominal lumens§ 4150K CRI 60 + Cool white	3000K CRI 50 + Warm white	3470K CRI 60 + white	6380K CRI 70 + Daylight	5000K CRI 90 + C50	4160K CRI 80 + Deluxe cool white	2990K CRI 70 + Deluxe warm white (soft white)	3200K CRI 80 + Optima 32™d	3570K CRI 80 + Natural	5500K CRI 90 + Vita-Lite™d
F42T6	24	Single Pin	1840	1880	1920	1620[b]		1300				
F64T6	38	Single Pin	2925	3000				2140				
F72T8	37	Single Pin	3025	3100		2700[b]						
F96T8	50	Single Pin	4025	4050[b]		3550[b]						
F48T12	40	Med. Pipin	3000									
F60T17	40	Mog. Bipin	2850									
F24T12	20	Single Pin	1150			940[c]						
F36T12	30	Single Pin	1940	2050[a]		1655						
F42T12	35	Single Pin	2350			1990						
F48T12	39	Single Pin	2890	2940	2890	2480		2070			1975	2120
F60T12	50	Single Pin	3580			3120						
F64T12	51	Single Pin	3830			3270		3180		3250	3030	2700
F72T12	55	Single Pin	4480	4600[b]	4600[b]	3830						3335
F84T12	67	Single Pin	5280			4480						
F96T12	75	Single Pin	6620	6270	6175	5280	4550	4470	4380	4750	4250	4475

* The life and light output ratings of fluorescent lamps are based on their use with ballasts that provide proper operating characteristics. Ballasts that do not provide proper electrical values may substantially reduce lamp life, light output or both.
† Percent of initial light output at 70% of rated life at 3 hours per start. Average for cool white lamps. Approximate values.
‡ Rated life under specified test conditions at 3 hours per start.
§ At 100 hours.
‖ These lamps can also be operated 120 and 300 mA.
When lamp is made by more than one manufacturer, light output is the average of all manufacturers submitting data.
[a] General Electric.
[b] Sylvania.
[c] Philips.
[d] Duro-Test.

Reprinted with permission from IES, *1993 Lighting Handbook.*

FIGURE 17-20   Instant starting fluorescent lamps

QUESTIONS

17-1. Why are fluorescent lamps used extensively in industrial and commercial projects?

17-2. What are the three types of light which may be projected by a luminaire onto a surface?

17-3. What is the difference between *localized* and *generalized lighting*, and where would you consider using each of them?

17-4. What is meant by the terms *room index* and *room ratio?*

17-5. How do the reflection factors of the space affect the number of luminaires required?

17-6. What must be considered when determining wall reflection factors?

17-7. What is meant by the *maintenance factor* when considering luminaires?

17-8. What does *coefficient of utilization* mean with regard to luminaires?

Design Exercises

17-9. Determine the total lumens and the number of luminaires required for the following problem:

Room: 24 ft × 50 ft

Fixture mounting height: 9 ft (above working plane), 1 ft 0 in. below ceiling

Working plane: 3 ft 0 in. above finished floor

Use: Laboratory, 120 fc

Ceiling reflectance: 70%

Wall reflectance: 50%

Floor reflectance: 20%

Maintenance factor: Clean, 12 months

Luminaires: Fluorescent, 40-W, instant-start, cool white, #29 (Fig. 17-19), 12 hours per start

17-10. Determine the total lumens and the number of luminaires required for the following problem:

Room: 17 ft × 25 ft

Fixture mounting height: 6 ft (above working plane), 6 in. below ceiling

Working plane: 2 ft 6 in. above finished floor

Footcandles desired: 75 fc, study hall

Ceiling reflectance:; 80%

Wall reflectance: 50%

Floor reflectance: 30%

Maintenance factor: Dirty, 24 months

Luminaires: Fluorescent, 40-W, instant-start, deluxe cool white, #34 (Fig. 17-19), 12 hours per start

CHAPTER 18

Security, Fire, and Smoke Systems

18-1 SECURITY

Except in a relatively few instances, building security has traditionally received little consideration during the design stage. However, the increase in crimes against businesses amounts to over $12 billion a year, and the increasing threats of violence from bombings and "shoot-outs" have resulted in significantly increased interest in security.

Building security must be approached in four basic ways: selection of materials used in the building, building design, guards, and electronic alarm systems.

Material Selection

Carefully select doors, windows, and their hardware to be certain that they will discourage intrusion. The windows should be designed so that the glass cannot be removed from the outside, and the hardware should be such that the window cannot be opened from the outside with a plastic card or wire. The doors in question should be metal, and locks should be selected which cannot be opened with a credit card. Also be certain that the door's hinge pins cannot be taken out from the outside of the building and the door removed. Maximum-security lock systems are available which have keys that cannot be duplicated on existing key-cutting machines; duplicate keys must be ordered from the manufacturer and are sent only to authorized persons. Obviously, these suggestions are not foolproof, but they do provide an economical beginning to building security.

Many building codes require that all doors which lead into stairways and exits be operable from both directions in case of fire or emergency. When such a door is not to be used except in an emergency, it may be wired to an alarm system which activates when the door is opened; large signs should be placed on such doors warning of the alarm system setup.

Buildings which are more prone to security problems because of the location or the type of occupancy should be built of materials which prevent easy entrance. For example, exterior wood frame walls are much easier to "open up" than poured concrete, concrete block, or brick walls. The roof assembly should also be carefully selected. It is easier to cut or "crowbar" an opening in a plywood roof than a concrete or gypsum roof.

Building Design

The designers of the building should pay particular attention to the locations of doors, windows, loading docks, and money-handling rooms. These areas should all be easy to view from other surrounding areas and be well lit (as should elevators and stairways). Ledges and exterior ornamentation which might allow people to climb up the side of the building should be avoided. Probably one of the best investments in terms of security is the provision of high levels of illumination throughout the project.

Guards

Guards are used to provide security in many buildings. It should be noted that seldom may a guard be compared with a policeman in terms of providing such security because the guards that are hired seldom have the training that policemen receive. Such guard jobs are generally low paying; and yet one survey has shown for over a third of those guards interviewed that the guard job was the best job available to them and that they had been unemployed before they took the guard job. Many of the guards hired range from those too inexperienced to do the job properly (for example, students who do homework on the job) to those who are not capable of handling the job (for example, some semi-retired personnel). It is interesting that those persons hired to protect the building and materials are the lowest paid personnel in many companies, perhaps making them susceptible to stealing or taking bribes for "looking the other way."

Neither guards nor electronic devices by themselves are the complete answer. The selective use of electronics, tied into central control points, reduces the need for guards. Using this approach, adequate money should be available to hire and train qualified guards.

Electronic Alarm Systems

For most projects, an electromechanical alarm system such as described here provides economical and effective security. Such systems have been used extensively in residences and in commercial, industrial, and institutional buildings. The setup of the system is quite simple, especially when it is installed in new construction. The monitoring system consists of the power source, which is connected by an electric circuit to all the doors and windows selected and to the alarm. Once the system is set, if the circuit is broken at any of the contact points (at a door or window), the alarm goes off. There are many types of installations and equipment available, depending on the equipment manufacturer and the desired security. The system may be connected to every door and window or only to a select group of doors and windows.

Control Units

The control unit used to activate and deactivate the circuit should be powered by the regular household power (120 V) available, with a stand-by battery power source to take over in case of a power failure. Many of the units have lights which indicate when the circuit is on. In addition, they may have a test alarm button so that the entire system can be checked.

To allow the occupant to leave and re-enter without sounding the alarm, an exit/entry control kit is installed. It includes an outside key-operated switch and an inside switch at the door. This type of control kit will sound the alarm if any attempt is made to disconnect the outside wallplate. Indicator lights on the wallplate turn on when the system is activated. Some exterior control kits are simply activated and de-activated by an outside key-operated switch, with none of the protective features mentioned.

When the outside control kit is used, a push-button interior alarm switch is installed. One inside alarm switch may be installed for the entire system, or several may be installed to operate individual doors and windows. The latter allows the opening of a door or window without having to de-activate the entire system.

Also available is a control kit with a time-delay exit and entry feature. This control delays the operation of the system, allowing the occupant to leave or re-enter the building without setting off the alarm. With this control kit, an outside key-operated switch is necessary.

Control kits should be mounted where they will be readily accessible to the adult occupants but not to any children or visitors—for example, in a closet, laundry area, bookcase, or cabinet.

Entry Detectors

The entry detectors usually used are either the plunger or the magnetic type.

The plunger-type detector is recessed in a hole about ¾ in. in diameter and 1¼ in. deep. It can be used on both doors and windows and is installed so that when the door or window is closed, the plunger is pushed in, and when the door or window is opened, the plunger pushes out, setting off the alarm.

The magnetic-type entry detector may also be used on both doors and windows. The two pieces of the detector are surface-mounted so that when the door or window is closed, they are in contact with each other, and when the door or window is opened, the magnets are separated, setting off the alarm.

Another entry detector often used is the floor-mat detector which will activate the alarm if stepped on. Common concealed locations for the floor-mat entry detector include under the rugs, inside all exterior doors, and just below windows.

Alarms

The alarm set off by the various systems may be a bell inside and/or outside the building, a bell inside and a horn outside, a horn inside and/or outside, or a horn with a light (beacon). In some locales, the alarm system can be connected directly to the police station. The type of alarm which will produce the best results should be used. When available, the best arrangement is to tie the alarm system into the police station. Generally, the next best arrangement is to have an inside bell or horn with an exterior horn and a light. The more noise the system makes, the more likely it is that neighbors will hear it and call police—and the more likely it is that whoever breaks in will leave without taking anything. To be effective, this type of system must depend on the cooperation of neighbors, so be certain that they understand what the alarm means and that they should call the police to report any attempted break-in.

Quite often, these alarm systems are combined with the fire/smoke alarms discussed in Sec. 18-6.

More elaborate electronic security systems utilizing ultrasonic sound, photoelectric and microwave devices, audio-activated devices, and closed-circuit television are also available. The advantages and disadvantages of these systems are compared in Fig. 18-1. Many of the systems are used in combination to achieve maximum control. Often, the integrated system is set up with a number of alarms that may warn of smoke, fire, and intruders. Obviously, the larger the building complex is, the more money that should be available for security systems.

Electronic Access

Electronic access control is being used by many large companies to control employee entrance into various areas. A commonly used access control requires insertion of a coded electronic "key" into a wall-mounted receptacle that decodes the "key" and activates the door lock only for the proper "key." This type of lock is much more secure than the standard cylinder lock used on most doors. Another control method requires the employee to punch a code number on a keyboard, and within several seconds, a picture of the person who is assigned that number is flashed on the screen for the guard to check. It operates by the same principle as the instant replay technique used on television and uses the same materials. A third access device requires a person to place a hand on a hand plate; the machine then measures such things as finger length, fingertip angle, and skin translucency and compares this information either with an identification card which is magnetically coated or with information which has been placed in the memory of the computer.

Type	Features	Application	Limitations
Ultrasonic High frequency (ultrasonic) sound, inaudible to human ears, completely saturates and is contained within the enclosed space to be guarded. Any slight movement or sound within the space will change the echo pattern and trigger an alarm.	Three-dimensional protection. Sound is inaudible. Detects all movement. Confined to enclosed space.	Banks, stores, plants, warehouses, office buildings, storage areas.	May be of limited effectiveness when placed near sound absorbent materials (e.g. carpet).
Photoelectric Invisible beam of infrared light. An alarm sounds when the beam is broken.	Up to 1,000-foot range. Interior or exterior models available.	Large areas or small, warehouse to doorway.	
Microwave Radar device which emits inaudible, measured radio signals. When person enters the field, the signal is interrupted and an alarm is set off.	Penetrates walls. Effective range to 10,000 sq ft.	Warehouses, storage areas, offices, banks, plants.	Movement outside the area being controlled may set off alarm.
Closed circuit television Cameras available with remote scanning, tilting, and zooming controls. Many areas may be monitored by one guard watching a group of receivers.	Low light level cameras available. Guards required cut in half.	Large complexes, hotels, warehouses, apartments, elevators, stairways, banks.	Cameras are costly. Due to high value of cameras, they are sometimes stolen.
Camera A surveillance camera set to take pictures automatically or on a signal.	Takes over 3,000 pictures before cartridge must be replaced. Signal may be a switch, foot plate, or any other electrically wired item.	Pilferage, vandalism, holdups, banks, loan companies, areas where money is handled.	Automatic setting could miss all the action if set to take photos too far apart. Expensive if photos taken too often. May provide photo of person but they then must be found.
Electromechanical Systems of floor mats, buttons, tapes, photocells, and screens that sound an alarm when activated.	Flexible, inexpensive system. Will deter unskilled intruders.	Schools, homes, warehouses.	Skilled intruder will easily avoid this type of security.
Audio All noise over a set level is picked up by a two-way microphone. Noises received above the level set off the alarm.	Guard may question intruder by microphone without being in physical contact with him.	Offices, warehouses, storage areas.	Loud noises from outside the area may set off alarm.

FIGURE 18-1   Security alarm system

18-2 FIRE AND SMOKE ALARM SYSTEMS

One of the greatest fears of any homeowner, businessowner, or director of an institution is fire. While the prime concern is always the loss of lives in a fire, many a business has never again risen from the smoke and flames. The threat of a fire destroying lives and property can be tremendously reduced by the proper installation of smoke detectors.

First, fire and smoke alarms are *not* the same thing. A fire detector is a device which can be preset to go off at a particular temperature. For example, most fire alarms for the interior of a residence are preset to go off when the air temperature goes over 135°F; technically, they are fire-heat detectors. A smoke detector is a device which has been preset to go off when a certain percentage of smoke accumulates (usually 2% to 4%). The smoke detector does not go off if it gets hot; its photoelectric cell activates the alarm only if it senses a 2% to 4% obscuration of light per foot. Similarly, the fire-heat detector does not activate if there is smoke, only if the heat goes above the preset temperature. Since most fire fatalities are due to smoke inhalation (even before the flames are widespread), the smoke detector offers the most protection and gives the quickest alarm in most installations.

Fire and smoke detectors may either be wired into the electrical system or be battery operated.

Those detectors that are wired into the electrical system should be designed to receive supplemental power from batteries in case of power failure. Such units should also have a small light which will go on if the battery fails, or some other type of device to notify the occupants that the battery must be changed. Fire and smoke detectors are often installed as part of the overall security system, as discussed in Sec. 18-5. When used in conjunction with the security system, different types of alarm signals are used, usually a steady signal when there is fire or smoke and a pulsating signal when there is forced entry. Such a setup means that neighbors should be made aware of the different signals and of what should be done if the alarm goes off. In some areas, it is possible to have the fire alarm connected to a central emergency office or to the fire station.

Battery-operated fire and smoke detectors should have a small light or other device to indicate that the battery is still functioning properly, as with the backup battery power source discussed above. They require no connecting wiring to the electrical system of the building, and this generally makes them ideal for installation when remodeling. While they have an alarm which will warn occupants of the building of a fire or smoke, they cannot be connected to an exterior alarm which might alert neighbors who could call the fire station, and they cannot be directly connected to the fire station.

In residences, the detectors are usually placed in a hallway (or other access area) leading to the bedrooms; in homes with bedrooms on the second floor, the units would be placed at the top of the stairway and near the kitchen. In other types of buildings, a careful check of the floor plan should indicate the best locations.

The installation of fire and smoke detectors is being required, especially in residences and institutions, by more and more building codes. Such codes may specify detector requirements and locations.

Standpipes and Hoses

Many large and tall buildings use water based on a standpipe and hoses system for fighting fires. The firefighters connect to the water hydrants in the street and pump the water through the siamese connect (Fig. 18-2) into the building where it can provide water to the hoses. Many large buildings have an overhead water tank to provide water during the crucial first minutes that a fire is fought. Typically the hoses are located at or near fire exit stairways, from which the firefighters enter the building.

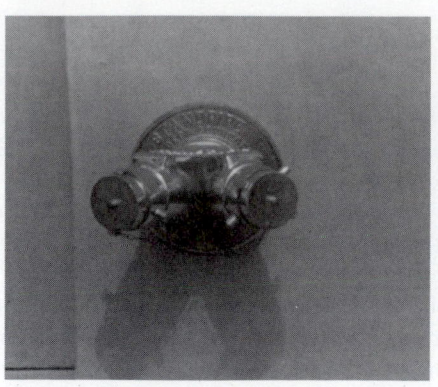

FIGURE 18-2   Stand pipe

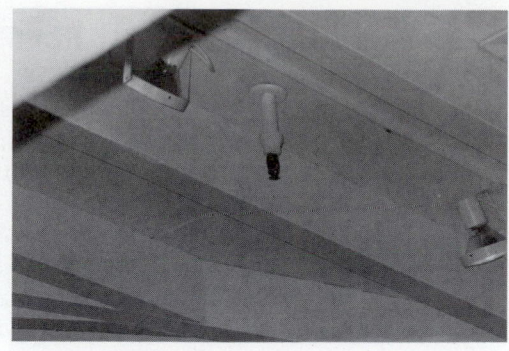

FIGURE 18-3   Sprinkler head

Sprinklers

Sprinklers (Fig. 18-3) are devices that open automatically to discharge water when excessive heat is detected. Typically the automatic sprinkler system consists of pipes placed in a horizontal pattern near the ceiling with sprinkler heads set at predetermined intervals or positions.

There are four common types of sprinkler systems: wet-pipe, dry-pipe, preaction, and deluge. The most common is the wet-pipe, whose pressurized water in the pipe and mains is released when the sprinkler head is activated. In the dry-pipe type, the pipes are filled with compressed air or nitrogen. When the sprinkler head is activated the compressed air escapes first, followed by water. The dry-pipe type is typically used in unheated buildings where there is danger that the water in the pipes would freeze and burst the pipes. Preaction is similar to dry-pipe except that the water first fills the pipe as an alarm is set off, providing an opportunity to extinguish the fire manually before the sprinklers open. This type is often used where the use sprinklers would cause extensive material or equipment damage, such as in stores and computer areas. Deluge systems allow for all sprinkler heads to go off at the same time. They are typically used where extremely rapid fire spread is expected.

Sprinkler heads are usually upright, pendant (pointing downward) and sidewall (pointing sideways). Most commonly used is the pendant type, which hangs below the pipes. This allows the piping to be concealed above the suspended ceilings with only the pendant head showing. Upright heads sit on top of the pipe and the entire system is exposed to view. They are commonly used in warehouses and in retrofitting older buildings. Sidewall sprinklers are often used in small rooms where they can throw a spray of water across the entire room. In this manner only one sprinkler is needed in the room.

Once activated, ordinary sprinklers continue to run until the main valve is closed manually. This can result in excessive water damage, one of the major causes of loss in many fires, in addition to wasting large quantities of water. Flow control sprinklers close automatically once the ceiling temperatures are reduced. Many of these sprinkler heads are reactivated if the temperature goes back up. They are not designed for use in dry-pipe systems, discussed next. Regardless of the type of sprinkler head, they must be replaced after they have been activated.

Water has some inherent disadvantages as a fire suppression system. First, water damages most of the building's contents and interior finishes; second, flammable oils tend to float on the water's surface and continue to burn; third, it conducts electricity and, finally, if it vaporizes into steam it may be harmful to the firefighters. Other methods may be considered when these factors are of major concern.

Other Fire Suppression Systems

Portable fire extinguishers can be used to put out most fires in their early stages. These extinguishers may contain mixtures of water, but they are also available with gases or dry chemicals. They are typically located along exit paths in the building. Portable fire extinguishers are labeled according to the class of fire they are meant to control.

Class 1A to 40A

The *A* indicates usually water, water-based agents, or multipurpose chemical agents. The numbers indicate the level of effectiveness in extinguishing fires, with 40 rated 40 times more effective than 1. This type of extinguisher is most effective on ordinary combustibles such as wood, paper, trash, and textiles.

Class 5B to 50 B

The *B* indicates flame-interrupting or smothering chemicals such as CO_2, foam, sodium, dry-base chemicals, or halogenated agents. The numbers indicate the minimum square footage of deep-layer fire it is capable of extinguishing. It is most effective on liquid petroleum, paint, and solvent fires.

Class C

The *C* indicates a non-electrical-conducting agent such as CO_2, sodium, potassium bicarbonate-base dry materials, or halogenated agents. It is most effective on electrical fires.

Class A:B:C

These are dry-chemical extinguishers that are considered multipurpose. They are usually filled with ammonium phosphate. They are not ideal for electrical fires since they leave a hard residue that is difficult to remove.

Class D

They are dry-powder extinguishers, usually filled with graphite or sodium chloride. They are used on combustible metals. Each extinguisher is designed for a specific metal, as noted on the nameplate.

Foams

Foam is very effective on flammable liquid fires and most popular in areas where fuel is likely to be, such as hangars. The foams are masses of gas-filled bubbles that are lighter than water and flammable liquids. They may be either air or chemical foams. They float on the surface of the burning liquids to smother the fire.

Carbon Dioxide (CO_2)

This gas is stored under great pressure as a liquid which turns to a gas when released. It smothers a fire by displacing the oxygen. As a result it is usually used in confined areas such as mechanical chases, unventilated areas, and display cases. Because it displaces the oxygen, it should not be used where there may be people or pets.

Halogenated Agents

These are liquids that turn to gas when released. They are halogenated hydrocarbons in which one or more of the hydrogen atoms has been replaced with halogen atoms, turning a highly inflammable gas into one that is nonflammable and can be used to extinguish flames. The most widely used is Halon 1301, commonly used in such areas as computer rooms where it can quickly extinguish a fire and yet do little harm to the contents of the room. It is also used on airplanes and museums, libraries, telephone exchanges, and kitchens.

QUESTIONS

18-1. Why has building security become increasingly important?

18-2. What types of electronic devices are used to provide building security?

18-3. Which of the systems in Question 18-2 is best suited to a residence, and why?

18-4. What is the difference between a fire detector and a smoke detector?

18-5. What problems may arise in hiring security guards for buildings?

CHAPTER 19

Sound Control

19-1 SOUND

This chapter discusses the proper design and installation of the mechanicals to be certain that they will not be a source of noise in the building. The control of acoustical problems is of prime concern, and the best results are achieved by anticipating the problems before they occur. First, the principles of sound transmission will be discussed, followed by the measurement of sound levels, and the architectural and construction elements which must be considered in reducing noise.

19-2 SOUND TRANSMISSION

Sound may be transmitted through the air (airborne) or through the structure (structure-borne). Planning is required to reduce the transmission of sound from the exterior to the interior and from room to room. Basically, in each space we want to keep the necessary sounds in and the unnecessary or undesirable sounds out.

Sound is not selective; it passes (transmits) through most building materials and any openings, no matter how small. As sound passes through building materials (walls, floors, and ceilings), its intensity is reduced, with the amount of reduction dependent on the types of materials and the construction used. Sound passes more readily through lighter, more porous materials than through heavy, dense, massive materials and assemblies of materials. The use of sound-absorbent materials on walls, ceilings, and floors will increase the transmission of sound through the surface; that is, it will make the room in which the sound originates quieter, but part of the sound will travel to the surrounding rooms.

19-3 DECIBELS

The decibel (dB) is used to measure the intensity of sound levels within the frequency range of human hearing. A chart showing relative sound levels in dB is shown in Fig. 19-1. From this chart, observe that a level of 25-40 dB is reasonably quiet, while noise levels of 60 dB and up become increasingly noticeable. The object is to reduce the noise to a reasonable level, not to eliminate it.

The important point to remember in dealing with decibels is that the apparent change in loudness varies greatly with actual change in sound level. A change of 3 dB is just barely noticeable to the listener, while a change of 5 dB is easily noticeable and a change of 10 dB makes a significant sound difference. A change from 50 to 60 dB makes the noise seem twice as loud while a change from 50 to 40 dB makes it seem half as loud.

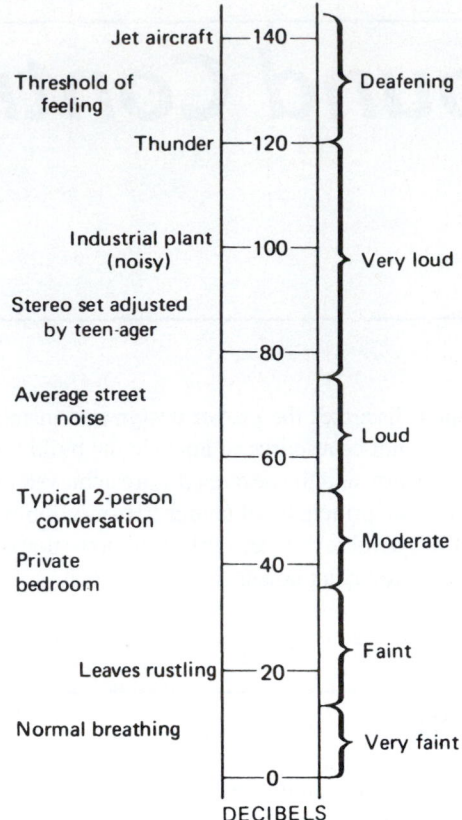

FIGURE 19-1   Decibels

19-4   ELEMENTS

Included in this section is a discussion of the various architectural and construction elements which must be considered to effectively control sound. Some of these elements fall outside of the mechanical designer's prime areas of concern and responsibility, but the designer should be aware of them and be certain that they are not overlooked.

Generally, the elements would be considered by the designer in the following order:

1. Surrounding environment
2. Arrangement and layout of rooms
3. Shape of rooms
4. Reflecting surfaces
5. Isolation of vibration
6. Isolation of impact
7. Isolation of sound

Surrounding Environment

The surrounding environment should be carefully checked before a building site is actually purchased and, once purchased, before planning room arrangements and wall materials. In addition, the zoning should be carefully checked to determine what types of businesses (and accompanying noise sources) may move into the surrounding area.

The location for a building should be chosen with the use of the building in mind. For example, a nursing home should not be placed under the approach pattern to an airport.

Once a site has been selected, exterior noise problems may be reduced by orienting the building on the site to reduce direct sound transmission and reflective sound from surrounding buildings (Fig. 19-2) and equipment. Another method of sound reduction is to shield the building from major noise sources using other buildings, barrier walls, and natural topography. Any barriers used, such as walls or other buildings, should be as close to the noise source and as high as possible.

In addition, the general layout and design must consider existing or possible future noises from surrounding areas. If there is an existing noise source—for example, noisy equipment used in an existing building—the general layout of the proposed building should place rooms which require quiet away from the existing noise and from noisy rooms. In the new building, perhaps the rooms for mechanical equipment may be placed closer to the noise source since they will not be affected by it. Figure 19-3 shows a suggested schematic solution to such a design problem.

Many times, the site itself and the project being designed necessitate that the quiet areas must be located adjacent to a noise source. The noise can still be reduced by placing the solid walls on the side of the building facing the noise source and by placing the windows on the side facing quieter areas (Fig. 19-4). Still, the designer must be careful that sound doesn't enter through the windows by bouncing off an adjacent wall (Fig. 19-5).

Arrangement and Layout of Rooms

Noisy rooms should be separated from quiet rooms by as great a distance as possible. The building layout should be designed so that rooms which are not as susceptible to noise, such

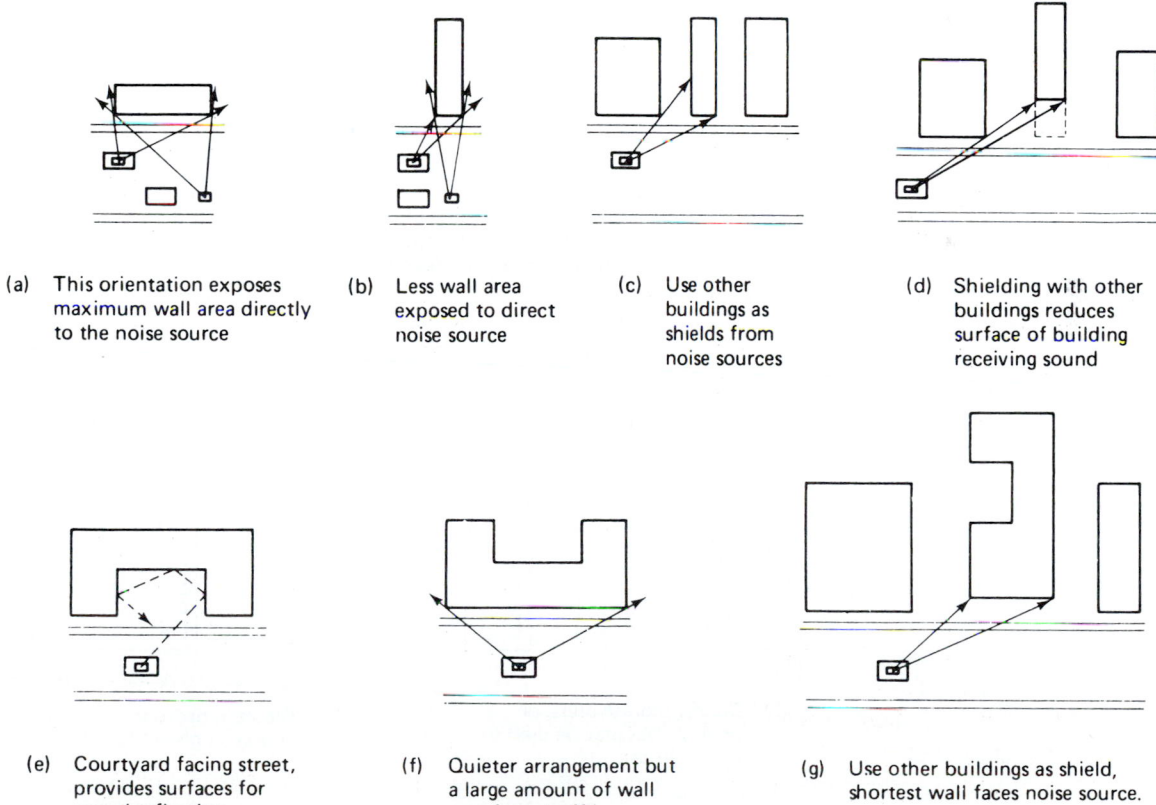

(a) This orientation exposes maximum wall area directly to the noise source

(b) Less wall area exposed to direct noise source

(c) Use other buildings as shields from noise sources

(d) Shielding with other buildings reduces surface of building receiving sound

(e) Courtyard facing street, provides surfaces for sound reflection

(f) Quieter arrangement but a large amount of wall area faces traffic.

(g) Use other buildings as shield, shortest wall faces noise source.

FIGURE 19-2   Building orientation

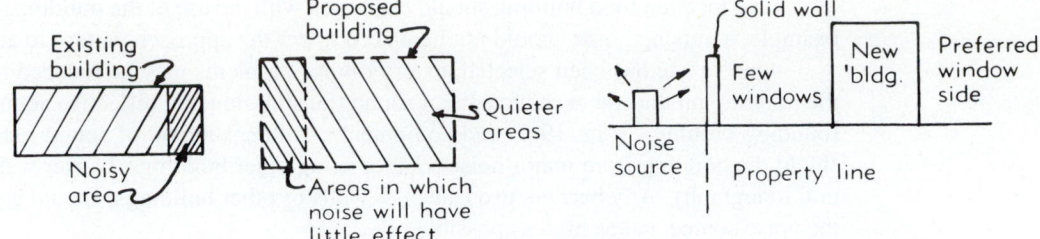

FIGURE 19-3   Planning sound zones

FIGURE 19-4   Sound shields

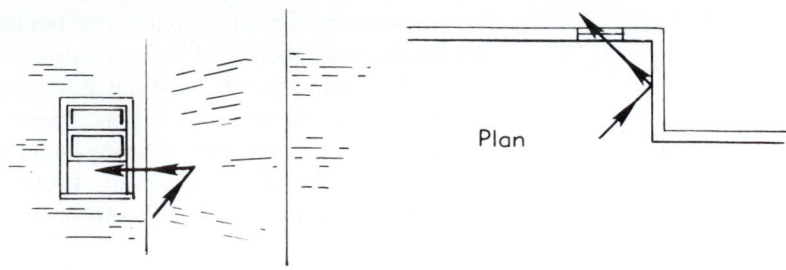

FIGURE 19-5   Sound reflection

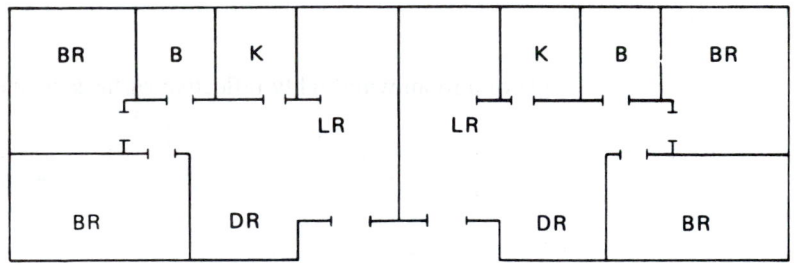

By mirroring (flopping over) the plan an approximately
equal noise level on each side of the wall for both
apartments is obtained.

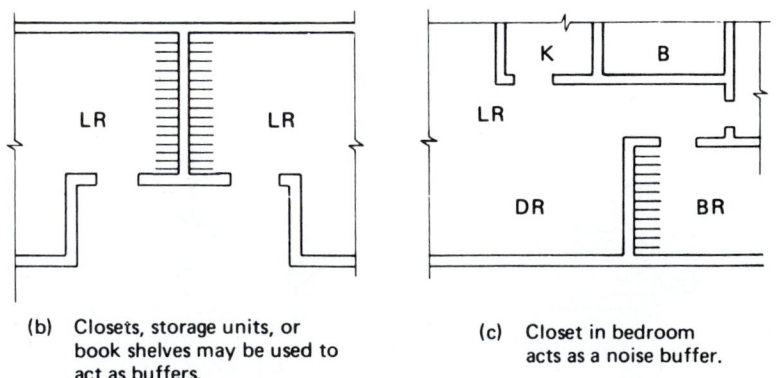

(b) Closets, storage units, or
book shelves may be used to
act as buffers.

(c) Closet in bedroom
acts as a noise buffer.

FIGURE 19-6   Noise buffers

as closets and corridors, act as buffers or baffles between those areas which contain noise sources and those which require quiet (Fig. 19-6). The rooms from which the noise will originate should be located wherever noise from exterior sources may be expected. Quiet areas should be located as far as possible from exterior noise sources. As an example of separation of rooms, a conference room should not be placed next to manufacturing areas or even noisy business machine areas or secretarial pools.

The travel of sound between rooms may also be controlled by avoiding air paths in the placement and design of doors and windows (Fig. 19-7).

Shape of Rooms

The room proportions used will affect the sound reflection within the space. Room shapes to be avoided are long, narrow rooms or corridors with high ceilings and rooms that are nearly cubical. Each of these conditions will cause excessive reverberation (sound reflection).

Ceilings which are domed or vaulted (or any concave surfaces) tend to focus the sound, causing it to be distorted. Large auditoriums with low ceilings also create a situation in which it is difficult for some of the audience to hear. Concave surfaces and large flat surfaces may be broken with splayed areas in order to diffuse the reflection of sound and to direct the sound as desired.

Reflecting Surfaces

All walls, floors, ceilings, and furnishings have certain characteristics which control the amount of sound they will reflect and absorb. If the surfaces in a room tend to be highly reflective, the room will probably seem loud, and it may even have a slight echo. An example of a room with highly reflective surfaces is one with plaster ceilings and walls and a terrazzo floor. The highly reflective qualities of a room can be controlled by using materials which absorb a large amount of the sound, such as carpeting, heavy drapes, and stuffed, upholstered furniture. Acoustical tile is also commonly used when a highly absorbent material is desired. Acoustical tile is most frequently used on ceilings, but it is an economical wall covering also.

Isolation of Vibration

Since the structural elements of the building readily transmit sound throughout the building, no noisy or vibrating equipment should be attached to them. Equipment within the building, such as heating units, pumps, and motors, should be mounted on resilient ma-

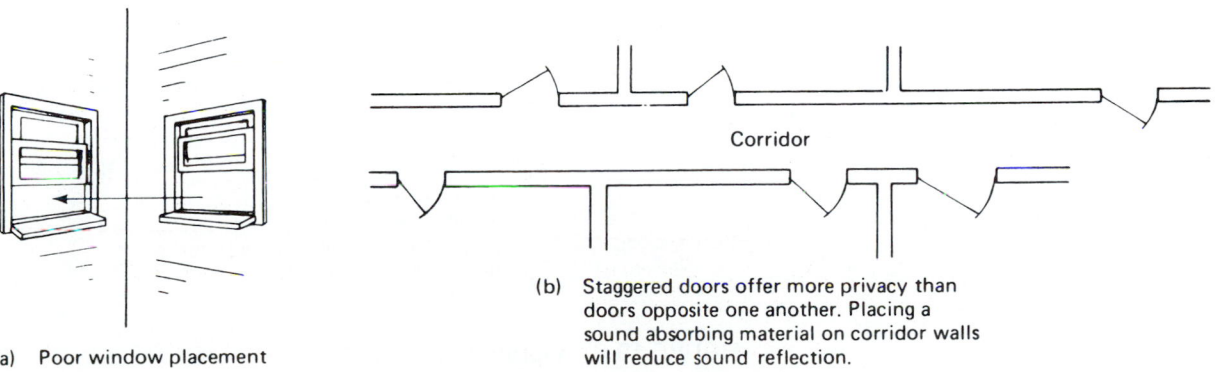

(a) Poor window placement

(b) Staggered doors offer more privacy than doors opposite one another. Placing a sound absorbing material on corridor walls will reduce sound reflection.

FIGURE 19-7 Doors and windows

chinery mounts and bases, which are available. Soft, resilient subfloor materials are often used under the wearing surface of the floors and under the equipment to reduce the transfer of vibration from the equipment to the structure.

Equipment located in the basement of a building may be placed on resilient mounts and a "pad" of concrete which is isolated, by expansion joints, from the rest of the floor. This is an example of discontinuous construction which is very effective when it can be used.

Vibration noises from equipment are also transmitted through pipes, ducts, or other similar conductors. The transmission of noise from the equipment, through these noise conductors, and into the various rooms may be reduced by providing a break where they leave the equipment. Ducts and certain types of pipes may be "broken" with flexible couplings and resilient connections (Fig. 19-8). Water pipes should be provided with expansion valves, expansion tanks, and air chambers throughout the system to reduce water hammering and knocking. Noise resulting from expansion and contraction of the water pipes (usually a creaking and groaning sound) may be controlled by using expansion joints in the line.

Isolation of Impact

Impact sound is that sound caused when one object strikes another. Typical examples of impact sound are footsteps, falling objects (such as shoes, toys, and machine parts), and hammering. When a structure is rigid and continuous, the sound easily travels through it. Impact noise may be controlled most effectively by using absorptive materials (such as carpeting), by isolating the noise sources (using discontinuous construction, (Fig. 19-9), and by reducing the possibility of flanking sounds (Fig. 19-10). The introduction of a masking or background sound (Sec. 19-5) should also be considered, but its effectiveness is limited when the sounds to be covered over are loud.

Isolation may be accomplished by separating the surface which will receive the impact from the structure supporting it. Floors may be isolated from the structure (Fig. 19-11) by using resilient subflooring and underlayment materials. Walls may be isolated from the structure by mounting the finish materials on resilient channels. The ceilings and walls in the surrounding spaces may also be mounted on resilient channels. Carpeting with a thick pad below it is one of the best absorptive materials in terms of reducing impact noises.

Flanking paths provide a channel through which the sound may easily travel. Rigidly connected electrical conduit, pipes, and air ducts are excellent examples of flanking paths.

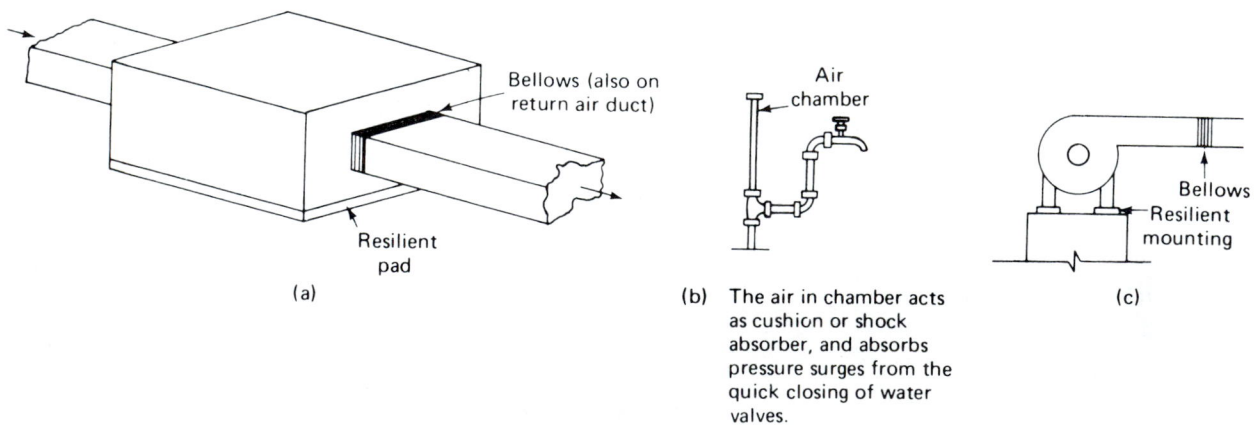

(a)

Bellows (also on return air duct)

Resilient pad

(b) The air in chamber acts as cushion or shock absorber, and absorbs pressure surges from the quick closing of water valves.

Air chamber

(c)

Bellows
Resilient mounting

FIGURE 19-8   Vibration

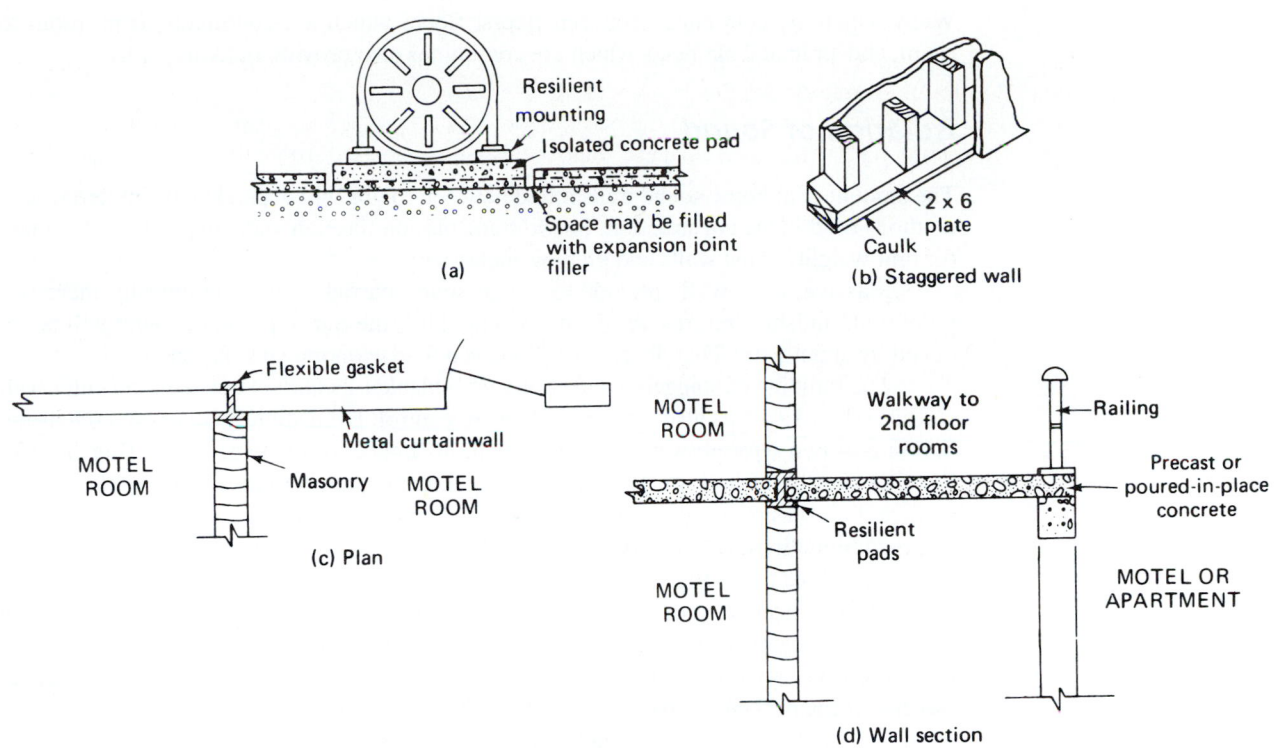

FIGURE 19-9 Discontinuous construction

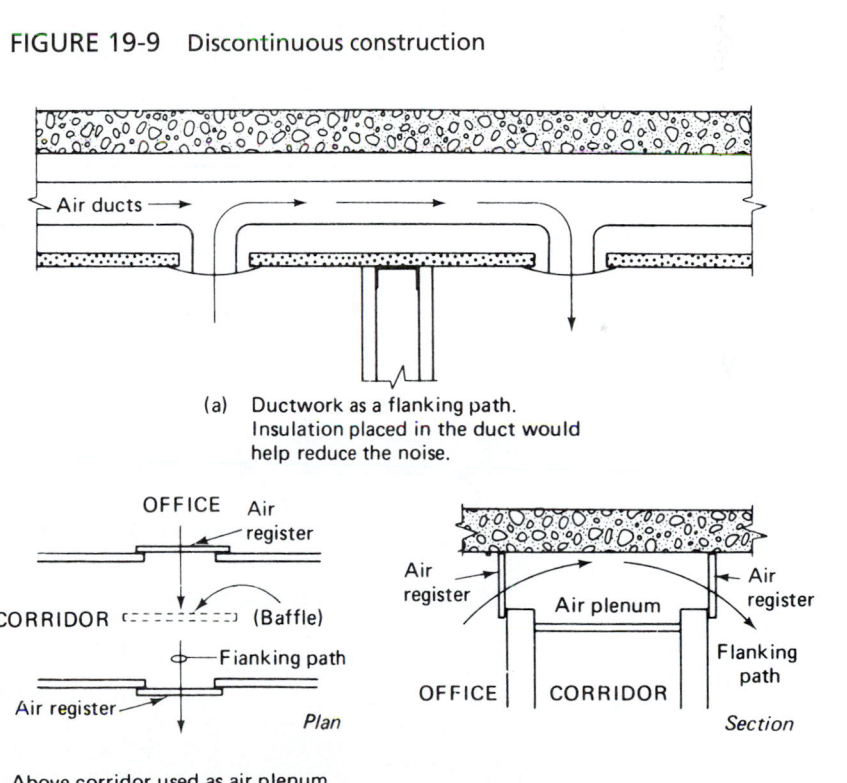

(a) Ductwork as a flanking path.
Insulation placed in the duct would
help reduce the noise.

Above corridor used as air plenum

(b) Flanking path may be broken by:
    1) Offset (staggered) registers and place insulation on
        walls of plenum.
    2) Place baffle between the registers.

FIGURE 19-10 Flanking paths

Walls which are continuous between floors, floors which are continuous from room to room, and structural elements which are continuous also provide flanking paths.

Isolation of Sound

The amount of airborne sound transmitted between rooms will depend on the materials and methods used in the construction. Sound transmission through walls depends on the mass (or unit weight) of the walls and on their inelasticity.

Massive, thick walls provide excellent sound barriers, but economically there is a point of diminishing returns. As the mass is doubled, the transmission of sound will be reduced by about 5 dB. Therefore, mass alone is not an economical solution.

The building of staggered stud walls with blanket insulation between the outer wall surfaces (Fig. 19-12) is also a method of sound control. Even slightly more effective is the building of two separate walls with no structural connections between them (Fig. 19-13). Another control method involves the use of resilient channels to mount the wall-covering surface, thus dissipating the sound energy in the channels. These methods also have their practical limitations. For any given mass, as the air space doubles, the transmission of sound is reduced about 5 dB (Fig. 19-14).

The practical, economical limit of sound reduction through walls is about 50 dB. With the use of background masking noises, this reduction is adequate for most conditions. At the same time, great care should be taken to eliminate as much sound transmission as possible through flanking paths, air ducts, soffits, and items placed back to back, such as medicine cabinets, electrical outlets, and lighting fixtures. Sound transmission through doors and door arrangements and through windows and window arrangements must also be carefully considered in achieving noise control.

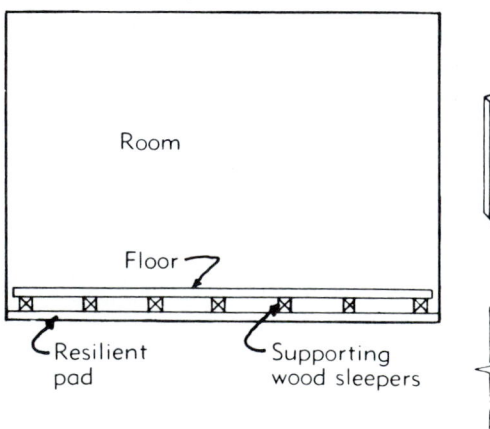

FIGURE 19-11   Isolated floors

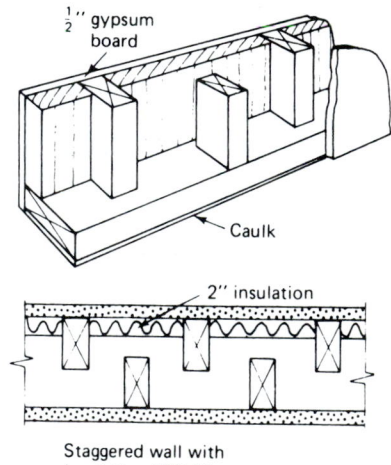

Staggered wall with insulation STC 50

FIGURE 19-12   Staggered wall

19-5 BACKGROUND NOISES

An effective means of reducing the awareness of sound is to use background noise within the room. The illustrations in Fig. 19-15 show a typical example of the effect of background noise in masking undesirable sound. Many stores and restaurants use piped-in music to

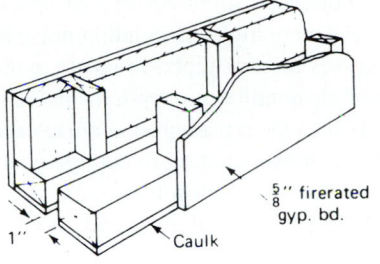

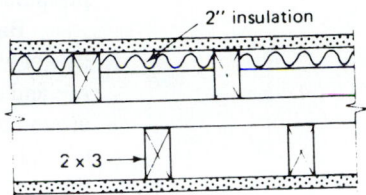

Double wall
with insulation
STC 51

FIGURE 19-13   Double wall

Air space	Sound transmission loss (decibels)
$\frac{3}{4}''$	30
$1\frac{1}{2}''$	35
$3''$	40
$6''$	45
$12''$	50

FIGURE 19-14   Air space

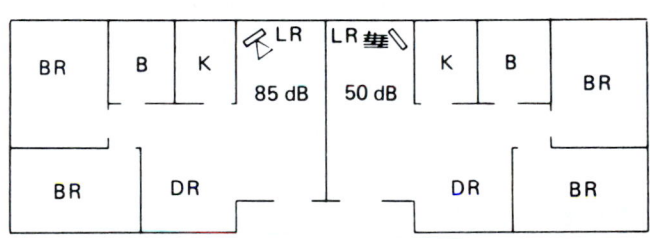

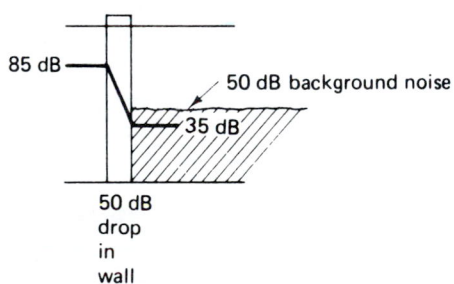

Sound inaudible — the level of the sound
must be brought at least 10 dB below
the background noise to make it inaudible.

FIGURE 19-15   Background noise

mask the noise of people working, traffic passing, and multiple conversations. Background noises are often introduced into very quiet "dead" rooms—for example, sleeping areas in nursing or rest homes or in residences on busy streets—where any noise, such as coughing, would be disturbing. They also provide a measure of privacy for speaking and telephone communications.

Acceptable levels of background noise range from 25 to 35 dB in bedrooms; 30 to 40 dB in living rooms, offices, and conference rooms; and 35 to 45 dB in large offices, reception areas, and secretarial areas. When background noises allowable to such relatively high decibel ratings are combined with effective transmission ratings for walls and floors, it is possible to tolerate relatively high surrounding noise levels.

Background noises include piped-in music, radios, records, or even the noise made by hot air heating and air-conditioning systems in the building. The blowers on most heating and cooking units provide a background noise capable of masking many of the noises around us. While many blowers are set to automatically shut off when the thermostat does not call for heat or cooling, it is worth the small extra cost to have these blowers on all of the time to provide both background noise and air circulation.

19-6 SOUND RATINGS

Materials and assemblies of materials are rated by a variety of different methods, each with its own meaning. The most common ratings are Sound Transmission Class (STC), Average Transmission Loss (TL), Noise Reduction Coefficient (NRC), Sound Absorption Coefficient at One Pitch (SAC), Impact Noise Rating (INR), and Impact Insulation Class (IIC).

Sound Transmission Class (STC)

This rating represents the transmission loss performance of the construction over a broad range of frequencies. The results are plotted, and a standard curve based on human responses at the various measuring frequencies is used to determine the rating given. The higher the rating, the better the construction in terms of reducing the sound. This rating is considered one of the most dependable in measuring overall performance.

Average Transmission Loss (TL)

The transmission loss performance of the construction is tested over a range of nine different frequencies. The results of these tests, at each frequency, are averaged together for a transmission loss rating. This does not take into account the fact that some people are more sensitive to noise in some of the frequency ranges than in others. It is possible that the construction does not provide the transmission loss desired at the proper frequencies.

Noise Reduction Coefficient (NRC)

This rating is used to measure the ability of a material to absorb sound rather than reflect it. Part of this absorbed sound is "killed" within the material, and part of it is passed through to the other side of the material. The NRC rating—which averages the absorption coefficients of the middle frequencies, rounded off to the nearest 5%—can theoretically range from 0 to 1, where a perfectly reflective material is rated at 1. If a material has an absorption rating of 0.25, the material will absorb 25% of any sound which strikes it and reflect 75% of the sound. When selecting materials, keep in mind that a difference of 0.10 in the NRC rating is seldom detectable in a completed installation.

Sound Absorption Coefficient at One Pitch (SAC)

This rating for sound absorption ability represents the peak performance for the material. It is not a representative rating of the material over the range of frequencies people hear. The SAC rating ranges from 0 to 1, similar to NRC.

Impact Noise Rating (INR)

This system is used to rate the ability of floor-ceiling assemblies to resist the transmission of impact sound. Ratings are based on plus (+) and minus (−) ratings from a standard curve, with a plus rating indicating a better-than-standard performance and a minus rating indicating a less-than-standard performance. The standard (0) is based on average background noises that might exist in typical, moderately quiet, suburban apartments. An INR of +5 indicates that the assembly averages 5 dB better than the standard. Where louder background noises are found, as in urban areas, a minus INR (to about −5) rating may be used without any detrimental effects. In quieter areas, an INR rating of +5 to +10 would be used since not as much background noise is available to mask the sound. The disadvantages of this rating system are its lack of relationship to the STC ratings and the negative values, which tend to cause confusion in the use of INR.

Impact Insulation Class (IIC)

This system is also used to rate the ability of floor-ceiling assemblies to resist the transmission of impact sound. IIC ratings are expressed in decibels and are more closely related to the STC ratings for airborne sound transmission than is the INR system. The IIC rating of any given assembly will be about 51 dB higher than its INR rating; therefore, an IIC rating of 61 would be comparable to an INR of +10. It should be noted that deviations can occur, and individual test data should be checked.

QUESTIONS

19-1. What is *sound transmission*?

19-2. If the sound level changes from 42 to 45 dB, how noticeable would it be to a person in the same room? What if the sound level changes from 42 to 52 dB?

19-3. How can the surrounding environment affect sound control?

19-4. What methods may be used to control and isolate vibration noises in buildings?

19-5. How can building orientation affect the sound level inside the building?

19-6. What is a *flanking sound path*, and how may this problem be avoided?

19-7. Define background noise. How is it used in controlling objectionable sound?

19-8. How can background noises be introduced into a space?

19-9. Why should the surrounding environment be checked before the building is designed?

19-10. How can the surrounding noise environment affect the design of a building?

APPENDIX A

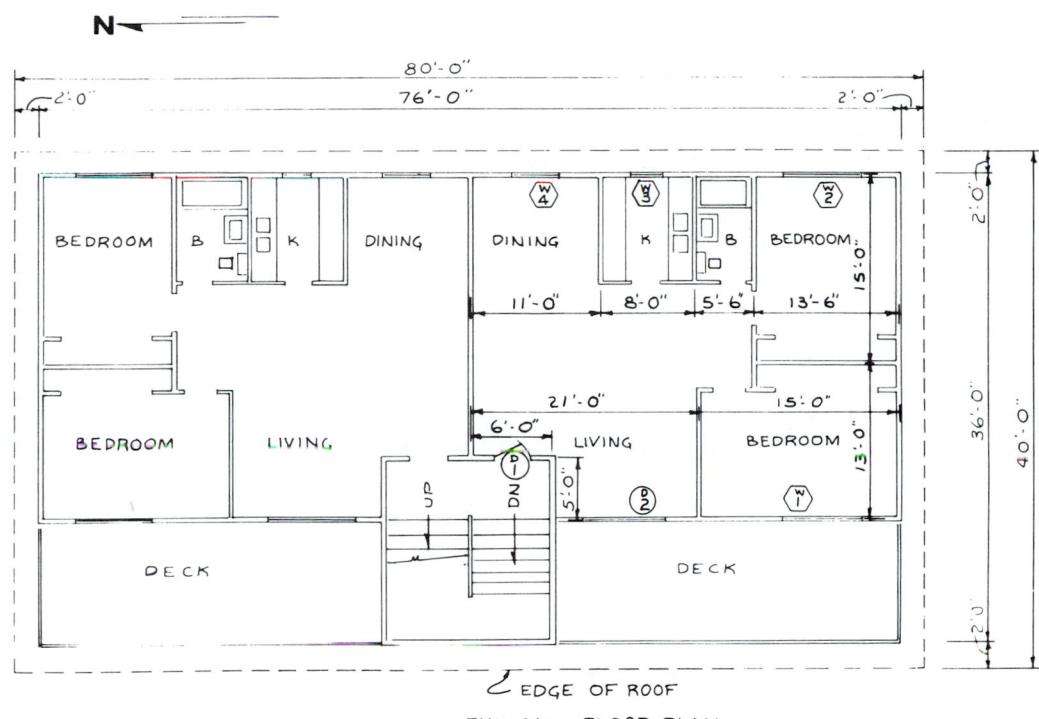

N ←

TYPICAL FLOOR PLAN

EDGE OF ROOF

Design data

Apartment building, 4 story
Floor to floor height: 10'-0"
3-in. service main
Street main pressure: 50 psi
2 exterior hose bibs
Street main to riser distance: 60 ft.
Building drain slope, ½ in per foot.
Flat roof
Occupancy: 2.5 persons per apartment
Heat loss values (U)
    walls 0.13
    roof 0.71
    windows and doors 0.61
Cooling values (HTM)
    walls 3.1
    roof 2.9
Infiltration: Average
D1-1¾ in. steel door, mineral wool core, steel fasteners, no
    thermal break
D2-Aluminum with thermal break, metal, double glazing, ¼-in
    air space.
Windows–same construction as D2.

SCHEDULE		
NO.	SIZE	TYPE
D1	3'-0" × 6'-8"	—
D2	6'-0" × 6'-8"	ALUM.-SLIDING
W1	5'-0" × 4'-0"	—
W2	5'-0" × 3'-0"	—
W3	3'-0" × 4'-0"	—
W4	4'-0" × 4'-0"	—

Electric

Air Conditioning, 2,000 watts
Electric Range, 10,500 watts
Lighting 3 watts per sq. ft.
Water heater, 3,500 watts
Dishwasher, 1000 watts
No clothes dryer

APPENDIX B

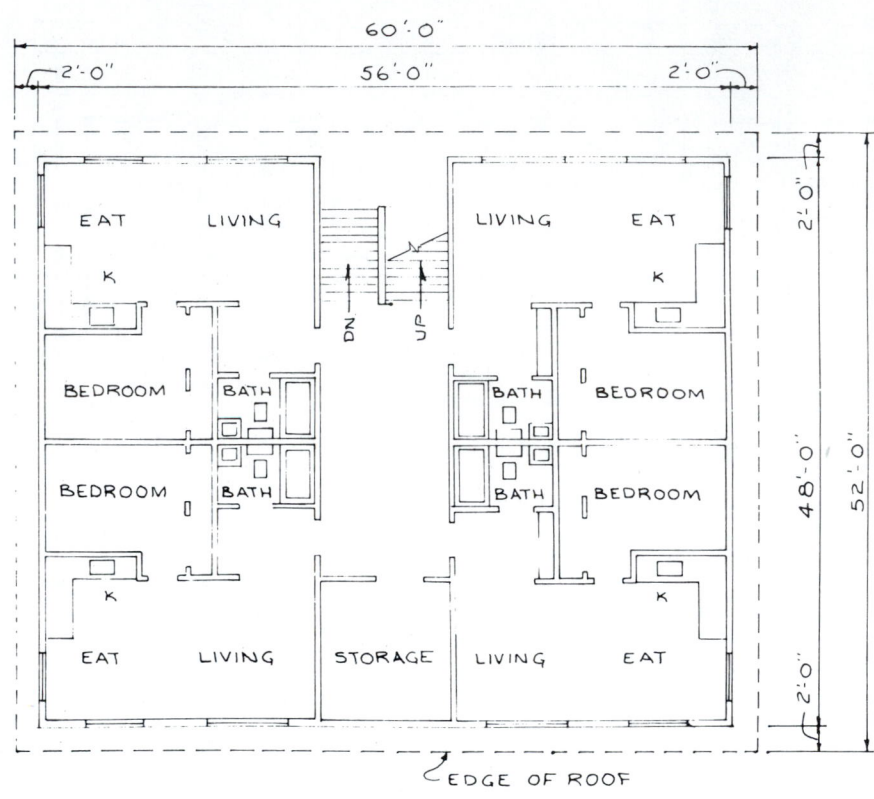

EDGE OF ROOF

TYPICAL FLOOR PLAN

Design data

Apartment building, 3 story
Floor to floor height: 9'-0"
3-in. service main
Street main pressure: 55 psi
3 exterior hose bibs
Street main to riser distance: 85 ft.
Flat roof
Occupancy: 1.5 persons per apartment
Food disposal units in each apartment

APPENDIX C

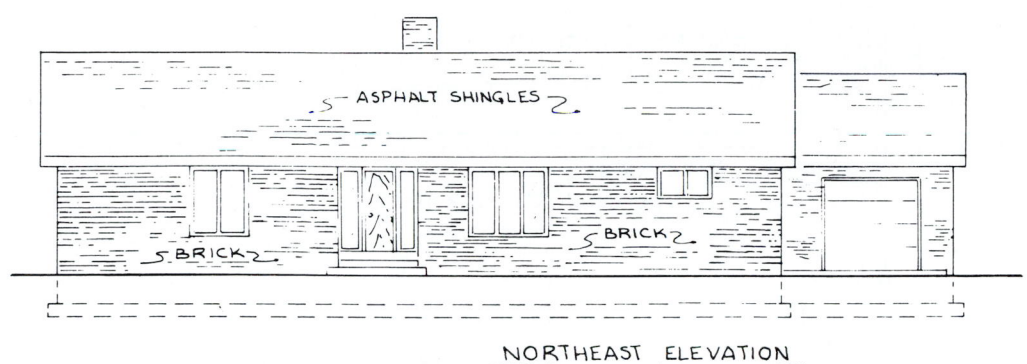

NORTHEAST ELEVATION

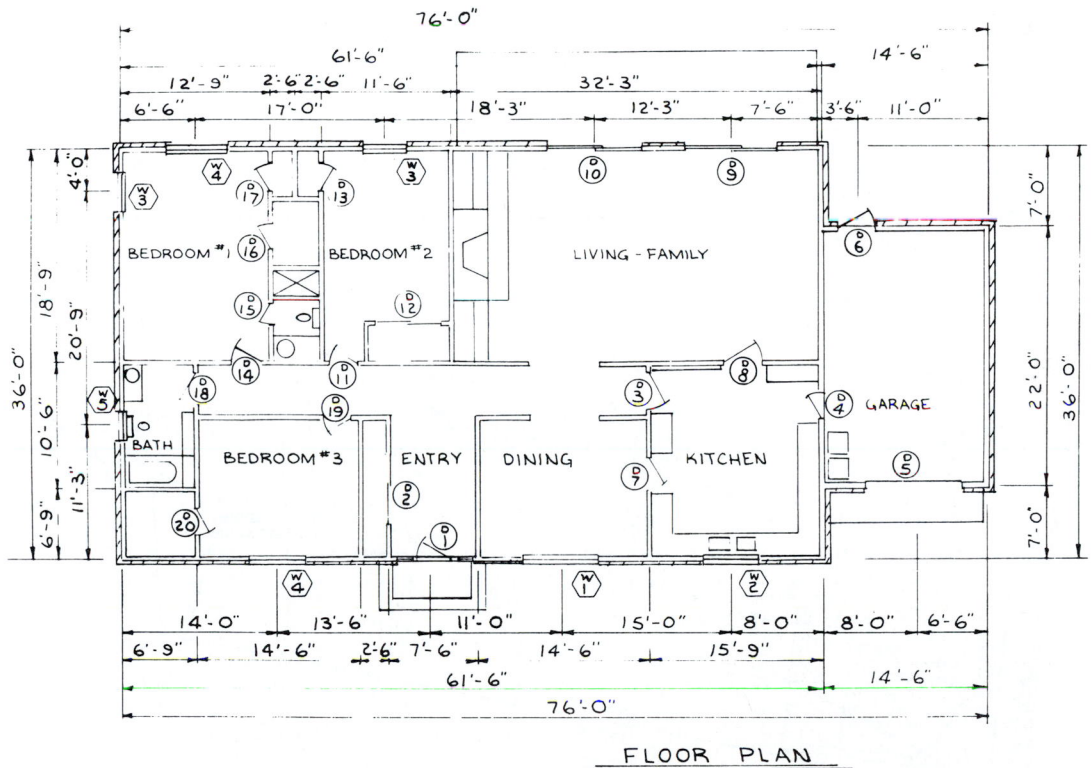

FLOOR PLAN

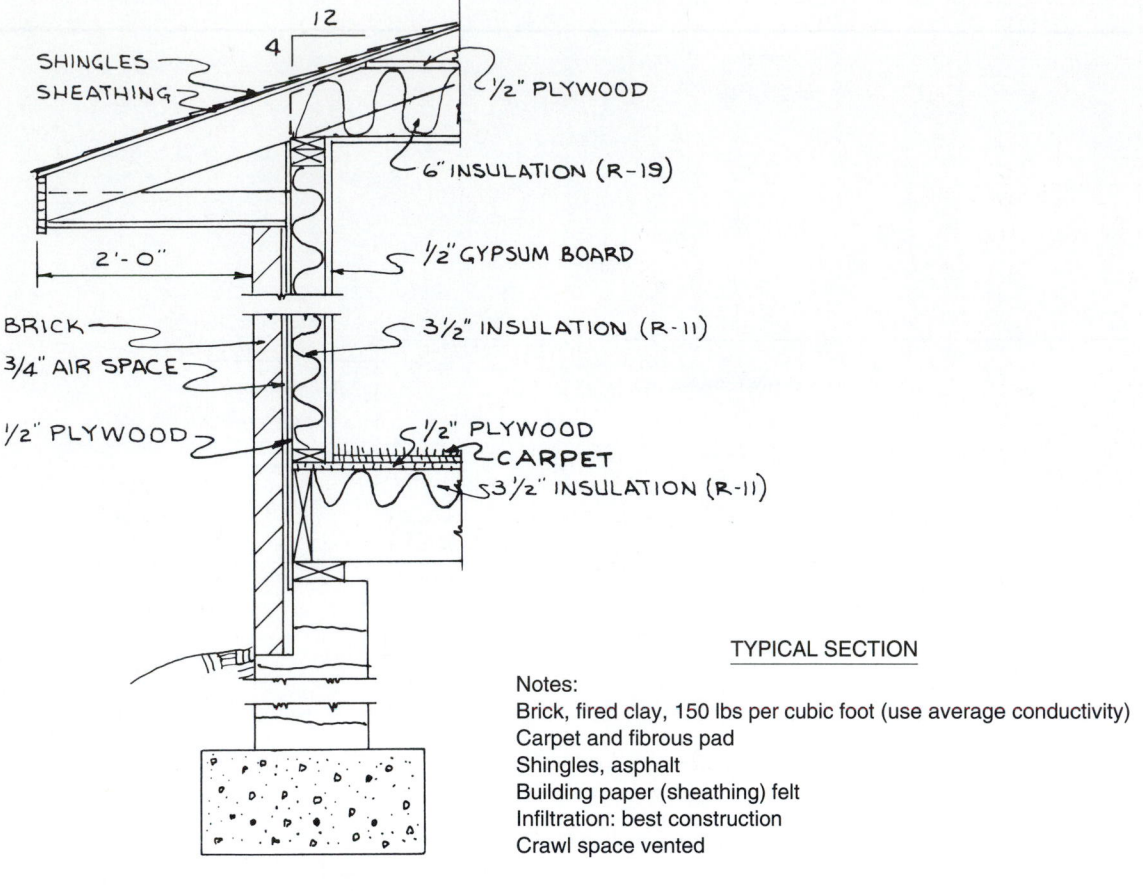

SHINGLES
SHEATHING

12
4

½" PLYWOOD

6" INSULATION (R-19)

2'-0"

½" GYPSUM BOARD

BRICK

3½" INSULATION (R-11)

¾" AIR SPACE

½" PLYWOOD

½" PLYWOOD
CARPET
3½" INSULATION (R-11)

TYPICAL SECTION

Notes:
Brick, fired clay, 150 lbs per cubic foot (use average conductivity)
Carpet and fibrous pad
Shingles, asphalt
Building paper (sheathing) felt
Infiltration: best construction
Crawl space vented

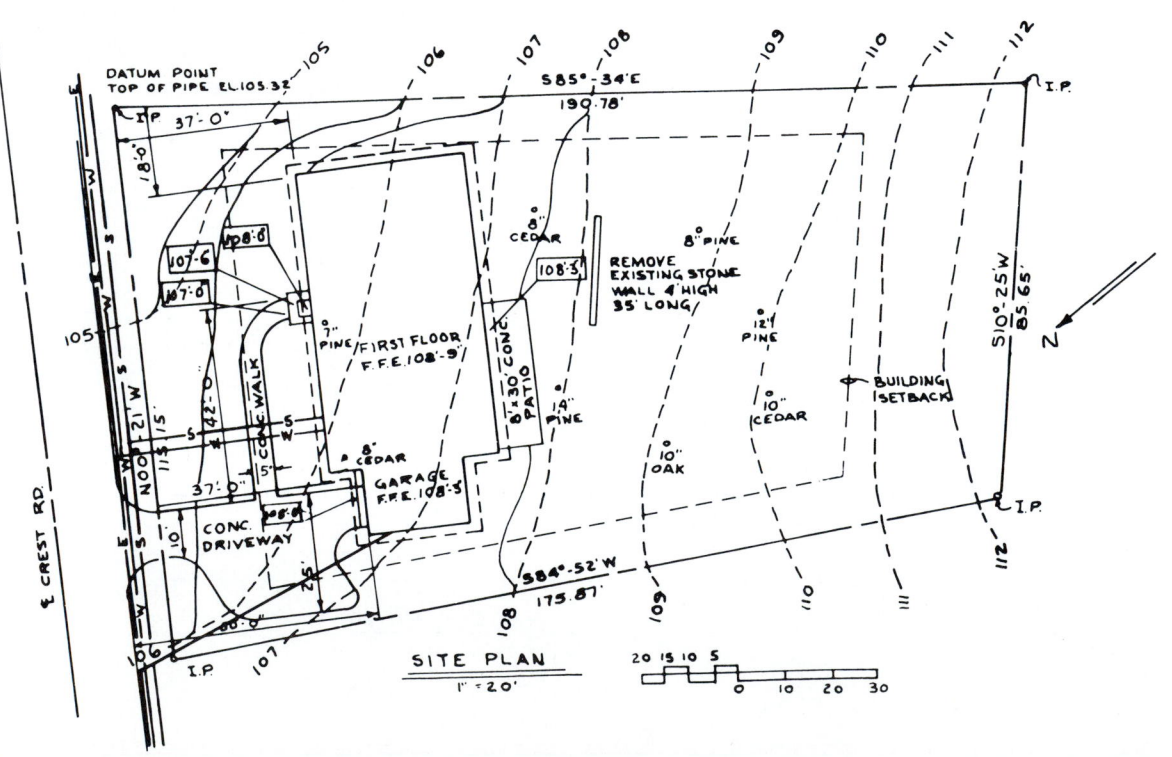

DATUM POINT
TOP OF PIPE EL.105.32

S85°-34'E
190.78

I.P.

I.P. 37'-0"

8" CEDAR

8" PINE

REMOVE
EXISTING STONE
WALL 4' HIGH
35' LONG

108.3

107.6

107.0

105

9" PINE / FIRST FLOOR
F.F.E.108.9

6'x30'CONC
PATIO

12"
PINE

14"
PINE

10"
CEDAR

BUILDING
SETBACK

S10°-25'W
85.65

N

8" CEDAR

GARAGE
F.F.E.108.5

10"
OAK

37'-0"

CONC.
DRIVEWAY

108

S84°-52'W
175.87

109

110

111

112

I.P.

E CREST RD

I.P.

106

107

SITE PLAN
1" = 20'

20 15 10 5
0   10   20   30

WINDOW SCHEDULE		
NO.	SIZE	TYPE
1	6'-0" × 5'-0"	CASEMENT
2	4'-0" × 3'-0"	''
3	3'-0" × 4'-0"	''
4	4'-0" × 4'-0"	''
5	2'-0" × 2'-0"	''

INSULATING GLASS

DOOR SCHEDULE		
NO.	SIZE	REMARKS
1,4,6	3'-0" × 6'-8"	
9,10	6'-0" × 6'-8"	
5	7'-0" × 8'-0"	
2,12	2-2'-6" × 6'-8"	
3,7,8,11 12,14,19	2'-8" × 6'-8"	
13,15,16 17,18,20	2'-4" × 6'-8"	

Design data
Occupancy: 2 adults, 4 children
Design temperature:
    heating: 10 F, 70 F
    cooling: 95 F, medium
Draperies, except kitchen
Roof, dark

Windows—double glazing, ½ air space, operable, metal edge of glass, aluminum with thermal break.

DOORS—

D1—1¾" steel door, mineral wool core, steel stiffeners, no thermal break

D4—1⅜" hollow core

D9 & D10—Double glazing, ½" Airspace, metal edge of glass, double door, aluminum with thermal break

Entry glass—double glazing, ¼" air, metal edge of glass

APPENDIX D

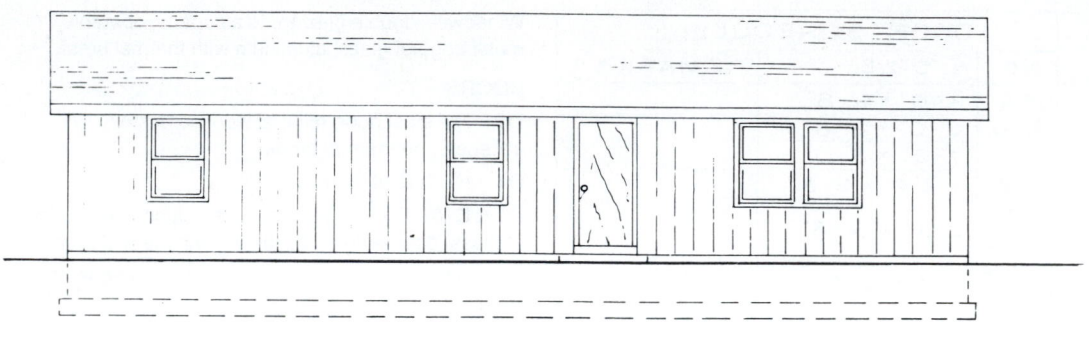

FRONT ELEVATION

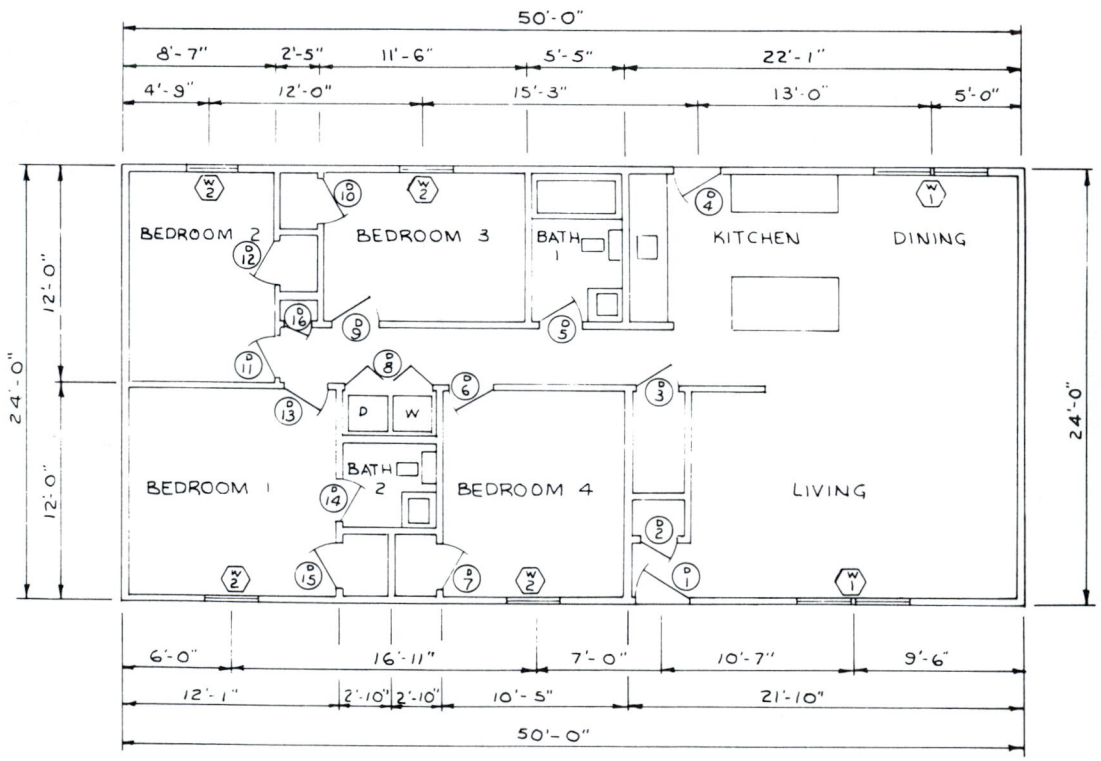

FLOOR PLAN

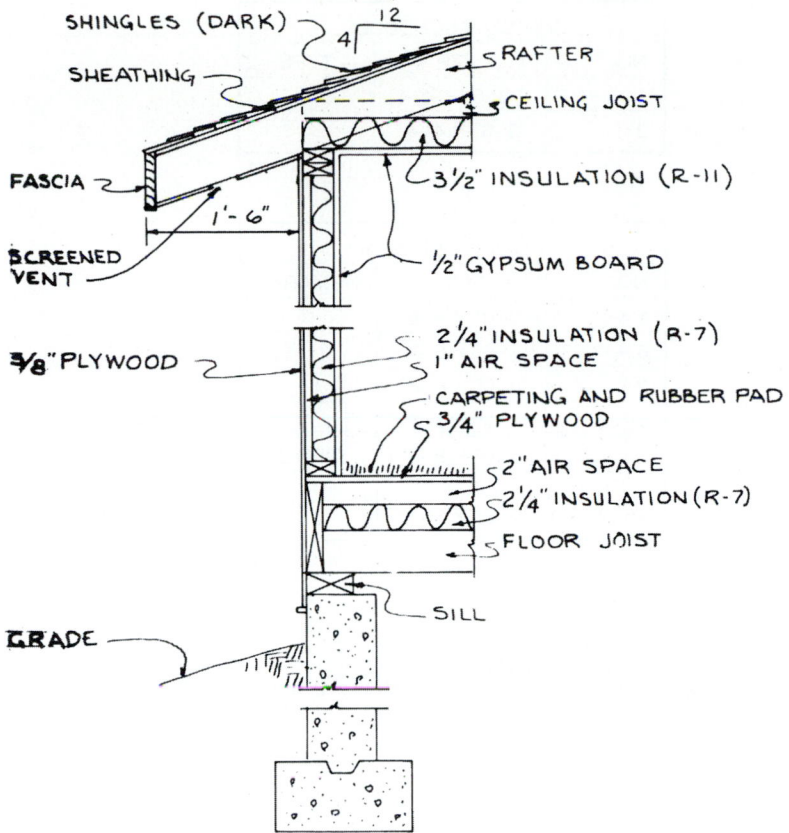

SHINGLES (DARK)
SHEATHING
4 ⎡ 12
RAFTER
CEILING JOIST
FASCIA
3½" INSULATION (R-11)
SCREENED VENT
1'-6"
½" GYPSUM BOARD
⅜" PLYWOOD
2¼" INSULATION (R-7)
1" AIR SPACE
CARPETING AND RUBBER PAD
¾" PLYWOOD
2" AIR SPACE
2¼" INSULATION (R-7)
FLOOR JOIST
SILL
GRADE

TYPICAL WALL SECTION

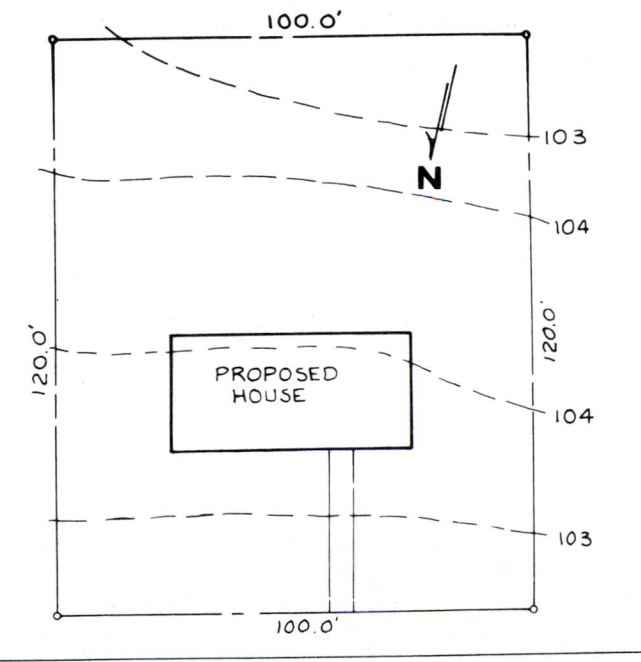

100.0'
103
N
104
120.0'
120.0'
PROPOSED HOUSE
104
103
100.0'

BELL VIEW RD.

SITE PLAN

WINDOW SCHEDULE		
NO.	SIZE	TYPE
1	2-3'-6" × 4'-4"	
2	3'-6" × 4'-4"	

DOOR SCHEDULE		
NO.	SIZE	REMARKS
1,4	3'-0" × 6'-8"	
6,9,11,13	2'-6" × 6'-8"	
2,3	2'-0" × 6'-8"	
16	1'-6" × 6'-8"	
8	5'-0" × 6'-8"	BIFOLD
5,7,10, 12,14,15	2'-4" × 6'-8"	

Design data
Location: Richmond, Va
Occupancy: 2 adults, 3 children
Food disposal
No window shading
Roof, dark
Infiltration, average
Windows—Aluminum with thermal break, operable, metal,
              double glazing, ¼ in. air space.
D1—1¾ in. solid core flush doors
D2—1¾ in. panel door with 1⅛ in. panels

Electrical
   Lighting 3 watts per foot
   Water heater 3800 watts
   Clothes dryer, 4400 watts
   Dishwasher 1,000 watts
   Range 11,700 watts
   Air Conditioner 2000 watts

APPENDIX E

Abbreviations, SI Units, and Conversions

TYPICAL ABBREVIATIONS

British thermal units	Btu
British thermal units per hour	Btu/h, BTUH
cubic feet	cu ft, ft3
cubic feet per minute	cfm
cubic meters	cu m/ m3
feet	ft
gallons	gal
gallons per minute	gpm
grams	g
inches	in
kilograms	kg
kilopascals	kPa
kilowatts	kW
kilowatt-hours	kWh
liters	L
liters per second	L/s
meters	m
millimeters	mm
square feet	sq ft, ft2
square inches	sq in, in2
square meters	sq m, m2
watts	W
watts per square meters	W/m2

CONVERSION FACTORS

Multiply	By	To Get
Btu/h (BTUH)	0.2928	watts
Btu/h/ft2	3.152	watts per square meter
Btu/hFft2	5.673	watts per kelvin per square meter
cubic feet	0.028	cubic meters

cubic feet	28.32	liters
cubic feet per minute	0.472	liters per second
cubic meters	35.32	cubic feet
feet	0.305	meters
feet	304.8	millimeters
gallons	3.785	liters
gallons per minute	0.06308	liters per second
inches	25.4	millimeters
liters	0.2642	gallons
liters per second	2.119	cubic feet per minute
liters per second	951	gallons per hour
liters per second	15.85	gallons per minute
meters	3.281	feet
millimeters	0.039	inches
pounds of force per square inch	6.895	kilopascals
square feet	0.0929	square meters
square inches	645.2	square millimaters
square meters	10.76	square feet
watts	3.1412	British thermal units per hour
watts per square meter	0.317	British thermal units per hour

Index